Health Care Outcomes
Collaborative, Path-Based Approaches

Suzanne Smith Blancett, EdD, RN, FAAN
Editor-in-Chief
Journal of Nursing Administration
Bradenton, Florida

Dominick L. Flarey, PhD, MBA, RN, CS, CNAA, FACHE
President
Dominick L. Flarey & Associates
and
The Center for Medical-Legal Consulting
Niles, Ohio

AN ASPEN PUBLICATION®
Aspen Publishers, Inc.
Gaithersburg, Maryland
1998

The authors have made every effort to ensure the accuracy of the information herein. However, appropriate information sources should be consulted, especially for new or unfamiliar procedures. It is the responsibility of every practitioner to evaluate the appropriateness of a particular opinion in the context of actual clinical situations and with due considerations to new developments. Authors, editors, and the publisher cannot be held responsible for any typographical or other errors found in this book.

Library of Congress Cataloging-in-Publication Data

Health care outcomes: collaborative, path-based approaches / [edited by] Suzanne Smith Blancett, Dominick L. Flarey.
p.   cm.
Includes bibliographical references and index.
ISBN 0-8342-1137-8 (alk. paper)
1. Outcome assessment (Medical care)   2. Medical protocols.
3. Critical path analysis.   4. Medical cooperation.   I. Blancett,
Suzanne Smith.   II. Flarey, Dominick L.
[DNLM: 1. Critical Pathways.   2. Outcome and Process Assessment
(Health Care)   W 84.7 H434 1998]
R853.087H4   1998
362.1—dc21
DNLM/DLC
for Library of Congress
98-9760
CIP

Orders: (800) 638-8437
Customer Service: (800) 234-1660

About Aspen Publishers • For more than 35 years, Aspen has been a leading professional publisher in a variety of disciplines. Aspen's vast information resources are available in both print and electronic formats. We are committed to providing the highest quality information available in the most appropriate format for our customers. Visit Aspen's Internet site for more information resources, directories, articles, and a searchable version of Aspen's full catalog, including the most recent publications: **http://www.aspenpub.com**
Aspen Publishers, Inc. • The hallmark of quality in publishing
Member of the worldwide Wolters Kluwer group.

Editorial Services: Lenda P. Hill
Library of Congress Catalog Card Number: 98-9760
ISBN: 0-8342-1137-8

Printed in the United States of America

1 2   3   4   5

Table of Contents

22—Case Managing To Maximize Quality and Cost Outcomes: A Total Hip Replacement Pathway 289

Mary G. Nash, Cynthia Barginere-Urquhart, Lisa Karen Brown, and John M. Cuckler

23—Outcomes-Based Practice: A Lumbar Laminectomy Model 301

Terri Hawkins Pigg

24—Total Joint Replacement Outcome Improvement 317

Shelly C. Anderson, Patricia A. Soper, Shannon Ericson, and Danny Gurba

Contributors

Janice Steele Allwood, MS, ARNP, CS
Tampa Bay Regional Burn Center
Tampa General Healthcare
Tampa, Florida

Shelly C. Anderson, RN, MSN, MBA
Nurse Manager
Clinical Nurse Specialist
Orthopaedics
Saint Luke's Hospital of Kansas City
Kansas City, Missouri

Carol Ansley, RN, MSN
Director
Clinical Quality Management
Alta Bates Medical Center
Berkeley, California

Joyce Arcus, RN, MSN
Facilitator
Care Management
Duke University Medical Center
Durham, North Carolina

Susan J. Ashcraft, MSN, CCRN
Clinical Nurse Specialist
Coordinator
Neuroscience Development
Redding Medical Center
Redding, California

Dawn A. Bailey, BSN, MAOM, RN
Director
Patient Care Services
Internal Medicine/Hypertension/
  Nephrology Nursing
Division of Nursing
Cleveland Clinic Foundation
Cleveland, Ohio

Cynthia Barginere-Urquhart, RN
Director
Case Management
University of Alabama, Birmingham
University Hospital
Birmingham, Alabama

Roxelyn G. Baumgartner, MS, RNC
Adult Nurse Practitioner
Case Manager
Urology
Vanderbilt University Medical Center
Nashville, Tennessee

Suzanne Smith Blancett, EdD, RN, FAAN
Editor-in-Chief
Journal of Nursing Administration
Bradenton, Florida

Lisa Karen Brown, RN, MSN
Clinical Care Coordinator
Orthopedics
University of Alabama, Birmingham
Birmingham, Alabama

Sue Brown, BSN
Director
Perioperative & Women's Services
Saint John's Medical Center
Santa Monica, California

Maureen Bueno, PhD, RN, CNAA
Assistant Vice President
Center for Disease Management and
  Clinical Outcomes
Robert Wood Johnson University
  Hospital
Assistant Professor
Department of Medicine
Robert Wood Johnson Medical
  School
University of Medicine and Dentistry
  of New Jersey
New Brunswick, New Jersey

Janice K. Bultema, MSN, RN, CNAA
Executive Director
Transition Planning
Northwestern Memorial Hospital
Chicago, Illinois

Cathy A. Campbell, BSN, RN, CNRN
Nurse Associate
Department of Neurological Surgery
Henry Ford Hospital
Detroit, Michigan

Virginia C. Campbell, RN, PhD, CNA
Board of Directors
Florida Health Sciences Center
Tampa, Florida

Lisa R. Cohen, MS, RN, CCRN
Clinical Nurse Specialist
Neuroscience Patient Care Service
Henry Ford Hospital
Detroit, Michigan

Norma J. Cole, MSW
Research Social Worker
Northwest Regional Spinal Cord
    Injury System
Department of Rehabilitation
    Medicine
Harborview Medical Center
University of Washington
Seattle, Washington

Susan Mott Coles, RN, MSN, AOCN
Clinical Nurse Specialist
Case Manager
Surgical Oncology, General Surgery,
    Urology
Center for Clinical Evaluation
Saint Thomas Health Services
Nashville, Tennessee

Marcia A. Colone, MS, PhD, MSSW
Director
Case Management
Northwestern Memorial Hospital
Chicago, Illinois

John M. Cuckler, MD
Division Director
Orthopaedic Surgery
University of Alabama, Birmingham
University Hospital
Birmingham, Alabama

Kim Curry, RN, MBA, PhD
Nursing Program Specialist
Hillsborough County Health
    Department
Tampa, Florida

Yvette De Jesus, RN, MSN
Clinical Coordinator of Disease
    Management Program
M.D. Anderson Cancer Center
Houston, Texas

Shannon Ericson, MS, PT
Physical Therapist
Saint Luke's Hospital of Kansas City
Kansas City, Missouri

Dominick L. Flarey, PhD, MBA, RN, CS, CNAA, FACHE
President
Dominick L. Flarey & Associates and
    The Center for Medical-Legal
    Consulting
Niles, Ohio

Nanette B. Fricke, BS, PT
Physical Therapist
Harborview Medical Center
University of Washington
Seattle, Washington

Anita Gottlieb, MA, RNC, CPHQ
Quality Improvement Coordinator
Arkansas Children's Hospital
Little Rock, Arkansas

Danny Gurba, MD
Orthopedic Surgeon
Dickson Diveley Midwest
    Orthopaedic Clinic
Saint Luke's Hospital of Kansas City
Kansas City, Missouri

Sharon E. Harpootlian, BS, MBA
Senior Data Analyst
Utilization Care Team Department
Henry Ford Hospital
Detroit, Michigan

Lynne Hedrick, BSN, MSN
Vice President
Nursing
Nursing Administration
The Children's Hospital
Denver, Colorado

Patti Higginbotham, RN, CPHQ
Director
Quality Improvement
Arkansas Children's Hospital
Little Rock, Arkansas

Diane L. Huber, PhD, RN, FAAN
Associate Professor
College of Nursing
The University of Iowa
Iowa City, Iowa

Louisa Kan, RN
Clinician
M.D. Anderson Cancer Center
Houston, Texas

Roey Kirk, MSM, CHE
President
Roey Kirk Associates
Healthcare Management Consultants
Miami, Florida

Michael Koch, MD
Associate Professor
Vice Chairman
Department of Urologic Surgery
Vanderbilt University Medical
    Center
Nashville, Tennessee

Sharon W. Lake, RN, MSN
Neonatal Case Manager
Department of Nursing
University of Kentucky Children's
    Hospital
Lexington, Kentucky

Kathleen M. Lewis, BSN, MSN
Nurse Manager
Thoracic/Cardiovascular Surgery
Thoracic Medicine
M.D. Anderson Cancer Center
Houston, Texas

David G. Litaker, MS, MD, FACP
Assistant Professor
Department of Internal Medicine
The Cleveland Clinic Foundation
Health Sciences Center of the Ohio
    State University
Cleveland, Ohio

Debra J. Livingston, MS, RN, CS
Director
Norman and Ida Stone Institute of
    Psychiatry
Northwestern Memorial Hospital
Chicago, Illinois

Maura MacPhee, RN, MSN, MS
Clinical Nurse Specialist
The Children's Hospital
Denver, Colorado
Instructor
University of Northern Colorado
Greeley, Colorado

Barbara J. Maggio, MSN, RN, C, CNS
Clinical Nurse Specialist
Pediatrics & Adolescents
Robert Wood Johnson University
    Hospital
New Brunswick, New Jersey

Karen March, RN, MN
Clinical Nurse Specialist
Neurosciences
Harborview Medical Center
University of Washington
Seattle, Washington

Anne Milkowski, RN, MS
Clinical Nurse Specialist
Neonatal Intensive Care Unit
All Children's Hospital
St. Petersburg, Florida

Lorraine C. Mion, PhD, RN
Director
Department of Nursing Research
Division of Nursing
The Cleveland Clinical Foundation
Cleveland, Ohio

Mae Taylor Moss, RN, MS, MSN, FAAN
Principal
Moss Management, Inc. International
Houston, Texas

Mary G. Nash, RN, MBA, PhD
Associate Executive Director
Chief Nursing Officer
University of Alabama, Birmingham
University Hospital
Birmingham, Alabama

Lynne Nemeth, RN, MS
Project Manager
Clinical Pathways and Coordinated
    Care
Harborview Medical Center
Patient Care Services
University of Washington
Seattle, Washington

Merrilee Newton, RN, MSN
Director
Quality and Clinical Resources
Alta Bates Medical Center
Berkeley, California

Marilyn Oermann, PhD, RN, FAAN
Professor
College of Nursing
Wayne State University
Detroit, Michigan

Jack S. Olsen, RN, MBA
Nurse Manager
Inpatient Orthopaedics
Harborview Medical Center
University of Washington
Seattle, Washington

Felicia Olt, RN, JD
Senior Quality Improvement
    Coordinator
Beth Israel Health Care System
New York, New York

Claire Paras, RN, MBA
Director
Interdisciplinary Practice Guideline
    Program
Department of Healthcare Quality
Beth Israel Deaconess Medical Center
Boston, Massachusetts

Barbara E. Parlotz, MSW
Supervisor
Social Work Department
Harborview Medical Center
University of Washington
Seattle, Washington

Patti Pattison, RN, MSN
Director
Nursing Administration
Professional Support Services
Alta Bates Medical Center
Berkeley, California

Terri Hawkins Pigg, RN, MSN, CNRN
Neuroscience Clinical Nurse
    Specialist/Case Manager
Saint Thomas Health Services
Nashville, Tennessee

Jan Randall, RN, MSN
Director
Care Management
Duke University Medical Center
Durham, North Carolina

Cary Robertson, MD
Associate Professor
Division of Urology
Department of Surgery
Duke University Medical Center
Durham, North Carolina

Christine Roeback, RN
Nurse Manager
Urology
Duke University Medical Center
Durham, North Carolina

Mark J. Rosen, MD
Chief
Pulmonary & Critical Care Medicine
Beth Israel Health Care System
New York, New York

Patricia Brita-Rossi, RN, MS, MBA
Nurse Manager
Nursing Program and Services
Beth Israel Deaconess Medical Center
Boston, Massachusetts

Tarek A. Salaway, MHA, MPH, MA
Clinical Reporting Specialist
Clinical Pathways and Coordinated
    Care
Harborview Medical Center
University of Washington
Seattle, Washington

H. Scott Sarran, MD, MM
Vice President
Medical Director
Chicago Partners
University of Chicago Health System
Crestwood, Illinois

Nancy Shendell-Falik, RN, MA
Assistant Vice President
Nursing
Robert Wood Johnson University
    Hospital
New Brunswick, New Jersey

Patricia A. Soper, RN, MBA
Chief Operating Officer
Saint Luke's Hospital of Kansas City
Kansas City, Missouri

Katherine B. Soriano, RNC, MS
Director
Pediatrics & Adolescents Units
Robert Wood Johnson University
  Hospital
New Brunswick, New Jersey

Patrice L. Spath, BA, ART
Health Care Quality and Resource
  Management
Consultant
Brown-Spath & Associates
Forest Grove, Oregon

Johnese Spisso, RN, MPA
Associate Administrator
Patient Care Services
Director of Nursing
Harborview Medical Center
University of Washington
Seattle, Washington

Mary Jane Strong, MS, RN, C
Clinical Nurse Manager
Norman and Ida Stone Institute of
  Psychiatry
Northwestern Memorial Hospital
Chicago, Illinois

Janet Taubert, RN, MSN, OCN
Oncology Clinical Nurse Specialist
M.D. Anderson Cancer Center
Houston, Texas

James K. Todd, MD
Director
Epidemiology and Inpatient Medicine
Professor
Pediatrics, Microbiology and
  Preventive Medicine
University of Colorado Health
  Sciences Center
Denver, Colorado

Becky Trella, RN, MSN
Director of Medicare Managed Care
Advocate Health Partners
Advocate Health Care
Oak Brook, Illinois

Linda M. Valentino, RN, MSN
Manager
Care Management
Beth Israel Health Care System
New York, New York

Nancy Wells, DNSc, RN
Director
Nursing Research
Vanderbilt University Medical Center
Nashville, Tennessee

Karol Wilson, PT, MPT
Physical Therapist
Harborview Medical Center
University of Washington
Seattle, Washington

Foreword

Among the myriad of changes in the health care industry, there (thankfully) remains one constant: a sincere desire to deliver high-quality care to patients. Nowhere is this more evident than in the broad enthusiasm with which critical pathways and outcomes maps have been embraced. Caregivers continue in their constant effort to provide the best care possible to their patients. Generally, when the right thing is not accomplished, it is either because they do not know what the right thing is (outcome), they do not know how to achieve the desired outcome (process), or there are insufficient resources and confusion about priorities (cost vs. quality).

I first learned about Critical Path Method (CPM) as a graduate student in 1974. Developed by engineers at the Du Pont Company in the 1950s for planning and control purposes, CPM was popular for two reasons: (1) key events were carefully scripted, forcing people to plan, and (2) the systematic scheduling of events within a process facilitated both direction and accountability. Even though it made sense and I liked its logic, I dismissed it as another hoop I had to jump through to graduate and shrugged it off, thinking, "I'll never use that." I discovered how wrong I was years later when I was on my way to work with a senior management team that was in the throes of implementing process improvement teams. In preparation for my work, I was reading *The Goal: A Process of Ongoing Improvement,*[1] a book of enormous popularity that has been credited with changing American business. As I got into the core of the book, bells and whistles started going off. I realized that process improvement, something I had been teaching and writing about for years, was in fact an organizationwide ap-

proach to systematically applying CPM (or, in our case, clinical pathways) to strategic business objectives (key clinical outcomes). Suddenly, systematizing the delivery of patient care did not seem so overwhelming.

Every process or pathway team I have ever worked with, in its quest to improve outcomes, has improved efficiency and, as a result, cost. In addition, planning to prevent problems greatly reduces failure costs (malpractice, rework, tarnished reputations, etc.). Planning ahead to identify outcomes and then designing processes to deliver on those outcomes always saves money. There is always a choice. The process can be maintained, improved, or blown up if it is beyond repair. Regardless of the process, the goal remains the same: to generate the desired outcome, using the easiest, most efficient process, the first time, and within cost constraints. Although it is difficult to ask caregivers, who have face-to-face contact with patients, families, and other customers on a daily basis, to think about costs, they are, in fact, best equipped to identify the maximum service available to patients within existing cost constraints.

Given the above, it is easy to understand the shift to path-based, collaborative practice and appreciate why it has gained popularity and become the preferred care delivery system. Savvy health care providers are breaking down barriers and removing obstacles to facilitate collaboration among functions, knowing that the effort will improve performance results for patients. Contributors to this book are part of a growing number of providers who have discovered just how well path-based collaborative practice will serve us well into the future. Readers will come away with a greater understanding of how path-based practice is transforming health care delivery and intensifying the need for collaboration. A timely work, this book provides the most current insights into today's systems of care delivery.

The potential value of the experiences within this text is huge. Whether you adopt a path as a whole or in part, use it for reference, or simply use it as a catalyst to start a discussion among members of your pathway team, the payoff will be worthwhile. Filled with a wealth of experienced-based knowledge, it gives readers the opportunity to learn from the creators' plans and experiences. As a reference, it is a valuable resource to clinicians who are seeking to do the right thing the right way. As a text, it becomes an off-site clinical experience. This book celebrates both the value of a team approach to improving the health of each individual patient and the importance of every individual who comes into contact with patients and their families.

The collection of work in this volume is impressive—and inspiring when one considers the sheer numbers of contributors, beginning with clinical providers and path-

way teams. Authors, teams, and individual caregivers have pooled their collective wisdom and participated in their own unique way. Congratulations to everyone who contributed and to Suzanne and Dominick, who had the courage and conviction to attack a project of this magnitude and importance in such a timely manner. Your contribution to the delivery of health care and the quality of the patient experience is immeasurable.

REFERENCE

1. Goldratt E, Cox J. *The Goal: A Process of Ongoing Improvement.* Croton-on-Hudson, NY: North River Press; 1986.

Roey Kirk, MSM, CHE
President, Roey Kirk Associates
Healthcare Management Consultants
Miami, Florida

Preface

During the last decade, no other industry was affected by change like the health care industry. We lived through unprecedented times of chaos, regrets, new beginnings, and the overall transformation of the health care delivery system. As we prepare to welcome the beginning of a new century, we realize that the transformation in health care is not over. January 1, 2000, has become a milestone for many predictions regarding the health care industry. How and if these forecasted events will come to pass is not yet known. Despite specific predictions, we believe that some premises are solid and undeniable: (1) transformation will continue in health care delivery well beyond the year 2000; (2) the acute care hospital will continue to develop collaborative partnerships resulting in the delivery of more and more care services outside its walls; (3) managed care will increase with capitation, becoming the primary reimbursement system; (4) the role of professional nurses will be focused on care and case management of populations; (5) our delivery system will be balanced between wellness and prevention care and disease management; and (6) health care delivery will be collaborative and outcomes based.

These first five trends mandate the sixth: that health care delivery will be collaborative and outcomes based. Though this trend is only now emerging, we believe that it will become the dominant theme across all health care settings. The massive changes that have occurred in the last two decades have prepared us for the reality of collaborative, outcomes-based practice.

Our knowledge of collaborative practices grew over the past several years as we worked with innovative health care leaders to bring you books on reengineering and care and case management. Evident in the writings of all those leaders who successfully reengineered processes and implemented case management were the concepts of collaboration, between institutions and among disciplines, and evaluation of outcomes. The emerging strength of these two concepts led to the development of this new work, *Health Care Outcomes: Collaborative, Path-Based Approaches*.

This book is the culmination of the changes that we have experienced over the last two decades. It is a collection of the most current and innovative presentations in path-based, collaborative practices. It takes us from the beginnings of case management and critical pathways in the 1980s to the new focus on measuring and managing outcomes in the late 1990s. In essence, this book tells the story of how we have taken innovations in health care delivery and created newer, more effective and efficient models and systems of care based on the management of care and the definition, analysis, and management of outcomes. We believe that it is a compilation of models and systems for the future of health care delivery.

As you read and study this book, you will be amazed at the ongoing themes of collaboration in health care delivery. After years of talk about collaboration, this book demonstrates the long-awaited realities of this practice. You will note that each case presentation is foundational in the use of pathways that demonstrate collaborative practice and provide means to identify and measure outcomes. You will realize the fine detail and the extraordinary work that teams of professionals must accomplish to realize cost-effectiveness and quality in patient care delivery. And you will have a greater understanding of how outcomes are the final determinants of care delivery and how they are paving the way for further change. This is an exciting work and one for which we are proud to have provided leadership.

We have grown in our knowledge and understanding of the dynamics of the new practice environment and how path-based, collaborative practice and the world of outcomes measurement and management have prepared us for the challenges of the new century.

We believe that you will find this book to be most valuable in understanding and preparing for meeting the demands of the next century. It provides a diverse compilation of knowledge and reference, as well as the "how to" of developing and enhancing your own collaborative, path-based practices and outcomes programs. It can be a valuable guide in leading your health care organizations forward in the next several years.

This book is a collaborative effort and experience in contributing to the advancement of health care delivery. We thank our contributors for their time, patience, dedication, and hard work in providing all of us with this timely reference. Our contributors are the emerging leaders in health care delivery; their institutions represent the most cutting-edge organizations of our time.

We thank and are indebted to key staff at Aspen Publishers. They are a dynamic group of dedicated professionals committed to publishing excellence. Their support and confidence in us will always be appreciated. We also thank Roey Kirk, well-known author and consultant in outcomes practice and management, for taking time from her very busy schedule to write the foreword to this book. She is a pioneer in the quality and outcomes movement, and this work would not have been complete without her input. We also thank our families and friends for their understanding and support through another massive publishing project.

As we think about how this book might make a small contribution to improving health care delivery, we remember two women whose life journeys ended just as we completed it: Mother Teresa of Calcutta and Princess Diana of Wales. May we learn from their examples and contributions how to care better for all humanity. Finally, we dedicate this book in the loving memory of J. Worthington Smith, whose life was taken unexpectedly at the beginning of this project.

Suzanne Smith Blancett, EdD, RN, FAAN
Dominick L. Flarey, PhD, MBA, RN, CS, CNAA, FACHE

■ Part I ■
Overview

The Evolution of Outcomes Management

Diane L. Huber and Marilyn Oermann

Of the many images of nursing, perhaps the most enduring is that of the "Lady with the Lamp." Reflecting the vision of Florence Nightingale caring for the troops during the Crimean War, this image has come to symbolize the caring presence of the nurse to the sick or injured and has been instrumental in building the trust and esteem that nurses hold in the public's view.[1] Yet nurses know that nursing is and can be much more than holding the lamp. In fact, one of the best kept health care secrets is that Florence Nightingale was the original nurse researcher and administrator who pioneered the systematic use of patient outcomes, in the form of mortality data, to improve health care.[2,3]

Nurses trace the history of outcomes initiatives back to Nightingale's work in the 1850s. However, nurses are not generally credited with either having an outcomes orientation or being the professionals with the best sets of skills to manage quality-of-care and cost-effectiveness initiatives within health care systems. Perhaps it is time to reclaim our roots and reemphasize the strategies used by Nightingale to decrease the mortality rate in a military hospital from 60 percent to almost 1 percent, using data collection, analysis, and care process improvement techniques specific to nursing care.[1] An argument can be made that the original nurse's role included outcomes management and that this has reemerged as a crucial aspect of contemporary health care.

By tradition, nurses are seen and often view themselves as caregivers. However, McClure identified the nurse role as having two dimensions: that of caregiver and that of care integrator.[4] Caregiver role functions are related to the augmentation of patients' dependency needs, comfort, education, therapeutics, and monitoring. The role of care integrator involves the linking, synthesizing, and integrating of the work output of multidisciplinary differentiated providers so that patient outcomes occur as needed and desired despite the complexity of the health care environment. Nurses possess the knowledge to perform as integrators and to effect change. This is the ultimate in outcomes: to coordinate and integrate care processes into a seamless continuum of cost-effective, high-quality health care that has a positive effect on the health of the public. Nurses do this by diagnosing patient and system problems, taking action to intervene, and evaluating by measuring and monitoring outcomes.

With such a crucial role for nurses, it is important to explore and build a knowledge base about outcomes: the meaning of related terms, an understanding of their historical development, and the relationships among outcomes measurement, outcomes management, quality initiatives, and performance issues. The purpose of this chapter is to build an understanding of outcomes and to chronicle the evolution of outcomes measurement and outcomes management.

DEFINITIONS

Fundamental to an understanding of outcomes and their measurement and management is a clear definition of terms. Definitions were culled from the literature for the following terms: *outcomes, outcomes measurement, outcomes management, pathways,* and *variance* (see Exhibit 1–1).

At the most basic level, outcomes are the results obtained from the efforts to accomplish a goal. *Outcomes* has been defined simply as "end results, or that which results from something"[2(p.158)] and as conditions to be achieved.[5] Lang and Marek noted that the complex construct of outcomes is usually defined as an end result of a

Exhibit 1–1 Definitions

Outcomes	End results, or that which results from something
Outcomes Measurement	Observing, describing, and quantifying indicators of outcomes
Outcomes Management	The multidisciplinary process designed to provide quality health care
Pathways	Using predetermined care activities and mapping timelines to form a guide to usual treatment patterns for a group of individuals with similar needs
Variance	The deviation or departure from the expected clinical trajectory

treatment or intervention.[6] They identified four components of outcomes: the outcome measure that represents the end result; the determination of when the end point occurs; the treatment or intervention used; and the identification of the problem, diagnosis, or population that gives rise to the treatment and is linked to the outcomes. Donabedian considered outcomes to be changes in the actual or potential health status of individuals, groups, or communities that could be attributed to either prior or concurrent health care.[7] Of the three measures of health care quality, structure, process, and outcomes, he considered outcomes to be the ultimate measure of care.[8]

Important to the definition of outcomes is an understanding of the multiple perspectives from which definitions are approached. While some definitions view outcomes globally as end results or changes in status, other definitions are more specific to a segment of health care. Outcomes, like the measures of the quality of health care, can be defined from medical, nursing, organizational, patient, or other perspectives.[9] For example, Lohr defined outcomes as the end result of medical care in terms of palliation, control of illness, cure, or rehabilitation of the patient.[10] She noted that the classic list of outcomes was death, disease, disability, discomfort, and dissatisfaction. Another medically focused framework is the Medical Outcomes Study (MOS), which identified clinical end points, functional status, general well-being, and satisfaction with care as the outcomes of physician care.[11] The MOS has been critiqued for lacking a multidisciplinary collaborative perspective.[12] Whether global or specific, outcomes really answer the basic question of whether the individuals, groups, or communities of concern were benefited by the care provided.[13]

Outcomes measurement has been defined simply as measuring the results of health care.[14] The measurement of outcomes encompasses activities of observing, de-

scribing, and quantifying indicators of outcomes.[15] This can be distinguished from outcomes monitoring, which is the repeated observation, description, and quantification of outcome indicators for the purpose of improving care.[15] Outcomes measurement draws upon the problem-solving and scientific inquiry processes. Outcomes measurement involves the following steps:

1. Determine the measures of interest (key elements or indicators).
2. Gather the necessary data.
3. Aggregate and analyze the data.
4. Interpret the results and subsequent actions.
5. Make changes.
6. Measure again to evaluate the effectiveness.[15]

Outcomes management relates to a variety of activities designed to use the data gathered from outcomes measurement and monitoring to continuously improve care toward an ideal achievement of outcomes.[15] *Outcomes management* has been defined as a multidisciplinary process designed to provide quality health care, decrease fragmentation, enhance outcomes, and constrain costs.[14,16] The core idea of outcomes management is the use of process activities to improve outcomes.

Pathways refers to critical paths or clinical pathways that form a structured multidisciplinary action plan outlining the critical or key events and activities and the expected outcomes of care for each discipline during each day of a care episode.[17] As introduced in the late 1980s by Boston's New England Medical Center, critical paths were designed for standardization of key events and time frames for a patient's hospitalization.[18] They are a method of planning and documenting care.[17] Pathways use predetermined care activities and map timelines to form a guide to usual treatment patterns for a group of individuals with similar needs. These paths of care to be delivered reflect the most appropriate path to take, with room for variance and corrections. Elements include client problems, expected outcomes, and intermediate outcomes or milestones per visit or day for an entire episode of care.[5]

Variance refers to a deviation or departure from the expected clinical trajectory. In the case of clinical pathways, the common sequencing of events is delineated for a similar group of individuals. Variance analysis is used to examine when and how an individual patient differs from the expected norm.[19] The goal is to reduce provider practice variations, minimize delays in treatment, and decrease resource use while maintaining or improving the quality of care.[20]

The above definitions form a context for understanding outcomes. Clearly, measurement of the quality of care revolves around the determination of outcomes and efforts to use this knowledge to improve the processes of care through outcomes measurement, monitoring, and management. Pathways and variance analysis are key strategies in shaping financial and clinical outcomes. An overview of the history of outcomes measurement and management gives a context for understanding the movement from traditional indicators to a contemporary focus on pathways and a major focus on outcomes.

HISTORY OF OUTCOMES MEASUREMENT

Dramatic changes in health care have become the norm in recent years. These changes have affected access to care, who provides care and in which settings, the structuring of health care organizations, and the costs of care. Many of these changes have arisen from the continuing need to control the costs of health care and from consumers' desire for information about the quality of care they are purchasing. From individual consumers through major corporations, the pressure of huge health expenditures has resulted in a call for greater accountability by health care organizations in providing quality and cost-effective care. Provider organizations need to demonstrate that their practices are sound and that their services are superior.[21]

Outcomes measurement provides a means of verifying the success of a provider's care in terms of predetermined outcomes. The focus on measuring outcomes is a response to the need to determine internally the quality of health care provided by an organization, to compare it with that of similar organizations, and to assess the cost-effectiveness of care. Ellenberg suggested that outcomes measurement is the direct result of the need to provide comparative databases on the effectiveness of treatment protocols, evaluations of health-related quality of life, and cost containment measures.[22] While the focus of early outcomes measurement was on the costs of health care, current efforts examine clinical, functional, patient satisfaction, and other types of outcomes. Outcomes measurement, therefore, has shifted from a more narrow focus on costs to encompass broader outcomes that are more clinically based and provide a measure of both the quality and effectiveness of care.

Factors Influencing Measurement of Patient Outcomes

A number of factors are creating the demand for outcomes measurement (see Exhibit 1–2). One is the tremendous variability in medical treatments in different geographic areas of the United States. Wennberg found in his landmark study comparing the cost and effectiveness of different invasive procedures for prostate surgery

Exhibit 1–2 Factors That Create Demand for Outcomes Measurement

- The tremendous variability in medical treatments in different geographic areas of the United States
- The need to measure the effectiveness of treatments with heterogeneous populations
- The priority given health care reform by the state and federal government, insurers, employers, and consumers
- The demand by purchasers for quantifiable information about the value of the health care dollars they are spending

that there was great variability in the procedures used for treating patients.[23] Less invasive techniques were found to be as successful as more invasive ones and were most cost-effective. The less invasive and less costly procedures, however, were not necessarily used by physicians. Other studies have documented variability in medical treatments for the same diagnosis. The related question is what effect these differences in treatments have on patient outcomes. The variability in medical practice has been one major impetus for outcomes measurement, fueled by questions among the public as to why such differences exist in certain geographic areas and sometimes across institutions in the same area.

Current efforts focus on determining the most appropriate and cost-effective medical treatments and translating these findings into practice. "The interest of payers and policymakers in evidence-based medicine clearly derives from a desire to identify sources of cost without benefit to free resources for other uses."[24(p.329)] Titler and Reiter suggested that the need to measure the effectiveness of treatments with heterogeneous populations is another factor underlying the movement toward outcomes measurement.[25] Patient outcomes data can be used to determine the effectiveness of medical treatments, nursing interventions, and care of other providers.

A third factor that has increased the demand for outcomes measurement is the priority given health care reform by the state and federal government, insurers, employers, and consumers. Yet demands for controlling costs of health care need to be balanced with a careful consideration of the quality of that care. Measurement of patient outcomes provides an opportunity to examine quality of care in relation to costs.

A fourth and related factor is that purchasers want quantifiable information about the value of the health care dollars they are spending. In particular, employers offering health care coverage as a benefit seek information about costs of care, services provided, and patient outcomes. Outcomes measurement provides quantifiable information for use by purchasers. One related issue, however, is the variety of definitions of quality and methodologies for measuring it, limiting the availability of comparable and reliable data.[26]

Significant Events in the Development of Outcomes Measurement

Nurses have always been concerned about patient outcomes and evaluating outcomes as a means of determining the effectiveness of care. Much of the philosophical basis for outcomes measurement is derived from the early work of Florence Nightingale. Nightingale used mortality statistics to portray the low quality of care provided to British soldiers during the Crimean War.[3] Through data collected by herself and her nurses on preventable deaths related to changes in sanitation, Nightingale developed compelling arguments for needed reforms. The reforms were effective in reducing the mortality rate at the Barrack Hospital in Scutari from 60 percent to approximately 1 percent.[27]

From Nightingale's time through the 1960s, patient outcomes data consisted mainly of mortality rates. Later, morbidity rates also were used for measuring the quality of care.[28] In nursing in the early 1960s, Aydelotte studied patient welfare as an outcome of nursing care.[29] Multiple outcomes measures were used, such as number of postoperative days, doses of medications, instruments to measure the patient's behavioral characteristics and physical state of health, and time the patient spent in certain activities. This was an important study in that the impact of nursing care on outcomes was examined.

The 1970s marked a significant period in outcomes measurement. The Joint Commission on Accreditation of Healthcare Organizations (Joint Commission) recommended the use of outcome criteria in nursing audits, and many health care organizations developed their own criteria for measuring the quality of their services.[30] A number of notable projects at this time influenced the development of outcomes measurement in nursing. Hover and Zimmer proposed five outcomes for measuring the quality of nursing care: the patient's knowledge of illness and treatments to be performed, knowledge of medications, skills, adaptive behaviors, and health status.[31] Horn and Swain developed an extensive set of outcomes measures to evaluate the quality of nursing care; a total of 348 outcome measures were then categorized based on Orem's self-care theory.[32] Daubert developed a patient classification system that included five levels of outcomes from recovery to terminal care.[33] These beginning projects provided a framework for outcomes measurement in nursing that extended beyond the traditional measures of mortality and morbidity.

In the area of community health, the Visiting Nurse Association of Omaha in the 1970s began development of the Omaha System. The intent of the Omaha System was to improve the agency's patient record system. As part of this system, the Problem Rating Scale for Outcomes was developed to measure three outcomes: (1) *knowledge,* the ability of the patient to remember and interpret information; (2) *behavior,* observable responses, actions, or activities of the patient; and (3) *status,* the condition of the patient in terms of signs and symptoms.[34] The Problem Rating Scale for Outcomes is an instrument for use by nurses and other health professionals to measure change in the client in terms of these outcomes. The testing of the Omaha System has continued from the beginning phase of development in early 1970s through the present. The System has evolved into a model for practice, documentation, and information management for use not only by home care and public health nurses but also by other health personnel.[35]

The Past 20 Years

The 1980s set the stage for continued emphasis on patient outcomes as a measure of quality of care. The use of outcome criteria to evaluate care paralleled the introduction of various mechanisms for controlling health expenditures, such as prospective payment, and related concern about the quality of care.[10] The Health Care Financing Administration (HCFA) introduced an initiative to measure the effectiveness and appropriateness of health care services for both Medicare and Medicaid programs.[36] This initiative included the development of a large database to monitor outcomes of certain treatments.

In 1989, the Agency for Health Care Policy and Research (AHCPR) was created with a commitment to examine the effectiveness of medical treatments for various conditions. One goal was to control Medicare spending by developing practice guidelines to enhance the quality and effectiveness of care. Clinical practice guidelines, developed from research and expert opinions, provide a blueprint for managing the care of patients. These research-based and standardized guidelines assist practitioners and patients in arriving at decisions about health care for specific clinical conditions.

Another significant impetus for outcomes measurement during the 1980s was the Joint Commission's Agenda for Change, approved in 1986, aimed at improving quality of health care through the accreditation process.[37] One important component of the Agenda for Change was the development of outcome indicators. An indicator is a quantitative measure of patient care that can be used to monitor and evaluate quality of care delivery.[37] Nurses can use such indicators to assess performance and evaluate patient outcomes. The indicators were developed by expert groups with nurse members and tested extensively, resulting in the development of the *Indicator Measurement System* (IMSystem) in 1994. Participating hospitals submit patient data and receive comparative reports with risk-adjusted information. On the basis of these reports, an organization can compare its actual rates with its predicted rates and with rates of other hospitals enrolled in the IMSystem.[34]

Outcomes Measures Sensitive to Nursing

From the 1980s to the present, nurses have been increasingly interested in identifying patient outcomes to measure the effectiveness of nursing care. The focus of these studies has been on describing outcomes sensitive to nursing care.

A number of these studies have been in home care. Lalonde identified seven outcome measures for patients receiving home care: taking prescribed medications, symptom distress, discharge status, caregiver strain, functional status, physiological indicators, and knowledge of health problems.[38] Similarly, Rinke attempted to categorize outcomes for home health care into five areas: physical, behavioral, psychosocial, knowledge, and functional.[39] During this same period, an outcome-based home care quality assurance program was developed in Alberta.[40] The program included both client and family outcomes important for monitoring in home care. The outcomes included pain management, symptom control, physiological status, activities of daily living, instrumental activities of daily living, well-being, goal attainment, knowledge and ability to apply that knowledge, patient and family satisfaction, family strain, and home maintenance.

Marek described types of outcomes appropriate for measuring the effectiveness of nursing care: physiological, psychosocial, functional, behavioral, knowledge, home functioning, family strain, safety, symptom control, quality of life, goal attainment, patient satisfaction, cost and resource utilization, and resolution of nursing diagnoses.[2,30,41] While there is still no consensus as to outcomes to measure, these early studies provided a beginning point for many of the current outcomes measurement projects in nursing.

The measurement of outcomes in long-term care also began during the 1980s. The Omnibus Budget Reconciliation Act of 1987, with the goal of improving the quality of care in nursing homes, mandated the development of outcomes measures for long-term care.

An extensive research program to develop and test outcome measures for home care was initiated by Shaughnessy in the 1980s, resulting in the standardized Outcomes and Assessment Information Set (OASIS) for

home health care. In an early study, Shaughnessy examined the outcomes of Medicare nursing home and home health patients. From 1989 through 1994, a series of studies was conducted to develop and test outcome-based measures of quality in home care.[42] The goal of the research was to develop outcome measures that home health care agencies could incorporate into their quality improvement programs. In this way, a partnership would be established between home care agencies and the Medicare program for collecting and processing information to improve patient outcomes, improve agency performance, and enhance the efficiency of the Medicare approach to quality assurance.[43] The new Medicare Conditions of Participation for certified home care agencies include a requirement that they collect outcomes data using OASIS.[44]

Outcomes Measures of the Quality of Health Care

The 1990s marked a period of intense scrutiny of the quality of services provided by health care agencies. Many proposals were developed for measuring the quality of health care and reporting data about service, quality, and cost. Some proposals emphasized practice guidelines, such as the AHCPR guidelines; more rigorous and outcome-based reporting requirements came from bodies such as the Joint Commission and HCFA; new standards emerged for health plan reporting, such as the Health Plan Employer Data and Information Set (HEDIS); and individual health care organizations developed their own systems for documenting health care quality in report cards and instrument panels.

The AHCPR clinical practice guidelines were established to assist practitioners and patients to arrive at decisions about appropriate health care for specific clinical conditions.[45] They also were intended to serve as quality review criteria and to set quality improvement goals.[46] To date, AHCPR has published 19 clinical practice guidelines. The focus now is on implementation of the guidelines by practitioners; nurses have assumed an important leadership role in implementing the guidelines in practice.[47] Brown et al. suggested that tailoring the guidelines to local conditions may be critical for implementation of them.[46] This creates a paradox for practitioners attempting to use the guidelines in that their validity may be compromised by modifying them, yet without this modification the guidelines may not be used by practitioners.

While many nurses are using the AHCPR guidelines in their own practice, others are developing practice guidelines at the local level. The intent of these locally developed guidelines is similar to that of the national ones; to assist nurses and patients in making informed decisions about health care. Dean-Baar suggested that the guidelines provide approaches to managing client conditions.[48] In addition, the guidelines may be used to evaluate patient care, identify future care needs, identify potential variations in nursing practice, reduce costs, and educate staff.[49,50]

The 1990s were a time in which groups such as the Joint Commission and HCFA shifted their attention to developing quality indicators that emphasized outcomes rather than the structures and processes of care. The Joint Commission's IMSystem is a performance measurement system for voluntary participation by health care organizations. An indicator is a valid and reliable quantitative process or outcome measure related to one or more dimensions of performance.[51] Indicators indicate an element of the process being measured or an outcome of that process. Traditionally, accreditation agencies such as the Joint Commission surveyed organizations' structure and process capabilities rather than their outcomes. Today, accreditation bodies focus on measures of actual performance.[52]

In the initial development of the IMSystem, the goal was to design an outcome-focused performance assessment to assist organizations to improve the quality of their care. Following extensive testing, indicators were gradually added to the IMSystem for a total of 33 indicators.[52] In 1995, the focus of the IMSystem shifted to create a broader group of performance indicators including those developed by health care organizations that would meet predetermined criteria.

Patient outcomes data are increasingly important in evaluating the quality of care provided by home health agencies. HCFA has focused its attention on measuring patient outcomes rather than organization and provider activities.[53] As indicated earlier, Medicare's OASIS data set is outcomes based. OASIS is designed to measure outcomes for adult home care patients; home care agencies will receive three annual outcome profile reports comparing their patient outcomes to a national sample of patients as well as to monitor their own improvement over time. Three types of outcomes are included in OASIS: (1) *end result,* a change in the patient's health status between two or more points in time; (2) *intermediate,* a change in patient behavior, affect, or knowledge that might affect end-result outcomes; and (3) *types of health care utilization,* such as hospital admission.[42,53]

HEDIS was developed by the National Committee for Quality Assurance to provide managed-care organizations with a standardized system for measuring quality performance indicators and for reporting this information. In addition, it allows organizations to monitor improvement activities over time, provides information to employers and purchasers for comparing managed-care organizations, and enables organizations to determine priorities for prevention.[54] There are a number of versions of HEDIS; the most recent version, 3.0, is intended for collecting data on commercial and Medicare and Medicaid risk populations. HEDIS 3.0 focuses more on outcomes than earlier versions. The performance indicators are categorized into eight domains: effectiveness of care, access and availability of care, satisfaction, health plan stability, use of services, costs of care, informed health care choices, and health plan descriptive information.[55] HEDIS provides a systematic measurement process, thereby enabling health plans to compare their outcomes nationally.

Report cards provide information about the quality and costs of health care to meet the needs of consumers and purchasers. While report cards are a positive response to the need for comparative information, there may be problems with the validity of the data gathered.[56] Uses of report cards include providing guidance as to which providers of care achieve the best clinical, functional, and satisfaction outcomes at lowest cost; holding providers accountable for achieving outcomes and maintaining costs; identifying opportunities for improvement; and identifying benchmarking sources.[56] The information in report cards, however, must be clear as to the outcomes that are measured and must be valid and reliable. Some health care organizations issue report cards on themselves to demonstrate their quality, costs, and services and how well they are scoring in these areas.

Nelson et al. recommended designing instrument panel data collection systems to feed directly into report cards.[56] Instrument panel data have dual aims of learning about variation in performance in a system and meeting external information needs for purchasers. They provide for a balanced review of outcomes, such as clinical outcomes, functional health status, patient satisfaction, and costs.[56–58] One goal of instrument panels is to provide knowledge for improvement within the organization.

Nursing Outcomes Classifications

Efforts to describe and classify nursing-sensitive outcomes continue; these efforts are critical to measuring the impact of nursing care on patient outcomes. The Omaha System, which began development in the 1970s, includes standardized classifications for various conditions, interventions, and ratings of patient problems.[34] The Problem Rating Scale for Outcomes is designed specifically for measuring outcomes of care, knowledge, behavior, and status, thereby providing data for clinicians and administrators. Recent studies suggest the usefulness of the Omaha System in describing and quantifying nursing practice and in providing a systematic way of collecting client outcome data from diverse home health agencies.[59]

In 1989, the National League for Nursing and its Community Health Accreditation Program began its project to develop outcome measures of care quality for elderly patients receiving home care and to design a report card for collecting and reporting the data.[60] The outcomes are consumer based, such as knowledge and family support; clinically based, such as functional ability; and organization based, such as financial viability.

At the University of Iowa, work began first in the 1980s on developing a nursing intervention classification. In the 1990s, research was initiated to develop a comprehensive classification of nursing-sensitive patient outcomes. The Nursing-Sensitive Outcomes Classification (NOC) completes the third of the four patient-level nursing process elements of the Nursing Minimum Data Set. The NOC includes 190 patient outcomes sensitive to nursing interventions. Each outcome is labeled and defined and includes indicators, a measurement scale, and references.[61] The NOC also is significant because of its ability to be included in nursing clinical data sets in varied settings and in large regional and national databases used to assess the effectiveness of nursing interventions and inform policy makers.[62]

In 1994, the American Nurses Association began its project to identify quality indicators and measurement tools to measure the quality of nursing care in acute care settings. The indicators included outcome, process, and structure measures. Eight of the indicators originally identified were subjected to more intensive study because of their specificity to nursing quality and ability to be tracked.[63] The outcome indicators include nosocomial infection rate; patient injury rate; and patient satisfaction with nursing care, pain management, educational information, and care during the hospital stay.[63]

The 1990s will be remembered as a period that emphasized outcomes as indicators of quality. Continued efforts to determine the effectiveness of nursing interventions in achieving patient outcomes are critical. Without such efforts, the focus of outcomes measurement will remain decidedly on medical practice, with limited opportunity for nursing's contributions to be recognized.

HISTORY OF OUTCOMES MANAGEMENT

Outcomes management is the use of information collected through the measurement of outcomes to continually improve processes of care. Outcomes management is a multidisciplinary approach to promote quality health care, decrease fragmentation, improve patient outcomes, and control costs.[16] In outcomes management, data are collected and analyzed over a period of time, trends are identified, and decisions are made as to changes necessary to enhance patient outcomes and improve the effectiveness of care delivery.

In 1988, Ellwood proposed outcomes management as a "technology of patient experiences designed to help patients, payers, and providers make rational medical care-related choices based on better insight into the effect of these choices on the patient's life."[64(p.1551)] He proposed that outcomes management would enable health providers to analyze clinical, financial, and health outcomes, drawn from a national database, to identify relationships among medical interventions, outcomes, and cost. Ellwood envisioned outcomes management as a system that could be modified continuously and improved upon through advances in science, changes in patients' expectations regarding their care, and availability of resources.[64]

With this as a framework, one of the goals of an outcomes management program is to collect a standardized set of data on patient outcomes and then use this information to improve care processes, develop new research on effectiveness, and determine clinical policies.[65] Outcomes management has come to be recognized as a method for measuring and then improving performance on the basis of the results obtained.

Outcomes management provides a systematic approach to linking outcomes data with continuous quality improvement (CQI). The process begins with the measurement of outcomes, followed by continuous improvement of work processes to achieve better outcomes.[14]

Early quality assurance efforts focused mainly on resolving problems to improve system effectiveness, but there was limited concern as to whether these efforts actually improved patient outcomes. Standards of care were developed, and audits were initiated to ensure that there was compliance with these standards. The Joint Commission's Agenda for Change, however, required hospitals to develop systems for measuring quality of processes, structures, and patient outcomes. The perspective has now shifted away from quality *assurance* to continuous quality *improvement*. Houston and Miller described outcomes management as an essential part of CQI and other quality enhancement programs.[66]

Wojner developed an outcomes management quality model that clearly shows this relationship between outcomes management and quality improvement.[67] In phase 1, an interdisciplinary team of providers is organized to initiate the outcomes management process, outcomes are identified for measurement at specified points in time, and instruments for outcomes measurement are selected. In phase 2, interventions are created and tested, or structured care methodologies, such as critical pathways, protocols, and algorithms, are designed to standardize practices. Phase 3 involves the implementation of these methods to standardize practice within each discipline; this phase also includes the collection of data. During phase 4, data are analyzed, leading to the identification of opportunities for practice enhancement and new research questions. The process then recycles through phases 2 through 4. The continuous nature of outcomes management provides for cyclical measurement and practice improvement targeted at outcomes enhancement.[67]

Other types of improvement efforts are ongoing in individual health care organizations. Outcomes management, while similar in process to these, begins with identifying outcomes for measurement, then seeks to examine the processes and practices for achieving them. Through outcomes management, data gained from measuring outcomes provide the basis for improving processes and in turn achieving optimal patient outcomes.[68]

TRACKING AND EVALUATING OUTCOMES

It is no longer sufficient to put a clinical program, process, or delivery system in place and assume that all is well unless a sentinel event occurs. This is a strategy sometimes referred to as "shoot first and call whatever you hit the target." Today, consumers and payers expect providers to demonstrate effective, efficient, minimally costly, and outcomes-based care services. This is a strategy of identifying a target and then shooting. Thus, health care providers need to track and evaluate their care outcomes. Critical pathways and variance analysis are two major care management tools designed to meet outcomes management needs by structuring an identified target and making any necessary midcourse corrections.

Critical Pathways

Providers search for ways to have some assurance of providing the best care. Providing the best care at the least cost in a complex, multidisciplinary environment includes orchestrating the proper care components in appropriate time frames. This involves multidisciplinary teamwork to plan and develop a standardized written map of care activities and the sequencing of interventions. A critical pathway, also called a *clinical pathway* or *critical path,* is one major tool for outcomes tracking and evaluating. It is often used in conjunction with case management systems.

A critical pathway is a document designed to organize interventions and activities for an episode of care. It incorporates process and outcomes components. A critical path can form a standard of care or a care plan or both. The idea is to use critical thinking by all members of a care team to identify critical and predictable clinical incidents needed to achieve desired outcomes. Thus, the critical pathway is a practical form of practice accountability and documentation that can be used to track health outcomes, patient and provider activities, complications, and teaching/learning outcomes. Critical pathways may also be used as education tools, to prepare and orient patients before treatment, and to negotiate expectations and care roles with patients and families.[69]

Critical pathways can be thought of as protocols of interdisciplinary treatments that are based on professional standards of practice and placed in order on a decision tree.[69,70] Critical pathways should incorporate daily expected patient outcomes as subgoals. The daily subgoals form a system of incremental patient outcome targets for reaching the final goal that easily lend themselves to real-time planning and evaluating patient care. Critical pathways are now an accepted tool for clinical care planning.[71] Critical pathways also can be used to accomplish financial and systems goals. These can include the reduction of clinical practice variations, minimization of delays in treatment, and a decrease in resource use and costs.[20] Research has indicated that critical pathways are an important determinant of improved quality in a managed-care environment.[20] Further, the combination of case management, critical pathways, an outcomes database, and a report card approach to program evaluation was a synergistic method to improve care via secondary prevention of heart disease in one population studied.[71] Critical pathways, with their focused outcomes management design, are a key feature of improved quality-of-care efforts and deliberative outcomes management systems.

Variance Analysis

Variance analysis is a second major outcomes tracking and evaluation tool. A variance is anything that varies or differs. In relation to critical pathways, a variance occurs when there is a deviation from the standard, expected treatment path. When some aspect of care "falls off the path," the clinician monitoring and coordinating care delivery can note this, evaluate it, and take corrective action. Variance data can be aggregated over populations or groups to analyze trends having an impact on the larger scope of care delivery. Variance data analysis can be used to improve health care effectiveness, reduce risks, understand and strengthen provider interventions, modify care protocols, trigger research, and identify the impact of care processes on patient outcomes. Thus, provider decision making can be enhanced, and the real-time care of an individual patient can be improved or kept on target. Clearly, variance analysis is a key strategy for incorporation into CQI initiatives.

A variance, or departure, from a critical pathway can be either positive or negative. The direction is determined in relation to the desired outcome. For example, if reduced length of hospital stay is desirable, then any occurrence that contributes to lengthening stay time is considered negative. For individuals receiving health care, variances should be assessed concurrently with the care process. For aggregated populations, variances are combined and retrospectively analyzed. Variances can be derived from patients, practitioners, or systems.[72]

Variance data analysis begins with careful measurement of outcomes for population specificity, sensitivity, reliability, and validity. Outcomes may be clinical, functional status, financial, performance, or service quality related. For example, length of stay, patient satisfaction, patient knowledge, delays in treatment, errors, complications, preexisting conditions, social complications, or charge per case may be outcomes of interest. A time frame for data analysis needs to be chosen. For example, 100 percent of all records for a population can be monitored for a year, or data can be analyzed quarterly. Data for analysis will need to be selected from among the mass of data gathered. Variance data may only point to the need to gather additional data. Aggregated data, displayed via graphs, scatter plots, or other data display methods, can be examined for trends and patterns. Clinicians then can analyze the data by assessing the significance of variance patterns or sentinel events and the need to take corrective action. In this way, variance analysis becomes a key link in the CQI chain.

CONCLUSION

The use of outcomes measurement and management to magnify quality improvement and cost reduction activities remains the best strategy for genuine care improvement. These efforts also promise to help assuage the lingering public skepticism about managed care initiatives. Nurses have a major role to play in all aspects of outcomes measurement and management. The knowledge and insights about care management that nurses have to offer are central to effecting positive health care outcomes. Critical pathways and variance analysis are two essential tools. In the role of care integrator, nurses can demonstrate effective and efficient care contributions to vital health care processes of collaborative CQI and outcomes management that make a difference in the health care of individuals, groups, and communities.

REFERENCES

1. Kalisch PA, Kalisch BJ. *The Advance of American Nursing.* Boston: Little, Brown and Company; 1978.
2. Lang NM, Marek KD. The classification of patient outcomes. *J of Professional Nursing.* 1990;6:158–163.
3. Nightingale F. *Notes on Matters Affecting the Health, Efficiency, and Hospital Administration of the British Army.* London, England: Harrison & Sons; 1858.
4. McClure M. Introduction. In: Goertzen I, ed. *Differentiating Nursing Practice: Into the Twenty-First Century.* Kansas City, MO: American Academy of Nursing; 1991:1–11.
5. Peters DA. Outcomes: the mainstay of a framework for quality of care. *J Nurs Care Qual.* 1995;10(1):61–69.
6. Lang NM, Marek KD. Outcomes that reflect clinical practice. In: *Patient Outcomes Research: Examining the Effectiveness of Nursing Practice. Proceedings of the State of the Science Conference Sponsored by the National Center for Nursing Research.* NIH Pub. No. 93-3411. Washington, DC: NIH; 1992:27–38.
7. Donabedian A. *The Methods and Findings of Quality Assessment and Monitoring: An Illustrated Analysis.* Vol 3. Ann Arbor, MI: Health Administration Press; 1985.
8. Donabedian A. Evaluating the quality of medical care. *Milbank Q.* 1966;44 (3, part 2):166–206.
9. Gardner DL. Measures of quality. *Series on Nursing Administration.* 1992;3:42–58.
10. Lohr KH. Outcome measurement: concepts and questions. *Inquiry.* 1988;25(1):37–50.
11. Tarlov A, Ware J, Greenfield S, Nelson E, Perrin E, Zubkoff M. The medical outcomes study: an application of methods for monitoring the results of medical care. *JAMA.* 1989;262(7):925–930.
12. Kelly KC, Huber DG, Johnson M, McCloskey JC, Maas M. The medical outcomes study: a nursing perspective. *J of Professional Nursing.* 1994;10(4):209–216.
13. Shaughnessy PW, Crisler KS. *Outcome-based quality improvement: a manual for home care agencies on how to use outcomes.* Washington, DC: National Association for Home Care; 1995.
14. Nadzam DM. Nurses and the measurement of health care: an overview. In: *Nursing Practice and Outcomes Measurement.* Oakbrook Terrace, IL: Joint Commission on Accreditation of Healthcare Organizations; 1997:1–15.
15. Oermann M, Huber D. New horizons. *Outcomes Manage Nurs Prac.* 1997; 1(1):1–2.
16. Moss MT, O'Connor S. Outcomes management in perioperative services. *Nurs Econ.* 1993;11:364–369.
17. Cohen EL, Cesta TG. *Nursing case management: from concept to evaluation.* St. Louis: CV Mosby; 1993.
18. Zander K. CareMaps®: The core of cost/quality care. *The New Definition.* 1991;6(3):1–3.
19. Gardner K, Allhusen J, Kamm J, Tobin J. Determining the cost of care through clinical pathways. *Nurs Econ.* 1997;15(4):213–217.
20. Ireson CL. Critical pathways: effectiveness in achieving patient outcomes. *J Nurs Adm.* 1997;27(6):16–23.
21. Lansky D. The new responsibility: measuring and reporting on quality. *J Qual Improve.* 1993;19:545–565.
22. Ellenberg DB. Outcomes research: the history, debate, and implications for the field of occupational therapy. *Am J Occup Ther.* 1996;50:435–441.
23. Wennberg J. Outcomes research, cost containment and the fear of health care rationing. *N Engl J Med.* 1990;323:1202–1204.
24. Clancy CM, Kamerow DB. Evidence-based medicine meets cost-effectiveness analysis. *JAMA.* 1996;276:329–330.
25. Titler MG, Reiter RC. Outcomes measurement in clinical practice. *MEDSURG Nurs.* 1994;3:395–398, 420.
26. Wilson AA. The quest for accountability: patient costs & outcomes. *CARING.* 1996;XV(6):24–28.

27. Strodtman LKT. The historical evolution of nursing as a profession. In: Oermann MH. *Professional Nursing Practice.* Stamford, CT: Appleton & Lange; 1997:38.

28. Bergner M. Measurement of health status. *Med Care.* 1985;23:696–704.

29. Aydelotte M. The use of patient welfare as a criterion measure. *Nurs Res.* 1962;11(1):10–14.

30. Marek KD. Outcome measurement in nursing. *J Nurs Qual Assur.* 1989;4:1–9.

31. Hover J, Zimmer M. Nursing quality assurance: the Wisconsin system. *Nurs Outlook.* 1978;26:242–248.

32. Horn BJ, Swain MA. *Criterion Measures of Nursing Care Quality.* DHEW Pub. No. PHS78-3187. Hyattsville, MD: National Center for Health Services Research; 1978.

33. Daubert E. Patient classification system and outcome criteria. *Nurs Outlook.* 1979;27:450–454.

34. Martin KS. Nursing and patient care processes: nursing care outcomes measurement. In: *Nursing Practice and Outcomes Measurement.* Oakbrook Terrace, IL: Joint Commission on Accreditation of Healthcare Organizations; 1997:17–34.

35. Martin KS, Scheet NJ. *The Omaha System: Applications for Community Health Nursing.* Philadelphia: WB Saunders; 1992.

36. Roper WL, Winkenwerder W, Hackbarth GM, et al. Effectiveness in health care: an initiative to improve medical practice. *New Engl J Med.* 1988;319:1197–1202.

37. Nadzam DM. The Agenda for Change: update on indicator development and possible implications for the nursing profession. *J Nurs Qual Assur.* 1991;5(2):18–22.

38. Lalonde B. *Quality Assurance Manual of the Home Care Association of Washington.* Edmonds, WA: The Home Care Association of Washington; 1986.

39. Rinke L. *Outcomes Measures in Home Care: State of the Art,* Vol. 3. New York: National League for Nursing; 1988.

40. Sorgen LM. The development of a home care quality assurance program in Alberta. *Home Health Care Serv Q.* 1986;7(2):13–28.

41. Marek KD. Measuring the effectiveness of nursing care. *Outcomes Manage Nurs Prac.* 1997;1(1):8–12.

42. Shaughnessy PW, Crisler KS, Schlenker RE, et al. Outcome-based quality improvement in home care. *CARING.* 1995;XV(6):44–49.

43. Research update. Outcome-based quality improvement demonstration. *CARING.* 1996;XV(6):67.

44. Health Care Financing Administration. *Federal Register,* March 10, 1997;62(46):11004–11064.

45. Field MJ, Lohr KN, eds. *Clinical Practice Guidelines: Directions for a New Program.* Washington, DC: National Academy Press; 1990.

46. Brown JB, Shye D, McFarland B. The paradox of guideline implementation: how AHCPR's depression guideline was adapted at Kaiser Permanente northwest region. *J Qual Improve.* 1995;21:5–21.

47. Kaegi L. Nurses leading the charge to take national AHCPR guidelines into local settings. *J Qual Improve.* 1995;21:45–49.

48. Dean-Baar SL. Application of the new ANA framework for nursing practice standards and guidelines. *J Nurs Care Qual.* 1993;8:33–42.

49. Montgomery LA, Budreau GK. Implementing a clinical practice guideline to improve pediatric intravenous infiltration outcomes. *AACN Clinical Issues.* 1996;7:411–424.

50. Yoos HL, Malone K, McMullen A, et al. Standards and practice guidelines as the foundation for clinical practice. *J Nurs Care Qual.* 1997;11(5):48–54.

51. Joint Commission on Accreditation of Healthcare Organizations. Performance improvement tools for outcomes measurement. In: *Nursing practice and outcomes measurement.* Oakbrook Terrace, IL: Author; 1997:123–161.

52. Katz JM, Green E. *Managing Quality.* 2nd ed. St. Louis: CV Mosby; 1997.

53. Harris MD, Dugan M. Evaluating the quality of home care services using patient outcome data. *Home Healthcare Nurse.* 1996;14: 463–468.

54. Parisi LL. What is influencing performance improvement in managed care? *J Nurs Care Qual.* 1997;11(4):43–52.

55. National Committee for Quality Assurance (NCQA). *HEDIS 3.0,* vol. 2. Washington, DC: NCQA; 1997.

56. Nelson EC, Batalden PB, Plume SK, et al. Report cards or instrument panels: who needs what? *J Qual Improve.* 1995;21:155–166.

57. Nugent WC, Schults BA, Plume SK, et al. Designing an instrument panel to monitor and improve coronary artery bypass grafting. *J Clin Outcome Measure.* 1994;1(2):57–64.

58. Schriefer J, Urden LD, Rogers S. Report cards: tools for managing pathways and outcomes. *Outcomes Manage Nurs Prac.* 1997;1(1):14–19.

59. Martin KS, Scheet NJ, Stegman MR. Home health clients: characteristics, outcomes of care, and nursing interventions. *Am J Public Health.* 1993;83:1730–1734.

60. National League for Nursing (NLN). *Summary of Findings: In Search of Excellence in Home Care.* New York: NLN; 1994.

61. Johnson M, Maas M, eds. *Nursing Outcomes Classification (NOC).* St. Louis: CV Mosby; 1997.

62. Maas ML, Johnson M, Moorhead S. Classifying nursing-sensitive patient outcomes. *Image.* 1996;28:295–301.

63. American Nurses Association (ANA). *Nursing Quality Indicators.* Washington, DC: ANA; 1996.

64. Ellwood PM. Shattuck lecture—outcomes management: a technology of patient experience. *New Engl J Med.* 1988;318:1549–1556.

65. Kania C, Richards R, Sanderson-Austin J, et al. Using clinical and functional data for quality improvement in outcomes measurement consortia. *J Qual Improve.* 1996;22:492–504.

66. Houston S, Miller R. The quality and outcomes management connection. *Crit Care Nurs Q.* 1997;19(4):80–89.

67. Wojner AW. Outcomes management: from theory to practice. *Crit Care Nurs Q.* 1997;19(4):1–13.

68. Hoesing H, Karnegis J. Nursing and patient care processes: interdisciplinary care outcomes management. In: *Nursing Practice and Outcomes Measurement.* Oakbrook Terrace, IL: Joint Commission on Accreditation of Healthcare Organizations; 1997:35–62.

69. Huber D. *Leadership and Nursing Care Management.* Philadelphia: WB Saunders; 1996.

70. Simpson R. Case-managed care in tomorrow's information network. *Nurs Manage.* 1993;24(7):14–16.

71. Levknecht L, Schriefer J, Schriefer J, Maconis B. Combining case management, pathways, and report cards for secondary cardiac prevention. *J Qual Improve.* 1997;23(3):162–174.

72. Willoughby C, Budreau G, Livingston D. A framework for integrated quality improvement. *J Nurs Care Qual.* 1997;11(3):44–53.

■ 2 ■

Patient Care Outcomes: A League of Their Own

Dominick L. Flarey

Everyone is talking about them, many are measuring them, and the majority are continuing to define them. Patient care outcomes are a necessary component of the entire process of care delivery. As such, their measurement and management are becoming more and more imperative in this new era of health care. The concept of patient care outcomes is not new. Nursing in particular has been defining and measuring them since the birth of the profession. What is new is their importance in demonstrating the value of our care initiatives in today's managed-care environment.

What exactly are patient care outcomes? Patient care outcomes may be defined as the measurement and assessment of the status of the patient following health care interventions. While this definition may seem simple, the overall concept of measuring outcomes can be challenging. The new adage in managed care is that one must deliver a high-quality service at an affordable price, with good outcomes. Much debate exists as to what is considered high quality, what *affordable* means and by whose standards, and what constitutes good outcomes. Despite the ensuing debate, nursing is emerging as the profession most prepared to address the current issue of patient care outcomes. Other health care professions are looking to us to provide leadership and guidance with this new mandate.

WHY MEASURE PATIENT CARE OUTCOMES?

The first and most important reason to measure patient care outcomes is to provide health care professionals with evaluative feedback regarding the actual care

that is being provided. Each care or treatment modality has a particular goal for the patient. Thus, each activity will also have some potential effect on the patient's overall condition and status. Examining the outcomes of patient care reveals to us whether the patient is responding satisfactorily to the planned interventions and whether the patient is meeting preestablished benchmarks for his or her particular problem at specifically determined times in the course of care delivery.

Health care professionals need to assess patient care outcomes constantly to provide a level of quality care and intervene appropriately when patients are not meeting defined benchmarks for care progression. There are at least nine other important and compelling reasons to measure and manage patient care outcomes:

1. to demonstrate the rationale for choosing specific medical and nursing interventions
2. to provide benchmarks that can be used to evaluate subsequent care outcomes across settings, populations, and other demographics
3. to demonstrate to third-party payers and society as a whole the effectiveness of care delivery
4. to assist health care professionals in adequately defining the concept of "quality" in care delivery
5. to assist the health care team in developing an individualized plan of care based on continual assessments of outcomes
6. to assist health care professionals in developing a research-based practice that is rooted in actual clinical experiences in patient care
7. to develop and test theories of care delivery in the practice setting
8. to assist in placing a monetary value on patient care delivery related to specific outcomes

Source: Reprinted with permission from D.L. Flarey, Patient Care Outcomes: A League of Their Own, *Outcomes Management for Nursing Practice,* Vol. 1, No. 1, pp. 36–40, © 1997, Lippincott-Raven Publishers.

9. to test collaborative practice interventions and evaluate how synergy among disciplines leads to further enhanced care outcomes

While the above reasons for measuring outcomes are not all-inclusive, they do provide a compelling justification for nurses' continuing efforts to define the specific reasons for and benefits of measuring outcomes in patient care delivery. Nursing has carved a niche as a leader in defining and measuring care outcomes, especially those that go beyond the interest of medicine. From the first phases of our educational process to current practice at the bedside, to some extent we have always incorporated outcomes measurement in patient care delivery. Today, the spotlight shines brightly on outcomes measurement and management.

In planning for outcomes measurement, two decisions are what to measure and the "how to" of outcomes measurement and management. Deciding what to measure is based on what we *expect* our interventions to achieve for patients. Outcomes are what we *hope* to achieve for patients.

MEASURE METHODOLOGY

To more fully operationalize the concept of what we measure related to patient care outcomes, this chapter will use the patient condition of heart failure as an example. This particular condition was selected because it currently dominates as an issue in managed patient care due to its complexity, enormous costs, and functional impairment of patients. Heart failure is a frustrating and costly condition to treat; defining outcomes criteria for this population also is challenging.

The first step before actually defining what should be established as outcomes for measurement and management in patients who have heart failure is a methodology that will lead us to making correct choices. The methodology recommended here is based upon the premise of critical needs theory. With each of the major domains—physiologic, psychological, and cognitive—critical questions need to be answered or critical issues defined. Attention to this particular detail will lead to the most obvious outcomes that must be measured for the patient, based on his or her condition.

The starting place for this methodology is viewing the patient physiologically. At this level, a person's needs are related to physiologic functioning or homeostasis and survival. The most critical questions that need to be addressed here are:

- What is disrupting homeostasis?
- What body system or physiologic function is impaired?

- What objective and subjective criteria can be used to evaluate adequately the level of impairment experienced by the patient and the type and degree of response to interventions?
- Which disruption is most damaging to comfort and sustaining life?

On the basis of these questions and their answers, nurses can derive the particular outcomes that should be measured. Exhibit 2–1 demonstrates how this methodology can be used.

WHY WE MEASURE OUTCOMES

From Exhibit 2–1, we can easily identify what outcomes need to be assessed and what outcomes are the most important for evaluating effectiveness of treatment. On the basis of the above analysis, the following patient care outcomes are recommended for assessment from a physiologic perspective:

- degree of dyspnea
- utilization of oxygen
- degree of disability related to activities of daily living
- type of breath sounds
- functional status of the patient
- rate and rhythm of pulse
- color of skin and nail beds
- degree of chest pain
- presence of orthopnea
- degree of jugular venous distention
- degree of abdominal distention
- degree of peripheral edema
- evidence of digitalis toxicity
- adequacy of oxygen saturation

These are the most important outcomes related to patients who experience congestive heart failure. They may be assessed using the format of a nursing diagnosis and an expected outcomes statement. Exhibit 2–2 provides us with a simple example.

Once both a nursing diagnosis and an outcomes statement have been written, it is necessary to provide for assessment and documentation of the defined outcomes. Exhibit 2–3 provides a sample of an assessment related to the nursing diagnosis for congestive heart failure. This assessment should be completed at the time of patient discharge from an acute care hospital.

The assessment of patient care outcomes in Exhibit 2–3 provides a comprehensive, documented assessment of the critical indicators for outcomes measurement at the time of patient discharge. The rationale for providing a full, detailed outcomes assessment is that it

Exhibit 2–1 Patient Care Outcomes: Physiologic Needs

Critical Question	Response	Major Symptoms and/or Dysfunction
What is disrupting homeostasis?	Alteration in the pumping mechanism of the heart, leading to poor pulmonary perfusion, causing hypoxia, which compounds the alteration in the pumping mechanism of the heart	Fatigue Dyspnea Chest pain Adventitious breath sounds
What body system or physiologic function is impaired?	Cardiovascular system, pumping action of the heart, pulmonary system	Cardiac enlargement Pulmonary congestion Cardiac arrhythmia
What subjective/objective criteria can be used to evaluate adequately the level of impairment and the type and degree of response to interventions?	Subjective: feelings of dyspnea, degree of dyspnea, precipitating factors, degree of relief with oxygen, degree of relief from diuretics, degree of relief from cardiovascular drugs	Pulmonary congestion, low arterial P_{O_2}, resulting in inadequate tissue perfusion
Which disruption is most critical to sustaining life?	Poor or failed pumping action of the heart; pulmonary congestion; tissue anoxia, leading to cardiac arrhythmia	Dyspnea, bilateral rales, cyanosis, peripheral edema, pulsus alternans, orthopnea, cough, fatigue, confusion, jugular venous distention, abdominal distention

- provides for a comprehensive assessment of the patient at the time of discharge
- is the patient's profile of progression or lack of progression at the time of discharge from an acute care setting
- provides a means to evaluate treatment effectiveness over the course of the acute episode of care
- provides documentation of the effectiveness of treatment over the course of the acute episode of care
- conveys important patient information to other health care providers who will continue providing patient care services after discharge
- provides a vehicle for the collection and aggregation of data related to patient care
- provides internal benchmarks for analysis by health care providers

- is a means to guide continued treatment of the patient over a longer course of time
- provides critical information for the redesign of processes to improve future patient outcomes
- provides concrete information regarding treatment effectiveness for managed-care networks and regulatory agencies
- documents the outcomes of the nursing process
- provides for defensive documentation in instances of litigation
- provides a profile of the patient's health status at a particular point in time

A LEAGUE OF THEIR OWN

The assessment and documentation of patient care outcomes are more important in health care today than ever before. Consumers want to know the effects of treatment and are demanding a much higher level of quality from the care services they purchase. Patient care outcomes need to be firmly established for all of the diagnostic-related groups for which an organization provides care.

Defined outcomes should be a component of an outcomes pathway and should be documented at least once daily on the pathway. For the documentation of discharge outcomes, a dedicated outcomes assessment form, as provided in Exhibit 2–3, should be created and should be a part of the patient's permanent medical

Exhibit 2–2 Nursing Diagnosis and Outcomes Statement

Nursing Diagnosis: Gas exchange, impaired, related to pulmonary congestion and fluid in the alveoli.

Outcomes Statement: "The patient will have improved gas exchange, as evidenced by vital signs within normal limits for patient's age and condition, skin and mucous membranes without cyanosis or pallor, decreased dyspnea, and arterial blood gases within normal limits."[1(p.1175)]

Exhibit 2–3 Assessment of Patient Outcomes at Discharge

Dyspnea	No dyspnea at rest. Mild dyspnea noted after ambulating in hall. Relieved rapidly with rest.
Utilization of O_2	Used O_2 when eating and while giving self a bath. Not in use otherwise. Has O_2 at home.
Activities of daily living	Able to give self a bed bath, able to eat without assistance, able to dress self without signs or symptoms of respiratory or other distress.
Breath sounds	Bronchial, clear to auscultation.
Functional status	Self-care is good for condition. Alert, oriented, answers appropriately.
Pulse	92, slightly irregular. Rhythm strip shows persistent rare premature atrial contractions.
Color	Normal, no cyanosis, no jaundice.
Chest pain	No chest pain. Last episode was 2 days ago. Not using nitroglycerin.
Orthopnea	Sleeping with two pillows as prior to admission.
Jugular venous distention	None today. Last evident upon admission and prior to acute care treatment.
Abdominal distention	Girth is 38, which is preadmission girth. Down from 44 upon admission.
Peripheral edema	Mild, 1+ edema, nonpitting.
Response to digitalis	No evidence of digitalis toxicity. Digitalis level this morning in therapeutic range.
Oxygen saturation	O_2 saturation as of 22 hours ago was 96%. No physiologic evidence of significant change.

record. Information from this discharge outcomes form can be included in a large patient care outcomes database so that outcomes can be aggregated and analyzed over time. This aggregation of outcomes provides the basis for the evaluation of patient care outcomes on an organizationwide or systemwide level.

Critical information can be obtained from this aggregate of outcomes data. Reviewing the outcomes data over time can prompt the development of critical questions that can be posed to improve the quality of outcomes through process improvements. Some of these critical questions for the diagnosis of congestive heart failure are:

- How do our patients compare in their length of stay to patients of other organizations or health care systems?
- What percentage of patients develop digitalis toxicity?
- What percentage of patients can enjoy activities of daily living without significant assistance?
- What percentage of patients demonstrate resolution of adventitious breath sounds by the time of discharge?
- What other medical and nursing interventions or therapies might provide better patient outcomes?

While these questions are not all-inclusive, they do provide an example of how health care providers can use the aggregate information from patient outcomes data to develop newer, more innovative approaches to patient care. Answering these questions is not a easy task. Not all patients are the same; each comes to the acute care episode of care with a different degree of congestive heart failure, with different comorbidities, and with a different level of compliance with treatment. One of our greatest challenges today is adjusting for these discrepancies and problems in adequately assessing outcomes. Though the assessment of patient care outcomes is in its infancy, future applications to help adjust for patient differences will be more commonplace. Methods and models for the aggregation of data related to specific predetermined criteria such as age, sex, and cognitive functioning are being developed and will continue to be a major focus of effort.

Physiologic patient care outcomes are not the only measures that are important. Other individual patient-related outcomes may be equally important and should not be neglected. These outcomes are related to psychological status and cognitive status. With respect to psychological status, some outcomes for congestive heart failure to be measured include

- presence/absence of anxiety
- presence/absence of depression
- ability to cope with a chronic illness
- degree of family support
- effect of disability on mental health
- evidence of suicidal ideations
- ability to plan for the future
- motivation to take responsibility for the illness

Cognitive outcomes also are important because poor cognitive outcomes may prevent patients from achieving levels of wellness or improvements in their condition. A few cognitive outcomes that should be assessed in patients with congestive heart failure are

- the degree of understanding and the depth of knowledge that the patient has of the illness and self-care
- the degree of understanding that the family/support system has of the patient's illness
- the ability of the patient to care for self and plan care for self
- the degree of understanding of needed lifestyle changes

While this is not an exhaustive list of psychological and cognitive outcomes, it does provide a stimulus to think about the psychological and cognitive aspects of the patient's illness and how these can positively or negatively influence the course of illness and ultimate health status outcomes. Quality care means assessing all the dimensions of potential patient outcomes. Patient care outcomes are not exclusive to disease states and physiologic dysfunction. All three domains—physiologic, psychological, and cognitive—interplay to influence the overall health status of individuals. Therefore, parallel outcomes should be asssessed in each domain.

CONCLUSION

Defining patient care outcomes is a journey. We are just beginning that journey in health care along the road of patient care outcomes measurement and management. This is an exciting time in health care and one in which nursing has great opportunity to demonstrate its ability to positively affect the overall health of this nation.

As we continue to explore the realm of patient care outcomes, nurses will play more pivotal roles in defining what must be measured and, more important, what needs to be done to influence outcomes positively for patients in the future. We are more than likely to discover that nursing interventions, rather than some new fad in technology, have the most dramatic impact on patient care outcomes. Nurses have always known that nursing assessment and strong nursing care and interventions are what quality care is made of.

REFERENCE

1. Luckman, S. *Medical-Surgical Nursing: A Psychophysiologic Approach.* 4th ed. Philadelphia: WB Saunders; 1993.

■ 3 ■

The Future of Collaborative Path-Based Practice

Patrice L. Spath

Collaboration has many advantages when it is applied successfully in health care delivery. Collaboration builds an awareness of interdependence. When people recognize the benefits of helping one another and realize that it is expected, they will work together to achieve common goals. When people work together to achieve common goals, they stimulate each other to higher levels of accomplishment. Collaboration builds and reinforces recognition and mutual respect within a team. People have an opportunity to see the effect of their effort and the efforts of others on achievement. Most important, collaboration benefits the patient. For example, caregivers at the University of Massachusetts Medical Center, Worcester, Massachusetts, found that when surgeons, nurses, and anesthesiologists worked together, patients were more likely to receive appropriate laboratory tests before their elective surgery.[1]

The importance of collaboration among caregivers is not a new idea. In the early days of the nursing profession, Florence Nightingale stressed the concept of "system" rather than "self" in her book *Notes on Nursing.* Nightingale noted:

> How few men, or even women, understand what it is to be "in charge." To be "in charge" is not only carrying out the proper measures yourself but [also seeing] that every one else does so too. People who are in charge often seem to have a pride in feeling that they will be "missed," that no one [but themselves] can understand or carry out their arrangements, their system, books, accounts, etc. It seems to me that the pride is rather in carrying on a system . . . so that, in case of absence or illness, any body can understand . . . and all will go on as usual."[2(pp.24–25)]

Nightingale recognized that individual caregivers are part of the bigger system of health care. While each professional adds value to the process of patient care, that value is diminished when the system itself is dysfunctional.

A number of initiatives aimed at improving collaboration and the systems of care delivery have been undertaken since the days of Florence Nightingale. In the United States, the first care coordination efforts occurred in the public health and mental health sectors.[3] In hospitals, the nursing profession took the lead in promoting better collaboration among caregivers. This leadership was evidenced by initiatives such as the New Jersey Department of Health's Nursing Incentive Reimbursement Award Program, which was aimed at increasing nurses' accountability for the quality of patient care and enhancing interdisciplinary collaboration. Out of these efforts came nursing-oriented tools, such as critical paths, that helped to improve communication among hospital caregivers.[4]

In 1994, the Joint Commission on Accreditation of Healthcare Organizations changed the structure of all of its standards to emphasize the importance of collaboration. Standards are no longer contained in chapters that refer to individual departments or services. They are now divided into chapters referred to as "important functions"—which, according to the Joint Commission, are "goal directed, inter-related series of processes."[5(p.713)] To meet the intent of the standards, caregivers must jointly define patient care goals and work together to achieve those goals.[6]

In only a few short years since path-based practice was introduced by New England Medical Center in 1989, the concepts and tools have gained considerable support among caregivers in all settings. Today, it is hard to pick up a professional journal without finding one or more articles on clinical paths, team-based practice models, or similar concepts. However, the question of "What's next?" is still unanswered. Will clinical paths continue to be the primary document used to promote cooperation among caregivers? Are new tools on the horizon? Will a well-functioning system of care, with collaboration an important element in that system, remain just as important as it was in the days of Florence Nightingale? Without a crystal ball, this question can only be answered by an examination of today's trends.

This chapter describes four major issues that are likely to affect collaborative, path-based practice in the future: evidence-based medicine, reduction of process variation, continuum-of-care collaboration, and automated health information. The reason that each issue is significant is presented along with examples of how some of today's providers have already begun to respond.

EVIDENCE-BASED MEDICINE

The word "clinical path" initially brought images of cookbook medicine and other negative thoughts to many caregivers. However, practitioners are increasingly aware of their accountability in the application of research evidence to clinical practice.[7] For this reason, evidence-based medicine is becoming more important and clinical practice guidelines are becoming the tool of choice for stabilizing patient care processes. Clinical practice guidelines are defined as "systematically developed statements to assist practitioner and patient decisions about appropriate health care for specific clinical circumstances."[8(p.2)] Unlike the nursing-oriented paths originally designed for collaborative practice initiatives, clinical practice guidelines are developed using a rigorous methodological approach that includes a literature review to determine the strength of the evidence for each patient care recommendation in the guideline.

Clinical practice guidelines may be presented as *clinical algorithms*. These are written guides to stepwise evaluation and management strategies that include: (1) explicit descriptions of an ordered sequence of steps to be taken in patient care under specific circumstances; (2) required observations to be made; (3) decisions to be considered; and (4) actions to be taken.[9] They may also include *policy statements*. These are narrative descriptions of the recommended course of action in a particular clinical situation.

Clinical practice guidelines are not presented in a clinical path format. A clinical path is a collaborative practice tool that organizes, sequences, and times the major patient care activities and interventions of the entire interdisciplinary team for a particular diagnosis or procedure.[10] Clinical practice guidelines tend to focus on those aspects of diagnosis and treatment for which evidence is required.[11] Clinical paths include many interventions that do not require research evidence. These path recommendations are derived from clinicians' personal experiences or administrative "best-practices."

How Evidence-Based Medicine Will Affect Path-Based Practice

Rather than relying solely on personal observations, caregivers are now pooling research results and outcomes data to identify best practices for themselves and their institutions. The guideline movement is an effort to codify best practices and reduce the amount of inappropriate variation in patient care. Tomorrow's path-based practices are likely to include a combination of evidence-based tools, such as clinical algorithms and other guideline reminder systems, as well as clinical paths.

An organization that has already begun to move in this new direction is William Beaumont Hospital in Royal Oak, Michigan. The original clinical path initiative at William Beaumont in 1992 was based on the traditional nursing-oriented model.[12] While the original paths were approved by the medical staff, they primarily helped the nurses standardize their practices. The paths were not used by physicians and other caregivers. Therefore, in 1996, William Beaumont Hospital reinvented its path-based practices. Under the direction of Kay Beauregard, MSA, RN, Director of Clinical Management and Quality, and Steven C. Winokur, MD, Medical Director of Quality Improvement, the clinical pathway initiative was transformed into an collaborative, interdisciplinary process. Pathway project teams are now co-chaired by a physician leader appointed by the medical staff department chair and by a nurse or a professional from another department (e.g., dietary, respiratory). The pathway development process includes a review of current literature and industry "best practice" standards. Each pathway project results in three documents:

1. *Clinical Practice Guidelines.* This document outlines the treatment decisions for those aspects of patient care for which evidence or consensus is required. The guidelines are based on current research and, where research is lacking, physicians' personal observations. The guidelines are illustrated in a narrative fashion on a path-like format that resembles the hospital's overall clinical pathway form (see Exhibit 3–1 and Appendix 3–A for abbreviations key). In most instances, the narrative guidelines are accompanied by a clinical algorithm illustrating the key treatment decision steps (see Figure 3–1 and Appendix 3–A for abbreviations key).

Exhibit 3–1 Clinical Pathway for Congestive Heart Failure Clinical Practice Guidelines: Physician Guide

A pathway does not represent the standard of care. Clinical judgment may supersede these guidelines.

INITIAL EVALUATION

TEST	FINDING	SUSPECTED DIAGNOSIS
12 Lead ECG	Acute ST-T wave changes	Myocardial ischemia
	Atrial fibrillation, other tachyarrhythmias	Thyroid disease or heart failure due to rapid ventricular rate
	Bradyarrhythmias	Heart failure D/T low HR
	Previous MI	Heart failure D/T reduced left ventricular function
	Low voltage	Pericardial effusion
	Left ventriclar hypertrophy	Diastolic dysfunction
CBC	Anemia	Heart failure due to or aggravated by decreased oxygen-carrying capacity
UA	Proteinuria	Nephrotic syndrome
	RBCs or cellular casts	Glomerulonephritis
Serum creatinine	Elevated	Volume overload due to renal failure
Albumin	Decreased	↓ extravascular volume D/T hypoalbuminemia
T4, TSH	Abnormal T4, TSH	Heart failure D/T or aggravated by hypo/hyperthyroidism
2 DE	EF < 40% EF > 40%	Systolic dysfunction Diastolic dysfunction

MEDICATIONS

1. IV diuretics
2. Mg, K+ supps as needed
3. ASA if CAD, TIA, CVA hx
4. Hep SQ for DVT prophylaxis

Systolic Dysfunction
1. IV diuretics
2. Responding to diuretics without increasing BUN/CR?

YES—convert to oral diuretics, digitalize, ACE I

NO—consider IV inotropes, digitalize, vasodilator tx, consider swan to tailor tx

Diastolic Dysfunction
1. IV diuretics
2. Consider Nitrates, Ca channel blockers, and/or Beta blockers
3. Avoid Digoxin and most inotropes, use diuretics cautiously

ADMISSION CRITERIA

MCU/RMF
1. Mild/Mod DOE
2. VSS
3. +/- telemetry
4. Renal dose dopamine only
5. Dobutamine < 10 mcg not requiring titration

CPCU
1. Mild/Mod DOE
2. VSS
3. Continuous telemetry
5. +/- IV inotropes
6. R/O ischemia
7. +/- arrhythmias

CCU
1. Mod/Severe DOE
2. Unstable VS
3. 1:1 Nursing care required
5. +/- Swan ganz
6. Probable ischemia
7. Unstable arrhythmias

Discharge Medication Regime
1. ACE I
2. ASA if history of CAD, TIA, CVA.
3. Diuretics with K+ supplements as needed
4. Nitrates
5. Consideration of digoxin and Coumadin
6. Avoidance of Ca channel blockers if low EF

DISCHARGE GUIDELINES

1. Edema/rales at baseline for patient
2. Ambulating with minimal/no DOE
3. Afebrile, VS WNL for patient, no ECG changes for 24 hours, RH rhythms stabilized
4. Labs & CXR WNL
5. Diet instructions, low Na+ diet
6. Cardiac rehabilitation
7. Daily weight instructions
8. Early S/Sx of CHF and when to call doctors or 911 discussed
9. Discharge instructions reviewed

DAYS 1–5

EC
- 12 lead ECG
- pOx
- 2 DE
- IV diuretics
- Consult Physician
- TX HTN/Ischemia
- CXR

DAYS 1-5
- Adjust Rx
- Weight QD
- Educate
- Monitor labs
- Gradual ambulation according to CR guidelines
- Evaluate possible discharge Day 4

792B JUNE 97

Not a permanent part of the medical record. MISD please return to MQPM.

Confidential—Peer Review

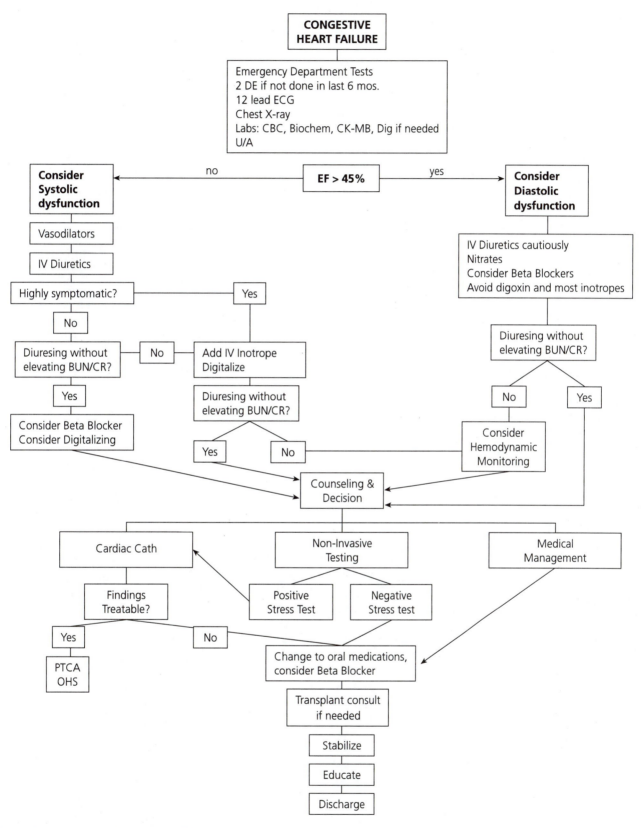

Figure 3–1 Congestive Heart Failure Clinical Algorithm. *Source:* Copyright © William Beaumont Hospital. Reproduction of the pathway is by permission of William Beaumont Hospital, Royal Oak, Michigan. All Rights Reserved.

2. *Physician Order Sheet.* All items that require physician orders are on this sheet (see Exhibit 3–2 and Appendix 3–A for abbreviations key). Separate medication order sheets may be developed for those patient groups that require numerous, complicated medication/IV orders.

3. *Clinical Pathway.* This document is used by the multidisciplinary team to coordinate patient care activities (see Exhibit 3–3 and Appendix 3–A for abbreviations key). It also serves as a tool for RN documentation, taking the place of the problem list or plan of care. The pathway includes common patient problems and expected outcomes and recommended nursing and clinical support service interventions. The physician-developed practice guidelines and elements on the order sheet are incorporated into the overall clinical pathway, but physicians are not expected to document on the clinical pathway.

The clinical practice guidelines and physician orders are approved by the affected medical staff departments and the medical care evaluation committee. The clinical pathway is approved by a work group of the hospital nursing practice committee. The director of pharmacy and, when appropriate, an infectious disease consultant also approve the pathway content.

William Beaumont's transition to an evidence-based, interdisciplinary approach to path-based practice has greatly enhanced the acceptance of pathways by all disciplines. During this transition, the caregivers at William Beaumont also learned an important lesson: one format for illustrating patient care recommendations cannot satisfy the needs of all disciplines. While the clinical pathway format adequately supports the nursing practice model, it is an unfamiliar paradigm for physicians. Physicians are accustomed to narrative clinical practice guidelines, clinical algorithms, and order sets. By broadening the available formats for illustrating path-based practice recommendations, the needs of all caregivers can be satisfied.

The importance of designing discipline-friendly reminder tools was also realized by Clinton Memorial Hospital Home Health. When physical therapists, speech therapists, social workers, and occupational therapists were asked to incorporate their treatment plans into the existing nursing care path model, the request was met with resistance.[13] It was necessary to allow these disciplines more latitude in designing paths that fit their unique care planning and treatment processes. Not only did these disciplines design a slightly different path format, but they chose to group their care paths into body system categories rather than diagnoses (e.g., upper extremity orthopaedic, nervous system).

CONTINUED NEED TO REDUCE VARIATION

Unnecessary or unintended variation in health care services can occur at the clinical decision-making level and in the delivery process itself.[14] Both types of variation have a potentially negative impact on health care costs and patient outcomes.

In the 1960s, the small area analysis studies of Dr. John Wennberg and colleagues revealed that where you live determines how you are medically treated.[15] Since that time, payers, clinicians, and providers have been working to reduce unnecessary variation in clinical decision making. As Berwick et al. pointed out in 1990:

> If Wennberg is right—and others have followed with similar findings—then the health care dollar is not only inflating, it is being spent largely in some colossal game of dice. The care patients receive apparently depends in large measure on who happens to be treating them. The evidence on variation makes simple trust in the quality of care seem naive.[16(p.8)]

Caregivers at Latter-Day Saints Hospital in Salt Lake City were among the first to substantiate the benefits of a stable patient care process.[17] In 1988, the hospital received a National Heart, Lung, and Blood Institute grant to test a new therapy for managing patients with adult respiratory distress syndrome (ARDS). A variant of traditional extracorporeal membrane oxygenation, the new therapy added equipment to simultaneously remove CO_2 and other waste products ($ECCO_2R$). A practice guideline for using $ECCO_2R$ grew out of this research project. After refining the guideline and by encouraging physicians to follow the guideline, the hospital was able to stabilize ventilator management for ARDS patients. Once treatment was stabilized, patient outcomes improved and costs were lowered.

Reducing variation in any process (clinical decision-making or the delivery process) involves two steps: (1) study the process as it now operates to identify potential sources of variation, and gather data to determine which activities are most likely to achieve desirable outcomes; and (2) stabilize the process by getting everyone to use the same procedures, materials, equipment, and so forth.[18] Clinical paths are a very popular tool for getting everyone to use the same procedures, materials, equipment, and so forth. However, the traditional path that focuses on the care provided to patients in particular diagnosis or procedure categories has limited use as a process stabilization tool. If everyone is being asked to follow the same process for a particular group of patients, then the group of patients must be clinically similar.

Exhibit 3–2 Congestive Heart Failure Physician's Order Sheet

ATTENTION PHYSICIAN:	This preprinted form is available to facilitate patient care.
	Order by checking the box in front of the item and/or completing the blanks where necessary.
	This is not for pharmacy orders.
	If additional items are needed, add them on the standard Doctors Order sheet (Form 280).

ADMIT ORDERS—CONGESTIVE HEART FAILURE

1. Admit to: ❏ MCU ❏ CPCU ❏ CCU ❏ _____

2. Condition: _____

3. Allergies: _____

4. Vital signs: • per unit protocol ❏ _____

5. Diet: • 2000 mg (2G) Sodium diet ❏ _____calorie ADA ❏ _____
    • Consult dietitian re: nutrition assessment and education
    ❏ Fluid restrictions _____ cc/day; ❏ IV _____ cc/day; ❏ PO _____ cc/day

6. Activity: • Bed rest, up in chair for meals & bedside commode for 24 hours then BRP and ambulate in room as tolerated Day 2

7. Nursing Treatments: • Cardiac Monitor (TMS) for MCU or Telemetry CPCU/CCU
    • I & O
    • Daily weight
    • CHF Teaching
    ❏ Foley catheter to dependent drainage, discontinue after 48 hours

8. Laboratory Tests:

• CBC (if not done in EC)	❏ IIT
• UA (if not done in EC)	❏ CKMB to total 3 (including EC labs)
• Magnesium Day 2	❏ Digoxin level
• Transition panel (if not done in EC)	❏ Lipid profile
• Electrolytes Q AM × 3 days	❏ PT
• BUN Q AM × 3 days	❏ PTT
• Creatinine Q AM × 3 days	❏ _____

9. Diagnostic Studies: • EKG (if not done in EC)
    • 2D Echo with Doppler (if not done in EC)
    ❏ CXR (_____), re: _____   ❏ portable
    ❏ MUGA

10. Respiratory Care: • O$_2$ per respiratory protocol

11. Consultation: ❏ Cardiologist Dr. _____   re: _____
    ❏ Dr. _____   re: _____

12. CPR Status Order: ❏ Full CPR (Call CPR Team)
    ❏ No CPR (CPR Team will not be called, no chest compressions will be initiated by unit staff in the event of cardiopulmonary arrest.)

_____ Pager No. _____
Attending Physician Signature

13. • Old medical records to nursing unit

Physician Assistant Signature	Date	Time	Physician Signature	Dr. Pager No.	Date	Time
Noted by Unit Secretary	Date	Time	Noted by RN		Date	Time

729A   JUNE 97 WP   Medical Record

Exhibit 3–3 Clinical Pathway, Congestive Heart Failure

A pathway does not represent the standard of care. Clinical judgment may supersede these guidelines.

CO-MORBIDITIES:

PATIENT PROBLEMS	EC	Admit Day 1	Day 2	Day 3	Day 4	Day 5
Alteration in fluid volume status R/T decreased CO.		VSS, I < O, diuresing AEB wt loss, dec. rales and edema.	Daily wts, diuresing, I < O.	Wt loss, I < O, rales and edema improving.	As day 3	VSS. At or near dry wt. Rales & edema at baseline.
Activity intolerance R/T decreased CO.		Tolerating minimal activity per CR level.	Tolerating activity with increased CR level.	Ambulating according to CR level.	As day 3	Ambulating without SOB. VSS.
Knowledge deficit R/T disease process.		Understands disease process, activity guidelines.	Understands disease, smoking cessation, daily wts. & S/S CHF.	As day 2.	Understands meds, dosages, indications, and side effects.	Reviews and understands discharge instructions.

Date:

	EC	Admit Day 1	Day 2	Decision Point	Day 3	Day 4	Day 5
LABS	___CBC ___BCP ___CK-MB ___BG if DM ___Dig level PRN ___UA	___PT/PTT ___UA if not done ___Eval need for TSH	___Lytes ___Renal Panel ___Mg ___Fasting LDL ___BG if DM as ordered	**MEDICAL MANAGEMENT continue**	___Lytes ___Renal Panel ___BG if DM as ordered	___Lytes ___RP if abnormal Day 3 ___BG if DM as ordered	___Repeat abnormal labs only
DIAGNOSTIC TESTS	___VS Q2 hrs and PRN ___CXR ___pOx or ABG ___12 lead ECG ___Telemetry ___2 DE if not recent eval	___VS per unit routine ___Admission weight ___I & O ___pOx ___Telemetry ___2 DE if not in EC	___VS per unit routine ___Weight ___I & O ___pOx ___Telemetry ___12 lead ECG	**NON-INVASIVE TESTING Schedule stress test for Day 3 or 4**	___VS per unit routine ___Weight ___I & O ___O₂ DC'd when sat >94% ___Telemetry	___VS per unit routine ___Weight ___I & O ___pOx with ambulation ___Consider DC telemetry	___VS per unit routine ___Weight ___DC I & O ___DC telemetry
TREATMENT	___IV diuretics ___Consider MS ___Treat HTN/ischemia ___O₂ per resp guidelines	___IV diuretics ___O₂ per resp guidelines ___Foley PRN ___Mg, K+ supps as needed ___Hep SQ for DVT prophylaxis ___**Medication considerations (See Physician Practice Guidelines)	___IV/PO diuretics ___Consider IVL ___O₂ ___***Medication considerations	**CARDIAC CATH Follow appropriate physician practice guide** ☐ **Pre-procedure checklist completed**	___IV/PO diuretics ___IVL ___O₂ ___***Medication considerations	___PO diuretics ___Adjust medications ___IVL	___PO diuretics ___DC IVL ___Eval need for home O₂
ACTIVITY	___CR level 1 ___Bedrest	___CR level 1 or 2 ___Bedrest ___Bedside commode	___CR level 3 ___BRP ___Chair for meals		___CR level 4 ___Amb in halls TID ___Chair for meals	___CR level 5 ___Amb in halls as tol	___CR level 6 ___Amb as tol
DIET	___2 gm Na	___2 gm Na	___2 gm Na		___2 gm Na	___2 gm Na	___2 gm Na
CONSULTS		___Cardiology ___Continuing Care	___Dietary ___PT if functional limitations		___Cardiac Rehab for Outpatient		
EDUCATION	___Orientation to hospital	___Intro to unit ___CHF teaching ___Activity instructions ___Diet, low Na/FR	___Smoking cessation ___Daily wt instructions ___S/Sx CHF & when to report	Tests/Procedures:	Handouts:	Medication Cards:	Discharge Checklist: ___When to call Dr or 911 ___Low Na Diet ___Daily weight ___Activity guidelines ___Medications ___Smoking Cessation ___Follow-up care ___Discharge instructions
RN SIGNATURE	RN:	RN:	RN:	RN:	RN:	RN:	RN:
SIGNATURE							
SIGNATURE							

7929 JUNE 97   N/A = Not applicable to patient   * = Variation; see Nurses/Progress Record for further information   MEDICAL RECORD

The challenge of defining a homogeneous group of patients is not easily overcome. Even the developers of clinical practice guidelines find it difficult to define clearly the population to whom their guideline applies. For example, thrombolytic therapy during acute myocardial infarction has been shown to save lives in many clinical trials.[19] Nonetheless, there is still confusion and debate about the effects of age, infarct location and extent, comorbid conditions, and time on responsiveness to thrombolysis. Even when the patient population is relatively homogeneous, all the factors of patient, disease, and therapy that physicians and other caregivers must consider in their clinical decisions cannot be built into a guideline or a clinical path.

How Variation Reduction Will Affect Path-Based Practice

Reduction of variation is likely to remain as a high priority among providers and caregivers. To get everyone to use the same procedures, materials, equipment, and so forth when treating a particular group of patients, the target population must be carefully selected. Developers of clinical paths are using several strategies to accommodate patient diversity better. Severity-adjusting patients in the same diagnostic or procedure group is one method used to account for the varying treatment requirements of a patient population. For example, pediatric patients with a diagnosis of asthma can require intensive care services or may be much less severely ill. The pediatric asthma evaluation scoring system shown in Table 3–1 is used by Good Shepherd Hospital in Barrington, Illinois, to severity-adjust patients who are placed on their pediatric asthma clinical path.[20] Illustrated in Exhibit 3–4 is their pediatric asthma path, which shows how the severity scoring system is used to determine which treatment plan is appropriate for the patient. At periodic intervals, the patient's current asthma severity score is calculated by the nursing staff and documented on the path. As the patient's severity score changes, so does his or her plan of care.

Some organizations design separate clinical paths for differing severity levels of patients admitted with the same diagnosis. For example, caregivers at Our Lady of Lourdes Medical Center in Camden, New Jersey, designed one clinical path for patients with congestive heart failure who required admission to an intensive care unit and a second path for congestive heart patients who did not require intensive care on admission.[21]

Another path-based practice strategy that can be applied to diverse patient groupings is the proactive approach being used at Pacific Medical Center in San Francisco. In 1991, Turley et al. applied the pathway method to patients with congenital heart disease. They found this population to be a very heterogeneous group of patients. For this reason, the traditional path was found to have limited value.[22] Therefore, a proactive model was developed. In the proactive pathway approach, the delivery system is responsive to the patient's performance, and the pathway can easily be altered to reflect his or her current treatment requirements. The patient's progression through his or her episode of care can be either slowed or accelerated on the basis of real-time feedback to the caregivers.[23]

Another tactic for overcoming the limitations of the traditional diagnosis/procedure-specific clinical path is to develop process-based or focused pathways. These paths illustrate the multidisciplinary process of care for a particular condition and are not specific to patients in a certain disease or procedure category. Examples of process-based clinical paths include those developed for wound care,[24] management of pressure ulcers,[25] organ donation,[26] comfort care of the dying patient,[27] management of patients with chronic pain,[28] and management of frail elderly patients.[29]

COLLABORATION ALONG THE CONTINUUM OF CARE

Improved collaboration among providers throughout an episode of care can have a positive impact on patient outcomes. However, the primary reason for today's systems integration is economic survival. Insurers, physicians, and hospitals are coming together to make money, protect their money, increase market share, protect market share, exert control over each other, and wrest control from the other side.

Table 3–1 Clinical Pediatric Asthma Evaluation Score (Good Shepherd Hospital, Barrington, Illinois)

	0	1	2
Pulse oximetry reading	95–100	85–94	<85
Color	pink on room air	pink on O$_2$	blue on O$_2$
Inspiratory breath sounds	normal	unequal	decreased to absent
Accessory muscles used (retractions)	none	moderate	maximal
Expiratory wheezes	none	moderate	marked
Cerebral function	normal	depressed or agitated	coma

Score of 0–2 is considered mild
Score of 3–8 is considered moderate
Score ≥ 9 is considered severe

Source: Copyright © Brown-Spath & Associates.

Exhibit 3–4 Pediatric Asthma Pathway (Good Shepherd Hospital, Barrington, Illinois)

	Day 1			Day 2			Day 3	
	Severe	Moderate	Mild	Severe	Moderate	Mild	Severe	Moderate
TESTS:								
Complete Blood Count	X	X						
Arterial blood gases if pulse oximetry < 90% saturation and patient on oxygen	X	X	X					
Theophylline level every 24 hrs if patient on aminophylline	X	X		X	X		X	X
Consider: Chest X-ray if febrile	X	X	X					
Electrolytes	X	X	X					
TREATMENTS:								
Albuterol Nebulizer	q 2-3 hrs. or prn	q 2–4 hrs. or prn	q 3–4 hrs.	q 3–4 hrs.	q 3–4 hrs.		prn	prn
Oxygen based on arterial blood gas or pulse oximeter results	X	X		X	X			
RESPIRATORY ASSESSMENT:								
Peak flow (Age ≥ 5 years)	q 4 hours	q shift	q shift	q shift	q shift		q shift	q shift
Pulse oximeter	Continuous	Continuous		Continuous	prn		prn	prn
Vital signs pre/post treatment	X	X	X	X	X		X	X
NURSING ASSESSMENT:								
Vital signs every 4 hours	X	X	X	X	X	X	X	X
Assessment every 4 hours	X	X	X	X	X	X	X	X
Asthma score	q 4 hrs	q 4 hrs	q 8 hrs	q 4 hrs	q 8 hrs	q 8 hrs	q 8 hrs	q 8 hrs
	Time/Score	Time/Score	Time/Score	Time/Score	Time/Score	Time/Score	Time/Score	Time/Score
	___ ___	___ ___	___ ___	___ ___	___ ___	___ ___	___ ___	___ ___
MEDICATIONS:								
Steroids	IV	IV	Oral or inhaler	IV	Oral or inhaler	Oral or inhaler	Oral	Oral
Antibiotics per chest X-ray, complete blood count, febril status	X	X	X	X	X	X		
Intravenous hydration if oral intake poor or vomiting	X	X	X	X	X			
Consider: Theophylline if indicated								
DIET/ACTIVITY:								
As tolerated	X	X	X	X	X	X	X	X
EDUCATION:								
Patient/Family teaching	X	X	X	X	X	X	X	X
Discharge instructions						X	X	X
DISCHARGE PLANNING:								
Assess home needs	X	X	X	X	X	X	X	X

Note: IV, intravenous; q, every; prn, as necessary.
Source: Copyright © Brown-Spath & Associates.

Caregivers are likely to see even more formal and informal linkages among service providers in the future. Much of this consolidation will continue to be economically driven. For example, in President Clinton's fiscal year 1998 budget, the Health Care Financing Administration (HCFA) announced plans to implement an integrated Medicare payment system for postacute services.[30] Administration officials have also discussed their intention to investigate methods for bundling Medicare payments for acute and postacute care. There is general agreement among HCFA officials and others that one consolidated (bundled) payment for a patient's entire episode of care would promote joint accountability among providers and ultimately lead to better patient outcomes.[31] However, considerable evaluation of differing bundled payment options must occur before HCFA can enact an "episode-of-care" payment system.

A second compelling argument for a seamless health care delivery system is that patients are demanding it. In the Pickwick/Commonwealth survey of hospital patients conducted in 1991, patients reported the following problems in making the transition from hospital to home:[32]

1. They are not receiving adequate clinical information during the hospitalization to make the transition to the community.
2. Their emotional and psychological needs are not adequately addressed.
3. They are not given enough control over their posthospital care choices.

In a national poll sponsored by the National Coalition on Health Care in December 1996, 79 percent of the 1,011 people surveyed agreed with the statement "There is something seriously wrong with our health care system."[33] Much of this concern relates to continuity-of-care issues. For example, researchers in Boston found that patients often feel that they have been provided inadequate information before hospital discharge regarding major elements of the postdischarge treatment plan, including medication and daily activities.[34]

Reimbursement and patient satisfaction will remain as two critical motives for improving collaboration among all providers along the continuum of care.

How Continuum-of-Care Collaboration Will Affect Path-Based Practice

The ultimate test for any organization will be its ability to integrate information systems and caregiver communication to deliver seamless, cost-effective quality health care services. Path-based practices are expanding to meet these needs. Clinical paths and other guideline re-

minder tools are being developed to help caregivers manage patients throughout an entire episode of care. These same collaborative practice tools are also being used in many disease management initiatives.[35]

For instance, after developing several clinical paths for their inpatient population, caregivers at St. Francis Medical in Trenton, New Jersey, embarked on a continuum-of-care program for patients with congestive heart failure (CHF).[36] This multidisciplinary effort produced several CHF treatment guidelines, which were displayed in many different formats: clinical pathways, clinical algorithms, and protocols. The format varied according to the intended users of the guideline and their intended purpose. For example, protocols describing appropriate management of CHF-related clinical situations were developed for use in the heart failure clinic and for telemanagement purposes. These protocols were developed jointly by the physicians, the medical residents, and an advanced-practice nurse.

Other components of the CHF program at St. Francis Medical Center include

- a heart failure clinic staffed by an advanced practice nurse who sees patients at regular intervals, monitors their progress, and suggests modifications to their plan of care as needed
- expanded cardiac rehabilitation services such that CHF patients are provided ongoing evaluations of their clinical status as well as an opportunity to interact with and be supported by staff and other patients
- telemanagement services that allow for 24-hour-a-day communication between the patient and his or her care providers
- a miniteam composed of community health specialists (nurse, social worker, dietitian, and other disciplines as needed) who complete the initial community-readiness assessment before the patient's discharge, validate the ongoing plan of care, and coordinate postdischarge referrals to relevant community agencies

Fully integrated health care delivery systems, such as Carondelet Health Care Corporation in Tucson, Arizona, are using a combination of clinical case managers, home care nurses, and community case managers to coordinate care for high-risk patient populations.[37] Tools such as pathways, guidelines, protocols, and physician order sets help to support the process.

AUTOMATED HEALTH INFORMATION

The health care industry spent $11.6 billion in 1996 to purchase products and services to support automated information systems efforts, a 16 percent increase over 1995. Expenditures in 1997 were expected to top the $13 billion mark.[38] Health care providers are spending these dollars to obtain information systems that can[39]

- serve all episodes of care wherever they take place within the organization
- provide immediate access to information at any site within the organization to all member and patient data for all episodes of care
- sort and analyze data in a variety of ways to support better clinical and business decisions
- replace the manual patient record with an electronic chart

The most direct contribution that information technology can make to collaborative path-based practice is to provide clinicians with better information for decision making, preferably at the point of care. This would enable clinicians to choose more effective services more quickly and help them avoid potentially tragic errors.[40] The range of available technologies to support this goal are vast and beyond the scope of this chapter. However, as these technologies become reality at the caregiver level, path-based practice is likely to change significantly.

For example, the health care community in Glens Falls, New York, is designing a computerized patient record system to augment their deployment of clinical practice guidelines, as well as to improve information flow.[41] This initiative is requiring a significant effort by physicians, hospital staff, and other entities involved in patient care to achieve a seamless flow of information through the continuum of patient care. Once this dream is realized, paper-based tools such as clinical paths will be replaced by point-of-care clinical decision support systems.

How Automated Health Information Will Affect Path-Based Practice

The goal of an automated health information system is to provide more complete and accurate information quickly to the clinician. Improving information flow and availability is expected to lead to better patient outcomes. Aside from the obvious continuity-of-care benefits from information linkages, clinical decision support systems are most likely to directly affect path-based practice.[42]

Clinical decision support systems will reduce or eliminate many of the inadequacies of today's paper-based tools. For example, patients may no longer need to be assigned to a particular category (diagnosis or procedure) before they can be placed on a clinical path. The patient's signs and symptoms, physical findings, test results, and background information will influence the suggested course of treatment. As the patient's condition changes, the computerized information system will analyze the new information about a patient and suggest an alternative plan of care. Computerization will also allow the development of complex treatment plans for patients with multiple comorbidities or complications, a logistical impossibility with paper-based clinical paths.

When implementing proactive pathways for their open-heart surgery patients, the cardiovascular caregivers at Pacific Medical Center soon realized that a paper-based system was too inflexible.[43] With the proactive pathway methodology, pathway recommendations and corresponding order sets needed to change as the patient's clinical condition fluctuated. The clinicians are currently in the process of developing a software program that will meet their needs.

Computerized information systems that interpret information about a patient using expertise captured in a computerized knowledge database are known as *expert systems.* Unlike traditional paper-based pathways, an expert system contains rules and decision algorithms that incorporate knowledge and judgment about the health problem at hand and alternative tests and treatments. These decisions are built into the expert-based system in the form of "if-then" rules as well as scoring algorithms, such as "If the patient's potassium is less than 3.0 mEq/dl and the patient is on digoxin, then the clinician should consider ordering potassium supplementation."[44(p.109)] These active care advice systems are designed to assist the clinician in performing diagnostic or therapeutic procedures (including pharmaceutical treatments) when the patient reaches certain stages in the process of care for a given health problem.

Active care advice systems may reduce the need for some of the surveillance people who now monitor the quality and appropriateness of care, such as utilization reviewers, case managers, and concurrent quality reviewers. Many automated information systems provide alerts to the clinician for situations such as potential adverse events (e.g., worsening of the patient's condition, based on the results of abnormal test results) and possibly inappropriate treatments (e.g., alerts regarding drug-drug allergies or drug-nutrient interactions).

To improve the appropriateness of resource use, active care advice systems can help clinicians make better informed decisions regarding testing or treatment options. For example, the clinician can be provided information about[45]

- likely conflicts or redundancies between a chosen test and others already ordered for the patient
- the results of previous tests on the patient that are like the one being ordered so that the clinician may reconsider whether the test really needs to be repeated

- the cost of a test or treatment ordered for the patient so that the clinician can reconsider whether it is really worth performing
- tests or treatments that would be less costly than the one ordered but equally effective in treating the health problem at hand

Pacific Applied Psychology Associates, Inc. (PAPA), an integrated behavioral health delivery system in northern California, is developing an automated active care advice system that will include clinical decision support tools that offer suggestions to the clinician regarding diagnosis, testing, and treatment.[46] The first component to be put in place is an expert system that will assist providers in making level-of-care decisions for psychiatric patients in crisis. The provider's treatment decision is electronically transmitted to a PAPA case manager to inform him or her of the case and the disposition decision made by the provider. The case manager responds via electronic mail with a confirmation of the services authorized.

Once a sufficient number of cases are entered into the system, PAPA ultimately hopes to be able to predict the treatment course, cost, and resulting outcomes for patients with different risk severity profiles. From this information, they expect to create pathlike tools that will further refine the clinical decision-making process and reduce the need for ongoing case manager oversight.

CONCLUSION

It is unlikely that Florence Nightingale could have envisioned how complex the system of health care would eventually become. Nonetheless, her advice about the importance of managing the "system of care" is more vital today than ever before. Collaborative, path-based practice is a critical element in systems management and is likely to remain important in the future. However, what we now recognize as the common components and tools of path-based practice are likely to change.

The evidence-based medicine revolution is causing a heightened awareness of the need for better bridges between research evidence and clinical practice. Tools such as clinical practice guidelines, pathways, algorithms, and order sets can help clinicians build these bridges. Those involved in path-based practice initiatives are advocating the design of discipline-friendly tools that all caregivers can feel comfortable using.

Reducing unnecessary variation in clinical decision making and the health care delivery processes can have many positive cost and quality benefits. Therefore, initiatives aimed at reducing variation are desirable. However, unlike manufacturing processes, the patient care process is affected by patient, disease, and therapy factors that

are difficult to anticipate and define prospectively. Therefore, clinical paths must be designed to accommodate desirable variation while eliminating unintended variation. Severity-adjusted pathways, proactive pathways, and focused pathways are paper-based solutions. Computerized health information techniques will offer clinicians even greater flexibility in designing a patient-specific plan of care.

The system of health care described by Florence Nightingale was confined to the hospital environment. Today's health care system stretches across many different provider sites. In response to financial and consumer demands, collaborative path-based practice is broadening its scope to cover the continuum of patient care. Case management, clinical pathways, and practice guidelines are vital elements of this integrated delivery system.

Paper-based patient care is quickly being replaced by automated health information technologies. The field of health informatics will change the way we deliver health care services by improving the retrieval, synthesis, organization, dissemination, and application of patient-reported, clinician-observed, and research-derived information. This trend is likely to have the greatest impact on collaborative, path-based practice. When clinicians in all sites of patient care have ready access to clinical decision support systems, the need for human intermediaries will decrease. Today's familiar pathway tools will evolve into computerized decision algorithms that prompt the caregiver with real-time "if-then" choices specific to the patient's current needs.

Two important elements of collaborative, path-based practice are not likely to change. As the delivery system becomes more complex and highly specialized, clinicians will find the micro-level issues of patient diagnosis and treatment increasingly challenging. This will leave them little time for the macro-level concerns, such as coordination among caregivers and provider sites. Individuals with clinical skills, human resource management talents, and organizational abilities will always be needed to address the macro-level problems.

The second element that will not go away is the need for the "human touch." It will always be necessary for clinicians to balance the needs of the individual patient against the findings of empirical research studies. There will never be an automated information system that can anticipate every clinical and psychosocial factor that must be considered in the decision-making process. Human compassion will also remain an important part of patient care. If clinicians are to meet patients' needs, they must treat both body and spirit.

The demise of collaborative, path-based practice will not occur in the foreseeable future. However, advocates may view the changes that will happen over the next few years as something other than what they perceive path-based practice to be all about. When facing the inevitable transitions, I encourage caregivers to ponder on the words of Alexander Graham Bell, who said, "When one door closes, another opens; but we often look so long and so regretfully upon the closed door that we do not see the one which has opened for us."

REFERENCES

1. Nardella A, Pechet L, Snyder LM. Continuous improvement, quality control, and cost containment in clinical laboratory testing. Effects of establishing and implementing guidelines for preoperative tests. *Arch Pathol Lab Med.* 1995;119(6):518–522.

2. Nightingale F. *Notes on Nursing.* London: Harrison; 1860.

3. Austin CD. History and politics of case management. *Generations.* 1988;Fall:7–10.

4. Zavorski LM, Taptich B. Creating a multidisciplinary disease management initiative. In: Spath P, ed. *Beyond Clinical Paths: Advanced Tools for Outcomes Management.* Chicago: American Hospital Publishing; 1997:71–101.

5. Joint Commission on Accreditation of Healthcare Organizations. *1996 Comprehensive Accreditation Manual for Hospitals.* Oak Brook Terrace, IL: Joint Commission; 1996.

6. Pelling M. *Hospital Managers' Guide to Joint Commission Standards.* Forest Grove, OR: Brown-Spath & Associates; 1997:4.

7. Haynes RB. Some problems in applying evidence in clinical practice. *Ann NY Acad Sci.* 1993;703:210–215.

8. Field MJ, Lohr KN, eds. *Clinical Practice Guidelines: Directions for a New Program.* Washington, DC: National Academy Press; 1990.

9. Clinical Practice Guideline Panel of the Quality Management Institute and Education Center. *Clinical Decision Making Aids: Clinical Practice Guidelines/Clinical Pathways/Clinical Algorithms Position Statement, Version 1.* Durham, NC: Department of Veterans Affairs, Veterans Health Administration; August 1996:11. (unpublished manuscript)

10. Spath PL. VHA looks at clinical practice guidelines, pathways, and algorithms. *JAHIMA.* 1996; 67(6):44–46.

11. Woolf SH. *Interim Manual for Clinical Practice Guideline Development: A Protocol for Expert Panels Convened by the Office of the Forum for Quality and Effectiveness in Health Care.* (AHCPR-91-19) U.S. Dept. of Health and Human Services, Public Health Service, Agency for Health Care Policy and Research; 1991.

12. Mosher C, Cronk P, Kidd A, McCormick P, Stockton S, Sulla C. Upgrading practice with critical pathways. *Am J Nurs.* 1992;92(1):41–44.

13. Hawley S, Davis B. Mapping home care services. In: Spath P, ed. *Beyond Clinical Paths: Advanced Tools for Outcomes Management.* Chicago: American Hospital Publishing; 1997:129–144.

14. Goonan KJ. *The Juran Prescription.* San Francisco: Jossey-Bass Publishers; 1995:42–45.

15. Wennberg JE, Gittelsohn A. Variations in medical care among small areas. *Sci Am.* 1982; 246:120–134.

16. Berwick DM, Godfrey AB, Roessner J. *Curing Health Care: New Strategies for Quality Improvement.* San Francisco: Jossey-Bass Publishers; 1990.

17. James B. Implementing practice guidelines through clinical quality improvement. *Front Health Serv Manage.* 1993;10(1):3–37.

18. Spath PL. Critical paths: a tool for clinical process management. *JAHIMA.* 1993;64(3):48–58.

19. Lau J, Antman EM, Jiminez-Silva J, Kupelnick B, Mosteller F, Chalmers TC. Cumulative meta-analysis of therapeutic trials for myocardial infarction. *N Engl J Med.* 1992;327(4):248–254.

20. Spath PL. *Mastering Path-Based Patient Care.* Forest Grove, OR: Brown-Spath & Associates; 1995:84–85.

21. Spath PL. *Mastering Path-Based Patient Care.* Forest Grove, OR: Brown-Spath & Associates; 1995:86–87.

22. Turley K, Tyndall M, Turley K, Woo D, Mohr T. Radical outcome method: a new approach to critical pathways in congenital heart surgery. *Circulation.* 1995;92(9) suppl II: 245–249.

23. Turley K, Turley K. Reducing length of stay and improving outcomes. In: Spath P, ed. *Beyond Clinical Paths: Advanced Tools for Outcomes Management.* Chicago: American Hospital Publishing; 1997:163–178.

24. Tallon R. Critical paths for wound care. *Adv Wound Care.* 1995;8(1):26, 28–34.

25. Mosher CM. Putting pressure ulcers on the map. *J Wound Ostomy Contin Nurs.* 1995;22(4):183–186.

26. Morningstar L. Organ donor path requires more work with families. *Hosp Case Manage.* 1994;2(8):140–142.

27. Morningstar L. Comfort care path strengthens support for terminally ill. *Hosp Case Manage.* 1995;3(9):139,142.

28. Morningstar L. Reusable chronic pain path account for multiple patient visits. *Hosp Case Manage.* 1994;2(9):151–154.

29. Paynter J, Ambrose K, Dolan K. Integrating geriatric evaluation and management with a multidisciplinary care planning process. In: Spath P, ed. *Beyond Clinical Paths: Advanced Tools for Outcomes Management.* Chicago: American Hospital Publishing; 1997:103–127.

30. Medicare Program; Changes to the Hospital Inpatient Prospective Payment Systems and Fiscal Year 1998 Rates; Proposed Rule. 42 CFR Parts 412, 413, and 489. *Fed Regis:* June 2, 1997. Volume 62, Number 105, Page 29901–29951.

31. Scanlon W, Dowdal T. *Medicare Post-Acute Care: Cost Growth and Proposals to Manage It Through Prospective Payment and Other Controls* (GAO/T-HEHS-97-106). Government Accounting Office testimony before the Committee on Finance, US Senate, 04/09/97.

32. Gerteis M, Edgman-Levitan S, Daley J, Delbanco TL. *Through the Patient's Eyes: Understanding and Promoting Patient-Centered Care.* San Francisco: Jossey-Bass Publishers; 1993:207–211.

33. International Communications Research. *How Americans Perceive the Health Care System: A Report on a National Survey.* January 1997 [WWW Document] URL http://www.nchc. org/perceive.html.

34. Calkins DR, Davis RB, Reiley P, et al. Patient-physician communication at hospital discharge and patients' understanding of the postdischarge treatment plan. *Arch Intern Med.* 1997;157(9): 1026–1030.

35. Dearing G. Standardized disease management improves processes of care. *Outcomes Measure Manage.* 1995;6(5):1–2.

36. Zavorski LM, Taptich B. Creating a multidisciplinary disease management initiative. In: Spath P, ed. *Beyond Clinical Paths: Advanced Tools for Outcomes Management.* Chicago: American Hospital Publishing; 1997.

37. Mahn VA, Spross JA. Nurse case management as an advanced practice role. In: Hamric A, Sprass J, Hanson C, eds. *Advanced Nursing Practice: An Integrative Approach.* Philadelphia: WB Saunders; 1996:445–465.

38. Monahan T. The business of healthcare information technology. *Healthcare Inform.* 1997;8:8.

39. Dorenfest SI. A look behind the rapid growth in healthcare IS. *Healthcare Inform.* 1997;7:44–47.

40. Horn SD, Hopkins DSP. Introduction. In: Horn SD, Hopkins DSP, eds. *Clinical Practice Improvement: A New Technology for Developing Cost-Effective Quality Health Care.* New York: Faulkner & Gray; 1994:1–5.

41. Anderson DJ, Freire CW, Hale P. Designing information systems for disease management. In: Spath P, ed. *Beyond Clinical Paths: Advanced Tools for Outcomes Management.* Chicago: American Hospital Publishing; 1997.

42. Connelly DP, Bennett ST. Expert systems and the clinical laboratory information system, *Clinics Labora Med.* 1991;11(1):136–138.

43. Turley K, Turley K. Reducing length of stay and improving outcomes. In: Spath P, ed. *Beyond Clinical Paths: Advanced Tools for Outcomes Management.* Chicago: American Hospital Publishing; 1997:163–178.

44. Gibson RF, Middleton B. Health care information management systems to support CQI. In: Horn SD, Hopkins DSP, eds. *Clinical Practice Improvement: A New Technology for Developing Cost-Effective Quality Health Care.* New York: Faulkner & Gray; 1994.

45. Congressional Office of Technology Assessment. Applications of Clinical Decision Support Systems. Appendix C in *Bringing Health Care On-line: The Role of Information Technologies* (GPO pub. no. 052-003-01433-5). Washington, DC: Government Printing Office; September 1995:53–54.

46. Pigott E, Alter G, Heggie DL. Linking expert systems to outcomes analysis. In: Spath P, ed. *Beyond Clinical Paths: Advanced Tools for Outcomes Management.* Chicago: American Hospital Publishing; 1997:253–261.

■ Appendix 3–A ■
Abbreviations Key

2D = Two-dimensional echocardiogram
2 DE = Two-dimensional echocardiogram
ABG = arterial blood gases
ACE = ACE I = ACE Inhibitor = angiotensin converting
  enzyme inhibitor
ADA = American Dietetic Association
AEB = as evidenced by
Amb = ambulate
ASA = aspirin
BCP = biochem panel
BG = blood glucose
BP = blood pressure
BRP = bathroom privileges
BUN = blood urea nitrogen
Ca = calcium
CAD = coronary artery disease
CBC = complete blood count
CCU = cardiac care unit
CHF = congestive heart failure
CK-MB = creatine phosphokinase myocardial bands
CO = cardiac output
CPCU = Cardiac Progressive Care Unit
CPR = cardiopulmonary resuscitation
CR = creatinine or cardiac rehabilitation
CVA = cerebrovascular accident
CXR = chest X-ray
DC = discharge, discontinue
DE = Two-dimensional echocardiogram
dec. = decreased
Dig = digoxin
DM = diabetes mellitus
DOE = dyspnea on exertion
D/T = due to
DVT = deep vein thrombosis
EC = emergency care

ECG = electrocardiogram
EF = ejection fraction
Eval = evaluation
FR = fluid restriction
Hep = heparin
HR = heart rate
HTN = hypertension
hx = history
K+ = potassium
I<O = intake less than output
I&O = intake and output
IIT = intense insulin therapy
IV = intravenous
IVL = intravenous lock
labs = laboratory tests
LDL = low-density lipoprotein
MCU = Medical Cardiac Unit
meds = medications
mg = magnesium
MI = myocardial infarction
MISD = Medical Information Services Department
MQPM = Medical Quality Program Management
MS = morphine sulfate
MUGA = multigated acquisition
Na+ = sodium
O_2 = oxygen
OHS = open-heart surgery
PO = oral
pOx = pulse oximetry
PRN = as needed
PT = physical therapy or prothrombin time
PTCA = percutaneous transluminal coronary angioplasty
PTT = partial thromboplastin time
Q = every
QD = every day

RBC = red blood cell count
RMF = regular medical floor
R/O = rule out
RP = renal panel
R/T = related to
SOB = shortness of breath
SQ = subcutaneous
S/S = signs/symptoms
S/Sx = signs/symptoms
supps = supplements
TIA = transient ischemic attack

TID = three times a day
TMS = telemetry monitor service
Tol = tolerated
TSH = thyroid-stimulating hormone
TX = treatment
UA = urinalysis
U/A = urinalysis
VS = vital signs
VSS = vital signs stable
WNL = within normal limits
Wts = weights

■ 4 ■

Developing Better Critical Paths in Health Care: Combining "Best Practice" and the Quantitative Approach

Dawn A. Bailey, David G. Litaker, and Lorraine C. Mion

As managed care and capitated payment plans become increasingly prevalent in today's health care marketplace, health care providers are continually confronted with the need to improve quality and reduce the costs of their services.[1–3] In today's environment, the economic outcomes of care are scrutinized and monitored as closely as clinical outcomes for patients. To remain competitive and viable, provider organizations must deliver effective care while maintaining rigorous control of costs and resource utilization. All provider disciplines, in particular nursing and physician staff, share in the responsibility for managing costs in response to competitive price structures. To achieve this objective, providers must have a better understanding of (1) the factors associated with higher costs of care relevant to the population under their care and (2) the management strategies that promote the best patient outcomes while controlling or reducing these costs.

Nursing services account for a significant portion of total hospital costs.[4–6] Thus, controlling nursing care costs and identifying their determinants within specific patient populations is of particular relevance to nurse managers.[4,7–9] The examination of nursing costs in the hospital setting, however, represents a challenge for a variety of reasons. Inconsistent definitions for nursing care services and nursing costs, differences in measures of patient illness, and differences in hospital characteristics such as size, type, length of stay, staffing ratios, personnel mix, and average nursing salaries complicate efforts in this area.[4,9]

Despite these difficulties, total hospital costs, including the costs of nursing services, must be carefully monitored and controlled. Clinical pathways or critical paths are one of the management strategies most frequently cited as a means of controlling and reducing care costs while pro-

moting optimal patient outcomes, such as improved physical functioning and enhanced patient satisfaction with care.[1,2,7,10–16] In use for more than 10 years, critical paths attempt to standardize and coordinate the interdisciplinary care that is typically provided to a group of patients by diagnosis or procedure. With a focus on quality and efficient resource utilization, critical paths map out the progression of care activities and procedures that are routinely recommended and expected to meet the challenge of providing optimal outcomes for a particular patient cohort.[15,17,18]

As a clinical management strategy, critical paths are reported to be helpful in accomplishing this goal.[2,7,10,12,14] The way that care paths diagram or display the types, frequency, and sequencing of health care resources allows health care managers to more readily identify the achievement of patient care outcomes in relation to the processes of care. Further, standardization of care processes via critical paths may reduce the costs associated with practice variations and lead to greater satisfaction with medical care.[13] Given the dynamic state of today's health care delivery and reimbursement systems, it is imperative that these clinical management strategies, developed in part to manage costs of care, and based on the prevailing features of a hospitalized population and health care reimbursement plans, be evaluated and refined on an ongoing basis.[3,7,10,12,16,18,19]

The purpose of this chapter is to present one strategy used to determine key factors associated with routine nursing care costs for patients who have undergone major orthopaedic surgery as a means of identifying deficiencies in an existing critical path. These data are then used as the basis for refining the existing critical path, and they provide continuous improvement in delivering more effective health care.

METHODS

Subjects and Setting

Five hundred consecutive patients were seen in the Department of General Internal Medicine at the Cleveland Clinic Foundation between August 23, 1994, and June 30, 1995, for preoperative assessment before major elective surgery. A subsample of 176 patients scheduled to undergo lower extremity joint replacement (DRG 209) or back surgery (DRGs 214 and 215) who would receive nursing care using an existing orthopaedic critical path were identified for these analyses. Eligible patients were at least 50 years old, could understand and speak English, and were scheduled for elective surgery with an expected inpatient length of stay greater than two days. Exclusion criteria included patient or physician refusal to participate.

Variables and Procedures

Routine nursing cost was used as the major outcome variable. At this institution, direct nursing costs are estimated per patient on a daily basis and are a function of the nursing workload and consumption of noncharge-able supply items. Nursing workload is measured daily via the MEDICUS system[20] and is downloaded into the hospital's financial database. Using data from this system, as well as supply utilization data from the Pyxis Corporation SUPPLYSTATION® system,[21] an estimate of the direct and indirect nursing costs is made for each patient on each day of his or her inpatient stay.

Using previous reports in the literature as a guide, selected information was collected preoperatively and postoperatively on variables that were thought to be associated with higher nursing care requirements and subsequent costs. Patients were seen for preoperative medical consultation in the Department of General Internal Medicine, and informed consent was obtained. In a 15-minute interview, the features assessed at this preoperative evaluation included demographics, a current medication list, a brief medical history highlighting the presence of either chronic or acute illnesses, smoking habits, previous psychological or neurologic diseases, history of delirium, and results of preoperative electrolyte, blood count, urinalysis, and liver function testing when available (Exhibit 4–1). Medication class (e.g., nitrates, oral hypoglycemics, corticosteroids, antidepressants) was determined for each patient as a dichotomous variable (medication used within one week of the preoperative evaluation) rather than by dosage or frequency. Self-reported attitudes on alcohol use were determined using

Exhibit 4–1 List of Variables Assessed for Correlation with Routine Nursing Care Costs

Demographic
Age (<70, 70+)
Gender
Race (white, nonwhite)

Clinical
Mortality rate[a]
Smoking history
Alcohol use[b]
Neurological diagnosis
Psychiatric diagnosis

Function
Physical function preadmission[c]
History of delirium
Preoperative cognitive function[d]
Postoperative delirium[d,e]

Pre- and Postoperative Medication Class (Used/Not Used)
Steroid
Benzodiazepine
Narcotic
Antihistamine
Antidepressant

Pre- and Postoperative Laboratory Indices
Sodium (<130, >150 mmol/L)
Potassium (<3.0, >6.0 mmol/L)
Glucose (<60, >300 mmol/L)
Hematocrit (<30%, ≥30%)
WBC (<12,000, 12,000+)

Hospital Variable
ICU care
Hospital length of stay

Note: ICU, intensive care unit; WBC, white blood cell count.
[a]Charlson Comorbidity Index[25] (5 or greater indicates increased risk of mortality).
[b]CAGE Questionnaire,[22] range 0 to 4.
[c]Specific Activity Scale[23] (SAS), range 1 to 4.
[d]Telephone Interview for Cognitive Status[24] (TICS), range 0 to 38.
[e]Confusion Assessment Method[30] (CAM), present or absent.

the four-item CAGE questionnaire[22] and an additional question about patients' perception of the effect of alcohol consumption on their health. A positive response to any of these five questions was interpreted as indicative of alcohol abuse. A standardized assessment of functional and cognitive status was also performed for each individual using the Specific Activity Scale[23] and the Telephone Interview for Cognitive Status (TICS)[24] score, which does not require written skills. Comorbid conditions were quantified using the Charlson Comorbidity Index.[25] Although events during the intraoperative course may have contributed significantly to subsequent events during the remaining inpatient stay, and thus to higher nursing costs, data on these factors were not collected.

Patients were evaluated daily in the postoperative period by trained clinical interviewers. This assessment included a brief standardized patient interview for cognitive status and level of pain and a review of both the nursing record and the patient's medical chart from postoperative days 1 through 4 (or until discharge) for reports of complications such as delirium, pulmonary embolism, myocardial ischemia, or pneumonia. Selected laboratory indices (serum sodium, potassium, glucose, white blood cell count, and hematocrit) were identified through review of the laboratory computer, and medication usage, as noted in the medication administration record, was also obtained.

Data Analysis

Data were entered using EPI-INFO, Version 5.01b,[26] and were analyzed with SPSS/PC+, Version 4.1.[27] Descriptive statistics were calculated on all variables. Independent variables were dichotomized to reflect abnormal/normal or at-risk/not-at-risk status, and univariate comparisons of nursing costs were made using Student's t test or the Mann-Whitney U test as dictated by the distribution of data. Two-tailed p values ($p < 0.05$) were used as the threshold for statistical significance throughout. Finally, independent variables were entered into an exploratory stepwise multiple regression model to determine factors associated with higher routine nursing costs.

RESULTS

Characteristics of the Study Sample

Of the 176 patients forming this sample, 93 (53 percent) were females, 151 were Caucasian (86 percent), and the mean age was 67.6 (±8.0) years. More than three-quarters of study participants were admitted for lower extremity joint replacement (84 percent), with the remainder undergoing back surgery. Most patients (97 percent) had SAS[23] scores of less than 4 and a Charlson Comorbidity Index[25] of less than 5 (97 percent), indicating relatively good levels of physical function and overall health. Preoperative cognitive function reflected by the TICS[24] ranged from 17 (significant impairment) to 38 (unimpaired), with a sample mean of 31, indicating minimal impairment. Patients incurred an average routine nursing cost of $1,070.76 (±603.39), with a mean length of stay of 4.89 (±2.6) days and mean intensive care unit (ICU) length of stay of 0.05 (±.33) days.

Determinants of Nursing Costs

Independent variables, with the exception of length of stay, were dichotomized to reflect clinical abnormality or increased risk. Routine nursing care costs were compared in the presence/absence of each variable and are presented in Table 4–1. The only variable associated with higher costs in this univariate analysis was the development of postoperative delirium: $1,323 (±559.44) for those who developed delirium versus $1,036 (±603) for those who did not ($t = 2.06$, $p = 0.04$). The Pearson correlation coefficient demonstrated a strong association between length of stay and routine nursing costs ($r = 0.95$, $p < 0.001$).

An exploratory stepwise regression model was then developed using nursing costs as the dependent variable and all preoperative and immediate postoperative features as independent variables. Five variables remained as independent factors associated with higher nursing costs: use of antipsychotic medication postoperatively, postoperative delirium, poor preoperative functional status, ICU stay, and increased length of stay. The use of antipsychotic medications, common in the management of postoperative delirium, was included in this exploratory model even though no significant correlation between the two was noted in univariate analysis. Additionally, this five-variable model accounted for nearly 94 percent of the variation observed in this data set for routine nursing costs. Not surprisingly, however, much of the adjusted R^2 resulted from the inclusion of length of stay in the model.

Table 4–1 Variables Associated with Higher Nursing Costs ($N = 176$)

	Costs	
Variable	Variable Present Mean (± SD) (N)	Variable Absent Mean (± SD) (N)
Postoperative delirium*	$1,323 (±559.44) (N = 21)	$1,036 (±603) (N = 155)
Feels guilty regarding drinking**	$942 (±241) (N = 10)	$1,078 (±618) (N = 166)
Postoperative steroid use**	$768 (±314) (N = 9)	$1,087 (±611) (N = 167)
Preadmission benzodiazepine use**	$1,635 (±1,380) (N = 11)	$1,033 (±500) (N = 165)

*$p < 0.05$. **$p < 0.15$.

DISCUSSION

Although many examples exist concerning the effectiveness of critical paths as a means to control patient care cost and enhance quality, the limitations associated with these paths, as well as suggestions for strengthening and improving this approach to patient care management, have been reported.[13,15,16,28] One of the most significant criticisms of these clinical management tools is that they are rarely evidenced based.[13,16,28] Establishing critical pathways using rigorous scientific methodology, however, requires labor-intensive, time-consuming procedures. The prevailing method of critical path development, by contrast, is the use of internal, institution-specific "expert knowledge" of the health care professionals to devise the approaches for patient care.[13,18] As such, critical paths reflect organizational culture and map out the current provider practice patterns specific to those care providers.[2,13,28] Although critical paths should be evidenced based, the advantage to this method of critical path development—one that is adapted to local practice—is enhanced clinician adherence.[16,29] In this study, we demonstrated a feasible approach that combines aspects of both methods and results in an institution-specific critical path that is based on clinical data and systematic evaluation.

Figure 4–1 displays a suggested process in which critical paths are developed using both local consensus and scientific processes. By using existing databases, the critical path can be evaluated for care processes (e.g., proportion of processes provided to that recommended) as well as selected patient care outcomes such as function, length of stay, complications, and cost. On the basis of these results, the critical path can be reviewed and modified as needed. The revised critical path can then be reevaluated. Thus, critical paths can be a useful tool within the hospital's existing continuous quality improvement program and as a management approach to cost containment.

We examined easily obtained patient variables before and during hospitalization and their association with nursing costs of care in an effort to identify the key determinants of nursing service utilization. Nursing cost, only one of many potential outcomes, was used as the criterion to evaluate the existing critical path. Although the focus was on cost, this same process could have been applied to existing critical paths using patient satisfaction data or other clinical endpoints, such as readmission rates. Cost data are readily available at this institution, and careful management of costs of care is both an institutionally relevant goal and a key job responsibility of nurse managers.

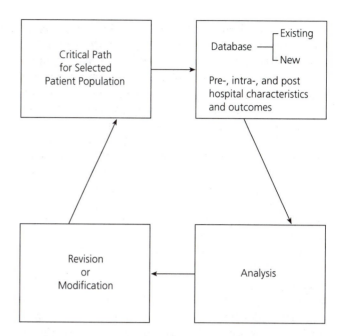

Figure 4–1 Strategy for Evidence-Based Critical Path Development.

Using this approach, the univariate factors associated with increased nursing care costs for this patient cohort were postoperative delirium, feelings of guilt associated with alcohol consumption, postoperative steroid use, and preoperative benzodiazepine use. When multiple factors were examined simultaneously, several of the factors remained as independent predictors of routine nursing costs. These included the use of antipsychotic medications, the development of postoperative delirium, and poor preoperative functional status. Previous reports have documented that postoperative delirium and poor functional status are associated with increased costs and longer lengths of stay.[31] Although no particular association between postoperative steroid use and cost of care has been described, one might speculate that these medications are used in circumstances associated with higher complication rates (e.g., chronic obstructive pulmonary disease exacerbation, chronic steroid dependence). Similarly, preoperative benzodiazepine use and feelings of guilt associated with alcohol consumption may reflect psychological comorbid conditions that affect nursing utilization.

Introducing elements into the critical paths that address these factors should result in beneficial changes in nursing utilization, hence cost. For example, nursing interventions aimed at the early identification of factors leading to delirium in these patients should be highlighted in the postoperative period. Nursing interventions on the critical path aimed at these factors would include preoperative identification of patients at risk, the development of specific interventions aimed at ameliorating or preventing the onset of delirium, and more intensive family involvement in daily care activities to assist in optimizing patients' cognitive status.

For patients with poor preoperative functional status, the critical path should delineate specific measures and interventions aimed at reducing further functional decline in these elderly patients. These interventions should be initiated in the preoperative phase and should include aggressive physical therapy and increased family involvement in maximizing the patient's functional ability. This study and others have demonstrated that poor functional status among elderly hospitalized patients accounts for higher total hospital costs.[31] Patients with impaired functional ability require more nursing care resources; thus, it would appear that the impact of the nursing costs associated with these patient characteristics accounts for the higher hospital costs.

Once these patient characteristics are identified as playing a role in increasing the nursing care costs of this patient population, it is a relatively simple task to revise and refine the critical path currently employed for this patient group. Revisions include the early preoperative identification of patients at risk and the specific interventions geared toward ameliorating the impact of these factors on nursing costs. Once completed, the revised critical path should again be reevaluated, with a new cohort of patients, to determine if the characteristics were amenable to treatment and if the nursing care costs were actually decreased. The data needed to make these determinations are readily obtainable from a computerized medical record. Indeed, ongoing evaluation and revision of all critical paths should be a part of every organization's continuous quality improvement processes. If critical paths are to be an effective means of controlling costs while promoting optimal patient outcomes, practitioners must consistently strive to build scientifically based findings into their refinements of these important documents.

CONCLUSION

In today's environment, nurse managers are increasingly confronted with the need to control nursing care costs. Identifying the determinants of these costs, therefore, is of particular relevance. Although critical paths have been cited frequently as a means for controlling and reducing costs, they are rarely evidenced based. Instead, they are developed based on local, institution-specific, "expert knowledge." We have demonstrated a feasible approach

that combines aspects of organizational culture and local practice patterns with systematic evaluation of clinical and cost outcomes, in the development of critical paths. Results from the systematic evaluations are then used to revise or modify the existing critical path. In this way, health care managers may have increased confidence in the validity of their locally developed critical paths' effectiveness in managing clinical and cost outcomes.

REFERENCES

1. Aspling DL, Lagoe RJ. Benchmarking for clinical pathways in hospitals: A summary of sources. *Nursing Economics.* 1996;14(2):92–97.

2. Clark CM, Steinbinder A, Anderson R. Implementing clinical paths in a managed care environment. *Nursing Economics.* 1994;12(4):230–234.

3. Low A. Reducing variation in patient care. *J Nurs Adm.* 1996;26(1):14–20.

4. Wilson L, Prescott PA, Aleksandrowicz L. Nursing: a major hospital cost component. *Health Services Research.* 1988;22(6):773–796.

5. McCormick B. What's the cost of nursing care? *Hospitals.* 1986;November 5:48–52.

6. Walker D. The cost of nursing care in hospitals. *J Nurs Adm.* March 1983:13–18.

7. Sovie MD. Tailoring hospitals for managed care and integrated health systems. *Nursing Economics.* 1995;13(2):72–83.

8. Buerhaus PI. Economics of managed competition and consequences to nurses: Part II. *Nursing Economics.* 1994;12(2):75–80.

9. Eckhart JG. Costing out nursing services: examining the research. *Nursing Economics.* 1993;11(2):91–98.

10. Crummer MB, Carter V. Critical pathways—the pivotal tool. *Journal of Cardiovascular Nursing.* 1993;7(4):30–37.

11. Lang M, Kolowich A, Glasheen J, Couch J, Anderson DH. *Critical pathways.* Santa Barbara, CA: COR Healthcare Resources; 1995.

12. Lumsdon K, Hagland M. Mapping care. *Hospitals & Health Networks.* 1993;October 20:34–40.

13. Anders RL, Tomai JS, Clute RM, Olson T. Development of a scientifically valid coordinated care path. *J Nurs Adm.* May 1997;27(5):45–51.

14. Blegen MA, Reiter RC, Goode CJ, Murphy RR. Outcomes of hospital-based managed care: a multivariate analysis of cost and quality. *Managed Care.* November 1995;86(5):809–814.

15. Pearson SD, Goulart-Fisher D, Lee TH. Critical pathways as a strategy for improving care: problems and potential. *Ann Intern Med.* December 15, 1995;123(12):941–948.

16. Merritt TA, Palmer D, Bergman DA, Shiono PH. Clinical practice guidelines in pediatric and newborn medicine: implications for their use in practice. *Pediatrics.* January 1997;99(1):100–114.

17. Coffey RJ. An introduction to critical paths. *Quality Management in Health Care.* 1992;1(1):45–54.

18. Ebener MK, Baugh K, Formella NM. Proving that less is more: linking resources to outcomes. *J Nurs Care Qual.* 1996;10(2):1–9.

19. American Pain Society Quality of Care Committee. Quality Improvement Guidelines for the treatment of acute pain and cancer pain. *JAMA.* December 20, 1995;274(23):1874–1880.

20. Hegyvary ST, Hausman RK, Kronman B, Burke M. *User's Manual for Rush-Medicus Nursing Process Monitoring Methodology.* Chicago: Technical Reports; 1979.

21. Pyxis Corporation. Pyxis Corporation SUPPLYSTATION®. San Diego, CA: Pyxis Corporation [Producer and Distributor].

22. Ewing JA. Detecting alcoholism: The CAGE questionnaire. *JAMA.* 1984;252:1905–1907.

23. Goldman L, Hashimoto B, Cook EF, Loscalzo A. Comparative reproducibility and validity of systems for assessing cardiovascular functional class. Advantages of a new specific activity scale. *Circulation.* 1981;64:1227–1233.

24. Brandt J, Spencer M, Folstein MF. The telephone interview for cognitive status. *Neuropsych Neuropsychol Behav Neurol.* 1988;1: 111–117.

25. Charlson ME, Pompei P, Ales KL, MacKenzie CR. A new method of classifying prognostic comorbidity in longitudinal studies: Development and validation. *J Chronic Dis.* 1987;40:373–383.

26. Dean AG, Dean JA, Burton AH, Dicker RC. *EPI-INFO, Version 5: a word processing, database, and statistics program for epidemiology on microcomputers.* USD, Inc. Stone Mountain, GA; 1990.

27. SPSS/PC+, version 6.1. SPSS, Inc. Chicago, IL.

28. Gorbien MJ. Clinical pathways: too hard a course for complex patients? *Continuum.* 1995:1–6.

29. Gray JAM. *Evidenced-based HealthCare. How to make health care policy and management decisions.* New York: Churchill Livingstone; 1997.

30. Inouye SK, van Dyck CH, Alessi CA, Balkin S, Siegal AP, Horwitz RI. Clarifying confusion: the confusion assessment method. *Ann Intern Med.* 1990;113:941–948.

31. Covinsky KE, Justice AC, Rosenthal GE, Palmer RM, Landefeld, CS. Measuring prognosis and case mix in hospitalized elders: the importance of functional status. *J Gen Intern Med.* April 1997;12: 203–208.

■ Part II ■
Respiratory

■ 5 ■

Community-Acquired Pneumonia Outcome Study: The Evolution of a Clinical Pathway

Linda M. Valentino, Felicia Olt, and Mark J. Rosen

Contemporary health care organizations and providers are challenged by rapidly changing reimbursement systems to reinvent or reengineer traditional patient care delivery. Health care providers face the daunting responsibility of demonstrating that patient care delivery is of high quality, cost-effective, satisfying to patients and families, and resource efficient.[1] Many health care organizations use clinical pathways to manage patient care processes and to demonstrate patient outcomes.[2] The use of clinical pathways was begun at Beth Israel Health Care System in the early 1990s. These clinical pathways were named *multidisciplinary action plans* (MAPs®). A Community-Acquired Pneumonia MAP® (Exhibit 5–A–1 in Appendix 5–A) was developed in 1991 to serve as a daily timeline guide for care of patients. This tool was intended to be used by medical, nursing, and ancillary staff. MAPs® served as daily care plans for the nursing staff. Although the MAP® was a tool developed for multidisciplinary use, it was seen by the multidisciplinary care team as a nursing care plan. Even though the MAP® was a permanent part of the patient's medical record, other disciplines rarely used it as a source of care planning and documentation.

In early 1994, a length-of-stay (LOS) initiative based on categories of disease management was undertaken to reduce overall hospital LOS by two days. A pulmonary team was established that met over a two-year period to address high-volume diagnosis-related groups (DRGs). Community-acquired pneumonia DRGs 89 and 90 were two of the high-volume DRGs that became the focus of the pulmonary team's work. The team also focused on the care of patients with bronchitis, asthma, chronic obstructive pulmonary disease (COPD), and tracheotomies and of patients requiring mechanical ventilation. The multidisciplinary team developed a Community-Acquired Pneumonia Clinical Guideline© based on case study, review of scientific literature, and the opinion of clinical experts (Exhibit 5–1). In addition, the pulmonary team, as well as other LOS teams, sought help from administration and the admitting office to cluster patients with similar medical and surgical diagnoses on the same inpatient clinical units. These units became known as *clustered units*. Whenever possible, pulmonary patients and those with a medical diagnosis of community-acquired pneumonia were clustered on the same in-patient unit. The Community-Acquired Pneumonia Clinical Guideline© highlights the purpose, preadmission considerations, treatment and management, and discharge indications for the hospitalized patient with severe pneumonia or less severe pneumonia. The development of key-point clinical guidelines was endorsed by senior management as a strategy to reduce LOS. This initiated an organizational effort to use clinical tools to reduce LOS. Crucial to the use and ultimate success of clinical guidelines was establishing "buy-in" of the physician group for the use of these clinical tools. The introduction of the MAP® in previous years had not had the same level of endorsement by administration, nor had the economic environment of health care been as competitive. The MAP® was largely viewed as a nursing tool, and the day-by-day format was viewed by some physicians as "cookbook" medicine. The MAPs® continued to be used in some clinical areas, but new MAPs® were not being developed. The Community-Acquired Pneumonia Clinical Guideline© replaced the MAP®. It was published in April 1995 and implemented soon thereafter. This chapter discusses the evolution of a community-acquired pneumonia clinical pathway. It outlines our experience and response to the dynamic changes in the health care environment.

Exhibit 5–1 Community-Acquired Pneumonia Clinical Guideline

BETH ISRAEL MEDICAL CENTER	CLINICAL GUIDELINE
TITLE: COMMUNITY-ACQUIRED PNEUMONIA	# P-2

PURPOSE:	• Initial selection of antibiotics. • Conversion of IV to oral antibiotics.
PRE-ADMISSION CONSIDERATIONS:	Many patients, especially young adults or adolescents without underlying medical conditions, can be treated as out-patients with oral erythromycin. Clarithromycin (Biaxin) or azithromycin (Zithromax) are more expensive alternative oral drugs; recommended in COPD patients.

TREATMENT/ MANAGEMENT:	**Hospitalized patient; less severe pneumonia**
	• Initial intravenous antibiotic choices:
	    cefuroxime (and erythromycin if Mycoplasma, Legionella or Chlamydia suspected)
	    ceftazadime or ticarcillin/clavulanate (Timentin) in place of cefuroxime for nursing home admission. If aspiration (anerobes) suspected, ticarcillin/clavulanat to be considered.
	
	Hospitalized patient; severe pneumonia
	Defined as one or more of the following: respiratory rate > 30/min, $PaO_2/FIO_2 < 250$, requiring mechanical ventilation, bilateral/multiple lobe involvement, increasing infiltrate by > 50% in 48 hours. Presence of shock, oliguria, requiring pressors.
	
	• Initial intravenous antibiotic choices:
	    cefuroxime *or* ceftazadime *or* ticarcillin/clavulanate
	                           AND
	    vancomycin *plus* erythromycin
	Note: Nafcillin to replace vancomycin for non-resistant staph whenever possible. Modify antibiotic choices per culture and sensitivities.
	
	Convert from IV to PO antibiotics when:
	• sustained decrease in temperature of 1 degree Fahrenheit × 24 hours
	• decrease in leukocytosis
	• improved pulmonary signs/symptoms
	• able to tolerate oral medication
	Note: This usually occurs day 2–5 of hospitalization in less severe pneumonia.

DISCHARGE INDICATIONS	**Discharge from Hospital:** May occur within 24 hours of switch to oral antibiotics providing no deterioration or other reason for continued hospitalization. *Note: This may occur day 3–6 of hospitalization in less severe pneumonia.* *Note: Chest radiography within first week on an improving patient is not necessary because the radiographic findings typically lag behind.*

PATHWAY

The development and implementation of the Beth Israel Health Care System Community-Acquired Pneumonia Clinical Guideline© started in 1995. It has continued over the last two years and in 1997 was revised into a critical pathway. The Community-Acquired Pneumonia Clinical Guideline© was developed by an interdisciplinary team that included physicians, registered nurses, social workers, a respiratory therapist, and a pharmacist. The members of this team represented several departments in the health care system: quality improvement, utilization management, case management, data analysis, and administration. The clinical guideline was developed concurrently with the previously described hospital LOS initiative. Once community-acquired pneumonia was identified, patients diagnosed with it were appropriate for a reduction in LOS. The pulmonary team conducted a retrospective chart review of 32 patients to obtain baseline data. Of these 32 patients, 18 were classified as DRG 89, and 14 patients were classified as DRG 90. All patients included in this review were admitted through the emergency department. All patients with a diagnosis of human immunodeficiency virus (HIV) or acquired immune deficiency syndrome (AIDS) were excluded from the study. The purpose of this chart review was to delineate the care delivered to a population of patients with community-acquired pneumonia admitted to Beth Israel. A data collection tool was developed by the pulmonary team. The team collected and analyzed the patient data. These data were then compared with findings of similar patient types in current medical literature. By considering these findings and current clinical practice, the clinical guideline was developed. In addition to the members of the original LOS team, this draft was reviewed for appropriateness by internal clinical experts in the departments of medicine, nursing, and social work.

The clinical guideline format is a template developed by the Beth Israel Health Care System tools committee, which is responsible for the oversight of all clinical tool development within the Beth Israel Health Care System. The clinical guideline template is structured to include "key points" of clinical care. The elements of this template include a purpose statement, preadmission and admission considerations, medical management, and discharge indications. Integrated within each section of the clinical guideline are the interdisciplinary aspects of care. Separate discipline-specific interventions are not outlined in the template. Clinical guidelines are not a documentation source; therefore, they are not a part of the permanent patient medical record. The guideline was intended to be a clinical reference tool to guide staff in the most optimal care interventions that would contribute to a reduction in LOS. The goals established for the development and implementation of a clinical guideline program were to decrease LOS, decrease cost, improve clinical outcomes, and integrate interdisciplinary patient care. The clinical guidelines developed by the LOS teams were introduced to providers within the health care system along with a restructured care and case management system. Case managers were identified as members of the health care team along with physicians, who would both select patients appropriate for the clinical guideline implementation. In addition, the case managers would be primarily responsible to gather variance data from the clinical guidelines, patients, practitioners, and delivery system. Once a new clinical guideline was completed and reviewed by the tools committee, it was published in several internal Beth Israel publications. The physician staff and all clinical department heads were given a published copy of the clinical guideline in the medical staff bulletin. Each in-patient unit maintains a copy of a binder of all clinical guidelines, and staff are notified of new or revised guidelines through unit managers and case managers.

CARE MANAGEMENT

The system of care management employed at Beth Israel has also evolved since its inception in 1990. The Beth Israel Health Care System is a tertiary care facility composed of three hospital sites and over 1,000 beds. The care management model implemented in July 1995 was the beginning phase of an organizationwide restructuring program of patient care delivery. This newly designed program evolved out of the LOS initiative begun a few years before. The Beth Israel care management system is defined as an interdisciplinary clinical care system supported by unit-based leadership teams. The leadership team guides the delivery of services through a patient-centered care philosophy. The leadership team is responsible for directing the cost and quality outcomes for a service area or clustered unit. The care management model is used in the acute care setting. However, this care management model is currently being expanded for the integrated health care system. The leadership team is composed of a manager of patient care services, a case manager, a liaison physician, and a social worker. Unit staff are invited to team meetings to discuss patient care issues and quality-of-care outcomes. In addition, there may be several consultants to the leadership team. For example, a play therapist may be a regular consultant to the pediatric leadership team. Quality improvement staff are permanent consultants to each team and act as facilitators. They provide leadership teams with support for interpretation of cost, quality, satisfaction, and process data. In addition, they facilitate communication of quality-related information from the leadership team to the hospitalwide quality improvement program.

Several processes composing the care management system were also restructured. Reengineering of the discharge planning process was an integral part of setting the foundation of the newly created care management system. This new system altered the roles of social workers and staff nurses in the discharge planning process. High- and low-risk criteria for discharge planning were developed. Upon admission of a patient to the acute care setting, the registered nurse completes an assessment of discharge needs. The registered nurse determines if the patient's discharge plan is high risk and therefore managed by the social worker or low risk and managed by the registered nurse. A forum for discussion about patient discharge needs was also established with the implementation of formal twice-weekly interdisciplinary discharge rounds. In addition, a formal program of daily interdisciplinary patient rounds was implemented. The care management program was initiated on 13 medical and surgical units at one hospital site. Plans for rollout of the program for all Beth Israel patient care units were to be completed over the next year. But, at the same time, a reengineering project for all patient care processes was begun, and a decision was made to implement care management with the entire reengineering process.

CASE MANAGEMENT

Case managers are employed in this system of care management. The case manager in most instances is a registered nurse prepared at the master's level who coordinates care for a select group of patients. In some instances, the specific case management model definition is dependent on the population of patients served. For example, the model uses social workers as case managers for patients in psychiatry and the AIDS program. The case management model used at Beth Israel follows the basic definition of case management approved by the Case Management Society of America: "Case management is a collaborative process which assesses, plans, implements, coordinates, monitors, and evaluates options and services to meet an individual's health needs through communication and available resources to promote quality cost-effective outcomes."[3(p.8)] The case manager role was initiated with the implementation of care management. Case managers were recruited from an internal group of clinical nurse specialists (CNSs). The choice to use master's-prepared nurses was consistent with the philosophy that as health care delivery systems are radically altered, the skills of advanced-practice nurses will be needed to link patient needs with available resources.[4] A case manager was assigned to all of the medical surgical units except one medical unit.

The case managers received training on managed care, patient selection, variance data collection, and the new discharge planning process. A case manager was assigned to the clustered unit for pulmonary patients. This case manager was a CNS with a strong specialty background in respiratory management. Case managers are responsible for collecting variance data on patients who are case managed. The case manager coordinates care so that patient outcomes can be achieved by preventing, wherever possible, patient, practitioner, and system variances. The case managers choose patients on the basis of case management selection criteria and review all new admissions for potential case management needs. At the same time the case manager, along with the physician, chooses an appropriate clinical guideline for the patient. The case manager and other members of the interdisciplinary team use the clinical guideline to direct the care of patients.

Case managers are responsible for collecting variance data on the clinical guideline. They participate in unit rounds, including daily interdisciplinary rounds and discharge planning rounds. Along with the unit-based leadership team, they are responsible and accountable for the monitoring of fiscal, quality, and clinical outcomes for both individual patients and the specific clustered patient population.

OUTCOMES

Measuring outcomes as a management strategy began appearing in the health care literature in the late 1980s. The process of outcomes management is driven by the processes or interventions that produce patient outcomes. The use of instruments like pathways serve two functions: to set forth a standard of practice and to provide a structure by which outcomes can be measured.[5] Since the implementation of the Community-Acquired Pneumonia Clinical Guideline©, several measurement studies have been performed. These studies attempt to measure variations in practice patterns, effectiveness, and patient outcomes. The care management program at Beth Israel Health Care System and in particular the case managers on the clinical units have been instrumental in the implementation of this clinical guideline in cooperation with the physician staff. From postimplementation of the clinical guideline to the present, the community-acquired pneumonia tool has evolved, and so has the outcomes measurement of the use of this tool. The results of our measurement studies will be discussed, as well as the present initiatives underway to revise the format of the clinical guideline to a clinical pathway to enhance the measurement of variance and outcomes.

INITIAL VARIANCE ANALYSIS

Variances in this case management system as well as others occur when a patient's hospital course differs from the care outlined in the clinical tool.[6] Types of variances are related to patient/family, health/illness, caregiver/provider, and process/environment. The monitoring strategy for the pneumonia clinical guideline developed by the pulmonary LOS team was never implemented. Initially, the physicians on the team, including the pulmonary fellow, were to review medical records retrospectively to measure the use and impact of the guideline. In response to physician feedback, it was decided that since the development of the guidelines was part of an institutional initiative, the organization would assume the measurement and monitoring process of guidelines. Therefore, it was decided that for the months of October and November 1995, case managers would collect data for patients on guidelines on a variance flowsheet (Exhibit 5–2). Data collection was limited to the

Exhibit 5–2 Variance Data Collection Flowsheet

Patient Variance Flowsheet

UNIT: _____

Transferred From: _____

                Date: _____

Transferred To: _____

                Date: _____

Guideline #: _____

MAP #: _____

Case Managed: Yes No Partially

Patient's Social Support: Adequate None Identified

Date:	Variance from Expected Course	Code

Completed by:

Date:

Source: Copyright © Beth Israel Health Care System.

pneumonia guideline, since the MAP® developed in 1991 was not being used. Data on the use of the Community-Acquired Pneumonia Clinical Guideline© were collected for 30 patients, and 27 of those patients were admitted to the pulmonary cluster unit. Data were also collected for the other pulmonary guidelines, asthma and COPD, on another 50 patients. Thus, variance data were collected on a total of 77 patients on the pulmonary unit. This number is important because the aggregation of the variance data is based on the total number of patients on guidelines on each unit. A code was assigned for each

guideline and variance collected to facilitate analysis of these data. For the pulmonary unit, the five top variances were medical status changes, other patient variance, secondary diagnosis with admission, other practitioner variance, and patient feeling unready for discharge. A comparison of these data to the overall data collected by the case managers for all guidelines and MAPs® showed that three of the pulmonary units' top five variances were the same for the institution. In descending order of frequency, the top three variances were medical status changes, other practitioner variance, and other patient variance. What exactly did these data tell us? We realized that capturing variance data and grouping it by patient, practitioner, and system was too general and did not answer the question: "How do we measure use and effectiveness of the guideline and its impact on patient care?"

COMPARATIVE CLINICAL GUIDELINE STUDY

In the ongoing process to measure the use and impact of the Community Acquired Pneumonia Clinical Guideline©, plans for a retrospective study comparing patients before and after guideline use were undertaken. Although data were collected before development of the guideline, it was felt that the same data collection tool should be used in comparing pre- and postguideline data. Therefore, a tool was developed (Exhibit 5–3) to capture information based on the Community-Acquired Pneumonia Clinical Guideline©. First-quarter 1994 and 1996 medical records were reviewed by a team of pulmonary specialists that included physicians, fellows, nurses, and care management and quality improvement staff. Records of patients with an admitting and principal diagnosis of pneumonia (ICD 486) with LOS between 4 and 20 days, an age greater than 17, and non-AIDS/HIV DRG were reviewed. The review was based on a random sample of patients with simple pneumonia. Patients diagnosed with severe pneumonia were excluded from the study. A total of 55 medical records met the above criteria and were reviewed. Thirty-three records were reviewed for 1994 and 22 for 1996. The review focused on clinical and cost outcomes. In addition to the indepth review of the 55 records, population data on LOS, clustering, and readmission rates for the full years 1994 and 1996 were compared.

Figure 5–1 represents population data showing that average LOS (ALOS) from 1994 to 1996 decreased by 1.2 days and median LOS by 2.0 days. A causal connection may be made between decreased LOS and implementation of the guideline. Clustering population data of patients with similar medical conditions placed on designated units were also reviewed. The graph in Figure 5–2 shows the units from which patients with pneumonia were discharged from 1994 to 1996. In 1994, pneumonia patients were widely distributed throughout the hospital, whereas the postclustering data show the majority of patients clustered on two units: 5L (pulmonary unit) and 7L (geriatric unit). The theory of clustering patients is based on the premise that the greater the volume of like patients treated on the same unit, the greater the likelihood that patients will receive cost-effective, efficient, quality care. This is another factor contributing to a decrease in LOS and an impact on quality of care.

Population data on readmission rates for 1994 and 1996 were also analyzed. The readmission rate for pa-

tients readmitted within 15 days with the same or related condition remained constant at 3 percent for both years. Comparison of Beth Israel data with benchmarking data collected by the Island Peer Review Organization (IPRO) in New York shows that Beth Israel readmission rates are within acceptable range. The IPRO study looked at 1,175 patients, with a readmission rate of 2 percent for patients readmitted solely for pneumonia and 8 percent for patients readmitted for any other cause. Therefore, we can probably conclude that the use of guidelines and clustering of the patients can reduce LOS with no impact on readmission rates. This further supports the theory that patients can be discharged earlier on the basis of criteria set forth in the guideline without having their care compromised by this early discharge.

An in-depth chart review also undertaken as part of this study looked at the antibiotic usage, time frames for switching from intravenous (IV) to oral (PO), and drug utilization costs. A reduction in the total number of antibiotics ordered by physicians was noted. In 1994, 11 different types of antibiotics were ordered. In 1996, four different types of antibiotics were ordered, with an increase in the use of cefuroxime by 26 percent. The change in use of PO antibiotics was not as drastic: the total number of PO antibiotics ordered by physicians was reduced from seven to six types, with Ceftin used 55 percent of the time. There was improvement in the switch from IV to PO antibiotics, from 34 percent being switched within four days in 1994 to 60 percent being switched within four days in 1996. As stated on the clinical guideline, the switch from IV to PO antibiotics did occur within two to five days of hospitalization. The Beth Israel rate for conversion from IV to PO antibiotics was higher than that in benchmarking data analyzed by IPRO in New York, which showed only 28 percent of patients switched from IV to PO antibiotics between October 1994 and September 1995. Following this study, the IPRO recommended that 62 percent of patients who met the clinical criteria should be switched within four days.[7] Beth Israel has met the benchmarking standard, with 60 percent of patients switched within four days. Beth Israel Health Care System providers are revising the criteria to include considering the switch by day 2 if the patient is tolerating PO fluids. We are continuing to monitor this switch on the basis of the guideline criteria, in addition to participating in the upcoming IPRO benchmarking study.

Last, in this comparative study, we looked at cost outcomes, particularly drug utilization costs. In 1994, the average cost per case for antibiotics was $299.76, whereas in 1996, the cost was $222.00 per patient. The most dramatic reduction in cost was seen in IV antibiotics from 1994 to 1996, with a decline of $124.74 per patient. The cost for PO antibiotics increased by $46.98 per

Exhibit 5–3 Pneumonia Clinical Guideline Study

CLINICAL GUIDELINE REVIEW
COMMUNITY-ACQUIRED PNEUMONIA   Med Rec. #: _____

Circle one: Admit Source: Home   Nursing Home   Shelter/Homeless

DO NOT WRITE IN SHADED AREAS: **INDICATORS**	YES	NO	CODE	DATE	N/AP	N/AV	UNABLE TO DETERMINE
1. Was patient treated prior to admission with PO ATB for pneumonia?							
1a. If YES, Write code							
Other (Specify) _____							
2. Check all that apply for severe pneumonia: Respiratory Rate > 30 min.							
Requiring mechanical ventilation							
Bilateral/multiple lobe involvement							
Presence of shock, oliguria, requiring pressors.							
3. If none apply to question #2, then patient has pneumonia							
3a. If pneumonia, write code for initial IV ATB of choice							
Other (Specify) _____							
4. If any apply to question #2, then patient has SEVERE pneumonia							
4a. If SEVERE pneumonia, Write code for initial IV ATB of choice							
Other (Specify) _____							
5. When was IV ATB initiated?							
6. When was IV ATB discontinued?							
7. Was patient switched to PO ATBs? When?							
8. Name of PO ATB _____							
9. If patient was switched to PO ATB, check all that apply: Decrease in temperature by 1 degree F × 24 hours							
Decrease in Leukocytosis							
Improved pulmonary signs/symptoms							
Able to tolerate oral medication							
10. Was patient discharged within 24 hours following initial switch to oral ATBs?							
11. Reason for continued hospitalization following switch to PO ATBs (Code)							
Other (Specify) _____							
12. Disposition Status (Code)							
13. Was organism resistant?							
13a. If resistant, was ATB changed?							

N/AV = Not Available; N/AP = Not Applicable

Source: Copyright © Beth Israel Health Care System.

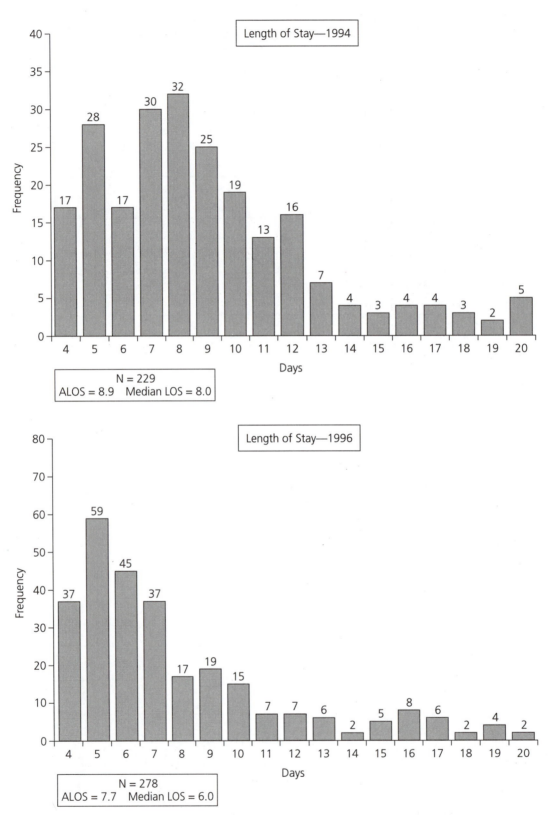

Note: The criteria used to define the population include an admit and principal diagnosis of pneumonia (486.xx), a length of stay between 4 and 20 days, an age of greater than 17, and a non-AIDS/HIV DRG.

Figure 5–1 Population Study, Full Years 1994 and 1996. *Source:* Copyright © Beth Israel Health Care System.

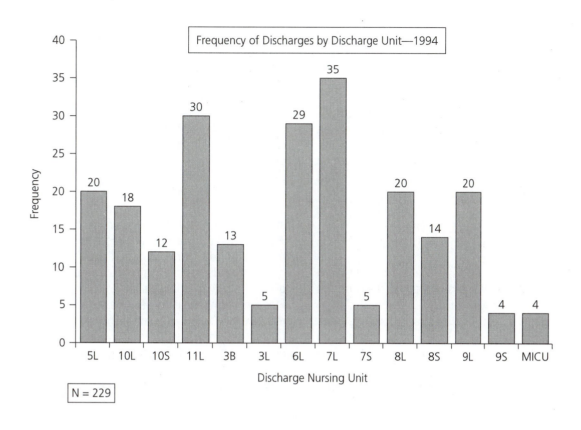

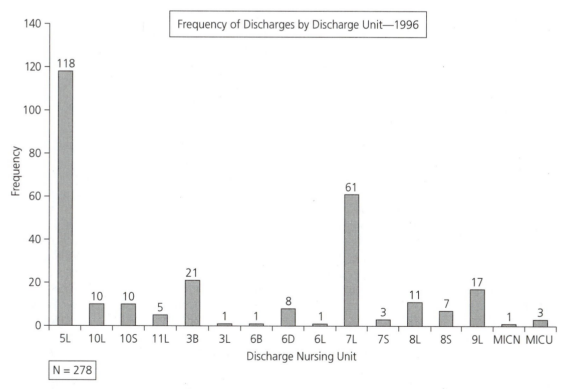

Note: The criteria used to define the population include an admit and principal diagnosis of pneumonia (486.xx), a length of stay between 4 and 20 days, an age of greater than 17, and a non-AIDS/HIV DRG.

Figure 5–2 Population Study, Full Years 1994 and 1996. *Source:* Copyright © Beth Israel Health Care System.

patient. A lower cost PO antibiotic is being investigated for use with pneumonia patients by Beth Israel Infectious Disease specialists. The findings from this study were significant because patient records were reviewed in depth, and general population data coded from the medical record were analyzed. This analysis yielded data to substantiate changes in patient care and subsequent revision of the clinical guideline. The next step in measuring patient outcomes and use of guidelines was undertaken by care management. A variance data collection system had to be put in place to collect data concurrently on clinical guidelines.

CONCURRENT DATA COLLECTION STUDY

A group of internal experts from the Beth Israel Health Care System met to determine how best to measure the use, effectiveness, and impact of MAPs® and clinical guidelines. A generic care management flowsheet was designed to collect these data. This tool was developed on the basis of the clinical guideline format. The flowchart developed was never implemented because it was determined to be too general to collect variance data. We did not want to make the same mistake twice by collecting data that were too general. The program coordinator of care management, with physician and nursing support, decided to develop a different strategy for variance data collection. A daily grid was developed from each clinical guideline (Exhibit 5–A–2 in Appendix 5–A). These data were collected concurrently by the case manager. The data collector assessed daily each key point on the clinical guideline for three possible outcomes. The data were coded as follows: "yes" represented that the key point on the guideline was met, "no" represented that the key point on the clinical guideline was *not* met, and "N/A" represented that the key point was not applicable to the patient. These data were collected from June 1, 1996, to August 31, 1996. A total of 32 clinical guidelines were implemented 777 times on 581 patients during this study period. The Community-Acquired Pneumonia Clinical Guideline© was fourth highest used when ranked with all clinical guidelines implemented. The community-acquired pneumonia guideline was used 71 times. Overall, this study concluded that 13 percent of all admissions to Beth Israel Health Care System had clinical guidelines implemented. On the pulmonary cluster unit, 37.8 percent of all discharges were on a pulmonary clinical guideline. A total of 61 patients on the pulmonary cluster unit were on the Community Acquired Pneumonia Clinical Guideline©. Twenty-three patients were coded upon discharge as DRG 89, with an ALOS of 7.8 days. Seven patients were coded upon discharge as DRG 90, with an ALOS of 3.6 days. These cases represent 43.4

percent and 58.3 percent of total cases admitted to be placed on this guideline, respectively. Thirty-five patients were coded upon discharge with DRG 89 and 90 were not on a clinical guideline; their ALOSs were 8.7 and 3.8 days, respectively. Initial intravenous antibiotic use was consistent with the clinical guideline in 87 percent of hospitalized patients on the pulmonary unit. Sixty-seven percent of patients had IV antibiotics converted to PO antibiotics by day 2, 3, 4, or 5 of hospitalization. Fifty-two percent of patients were discharged 24 hours after the switch from IV to PO antibiotics.

It was concluded that in most cases, the care of patients at Beth Israel Health Care System studied was consistent with the care outlined in the clinical guideline. The patients treated for community-acquired pneumonia on a clinical guideline had a lower length of stay than those not on the guideline. In addition, several recommendations were made to administration and the leadership teams. These recommendations were, first, to revise the clinical guideline to include interventions that were more specific to all members of the interdisciplinary team, such as nursing and psychosocial interventions. Although aspects of interdisciplinary care are contained in the clinical guidelines, they are not specific and therefore are not easily measurable. Second, the clinical guideline is not time defined. The time in which each key point was met varied. This was extremely difficult to capture in the data collection. It was recommended that the format of the clinical guideline be changed to a day-by-day format to facilitate data collection and to establish daily patient goals.

CONCLUSION

The Community-Acquired Pneumonia Clinical Guideline© has been revised into a day-by-day clinical pathway format (Exhibit 5–A–3 in Appendix 5–A). The clinical pathway will allow greater ease in the collection of data to measure variations in care. The case manager will continue to collect and analyze these data and report findings through the care management program. The care management program reports these data through the Beth Israel Health Care System quality improvement committee. Our method of data collection based on the criteria set forth in the clinical guideline has changed over time. It has been a learning process. We believe that the data collected have given us valuable information on how to proceed with improving care and guideline revisions. In addition, outcomes data collection has enabled us to substantiate to clinicians the rationale for changing the clinical guideline to a pathway format. Our data results show that, when possible, data should be collected concurrently so that when variances occur, they can be immediately addressed by providers.

Table 5–1 Length-of-Stay Comparison: Discharges for DRGs 89 and 90, 1993–1996

		1993		1994		1995		1996	
	DRG Code and Description	*Discharges*	*ALOS*	*Discharges*	*ALOS*	*Discharges*	*ALOS*	*Discharges*	*ALOS*
89	Simple pneumonia and pleurisy, age > 17, w/CC	423	11.1	332	9.9	398	9.1	409	8.3
90	Simple pneumonia and pleurisy, age > 17, w/o CC	96	6.2	78	6.0	84	5.2	107	4.4
	Total	519	10.2	410	9.2	482	8.4	516	7.5

Note: CC, comorbidity and complications.
Source: Copyright © Beth Israel Health Care System.

For the period of 1993 to June 1997, coded data, including overall ALOS and clustering data, have been extracted from charts of patients with DRGs 89 and 90. These data have been aggregated in Table 5–1, which shows that ALOS has gradually decreased from 10.2 days in 1993 to 7.5 days in 1996. The leadership teams, particularly the team on 5L, will continue to monitor LOS data and the use of the Community-Acquired Pneumonia Clinical Guideline©. The overall data for clustering do show a reduction in ALOS from 9.5 days to 8.1 days for the pre- and postcluster periods. Although specific 5L data are not available, we can see that clustering may be one of the many factors that affect ALOS and outcomes in general. We recognize that outcome data based on general population data (retrospectively coded information) such as average and median LOS and readmission rate are helpful as a means of identifying opportunities for improvement. Such data are not the end point but the starting point for identifying trends and documenting outcomes.

The implementation of the Community-Acquired Pneumonia Clinical Guideline© occurred in the context of many organizational changes. This makes it difficult to conclude which changes have had the greatest impact on the outcomes of patient care. However, outcomes measurement parameters have been described to include data collection related to financial, quality, variance analysis, and patient satisfaction.[1] The findings reported here include such data. Specific patient satisfaction data were not collected for patients with community-acquired pneumonia, but these data are collected on the pulmonary unit and reported quarterly to the leadership team. In the studies we have reported data collection, aggregation, and analysis for patients with community-acquired pneumonia have been conducted by the clinicians most closely associated with the care of these patients. We consider this a great success because it places the responsibility for improvement of patient care upon the providers, where it belongs.

REFERENCES

1. Strassner L. Critical pathways: the next generation of outcomes tracking. *Orthop Nurs.* 1997;March/April:56–61.

2. Wojner AW. Outcomes management: an interdisciplinary search for best practice. *AACN Clin Iss.* 1996;7(1):133–145.

3. Case Management Society of America. *Standards of Practice.* Little Rock: Case Management Society of America; 1995:8.

4. Porter-O'Grady T. Nurses as advanced practitioners and primary care providers. In: Cohen EL, ed. *Nurse Case Management in the 21st Century.* St. Louis, MO: Mosby-Year Book; 1996:10–20.

5. Wojner AW. Outcomes management: an interdisciplinary search for best practice. *AACN Clin Iss.* 1996;7(1):133–145.

6. Magdalena MA, Newton C. Managing variances in case management. *Nurs Case Manage.* 1996;1(1):45–51.

7. Silver A. Proceedings of Island Peer Review Organization: adult pneumonia: current concepts and new approaches. IPRO's recent quality improvement studies on pneumonia among New York state Medicare patients. Presented at Annual Meeting; March 26, 1997.

■ Appendix 5–A ■
Clinical Pathway Tools

Exhibit 5–A–1 Beth Israel Medical Center Multidisciplinary Action Plan

DIAGNOSIS: _____Community-Acquired Pneumonia_____

MD:

UNIT: _____Medicine_____

ADMISSION DATE: _____

DATE MAP INITIATED: _____

DRG #: _____79/80_____

_____

EXPECTED LENGTH OF STAY: _____4 Days (5th Day Optional)_____

PATIENT CARE MANAGER: _____

SOCIAL WORKER: _____

GOALS MUTUALLY SET WITH PATIENT AND/OR FAMILY

_____ YES

_____ NO   EXPLAIN _____

PATIENT ALLERGIES: _____

_____

_____

	YES/NO	DATE
DNR:		

HEALTH CARE PROXY
OR LIVING WILL: _____

continues

54

Exhibit 5–A–1 continued

DAY 1 of 5

MD: _____

DIAGNOSIS: _Community-Acquired Pneumonia_ _____

DATE: _____

MAP DOES NOT REPLACE MD ORDERS

PROBLEM	EXPECTED PATIENT OUTCOME/ DISCHARGE OUTCOME	NURSING INTERVENTIONS	ASSESSMENT/ INTERVENTION	
1. Impaired gas exchange	1A. Airway will be free of secretions. B. Absence of signs/ symptoms of respiratory distress. C. ≥ 90% O_2 Sat (if measured).	1 A. Respiratory assessment: auscultate breath sounds, rate pattern and depth of respirations, use of accessory muscles, skin color, capillary refill, vital signs, L.O.C. and affect. **Notify physician if condition worsens and anticipate need for intubation if condition deteriorates. B. Monitor ABGs, CXR reports, sputum C&S, gram stain results. C. Maintain HOB increased to facilitate clearing secretions. D. Encourage pt. cough unless cough is frequent and nonproductive. E. Administer IV fluids as ordered and encourage oral fluids. F. Note effectiveness of meds and inform MD if ineffective. G. Consider oral suctioning with Yankauer. H. Evaluate pt's. response to nursing interventions and revise prn.		
2. Infection	2A. Infection will be reduced as evidenced by afebrile, WBC, trending downward if high initially.	2 A. Assess patient for signs/symptoms of infection (especially monitor IV site for phlebitis): yellow or green sputum, temp., vital signs, increased WBC count, chills, sputum, C&S, gram stain. B. Notify MD for temp. > 100.3° F. C. Send specimens for culture as indicated. D. Provide antipyretic TX; ABX TX. E. Evaluate patient response to nursing interventions and revise prn.		
3. Activity intolerance	3A. Patient will progressively increase his/her ability to tolerate activity.	3 A. Inform patient of MD's imposed activity restrictions BRP, bedrest, and rationale to conserve energy. B. Assist pt. OOB to chair QD and assist with ADLs. C. Admin. analgesics, supplemental nutritional vitamins prn. D. Encourage pt. to assume frequent rest periods, sedentary diversional activities. E. Maintain O_2 delivery system (esp. nasal cannula) while pt. eats in order to prevent desaturation, SOB and increase patient's ability to eat. F. Evaluate patient's response to nursing interventions and revise prn.		

Note: ABGs, arterial blood gases; ABX, antibiotic treatment; ADLs, activities of daily living; BRP, bathroom privileges; C&S, culture & sensitivity; CXR, chest X-ray; HOB, head of bed; LOC, level of consciousness; OOB, out of bed; O_2 Sat, oxygen saturation; QD, every day; SAT, saturation; SMA, serum metabolic analysis; SOB, shortness of breath; TX, treatment; WBC, white blood cell count.

Exhibit 5-A-2 Community-Acquired Pneumonia Variance Data Collection Tool

Date started on guideline _____

*To Indicate transfer on grid

Admit Unit _____   Current Unit _____   Case Managed: _____ yes _____ no

Primary Diagnosis _____

Secondary Diagnosis _____   *Not a part of permanent record.

Admission Date: _____   Discharge Date: _____   Physician: _____   Case Manager: _____

BETH ISRAEL HEALTH CARE SYSTEM TITLE:	Pre-Admission	Date:___ Day:___ POD___	Date:___ Day:___ POD___	Date:___ Day:___ POD___	Date:___ Day:___ POD___	Date:___ Day:___ POD___
CLINICAL GUIDELINE: Community-Acquired Pneumonia # P-2						
PRE-ADMISSION CONSIDERATIONS:	Many patients, especially young adults or adolescents without underlying medical conditions, can be treated as out-patients with oral erythromycin. Clarithromycin (Biaxin) or azithromycin (Zithromax) are more expensive alternative oral drugs; recommended in COPD patients.					
TREATMENT/ MANAGEMENT:	**Hospitalized patient; less severe pneumonia** • Initial intravenous antibiotic choices: cefuroxime (and erythromycin if Mycoplasma, Legionella or Chlamydia suspected) ceftazadime or ticarcillin/clavulanate (Timentin) in place of cefuroxime for nursing home admission. If aspiration (anerobes) suspected, ticarcillin/clavulanat to be considered.					
	Hospitalized patient; severe pneumonia Defined as one or more of the following: respiratory rate > 30/min, PaO₂/FIO₂ < 250, requiring mechanical ventilation, bilateral/multiple lobe involvement, increasing infiltrate by > 50% in 48 hours. Presence of shock, oliguria, requiring pressors. • Initial intravenous antibiotic choices: cefuroxime or ceftazadime or ticarcillin/clavulanate AND vancomycin *plus* erythromycin *Note: Nafcillin to replace vancomycin for non-resistant staph whenever possible. Modify antibiotic choices per culture and sensitivities.*					
	Convert from IV to PO antibiotics when: • sustained decrease in temperature of 1 degree Fahrenheit × 24 hours • decrease in leukocytosis • improved pulmonary signs/symptoms • able to tolerate oral medication *Note: This usually occurs day 2–5 of hospitalization in less severe pneumonia.*					
DISCHARGE INDICATIONS:	**Discharge from Hospital:** May occur within 24 hours of switch to oral antibiotics providing no deterioration or other reason for continued hospitalization. *Note: This may occur day 3–6 of hospitalization in less severe pneumonia.* *Note: Chest radiography within first week on an improving patient is not necessary because the radiographic findings typically lag behind.*					

Source: Copyright © Beth Israel Medical Center. Reproduction of the pathway is by permission of Beth Israel Health Care System, All Rights Reserved.

Exhibit 5-A-3 Beth Israel Care System—Care Path

Community-Acquired Pneumonia—#P-2	Pre-Admission Considerations/Admission Criteria	Discharge Criteria	
Additional Care Path initiated: **Goal LOS:** Severe Pneumonia 2 days Less Severe: Ambulatory	1. Many patients, especially young adults or adolescents without underlying medical conditions, can be treated as outpatients with oral erythromycin. Clarithromycin or azithromycin are more expensive alternative oral drugs; recommended in COPD patients. 2. Admit to ICU for hypotension, severe hypoxemia, respiratory failure, or other severe comorbid conditions. 3. Admit for floor care for toxicity, failure to respond to appropriate outpatient treatment, pleural effusion, or isolation if immunosuppressed state.	1. Sustained decrease in temperature of one degree Fahrenheit × 24 hours 2. Decrease in leukocytosis 3. Improved pulmonary signs/symptoms 4. Able to tolerate oral antibiotics × 24 hours	Stamp Addressograph Name of Service Attending/House MD:

Plan	Day #1—Point of Entry Date:	Day #2 Date:	Day #3 Date:	Day #4 Date:	Day #5 Date:
Assessment Treatments Interventions	History and Physical Ausculate lungs, assess patient for SOB. Monitor respiratory status Determine if pneumonia is *Less Severe* or **Severe** **Severe pneumonia** defined as one or more of the following: respiratory > 30/min, $PaO_2/FIO_2 < 250$, requiring mechanical ventilation, bilateral/multiple lobe involvement, increasing infiltrate by > 50% in 48 hours. Presence of shock, oliguria, requiring pressors. • Monitor temperature • O_2 therapy, if indicated	Continue respiratory assessment Monitor temperature If on mechanical ventilation, consider weaning	Evaluation O_2 therapy ⇒⇒⇒	⇒⇒⇒	⇒⇒⇒
Tests Procedures	ABG, if indicated CBC, SMA, Chest X-ray	Monitor WBC Chest X-ray, if indicated	⇒⇒⇒	CBC, SMA, Chest X-ray, as indicated.	CBC prn
Medications	*Hospitalized patient; less severe pneumonia* Initial intravenous antibiotic choices: cefuroxime (and erythromycin if Mycoplasma, Legionella or Chlamydia suspected) ceftazadime or ticarcillin/clavulanatein (Timentin®) in place of cefuroxime for nursing home admission. If aspiration (anaerobes) suspected, ticarcillin/clavulanate to be considered. **Severe pneumonia:** Initial intravenous antibiotic choices: cefuroxime or ceftazadime or ticarcillin/clavulanate AND vancomycin plus erythromycin	Consider conversion from IV to PO antibiotics. **Severe:** ciprofloxacin HCL Less Severe: ampicillin/sulbactan (Augmentin®), clarithromycin, cefuroxime	Consider conversion from IV to PO antibiotics *Discharge home if converted to PO antibiotics on Day 3*	⇒⇒⇒	PO antibiotics
Fluids Nutrition Elimination	IV I&O	Consider Diet/tube feeding IV, I&O	⇒⇒⇒ DC IV with conversion to PO	⇒⇒⇒	⇒⇒⇒
Activity	HOB ↑, limited ambulation Bedrest (severe)	Ambulate as tolerated	⇒⇒⇒	Ambulates independently	
Consults	Consider Pulmonary and/or Infectious Disease consult				
Discharge Planning	Discharge Planning Assessment Tool (DPAT) Complete Referrals Assess Psychosocial, Spiritual Needs	⇒⇒⇒	ICU → floor if severe *Discharge* *Follow-up appointment*	⇒⇒⇒	Discharge Follow up appointment
Patient/Family Education	Assessment of learning needs, coughing and deep breathing. Discharge Teaching.	⇒⇒⇒	*Verbalizes understanding of medication use*	Verbalizes understanding of medication use	⇒⇒⇒

Note: ABG, arterial blood gas; CBC, complete blood count; COPD, chronic obstructive pulmonary disease; DC, discontinue; HOB head of bed; I&O, intake and output; ICU, intensive care unit; IV, intravenous; PO, oral; SMA, serum metabolic analysis; SOB, shortness of breath; WBC, white blood cell count.
Source: Copyright © Beth Israel Medical Center. Reproduction of the pathway is by permission of Beth Israel Health Care System. All Rights Reserved.

■ 6 ■

The Pneumonia CareMAP®: Improving the Process To Enhance Outcomes

Patti Pattison, Merrilee Newton, and Carol Ansley

With an increasing transition to capitation, the health care system has been continually challenged to find ways to streamline the cost of care while maintaining quality. These expectations for value have fostered a heightened focus on reducing variation in patient care and improving outcomes measurement. Clinical pathways, in conjunction with case management, have become important tools that have proven to be effective in standardizing care. This move toward appropriate standardization has helped reduce unnecessary variation in patient care and has promoted improved communication and coordination. Standardized tools such as clinical pathways define expected practices and promote consistency of care. The conscientious user of these tools is rewarded with improved satisfaction for both caregivers and patients, enhanced clinical outcomes and appropriate resource use.

One year after an aggressive implementation, the assessment of an 18-month project to institute multidisciplinary action plans (CareMAPs®) demonstrated significant underutilization and minimal integration into practice. This chapter describes the initial development project and the subsequent quality improvement process that resulted in the reimplementation of CareMAPs® with a major focus on the measurement of process and outcomes. The pneumonia CareMAP® will be used to illustrate the new material and clinical outcomes measurement.

THE FIRST EXPERIENCE

Alta Bates Medical Center is a 540-bed nonprofit acute and postacute community hospital located in Berkeley, California, serving a diverse population. CareMAPs® were first implemented in 1995 for use as a guideline to focus the health care team on delivery of high-quality, cost-ef-

fective care across a defined time line. A stepwise approach was taken to implement this original process:

1. formation of a CareMAP® oversight steering committee with multidisciplinary leaders and a nursing subcommittee
2. establishment of ad hoc multidisciplinary clinical study groups to develop diagnosis and procedure specific clinical recommendations, including pneumonia
3. development of an endorsement process for nursing/medical staff and other ancillary services
4. implementation of the CareMAP® after a pilot study
5. design of an ongoing evaluation process

Structure

The oversight steering committee was formed to outline the process for development and implementation of CareMAPs® throughout the organization. Initial physician concerns were focused on medical-legal issues and a sense of loss of autonomy related to "cookbook medicine." Discussion with legal counsel was helpful in allaying these concerns and validating CareMAPs® as clinical guidelines to be used at the discretion of the physician on the basis of the individual needs of the patient. However, a certain amount of apathy persisted, and only a core group of physician leaders actively supported and endorsed the process.

During the design process, the nursing subcommittee had lengthy discussions regarding the challenges of documentation using the conventional clinical pathway daily format to track patient progress, especially considering an increasingly abbreviated length of stay. The prescriptive nature of the day-by-day format offered little allowance for patients' uneven progress. With consideration of these issues, the design of a daily format was completed. Care outcomes were identified with key sequential interventions covering the entire course of the hospitalization. Finally, these interventions were incorporated into "care events," which included diagnostic tests, activity and safety, treatment/patient care, medications, diet/nutrition, patient/family teaching, discharge planning, and assessment/interventions/outcomes (see Exhibit 6–1).

Development

Diagnoses and procedures for CareMAPs® were generally selected using volume and cost data. Not only was pneumonia the second largest diagnosis, but there was a well-respected pulmonologist who was willing to sponsor and chair the project. The pneumonia clinical study group included representatives from nursing, respiratory therapy, pharmacy, and lab, as well as primary care and infectious disease physicians. A review of the literature and standards of practice in the community were considered, in addition to current practice patterns and clinical outcomes. Although no differences in physician-specific clinical outcomes were observed, there was substantial variation in the treatment of pneumonia patients, particularly in the areas of antibiotic usage, lab tests, X-rays, and length of stay (LOS). These areas were scrutinized closely to achieve improvements in delivery of care and resource use. The study group developed the CareMAP® and clinical practice recommendations. Physician orders and patient educational materials were not developed in this first attempt.

Implementation

The educational strategies developed by the nursing subcommittee, with input from the oversight committee, were essential to the implementation. A variety of educational modalities were used, including inservice education, staff meetings, posters, and classroom discussions. Educational efforts involved patient care staff, physicians, case managers, discharge planners, social workers, and other disciplines as appropriate. To support implementation further, a self-instructional manual was developed, *Instructions for Developing and Implementing a CareMAP,* which included guidelines for development, the policy and practice, and the tracking and approval form.

CareMAPs® were finally implemented in the organization after a small, reasonably successful pilot on a medical unit. Although a policy for ongoing evaluation and revision was established, there was little structure or clear account-ability for monitoring and evaluating the process. Hindsight demonstrates that there was a much more intensive focus on the design and development component than on the implementation and monitoring aspects.

DEFINE AND REDESIGN TO IMPROVE

As managed care continued to grow in the region, the organization once again pressed the need for cost-effective quality care, triggering a more intensive review of the CareMAP® project. A core group of quality professionals and nursing leaders came together to assess the current use. They verified the suspected underutilization through chart review and physician and staff anecdotal information. Over subsequent months, attempts were made to improve CareMAP® use through all of the usual routes—exhortations, communication, inservices, and admonitions—but with only minimal improvement. Ultimately, continuous improvement strategies were used to identify the root causes of ineffective use, improve the pathway process, and standardize practice.

CLARIFYING ROOT CAUSES

As a first step, discussions were held with several multidisciplinary focus groups. A constellation of issues related to poor utilization of the CareMAPs® was identified:

- Physician support was minimal.
- Nurses were marginally vested in the CareMAP® process due to several factors:
  1. challenges regarding the daily format
  2. confusion about the definition of variances
  3. inconsistency with documentation practices
- Case managers were not integral to the CareMAP® process.
- Physician order sets were not used to promote recommended practice.

Along with these staff and physician concerns, surveys consistently indicated that patients did not have a clear understanding of their illness or how to care for themselves at home. Survey scores were also low on patients' perception regarding coordination of care among inpatient providers.

With the addition of the case managers, the expanded core team developed a fishbone diagram to assist in identifying the root causes of underutilization (see Figure 6–1). There were no issues identified with the actual clinical content of the CareMAPs®. Instead, after analysis of the fishbone, the efforts for improvement focused on three themes: format, process, and responsibility.

Exhibit 6–1 Community-Acquired Pneumonia CareMAP®

Admit Date:	Target LOS: 5–6 days		Projected DC Date:		Actual DC Date:
CARE EVENTS	Date(s)	Date(s)	Date(s)	Date(s)	Date(s)
	Day 1	**Day 2**	**Day 3**	**Day 4**	**Day 5/6**
Unit	ER Admit/Med Surg unit ICU for hypotension/resp failure	Med/Surg unit	Med/Surg unit	Med/Surg unit	Med/Surg unit
Diagnostic Tests	CBC/Chem 7 Blood cultures × 2 Sputum cult w/Gram stain ABG on room air Chest X-ray		CBC	O_2 sat on RA if O_2 therapy used Chest x-ray if patient not improving	
Activity & Safety	Progressive ADL and ambulation Bedrest OOB/Amb	OOB/AMB BID	OOB/AMB TID →		
Treatments/ Patient Care	I&O/Maintain fluid balance → VS q 4 h → Resp isolation until organism identified O_2 therapy as ordered → Neb/meds/MDI as ordered for airway obstruction →				VS q shift
Medications	IV Cefuroxime 750 mg q8h → or IV Erythromycin 500 mg q6h Time started:			Convert to PO antibiotics	
Diet/Nutrition	Diet as tolerated →				

Patient/Family Teaching	Instruct on Resp Isolation
	Describe pneumonia, symptoms, expected treatment
				Pulm CNS for MDI instructions if home use planned	Review discharge plans:
					  Meds & Treatments
					  ADL
					  When to call MD
					  Follow-up appt.

Discharge Planning	Refer to PCC/social worker/DC Planner	RN/SW/DP plan w/MD for potential discharge plan		Discharge plan in place	
Assessment Interventions, & Outcomes	**HYPOXEMIA/TACHYPNEA** Note paO_2/SaO_2 and O_2 need → *Without hypoxemia on O_2* → Resp rate/distress: Assess SOB, cyanosis, change in LOC, restlessness, tachycardia, diaphoresis **INEFFECTIVE AIRWAY CLEARANCE** Assesses breath sounds: rhonchi, rales, wheezing, stridor Sputum Observation—consistency, color, amount Cup at bedside—change cup q 24 hours Maintain fluid balance → Notes CXR Results → Consider pulm toilet/therapy → **ACTIVITY LEVEL** Assess tolerance to ADLs → Progress activity: Bedrest/OOB/Ambulation Frequent rest periods		*Without resp distress on bedrest* Refer to Pulmonary CNS if respiratory distress persists *Assisted ADLs*		***No O_2 or home therapy initiated*** Without respiratory distress with ADLs ***No adventitious sounds Sputum decreases*** ***No/Minimal Edema Clears secretions independently Independent ADL or return to pre-hospital status***
Daily Review by:	11–7 _____, RN 7–3 _____, RN 3–11 _____, RN _____, MD	11–7 _____, RN 7–3 _____, RN 3–11 _____, RN _____, MD	11–7 _____, RN 7–3 _____, RN 3–11 _____, RN _____, MD	11–7 _____, RN 7–3 _____, RN 3–11 _____, RN _____, MD	11–7 _____, RN 7–3 _____, RN 3–11 _____, RN _____, MD

Note: ABG, arterial blood gas; ADLs, activities of daily living; AMB, ambulate; BID, twice daily; CBC, complete blood count; CNS, clinical nurse specialist; CXR, chest X-ray; DC, discharge; ER, emergency room; ICU, intensive care unit; I&O, intake and output; LOS, level of consciousness; MDI, metered dose inhaler; O_2 Sat, oxygen saturation; OOB, out of bed; PCC, patient care coordinator; RA, room air; SOB, shortness of breath; TID, three times a day; VS, vital signs.

Source: Copyright © Alta Bates Medical Center.

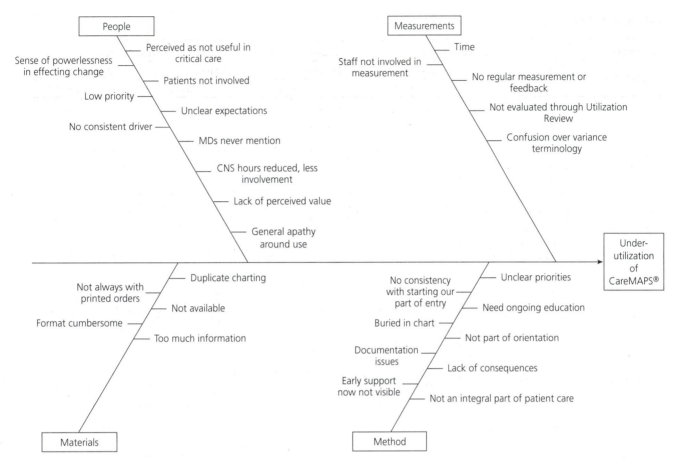

Figure 6–1 Fishbone Diagram on Causes of Low CareMAP® Usage. *Source:* Copyright © Alta Bates Medical Center.

Focused Improvements

Reformatting the CareMAP® was a greater challenge than initially anticipated. The focus groups were frequently consulted as the core team revised the forms to meet the identified parameters for content, ease of use, and consolidation. The revisions encompassed three key design changes: replacing the daily format with core concepts of care referenced as "essential elements of care," summarizing the clinical recommendations and listing team members on the reverse of the MAP® for reference, and adding documentation of variances and discharge outcomes. Additionally, the "stand-alone" MAP® was replaced with a packet that included

- CareMAP® with clinical recommendations
- physician order set
- nursing plan of care
- identifying stickers
- patient education "map of care"

(See Exhibits 6–A–1 through 6–A–4 in Appendix 6–A.)

An improved process for using the pathway and more clearly delineated responsibilities were identified and incorporated into the policy and practice. As rewritten, the staff registered nurse (RN) will evaluate appropriateness of the pneumonia CareMAP® and ensure placement of the packet in the patient's medical record at the point of entry. On the inpatient unit, the secretary will place the MAP behind the medical record "CareMAP®" tab. A sticker will be placed on the outside of the chart and the treatment rand to indicate the LOS target. The RN notes the orders, indicates the LOS target date, reviews the patient education material with the patient and family, and individualizes and initiates the nursing plan of care. This plan will become a part of the treatment rand, which is used during shift report. On rounds, the case manager will also verify appropriate initiation of the CareMAP®. Both the case manager and RN will identify and document variances to progress. Upon discharge, the RN will complete the actual LOS and discharge outcomes, ensuring that the patient understands the instructions for care at home (see Figure 6–2).

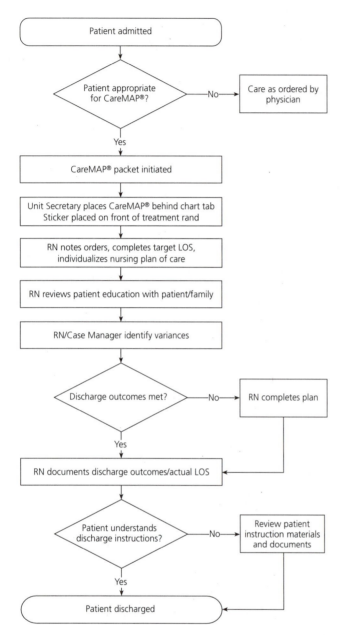

Figure 6–2 CareMAP® Flowchart. *Source:* Copyright © Alta Bates Medical Center.

Implementation

Although a major emphasis had been placed on the training and implementation plan with our first attempt, the core team agreed that a more intensive and prolonged effort was required to ensure success of the pneumonia CareMAP® process during the two-month pilot. Staff on pilot units were required to attend an inservice or complete a self-instructional module. The core team, along with extensive support from case management staff and clinical nurse specialists, was available for daily follow-up and questions or concerns. Weekly core team debriefing sessions were held, issues were discussed, and action plans were developed. Points of clarification were communicated to staff and physicians on the pilot units. As system or process problems surfaced, team members immediately tracked down root causes and made corrections.

Extensive data were gathered related to use of the MAP, process steps, and LOS. An additional element incorporated into this second implementation was an evaluation from CareMAP® customers. Discipline-specific tools were developed to capture input from RNs and physicians, and a "nationally normed" patient survey was mailed to all patients whose care was guided by the pneumonia CareMAP®. Review of the internal customer responses resulted in minor revisions. The patient survey demonstrated no noticeable improvement in the areas of education and care coordination. Findings were communicated to all levels of the organization, and, with final revisions to form and format, attention turned to measuring outcomes.

OUTCOMES MEASUREMENT

All too often, new organizational projects have been given high priority and implemented through the intense efforts of many, only to falter through loss of momentum without measurement and communication of results. Along with skillful implementation, effective outcomes measurement and monitoring were key to the renewed initiation of CareMAPs®. In this second effort, however, the core team was determined to be as attentive in measuring the implementation process as it had been in summarizing clinical outcomes. Plans were also made to provide frequent progress reports to all involved.

Process Outcomes

The new CareMAP® format and procedures required physicians, nurses, and case managers to make changes in established work routines. The core team recognized that to be successful, the new expectations for practice would need to be monitored, and regular feedback regarding progress would have to be given to the staff. This proved to be invaluable in acknowledging staff performance and identifying areas that needed continued effort. A concurrent data collection process was developed to monitor key processes for the project. The core team identified the measures, and a collection tool was created. By using the daily admission report, patients with confirmed or possible pneumonia were identified and added to the master patient list, and daily rounds were made to collect the information. A postdischarge review of the medical record was also done. Measurement focused on those elements fundamental to the change and included

- appropriate patients placed on MAP® (exclusions were admission to nonpilot units; non–community-acquired pneumonia, such as pneumocystis; and pneumonia ruled out)
- CareMAP® initiated, in chart behind new tab
- labels placed on chart and treatment rand
- physician order set used by admitting MD
- nursing plan of care added to treatment rand
- patient education materials given to and reviewed with patient
- documentation of actual discharge date and discharge outcomes completed by RN
- patient discharged by target LOS and, if longer, variance identified

This data collection was labor intensive, but the unit-specific summaries that were developed drove the weekly core team meetings. Areas needing more work were clearly identified and this enabled the team to provide specific direction to the nursing staff and case managers. During daily rounds to gather data, team members began to see signs of heightened attentiveness to the CareMAP® process: the data summaries were posted in the staff conference rooms, and case manager activities centered on the use of the MAP® in planning for dis-

charge. Conclusions from the measurement of these process outcomes included the following:

- Eighty percent of all pneumonia patients were admitted through the emergency department. Therefore, it was imperative that the new process be solidly in place at this point of entry.
- Seventy-six percent of all pneumonia patients managed by the MAP® were discharged home or to the hospital-based skilled nursing facility (SNF) by the target LOS of four days.
- There was 31 to 85 percent completion of expected documentation:
  1. 43 percent for discharge date
  2. 31 percent for patient outcomes, with wide variation between nursing units. This was a new documentation practice that was not fully integrated by the end of the eight-week pilot and will require continued monitoring. However, in those MAPs® with complete documentation, patient outcomes were met 100 percent of the time.
  3. 39 percent for signature of discharging RN
  4. 85 percent for variance when LOS greater than four days

(See Tables 6–1 and 6–2.)

Table 6–1 Pneumonia CareMAP® Pilot: Process Outcomes

		6N	4W/4E	Other	Total
Totals	Total admissions	30	28	16	74
	Exclusions	3	6	16	25
	Pilot cases	27	22		49
Admissions	ER admits	21	18		39
	Direct admits	6	4		10
	Total	27	22		49
	% ER admits				80%
Initiation CareMAP®/	Initiated at Point of Entry	15	14		29
CareMAP®		56%	64%		59%
Orders	Delay/incomplete	9	7		16
	No preprinted orders (PP)	8	2		10
	No MAP®/no PP orders	3	1		4
Discharges	Total discharges	27	22		49
	Length of stay				
	1 day	1	1		2
	2 days	6	3		9
	3 days	8	6		14
	4 days	5	7		12
	5 days	2	2		4
	6 days	2			2
	7 days	1	1		2
	8 days	1	2		3
	>8 days	1	1		2
	Discharged by target LOS	74%	73%		76%
	Total ALOS	3.9	4.2		4.0

Table 6–2 CareMAP® Completion at Discharge

	6N		4W/4E		Total	
Total Reviews	27	100%	15	100%	56	100%
Discharge date						
entered	16	59%	5	23%	21	43%
Patient outcomes						
All checked	12	44%	3	14%	15	31%
Some checked	3	11%	—	—	3	6%
None checked	12	44%	19	86%	31	63%
Signature present	14	52%	5	23%	19	39%
Variance required/						
present	5/7	71%	6/6	100%	11/13	85%

Source: Copyright © Alta Bates Medical Center.

Variance Analysis

Since the documentation of variances potentially affecting LOS was a new component of the CareMAP® process, the core team was particularly interested in reviewing the documentation in this section. A summary of variance "Reason Codes" collected from the MAPs® was completed following the pilot, with not unexpected results. Over 90 percent of all variances were related to patient condition: slow to progress, significant comorbidities, or a new clinical event. Although often viewed as frequent causes of variance, delays in discharge or hospital services accounted for less than 10 percent of variances noted.

In planning for the full implementation of the CareMAP® process on all nursing units, the core team decided to continue using the "Variance" section on the MAP®. However, there were concerns about the ongoing usefulness of the information provided by the variance process and the intensity of the documentation and data collection. During the pilot, copies of the MAPs® were made at discharge to allow summary and analysis of the data. Even for 50 patients, this was an intensive manual process, and one that the team could not support for hospitalwide implementation and evaluation. An equally complex process of printing multipart MAPs® and directing staff to route copies of all completed MAPs® to some central location for analysis was also rejected. Overall, the team felt that it needed more experience with variances to appreciate the potential value. Therefore, monitoring of more easily collected data on patient outcomes and average LOS (ALOS) will continue for now, with a plan to do more focused, in-depth reviews of the specific variance data when downward trends in desired outcomes are noted.

The core team was extremely gratified with the measurement of the process outcomes, notwithstanding the considerable time and effort spent in collecting the data. The summaries were invaluable in motivating all involved to persevere and clearly identified where efforts were needed.

CareMAP® Automation

As a result of the pilot project, the team recognized the need to identify those patients for whom a CareMAP® was instituted, preferably through automation in lieu of record review or collection of voluminous, duplicated copies. A unique charge for each MAP® has been initiated, along with an automated order set in the order communication system. This will allow monitoring to determine what percentage of each patient group had a MAP® initiated, as well as further detail about use patterns to focus on improvement.

Although the organization is moving toward an automated clinical documentation system that will incorporate on-line use of and documentation to the CareMAP®, it will be quite some time before this is in place. In the meantime, other available computerized data systems, including the hospital information and decision support programs, will be used to collect most of the indicator data that are currently monitored.

OVERSIGHT OF CAREMAP® MEASUREMENT AND MONITORING

As mentioned earlier, CareMAPs® and order sets were originally developed by ad hoc groups of clinicians who reviewed current and community practices to complete clinical guidelines. Once these were implemented, the groups usually stopped meeting, since they were not designed to have ongoing oversight for their respective CareMAPs®. Fortunately, the department of medicine/family practice had an active utilization management committee that assumed the oversight of all medicine CareMAPs®, and as the process improved, the group identified appropriate outcome indicators to monitor. However, there were no other utilization management committees, and there was very little physician ownership of other CareMAPs®.

Two years ago, the medical staff redesigned its quality activities to streamline the peer review process and promote continuous improvement. As a result, there are now five departmental, multidisciplinary clinical quality management committees that are responsible for the oversight of CareMAPs®, including development and approval, collaboration with implementation, and ongoing outcomes monitoring.

Current CareMAP® Indicators

The data set for each CareMAP® includes indicators common to all MAPs® and some indicators unique to each, which monitor desired changes in process/practice or patient outcomes. Common indicators include

- number of cases
- ALOS

- percentage of patients discharged alive by the target LOS (an excellent indicator to monitor practice patterns and use of the CareMAP®)
- average variable cost per case
- mortality rate
- readmission rate within 30 days (including readmission diagnosis)
- infection rate for surgical procedures

Unique indicators for pneumonia include

- percentage of cases using desired antibiotics
- hospital day that antibiotics are converted from intravenous to oral
- average chest X-rays per stay
- percentage discharged to hospital-based SNF

Clinical Outcomes for Community-Acquired Pneumonia

Overall, quality indicators (e.g., patients treated with recommended antibiotics, mortality, readmissions) have maintained or improved with the use of the CareMAP®, while measures of resource use (ALOS, variable cost per case) have shown reductions. Some important results include

- Percentage receiving recommended antibiotics increased from 85 percent to 100 percent.
- Readmissions within 30 days (all causes) decreased from 16 percent to 8 percent.
- ALOS for pilot cases managed on the MAP® was 3.9 days; for all pneumonia cases, ALOS decreased from 5.0 to 4.7 days, with a full-year forecast of 4.5 days for all cases.
- Percentage of patients discharged alive by the target four days increased from 52 percent to 64 percent in the past five months.
- Variable cost per case has declined 19 percent.

(See Table 6–3.)

During the quarterly reviews of the indicators, requests for follow-up may be made when results raise questions. When a four-year trend in the mortality rate for pneumonia showed a consistent downward trend, there were concerns as to whether the increased early discharge to the SNF was simply shifting the rate from the acute hos-

Table 6–3 Outcome Indicators: Pneumonia

		Prestudy FY 1990	1994 FY	1995 FY	1996 FY	1997 Q1
DRGs: 89 or 90						
Number of cases	All cases					
	Cases w/LOS < 15					
Average length of stay	All cases					
	Cases w/LOS < 15					
Variable cost per case	All cases					
	Cases w/LOS < 15					
Live discharge by CareMAP® LOS target	Live discharges					
Live discharges by CM LOS: all live discharges	Discharged by CM LOS					
CareMAP® LOS target = four days	Rate					
Mortality rate (Includes deaths in SNF/TRU following the acute stay)	Deaths					
	Rate					
Readmission rate (Within 30 days for any reason)	Readmits					
	Rate					
Focused Indicators						
Discharged to SNF	Cases					
	Rate					
	SNF ALOS					
Average number of chest X-rays per case	# of chest X-rays					
	Cases using X-rays					
	% of cases using X-rays					
	# X-rays per case					
% of cases using Cefuroxime, Erythromycin or Ceftriaxone	Cases using					
	Rate					

Note: ALOS, average length of stay; CM, CareMap®; DRG, diagnosis-related group; FY, fiscal year; LOS, length of stay; SNF, skilled nursing facility; TRU, transitional rehabilitation unit.

Source: Copyright © Alta Bates Medical Center.

pital to that service. This inquiry resulted in the inclusion of the SNF in the calculation of the mortality rate. Subsequent review continues to show a decline in the combined mortality rate for pneumonia patients.

Future Enhancements to Outcomes Monitoring

The core team is currently developing systems to add two additional measures to the CareMAP® monitoring process: (1) refining the existing process for the patient discharge satisfaction survey to regularly identify and sample patients using CareMAPs®, to determine if satisfaction is affected by the use of a MAP®; and (2) instituting use of a severity-adjusted database of northern California hospitals to compare several outcomes of the CareMAP® groups with other hospitals' outcomes, including ALOS and readmission and mortality rates.

CONCLUSIONS

Lessons Learned

After a careful review of the first experience, by using a continuous improvement process, the multiple causes of CareMAP® underutilization could be clustered into issues of format, process, and responsibility. This allowed the

team to effectively redesign all components of the CareMAP® system and focus energy on reimplementation. There are several implications for other providers developing or improving critical pathways:

- Consider options in the tool format—more global aspects of care rather than detailed daily care events.
- Focus intently on all process elements during implementation.
- Standardize staff education for both implementation and new staff orientation.
- Incorporate the clinical pathway into a comprehensive packet.
- Regularly and frequently collect outcome measures and communicate them to all constituents.

(See Table 6–4.)

Future Plans

The successful pilot of both a medical and a surgical CareMAP® has allowed the team to move forward in revising and implementing additional MAPs®. It is anticipated that the format will be extended to, and work equally well in, specialty areas such as obstetrics and rehabilitation. Each clinical quality management committee

Table 6–4 Summary of CareMAP® Improvements

	Original	Improved
Format	• Day-by-day format for listing "care events"	• Summary of "essential elements of care" (care concepts for the total episode of care)
	• No physician order set	• Physician order set
	• No documentation of variance and discharge outcomes	• Documentation of variances and discharge outcomes
	• No patient MAP®	• Patient education map of care
	• Nursing plan of care with an intervention focus on MAP®	• Plan of care coincides with "essential elements of care"
	• Singular MAP®	• CareMAP® packet includes MAP®, order set, plan of care, and patient education
Process	• Initiation on nursing unit	• Initiation at point of entry to hospital (usually ER or preop testing)
	• No unique identification of patients on MAP®	• MAP® stickers on chart and treatment rand
	• Limited documentation (signature of RN/MD noting review)	• Documentation of LOS, variances, and discharge outcomes
	• No specific patient educational materials	• Patient/family education facilitated by specific written materials
	• Staff education optional	• Staff education required by class or self-instructional manual
Responsibility	• Unclear roles and expectations	• Roles and responsibilities clearly delineated and monitored
	• Case managers not integrated in process	• Case managers instrumental in initiating MAP® and using for communicating with RNs/MDs
	• Inconsistent administration support from patient care and physician leaders	• Full organizational support
	• Minimal clinician ownership and monitoring	• Defined oversight and regular monitoring through quality committees

will annually develop CareMAP® packets for at least one diagnosis or procedure. The core team, once an interim improvement committee, will continue as a permanent part of implementation, ensuring not only the development of future CareMAPs® but also the continual assessment and improvement of the process.

Outcomes

Outcomes measurement has been described as an important, if not the most important, component of this CareMAP® improvement project. Measurement was instrumental in tracking the implementation process and monitoring patient outcomes and resource use. As the introduction of additional MAPs® is completed this year, with others to follow, the efforts to regularly measure and monitor must be sustained to

- identify and communicate achievements in order to reward efforts and enhance physician and staff support for the process
- provide direction for improved use
- evaluate the appropriateness of CareMAP® targets and clinical recommendations

In addition, the core team is particularly interested in evaluating the impact of deviating from the conventional pathway model, which includes detailed daily care events, and continuing to assess the usefulness of process-focused variance analysis.

Market forces have unquestionably prompted the initial endorsement of tools such as the CareMAP®, which hold the potential for clinical effectiveness. However, the sustained support of clinicians, administrators, and hospital boards will only come through credible measurement that demonstrates expected or improved clinical outcomes and resource use. The purchasers and consumers of health care, as well as review organizations such as the Joint Commission on Accreditation of Healthcare Organizations, are also leading the drive for this same information to assist in determining the value of major health care expenditures. The measurement of outcomes for specific patient populations, fostered by the clinical pathway process, is an excellent way to meet this imperative.

SUGGESTED READING

Ebener K, Baugh K, Mansheim-Formella N. Proving that less is more: linking resources to outcomes. *J Nurs Care Qual.* 1996;10(2):1–9.

Goode C. Impact of a care map and case management on patient satisfaction and staff satisfaction, collaboration, autonomy. *Nurs Econ.* 1995;13(6):337–348.

Horn S, Hopkins D. *Clinical Practice Improvement: A New Technology for Developing Cost-Effective Quality Health Care.* Faulkner & Gray's Medical Outcomes and Practice Guidelines Library; 1994:1.

Ireson C. Critical pathways: effectiveness in achieving patient outcomes. *J Nurs Adm.* 1997;27(6):16–23.

Lumsdon K, Hagland M. Mapping care. *Hosp Health Net.* 1993; 67(20):34–40.

Spath P. Path-Based Patient Care Should Build Quality into the Process. *J Healthcare Qual.* 1995;17(6):26–29.

Zander K. Evolving mapping and case management for capitation. Part III: Getting control of value. *New Definition. Cen Case Manage.* 1996;11(2):1–2.

■ Appendix 6–A ■
CareMAP® Revisions, Orders, and Plan of Care

Exhibit 6–A–1 Community-Acquired Pneumonia: Essential Elements of Care and Documentation of Variance and Discharge Outcomes

This CareMAP® is a generally suggested approach, but the decision whether or not to abide by it has to be based on the individual circumstances of a particular patient. Any decision to follow or modify the CareMAP® shall be within the discretion of the physician.	Target LOS: 4 days	Target D/C Date: _____
		(admit date + 4)
	Admit Date: _____	Actual D/C Date: _____

Essential Elements of Care—see pre-printed orders

Antibiotics	• Timely initiation of appropriate IV antibiotic therapy; change to oral antibiotic in 2–3 days
Airway Clearance	• Effective respiratory therapy and airway clearance in patients with airway obstruction
Activity	• Early ambulation and progressive ADLs
Education	• Disease process, symptoms, treatment
	• Discharge Instructions See reverse for summary of clinical recommendations

Factors/Variances Potentially Affecting LOS

Date	#	Description of Event	Reason Code	Signature	Resolved	Init.
	1					
	2					
	3					
	4					
	5					
	6					

Reason Codes	**Patient/Family**	**MD/Caregiver**	**Systems**	**Discharge/Transfer**
	A1 Patient condition	B1 Not done/delayed	C1 Service availability	D1 Placement unavailable
	A2 Availability/decision	B2 Order/Decision	C2 Lack of information/data	D2 Financial/Legal issues
	A3 Other	B3 Other	C3 Other	D3 Other

Discharge Outcomes **Met** **Not Met**
- Discharged by target LOS. _____ _____
- Patient has minimal/no respiratory distress at rest or with ADLs _____ _____
- Patient has decreasing/no:
   - productive sputum _____ _____
   - fever _____ _____
- Verbalizes understanding of disease process, management of self-care, and receives discharge information. _____ _____

If not met, indicate discharge instructions or plan of action: _____

_____

Discharging RN Signature _____

Note: ADLs, activities of daily living; LOS, length of stay.
Source: Copyright © Alta Bates Medical Center.

Exhibit 6–A–2 Management of the Patient with Community-Acquired Pneumonia: Clinical Recommendations

Recommendations of Pneumonia Clinical Study Group, February 1992; reviewed and revised January, 1997
Accepted by the Dept of Medicine/FP UM Committee, QA Committee February 1992
Study Chair:
Study Group Members:

Summary: Major Clinical Recommendations

Admission Workup

- Initial workup should include: –CBC with differential   –Chest X-ray   –Blood culture × 2
     –Chem 7   –O$_2$ saturation   –Sputum culture with Gram stain
     If TB suspected, AFB smear and culture × 3 and intermediate PPD

Level of Care

- Patients with pneumonia infrequently require admission to a critical care unit. Criteria that may suggest a need for critical care include respiratory failure and hypotension.

Chest X-rays

- Some patients may require only one chest X-ray, for admission evaluation, and most require no more than two. A second chest film may not be necessary during hospitalization if the patient's clinical condition improves and the patient will have adequate outpatient follow-up. A dramatic radiographic improvement during early treatment should not be expected. It takes an average of 6 weeks for a chest film to clear completely.

Respiratory Therapy

- An assessment by Respiratory Therapy should be completed in patients with airway obstruction and problems with secretions. If indicated, RT protocol for bronchodilator therapy and secretion management should be initiated.

Antibiotic Therapy

- Recommended IV antibiotics for initiation of therapy include one of the following 3 choices:
  –Cefuroxime 750 mg q8hrs,   OR   –Ceftriaxone 1 Gm QD,   OR   –Erythromycin 500 mg q6hrs (Penicillin allergy)
  Additional considerations: –severely ill patients, Erythromycin 500mg q6hrs *in addition* to one of above cephalosporins
     –suspected Legionella, Erythromycin 1 Gm q6hrs
- Initiation of antibiotics should begin within 1 hour of order and should not be delayed for sputum collection or results.
- Timely conversion from IV to PO antibiotics should be done, usually in 48–72 hours in uncomplicated cases.
  Indications include: –fall in WBC and decrease in fever (not necessarily afebrile)
     –able to take oral fluids and medications
     –fall in respiratory rate, able to clear secretions well
     –improvement in oxygenation (not with underlying lung disease)

Infection Control

- Patients with respiratory signs and symptoms should be placed in any single room and care givers should observe routine BSI procedures. Physician should evaluate the need for private room after 24–48 hours.
- If TB is suspected, the patient should be placed on Respiratory Isolation and admitted/moved to a negative pressure isolation room (NPIR). Contact Infection Control staff. See Algorithm for Respiratory Isolation.

Discharge Planning

- Consider discharge shortly after patient converts from IV to PO antibiotics. The following clinical conditions need NOT be met prior to discharge:
  –Afebrile   –Clear chest X-ray   –Normal WBC   –Normal ABGs*
  *Patients with underlying lung disease may have persistent hypoxia and can receive oxygen therapy at home.
- The SNF should be considered after 72 hours for patients needing continued skilled observation, IV therapy, PT or RT.

Note: ABGs, arterial blood gases; AFB, acid fast baccilli; BSI, body substance isolation; CBC, complete blood count; PPD, purified protein derivative; PT, physical therapy; RT, respiratory therapy; SNF, skilled nursing facility; WBC, white blood cell count.
Source: Copyright © Alta Bates Medical Center.

Exhibit 6–A–3 Community-Acquired Pneumonia: Nursing Plan of Care

| Plan of Care: | PNEUMONIA; COMMUNITY ACQUIRED |
| Target LOS: | 4 days |

Problems	Date	Intervention	Goal (Outcome)
Airway Clearance		1. Assess color and consistency of secretions. 2. Assess respiratory rate, pattern, and ability to clear airway. 3. Encourage turn, cough, and deep breathe. 4. Oxygen therapy PRN. _Signature_ 5. _____ _____ 6. _____ _____	Patient will demonstrate effective airway clearance as evidenced by the patient's ability to expectorate secretions and absence of labored respirations.
Activity		1. Assist patient to chair/ambulation, use portable O$_2$ if needed. 2. Group care activities to provide rest. _Signature_ 3. _____ _____ 4. _____ _____	Patient will demonstrate tolerance to activity as evidenced by minimal shortness of breath, vital signs within patient's normal values.
Knowledge Deficit		1. Perform readiness to learn prior to teaching and assess barriers to learning. 2. Provide early and frequent patient/family education. 3. Patient education materials given to patient/family and content reviewed: _Signature_ _____ _____ _____ _____	Patient verbalizes and/or demonstrates adequate knowledge of illness as evidenced by discussing accurate information related to healthcare problems, self-management and follow-up care.

Source: Copyright © Alta Bates Medical Center.

Exhibit 6–A–4 Community-Acquired Pneumonia: Patient Education Sheet

PNEUMONIA: WHAT YOU NEED TO KNOW

What is Pneumonia?

❑ Pneumonia is an infection in the lung

Some of the Common Signs and Symptoms of Pneumonia Are Listed Below. You May or May Not Develop All of These Symptoms.
❑ Sudden onset of shaking, chills, fever
❑ Discolored, thick phlegm/sputum
❑ Chest pain or discomfort worsens with deep breathing or coughing
❑ Shortness of breath or difficulty breathing
❑ Heart beating fast or strong

During your hospitalization . . .

❑ You will be hospitalized for approximately 4 days

❑ While you are in the hospital, tests may be ordered by your doctor to determine infection.

❑ Your care will include:
  ❑ Antibiotic medication (IV or pills)
  ❑ Oxygen if you are having difficulty breathing
  ❑ Adequate rest periods

❑ You can help by:
  ❑ Drinking large amounts of fluid
  ❑ Maintaining a nutritious, well-balanced diet
  ❑ Following your plan of care

Instructions to Remember When You Are at Home:

❑ Get plenty of rest

❑ Avoid smoking

❑ Ask your doctor for recommendations regarding flu immunizations

❑ Notify your doctor if:
  ❑ You feel more tired
  ❑ You develop a fever
  ❑ Your phlegm/sputum increases or becomes more discolored

For further information, please notify your nurse.

Source: Copyright © Alta Bates Medical Center.

7

Collaborative Efforts in Managing Patient Outcomes: The Pulmonary Resection Pathway

Janet Taubert, Kathleen M. Lewis, Louisa Kan, and Yvette De Jesus

The University of Texas M.D. Anderson Cancer Center is a comprehensive cancer center located in Houston, Texas. An academic and tertiary referral center, our institution treated approximately 52,300 patients with cancer in 1996.

In assessing, evaluating, and monitoring the quality of care received by patients, the health care industry is shifting its focus from individualized patient care to a broader, population-based perspective.[1] Gould and Schaeffer have identified oncology as a primary managed-care product for three reasons:

> (1) Cancer is an increasingly treatable disease, (2) costs associated with oncology treatment are already considerable, and likely to continue to increase with progress in medical technology, and (3) these costs cannot be entirely eliminated, but can be controlled by adopting treatment strategies that make cost effectiveness a priority for the delivery system."[2(p.182)]

In l996, in response to a changing health care system and the impact anticipated from managed care, M.D. Anderson Cancer Center established a disease management program. The goal of this program is to coordinate oncology services across the continuum of care to provide patient-focused, effective oncology services resulting in optimal clinical and financial outcomes. Toward this goal, the program works to develop practice guidelines, collaborative pathways, patient education material specific to pathways, and computer-based medical records. This program is also the foundation for case management and future physician profiling focused on measurements of quality and outcomes.[3] A practice guideline defines care across the continuum for a particular cancer type; it is viewed as an algorithm with branches based on the results of staging, procedures, and treatment (Figures 7–1 and 7–2). For example, the M.D. Anderson practice guidelines for both small-cell and non–small cell lung cancer outline the staging processes, the treatment plans based on diagnosis, and the follow-up therapy. Thus, in an individual with a lung cancer diagnosed as T1N0, a practice guideline would guide the evaluation and choice of therapy (e.g., pulmonary resection) and follow-up treatment. A collaborative pathway is an interdisciplinary tool describing expected events, interventions, and outcomes in an episode of care with a particular treatment modality, such as surgery, chemotherapy, or radiation therapy. For example, for patients undergoing surgery, the pathway identifies the length of stay, essential interventions, patient education, discharge planning, and predicted patient outcomes. Physician profiling will provide severity-adjusted data derived from outcome measurements and correlate these data with physician use of guidelines, pathways, and medical resources.

Approximately 3,000 new patients were seen and evaluated for treatment in M.D. Anderson's thoracic surgery clinic this past year. In this chapter, we will review the development and implementation of M.D. Anderson's collaborative pathway for pulmonary resection, with an emphasis on the use of the pathway and expected outcomes.

PULMONARY RESECTION AT M.D. ANDERSON CANCER CENTER

The standard treatment of stages I, II, and selected stage IIIA lung cancers is surgical excision. Up to 80 percent of patients with early-stage lung cancer (T1N0) can be cured with a surgical resection, and in stage II disease, the 5-year survival rate after surgical resection is approxi-

Non-Small Cell Lung Cancer

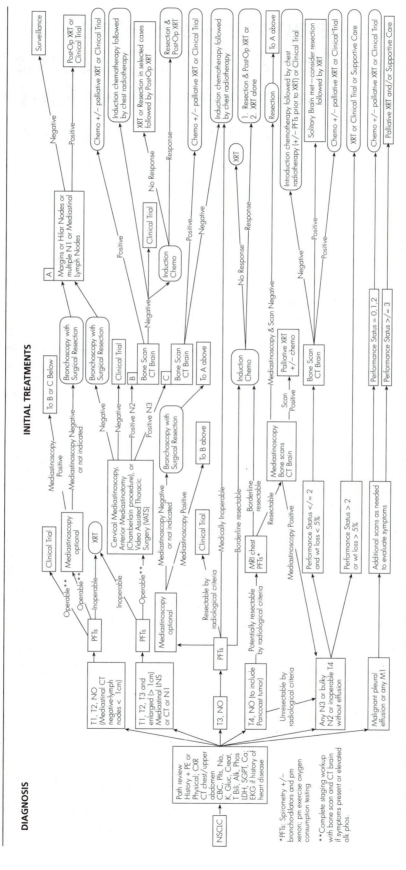

Figure 7–1 Non-Small Cell Lung Cancer. *Source:* Copyright © M.D. Anderson Cancer Center.

Notes: PFTs = Pulmonary Function Tests; CT = Cat Scan; XRT = Radiotherapy; NSCLC = Non–Small Cell Lung Cancer; PE = Physical Exam; CXR = Chest X-ray; T = Tumor; N = Node; M = Metastases; VATS = Video Assisted Thoracoscopy.

Small-Cell Lung Cancer

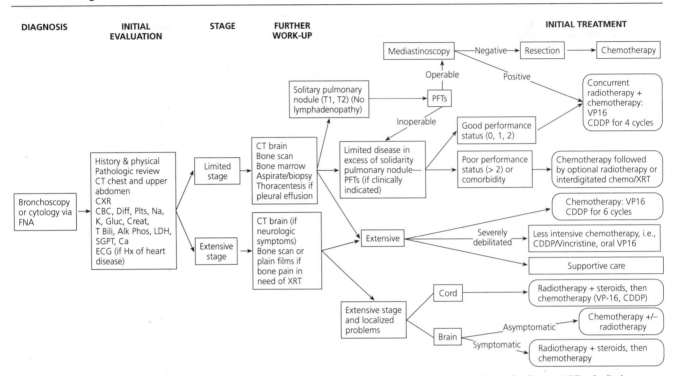

Notes: FNA = Fine Needle Aspiration; VP-16 = Etoposide; CDDP = Cisplatin; PFTs = Pulmonary Functions Tests; T = Tumor; XRT = Radiotherapy.

Figure 7–2 *continues*

mately 40 percent.[4] The surgical procedures classified under pulmonary resection include lobectomy, wedge resection, segmentectomy, and resection of pulmonary metastases. These are areas representing a high patient volume with fairly consistent postoperative outcomes. The majority of patients undergoing pulmonary resection at our cancer center fall into one of two categories: patients with a surgically resectable lung tumor and patients with a surgically resectable metastatic lung lesion. Primary sarcoma is the most common cancer type among patients in the latter group, but this group may also include patients with primary cancers of the breast, bladder, prostate, and testes.

Pulmonary resections in our institution are performed by our thoracic/cardiovascular surgeons; between 25 and 40 procedures are performed each month. The surgical procedure, postoperative interventions, and outcomes are very similar between patients. In view of this and the fact that our surgeons all follow a similar standard of practice, pulmonary resection was an ideal procedure for which to develop a pathway.

HOW WE DEVELOPED THE PATHWAY

In December 1995, a central working group consisting of a thoracic/cardiovascular surgeon, the nurse manager of the thoracic surgery inpatient unit, the oncology clinical nurse specialist for the inpatient/outpatient thoracic unit, a representative of the patient education department, and a representative from the disease management program was organized to devise a plan for development of the pathway and to suggest a timeline. The role of the disease management program representative initially was to guide the group through the phases of pathway development, focusing on the pathway format rather than the practice content. An important question we had to ask ourselves before developing the pathways was what procedures we were routinely performing that had not been scientifically validated. This practice was determined not to contribute to the patient's overall outcome. The pathway process allowed us an opportunity to review our practice related to current science and to correlate current practice with patient outcomes.[5]

From the thoracic/cardiovascular-attending physicians at M.D. Anderson, a surgeon was chosen to serve as the physician representative in our working group. This physician was responsible for facilitating communication between his surgical peer group and the pathway working group. We felt it was imperative that our physicians "buy into" the pathway development process up front to make it successful. Physicians' involvement in the devel-

Small-Cell Lung Cancer *continued*

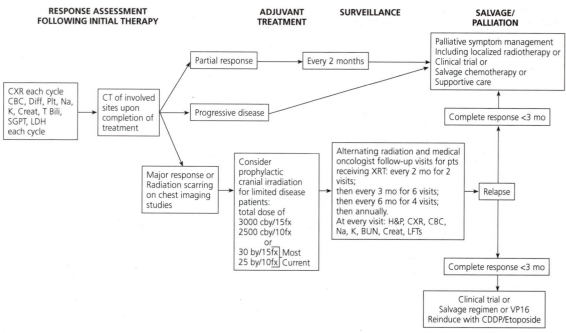

Notes: Fx = fractions; CDDP = Cisplatin; VP-16 = Etoposide; XRT = Radiotherapy.

Figure 7–2 Small-Cell Lung Cancer. *Source:* Copyright © M.D. Anderson Cancer Center.

opment and implementation of the pathways contributes to improved outcomes and provides opportunities for maintaining and ensuring quality.[6]

Format of the "Pathway Packet"

The format for pathways used at our institution was developed by the disease management program. Each "pathway packet" has six separate but corresponding components: the pathway, the informed surgical consent form, the physician order set, a record of outcomes, a record of alterations to the plan of care, and the corresponding patient educational material.

Pathway

The pulmonary resection pathway is divided into several categories: assessment, consultations, diagnostic tests, treatments, medications, performance status/activity, nutrition, teaching/psychosocial, discharge planning, and outcome criteria. Exhibit 7–1 provides a listing of the major interventions occurring on the day of surgery. Although the pulmonary resection pathway is viewed as a guideline for all patients undergoing pulmonary resec-

tion, care is always tailored to meet the individual patient's needs.

The pulmonary resection pathway covers the preoperative visit in the thoracic surgery outpatient clinic; the admission day, which is also the day of surgery; six postoperative days, with discharge expected by day 6; and one-month and six-month follow-up clinic visits. The expected length of stay was based on past practice, standards of our practice, and a review of the literature.

Informed Surgical Consent Form

The informed surgical consent form was designed to provide space for the individual surgeon to fill in the actual procedure scheduled to be performed, such as a right upper lobectomy, before obtaining an informed consent from the patient.

Physician Order Set

A physician order set was developed by our thoracic/cardiovascular surgeons and then evaluated by members of our pathway working group. The intent of the physician order set is to decrease the variation in physician or-

Exhibit 7–1 Pulmonary Resection Pathway for the Day of Surgery

MRN: _____

Name: _____

DOB: _____

Admit Date: _____

Path Entry: _____

Surgical Pathway

Site/Sub-site: Thoracic/Lung
Pathway: ThS3 Pulmonary Resection

Category of Care	Same Day Admit Surgery Order: O_2
Assessment/Eval	Monitor Bed
Consult	
Diagnostic Test	H & H Na, K, Cl, CO_2 (if blood loss >/= 200cc) CXR in PACU
Treatment	Incentive Spirometer q1h while awake; q3–4h during night InspirEase with Albuterol Inhalant q4h prn for bronchospasm CT to –20 cm suction O_2 4L, titrate down if O_2 Sat >93% NT suction prn w/red rubber cath Foley BSD Initiate Respiratory Tx per ABRC protocol if one of the following: a) O_2 saturation <93% on >40% O_2 b) NT suctioning required >q4h or c) Nebulizer treatments required >q4h d) Pts with concurrent conditions (i.e., pneumonia) requiring respiratory therapy interventions >/= q3h
Medication	Gentamycin 1 mg/kg IVPB 1 dose pre-op; 2 doses post-op and Cleocin 900 mg IVPB 1 dose pre-op; 2 doses post-op or Unasyn 1.5 mg IV q8h (if renal insuff) IV Fluids @ 75 cc/hr Analgesia Epidural or PCA Droperidol 0.625 mg IV q6h prn N/V Laxative of Choice Benadryl 12.5 mg IVPB prn itching Tylenol gr x PO q4h prn if temp >38 degrees Heparin 5000u SQ BID until discharge OR compression boots until ambulating TID
Performance Status/Activity	HOB @ 30 degrees Out of bed to chair 15 min
Nutrition	NPO Ice Chips post-op
Teaching/Psychosoc	Reinforce teaching Durable Power of Attorney for Healthcare Information given and/or reinforced
Discharge Planning	Needs assessment
Outcome Criteria	Discharge Criteria: stable vital signs; afebrile without significant pulmonary complications; incision line clean and without signs of infection; tolerating diet; activity appropriate for discharge
Followup Criteria	

Note: CXR = chest X-ray; PACU = Post Anesthesia Care Unit; CT = chest tube; NT = nasal tracheal; BSD = bedside drainage; ABRC = Assessment-based Respiratory Care; HOB = head of bed; PCA = patient controlled analgesia; N/V = nausea/vomiting.

Source: Copyright © M.D. Anderson Cancer Center.

ders throughout the patient's hospitalization. This order set addressed medications, diagnostic tests, X-rays, treatments, activity, and nutrition for patients undergoing a pulmonary resection. Elements of the physician order set correspond with the categories of care from the pathway. Order sets are preprinted and outline the entire course of hospitalization. The use of order sets allows for timely transcription of orders, which provides more time for delivering care. All that is required on the part of the physician or designee is to fill in the dates for things to be accomplished and sign the orders.

Record of Outcomes and Alterations to Plan of Care

A Patient Outcomes Documentation Record was developed to measure expected patient outcomes. This document consists of a list of expected outcomes and is filled out every shift, or three times per 24-hour period. The Outcomes Documentation Record was developed in collaboration with our institutional documentation committee to provide for continuity of documentation and to prevent double documentation by the staff. Documentation of the care for patients on the pathway is outcomes based and is complemented by a shift-to-shift assessment that uses documentation by exception. Daily patient outcomes are documented as met or unmet. We chose to refer to unmet outcomes as *alterations* to avoid the negative connotations that often accompany the term *variances.* Alterations should be viewed as neither positive nor negative. An example of an alteration is discharge of a patient sooner than the pathway predicts. Other alterations may result in an increase in the length of stay, additional costs, and/or decreased patient satisfaction. An example of such an alteration in our patient population would be inability to provide adequate pain control. That often leads to an ineffective cough, ineffective use of the incentive spirometer, and inability to ambulate, which in turn may result in an increased length of stay and decreased patient satisfaction.

Patient Education Material

As our pathway was in development, the patient education representative was working on an educational booklet entitled *Your Care Path—Pulmonary Resection.* This booklet was developed with input from our pathway-working group. The information in this booklet reflected the details outlined in the pathway. It covered preoperative needs; information related to the surgery; what one should expect each day postoperatively, including the importance of pain control, ambulation, use of the incentive spirometer, and control of nausea and vomiting; and discharge and home care instructions. All patients requiring a pulmonary resection review the educational booklet at their preoperative clinic visit; this ensures consistent preoperative teaching for all patients requiring a pulmonary resection. The preoperative educational preparation consists of the patient's and family's viewing of a video on thoracotomy that outlines care issues from the preoperative stage through the postopera-

tive stage, including home care. Care issues such as instruction in the use of an incentive spirometer, with a return demonstration and distribution of an incentive spirometer for take-home practice, and a discussion of smoking cessation are completed during the perioperative stage. An assessment by anesthesiology, which consists of discussion of anesthesia and the use, benefits, and side effects of an epidural catheter for pain control are also completed.[6] It is important to address pain control, ambulation, and use of the incentive spirometer, which are techniques for prevention of pulmonary complications following surgery. These issues are currently reviewed throughout each patient's hospital stay and are felt to directly affect length of stay, patient satisfaction, and utilization of resources.

Nursing Diagnoses

Nursing process is viewed as a uniform way to identify problems (nursing diagnosis), provide interventions, and define expected outcomes for patients with specific needs related to actual or potential high-risk problems.[7] These diagnoses provide the basis for the selection of nursing interventions chosen to achieve appropriate outcomes to deal with specific problems, such as pain or infection. We chose not to include nursing diagnoses within our pathway. Instead, we chose to address the diagnoses/patient problems through a collaborative interdisciplinary approach. Collaboration among the health care team members is essential to the achievement of patient outcomes. An example of this within our pathway is the diagnosis of pain. The issue of adequate pain control is addressed each day of our pathway. Maintenance of adequate pain control is achieved only through the collaborative efforts of the nursing staff, surgeons, and the physicians and nurses of our acute pain service in combination with our patients and families. The use of pathways provides a measurable, objective approach to providing patient care. There is a need to combine nursing process with pathway development to achieve our multidisciplinary approach to patient care.

Developing Consensus

After the pathway had gone through several revisions, the final version was presented to M.D. Anderson's thoracic/cardiovascular surgeons by the pathway-working group. Overall, the surgeons were in agreement with the pathway. One issue of controversy was the use of 5,000 units of heparin given subcutaneously twice a day versus the use of antiembolic compression boots. The risks, benefits, and alternatives for each of these treatments were discussed in depth, and the relevant literature was consulted. The decision was reached that one of these options must be initiated, although it would be left up to the discretion of the physician which intervention to choose. Once the working group accepted the pathway, it was then submitted to the institutional guideline review committee. The purpose of this interdisciplinary group is to review the pathway objectively. This objective review focuses on specific pathway interventions, such as the frequency of postoperative chest X-rays for the pulmonary resection patient, and makes appropriate recommendations for changing or approving the pathway as written.

Final Approval

As final approval for the pulmonary resection pathway was being obtained, plans were made to educate the staff of both the inpatient and outpatient thoracic surgery units in the use of the different elements of the pathway. Educational inservices were held involving surgical, nursing, and clerical staff in both areas and on all shifts. In addition, training was provided for members of the other disciplines involved in our pathway, clinical dietitians, case managers, and other disciplines assigned to our thoracic units.

Mechanisms for Revision of the Pathway

On the basis of documentation of daily outcomes and clinical evaluation, the pathway underwent a concurrent revision process. In addition, the pulmonary resection pathway was reviewed in depth for revisions of expected outcomes one year after implementation on the basis of a chart audit performed by the disease management program representative in collaboration with the clinical nurse specialist. On the basis of this audit, we were able to identify a number of areas of necessary revision within our pathway. For example, the presence of an air leak was added to our outcomes documentation, since this was a clinical issue that occasionally increased our length of stay. A need to reeducate our staff regarding complete and accurate documentation was identified. In a joint decision between the nurse manager and clinical nurse specialist, a plan was devised to achieve 100 percent compliance among the nursing staff in pathway documentation. We chose to include compliance with pathway documentation as a part of nurses' evaluation process. A 100 percent compliance with pathway documentation within a six-month period from their previous evaluation is required.

USE OF THE PATHWAY IN CLINICAL PRACTICE

Patient Enrollment

Patients undergoing a pulmonary resection are enrolled onto the pathway during their preoperative clinic visit; their primary clinic nurse performs this enrollment. Enrollment was initially performed manually by the clinic nurse, and an enrollment form was faxed to the disease management program, where the enrollment information was entered into a Lotus Notes software database. This allowed tracking of all patients enrolled. This software is not on-line documentation. The software (Cancer Manager) is used to develop pathways, enroll patients, and print their respective pathway documents ready for the chart and for the patient. Our institution is in the process of developing a computer-based patient record with pathways as a component. The use of computer-based patient record systems assists in the quality process of patient care and allows pathway monitoring to become a prospective rather than a retrospective review.[8] For such systems to be effective in the patient care setting, they must be user friendly and time saving for the staff.[9] The disease management personnel conducted one-on-one educational inservices with the clinic staff. In addition, this implementation of the computer-based system necessitated backup support from our computer services division, which provides technical support, such as updating the printer to print at high speed. Currently, the change to computer-generated copies of the pathways is being implemented and pilot tested on our inpatient thoracic surgery unit. In the future, the computer-based patient record with the collaborative pathway process will include on-line documentation.

Exceptions to Pathway Enrollment

At present, all patients undergoing a pulmonary resection are placed on this respective pathway. Exceptions to this are examined individually; it is understood through education by the clinical nurse specialist that any patient undergoing additional complex procedures in addition to a pulmonary resection will not be placed on this pathway. For example, a patient may undergo a procedure (thoracotomy) resulting in a combination between thoracic and cardiovascular surgery and/or neurosurgery, general surgery, or plastic surgery, such as a resection or a large soft tissue chest sarcoma with a flap placement, thus resulting in an additional plan of care and change in their overall outcomes. The oncology clinical nurse specialist in charge of managing the pathways meets with the clinic staff to assess these patients individually.

Clinical Nurse Specialist's Role in Pathway-Guided Care

The management of patients on pathways is dependent upon practitioners' clinical expertise with a select population, their knowledge and understanding of the evolving health care system, and their ability to collaborate and coordinate with all members of the health care team and to utilize critical thinking skills. The clinical nurse specialist, whose roles include direct patient caregiver, consultant, educator, researcher, collaborator, change agent, and clinical leader, is uniquely identified to manage this process.[10,11] The oncology clinical nurse specialist coordinates the care with all patients and uses the pathway as a tool for communication to all health care providers responsible for the plan of care. Her responsibilities include ensuring that

- patients are properly enrolled onto the pathway
- pathways are initiated postoperatively
- alterations to the pathway have been identified
- pathway alterations have had interventions initiated
- pathway documentation is complete
- the pathway is reassessed and revised as needed
- patient outcomes are monitored; trends are identified, and opportunities for change are implemented

Clinical Nurse's Role in Pathway-Guided Care

Adherence to the pathway and documentation of alterations and interventions specified by the pathway are the responsibilities of each primary clinical nurse caring for each patient. Nurses are accountable for identifying and planning for discharge needs, beginning on their preoperative visit; this role includes collaborating with the case manager. Additionally, their responsibilities include ensuring that:

- patients follow the pathway
- patients' alterations are addressed with the clinical nurse specialist/physician
- all alterations are addressed, and interventions are implemented promptly and appropriately
- patient outcomes and alterations are documented completely
- the pathway is utilized in the change-of-shift report
- the pathway as teaching tool is consistently reinforced with the patient

Physician's Role in Pathway-Guided Care

A major role that thoracic surgeons play in pathway-guided care is, first, to ensure that patients are treated according to the pulmonary resection pathway. This role is successful if the surgeons have ownership of the pathway. The ownership should start during the development and should be ongoing. An example of this ownership is demonstrated by their unanimous decision to use the same antibiotics postoperatively for the same time frame. Second, their daily documentation in the patient chart reflects the outcomes of the pathway, with emphasis on unmet expected outcomes. Third, in their daily collaboration with the patient's nurse, their focus of discussion is on expected outcomes that are unmet and on formulating a plan of care to reach the identified goals.

Case Manager's Role in Pathway-Guided Care

The role of the case manager begins with the identification of those patients on pathways through collaboration with the clinical nurse specialist. The case manager is responsible for the initial contact with the insurance carrier who provides coverage for the patient. To date, a portion of insurance carriers will accept pathways as an appropriate plan of care and approve hospital days on the basis of this plan. Failure to the insurance carrier to accept this as a plan of care usually results in multiple, often daily telephone calls between the case manager and insurance carrier, causing inappropriate utilization of time and resources and yielding frustration. The need for effective communication and further education with insurance carriers related to length of stay based on pathways cannot be overemphasized. The case managers play a vital role in the management of the disease process; our objective is the management of the disease versus event-driven treatment.[12] The case manager's focus is in three areas: (1) clinical management, patient compliance with pathways, and evaluation of outcomes; (2) cost-effectiveness, charges-associated management of the disease process; and (3) financial management dealing with reimbursement charges. An additional role of our case managers is providing discharge planning in collaboration with the nurses and physicians. Most commonly, the discharge needs of our patient population are setting up home oxygen, arranging home care nurse visits, and occasionally setting up assertive devices for ambulating.

Ongoing Documentation

After the patient is enrolled onto the pathway, continuing documentation consists of two forms: the Patient Outcomes Documentation Record (Exhibit 7–2) and the Alterations to Plan of Care Form (Exhibit 7–3). The pa-

Exhibit 7–2 Patient Outcomes Documentation Record

MRN: _____

Name: _____

DOB: _____

Admit Date: _____

Path Entry: _____

Patient Outcomes Documentation

Site/Sub-site: Thoracic/Lung

Pathway: ThS3 Pulmonary Resection

☑ Patient Met Criteria   *Criteria Unmet by Patient

		D	E	N
Day of Cycle				
Day of Week				
Date				
Category of Care	Same Day Admit Surgery			
Assessment/Eval				
Consult				
Diagnostic Test	CXR shows absence of pneumonia and proper CT placement.			
Treatment	Effective cough. Vital signs stable. CT drainage system patent with no evidence of air leak. Urine output >/= 30cc/hr.			
Medication	Remains free of s/s of infection. Pain level <5; adequate pain control without excessive sedation. Absence of N/V. Temp < 38C.			
Performance Status/Activity	Up to chair for 15 min.			
Nutrition	Taking ice chips post-op			
Teaching/Psychosoc	Demonstrates use of incentive spirometer. Durable Power of Attorney for Healthcare: (check one) 1. Copy of document in Medical Record. 2. Understands but document not brought to hospital			
Discharge Planning				

Date	Initials	Signature/Title	Date	Initials	Signature/Title	Date	Initials	Signature/Title

Note: CXR = Chest X-ray; CT = chest tube; N/ V = Nausea/vomiting; D = Day; E = Evening; N = Night.

Source: Copyright © M.D. Anderson Cancer Center.

Exhibit 7–3 Alterations to Plan of Care Form

MRN: _____

Name: _____

DOB: _____

Admit Date: _____

Path Entry: _____

Alterations to Plan of Care

Alterations:

1. Anemia
2. Activity
3. Airway obstruction
4. Aspiration
5. Bleeding
6. Cardiovascular complications
7. Confusion
8. Constipation
9. Decreased urinary output
10. Dehiscence
11. Diarrhea
12. Falls
13. Fatigue
14. Febrile
15. Fistula
16. Fluid/electrolyte imbalance
17. Hemodynamic instability
18. Ileus
19. Incontinence
20. Infection
21. Knowledge deficit
22. Mental status changes
23. Nausea/vomiting
24. Neuropathy
25. Nutrition
26. Obstruction
27. Pain
28. Phlebitis
29. Pneumonia
30. Psychosocial
31. Pulmonary complications
32. Self-care deficit
33. Sepsis
34. Seizures
35. Skin breakdown
36. Stomatitis
37. Thrombosis
38. Urinary retention
39. UTI
40. Wound infection
41. Other alteration (please indicate in chart below)

Date/Time	Alteration Number	Path Day if on Path	Supplementary Documentation Utilize this area to document additional information. Documented by D—Data, A—Action, R—Response, T—Teaching Assessment of alteration should continue every shift until resolved.

Every entry must be dated, timed, and signed.

Source: Copyright © M.D. Anderson Cancer Center.

tient outcomes documentation is viewed as an interdisciplinary form to be completed by the appropriate disciplines. This document is completed three times per 24-hour period, or every eight hours; the outcome is recorded as either met or unmet. All outcomes that are charted as unmet require further documentation on the Alterations to Plan of Care Form. Documentation on this form consists of identifying the alteration, the day of the pathway, and interventions and teaching implemented to correct the alteration. Within the supplementary documentation, information is charted using a focused charting format, DART, where D stands for data, A for action taken, R for response to the intervention, and T for teaching involved with the actions. An example of this charting is:

Data: Patient complains of a pain level of 8/10; patient has poor cough effort

Action: Acute pain service physician notified; patient bolused with 5 cc of Fentanyl epidural; epidural basal rate increased to 6 cc/hour

Response: Patient's pain level decreased to 4/10; will closely monitor patient's pain level and assess patient's cough effort

Teaching: Patient reinstructed on use of epidural pump; reinstructed patient on ability to push prn epidural medication every 10 minutes as needed

RESULTS OF PATHWAY IMPLEMENTATION

To determine the impact of pathway implementation, a baseline needs to be determined. Once the baseline is established, the impact of the pathway becomes evident. This section will review the various outcomes of pulmonary resection and how they have changed since the pathway was implemented. From February 1996 through July 1997, a total of 450 individuals were enrolled onto the pulmonary resection pathway. The median age of an individual undergoing this procedure was 58 years, with 92 percent of the population 36 years of age or older. The population was 62 percent male and 38 percent female.

Outcome Measurement in Oncology

The principle of outcomes is to identify ways to maximize quality while minimizing resource utilization; quality indicators are used to determine the most effective treatment plan for a specific condition or disease process.[13]

Cancer is currently one of the more costly disease processes to manage, accounting for more than 20 percent of all health-care costs.[14] Traditionally, academic institutions have not been financially challenged, due in part to grants, contracts, government appropriations, and research benefits. The changing health care system and the onset of managed care have challenged us to continue to provide quality care while dealing with a variety of contracted services. At M.D. Anderson Cancer Center, we view outcomes as three separate collaborative efforts: (1) *daily outcomes,* those driven by daily interventions; (2) *pathway outcomes,* those driven by length of stay, patient satisfaction, and utilization of resources; and (3) *disease outcomes,* those driven by complications, morbidity, and mortality. The major focus is our daily outcomes driven by daily interventions that directly affect our pathway and disease outcomes.

OUTCOMES DRIVEN BY DAILY INTERVENTIONS

Pulmonary Outcomes

Aggressive pulmonary care plays a major role within our pathway. An expected outcome we strive to achieve is that the patient will not develop pneumonia. The expected outcomes measured are directly related to the prevention of pneumonia: (1) the patient has an effective cough, (2) the patient performs the incentive spirometer effectively, (3) the patient is ambulating, and (4) the patient is afebrile. Some controversy grew between nursing and respiratory therapy staff as to who would assume responsibility for the delivery of pulmonary care. Before pathway implementation, pulmonary treatments were performed by both respiratory therapy and nursing, but not necessarily in a coordinated manner. Our decision was to have nursing staff be accountable for incentive spirometer treatments, oxygen saturation monitoring, ambulation, and weaning from oxygen therapy. However, as in many institutions, respiratory therapy has developed and implemented assessment-based respiratory care, a protocol in which the therapist is delegated the responsibility for assessing the patient and directing and ordering the pulmonary plan of care, including ordering medicated nebulizer treatments, chest physiotherapy, and incentive spirometry and determining oxygen needs. Conflict arose by having two different pathways addressing many of the same issues without a clear definition of the roles of the nurse and the respiratory therapist. Although both disciplines wished to achieve the same result—prevention and/or resolution of pulmonary complications—it was clear that the presence of two pathways resulted in duplication of work effort, unnecessary resource utilization, and confusion among the care providers. This controversy was addressed through meetings between respiratory therapy and our pathway-working group. There was compromising in these meetings, and the end result was to add onto our pathway trigger points for instituting and discontinuing assessment-based respiratory care. In return, respiratory therapy agreed to supply our thoracic inpatient unit with a unit-based therapist with the intention of helping to improve the collaboration between nursing/physician staff and respiratory therapy to achieve an appropriate plan of pulmonary care so that pneumonia/pulmonary and/or other complications would be prevented.

Nutritional Outcomes

Nutritional outcomes are monitored daily by our clinical dietitian. They begin with the patient's tolerating ice chips immediately postoperatively and go on to the patient's advancing from a liquid diet to a regular diet. The clinical dietitian collaborates with the nursing staff to identify patients at high risk for nutritional deficits, such as patients who are elderly or who have previously received chemotherapy or radiation and/or take oral supplements regularly. The outcome expected by discharge is that all patients' nutritional intake will be at least 50 percent. If intake is less than 50 percent, our clinical dietitian assesses the patient. The outcome of nutrition was an area that was consistently met before implementation of our pathway and continues to be met. There are clinical issues that may affect this outcome, such as nausea and vomiting. At present, a tracking process is being attempted. Nausea and vomiting were identified as quality indicators in the postoperative population. If these are controlled, the outcome of nutritional intake should not be altered. If they are not, they may lead to unmet outcomes that may affect length of stay.

Postoperative Cardiac Outcomes

The most common complication following a pulmonary resection is supraventricular arrhythmia, especially atrial fibrillation.[15] In an attempt to identify this common postoperative occurrence, we incorporated in our pathway, shown in Exhibit 7–1, the instruction of admitting the patient to a monitored bed. In collaboration with our thoracic surgeons, we chose to install monitors on our inpatient unit with capabilities for continuous cardiac and pulse oxymetry readings. Before the use of these monitors, patients with cardiac arrhythmias required a

transfer back to the intensive care unit to receive treatment for these arrhythmias. Implementation of using the monitors has shown a significant decrease in resource utilization by reducing time spent in the recovery and intensive care units; additionally, the cost incurred when using the cardiac monitor for continuous pulse oximetry has proven to be less than when utilizing pulse oximetry through respiratory therapy. An issue of patient and family satisfaction was identified when patients required a transfer back to the surgical intensive care unit and were limited in the time they could spend with their family members. Since the use of the cardiac monitors on the unit began, patients have been treated for their cardiac arrhythmias successfully and safely. The use of cardiac monitors required all of our nursing staff to attend cardiac interpretation classes and complete successfully an exam on cardiac interpretation. In addition, nurses are now attending advanced cardiac life support classes to obtain certification in this field. The effort to achieve this change was a collaborative effort between staff education, nursing, and physician staff. The overall outcome has been positive; nursing staff has accepted and is fulfilling a very challenging role. Physicians, nurses, and patients are satisfied with the result of treating arrhythmias on the unit without dealing with the transfer back to the intensive care unit. The institutional change was challenging and required commitment and resources, but in view of the outcome, it has proven its worth.

Other Postoperative Outcomes

With use of the pathway, combined with our provision of cardiac monitoring, most patients undergoing a pulmonary resection require three to six hours of recovery in the post anesthesia care unit before transfer to our inpatient area. Before this, patients generally spent one to two days in the surgical intensive care unit. At present, the patient is usually extubated, if clinically appropriate, immediately following surgery, and use of the incentive spirometer is reinforced at this time. The fact that extubation takes place immediately following surgery means that the patient is able to transfer to the floor with a cardiac monitor. A nurse and physician from our acute pain service team, in collaboration with the members of the primary team, address pain control during the postoperative period. The majority of patients undergoing a pulmonary resection have an epidural catheter placed for pain management, which remains in place until all chest tubes are removed. A physician and nurse from the acute pain service will follow the patient for pain control until the epidural catheter is removed; this practice requires close collaboration between the pain service team and our nursing staff in addition to the patient and family. Pain control is addressed each day of the pathway, utilizing a pain scale of 0 to 10, with conversion from epidural to oral pain medication on day 5 of the pathway or on the day that all chest tubes are removed. Our target outcome for pain control is that pain control is at a level of 5 or less, depending on the particular day of the pathway, without the excessive sedation present.

Ambulation is another postoperative outcome that is measured beginning on the day of surgery. As shown on our pathway, the patient is expected to begin ambulation on the day of surgery. This issue has required education and reinforcement with our patients, who were hesitant to begin ambulating immediately following the extensive surgery. An additional outcome that we monitor closely is oxygen requirements. We titrate the oxygen off soon after surgery, generally within three days postoperatively. Before our pathway, oxygen titration was generally a responsibility of respiratory therapy. With the use of our monitors with continuous oxygen saturation readings and the investment by the unit in portable oxygen saturation monitors, this role has been delegated to our nursing staff, thus reducing resource utilization. We believe that giving this responsibility to the nursing staff has led to better coordination and achievement of our patients' needs.

Teaching/Educational Outcomes

Preoperative teaching continued to be performed in the clinic setting after implementation of our pathway. Our expected patient outcomes were specific: (1) viewing of a thoracotomy video, (2) review of the educational booklet, (3) preoperative teaching, and (4) instruction with return demonstration on the use of an incentive spirometer. In addition, the nursing staff was given the responsibility of distributing information related to the durable power of attorney for health care. This was viewed not as an end-of-life discussion but simply as an opportune time to provide information to patients regarding designating an individual to speak for them if they were unable to. The decision to include this within our pathway was made in collaboration with our institutional ethics committee. The patient was given several options: to sign papers designating an individual to hold durable power of attorney for health care, to take the information home to read and express a decision when returning to the hospital for surgery, to bring a copy of an existing durable power of attorney for health care, or to reject the document. This outcome was also addressed

on the day of surgery to provide follow-up on the presence of such a document within the patient's chart. The nursing staff in the clinic area demonstrate an ease in distributing this type of information to patients and answering questions as appropriate; additionally, physicians and a social worker are present in the clinic to discuss further issues that may arise at the patient's request.

Discharge Teaching

Before implementation of our pathways, discharge teaching was generally initiated on the day of discharge. Since implementation of the pathway, discharge teaching begins on day 3 following surgery with a review of the educational booklet distributed preoperatively in the clinic. By day 6 or the day of discharge, the need is only to reinforce the teaching. We continue to have to reinforce the need for the nursing staff to begin teaching by day 3 along with reinforcing the other daily outcomes expected for the patient according to their plan of care. It is important that the patient and family members be aware at the beginning of their pathway what is expected of them along with their expected day of discharge will be so that they can set their expectations and have a support system available upon discharge. The case manager is also actively involved early in the pathway to identify patients who may require certain specific needs for discharge such as home oxygen.

Follow-up Clinic Visit

Because M.D. Anderson is a referral center, our patients are usually discharged back to their primary care physician, and a majority of our patients reside outside the state of Texas and out of the country. Patients are scheduled for one-month and six-month follow-up clinic visits, at which times patients are assessed to ensure that they have achieved and maintained full range of motion of their affected shoulder. Additionally, discussions between patient and physician are held regarding the surgery and any further plans for treatment if indicated on the basis of pathology results from the surgery.

OUTCOMES DRIVEN BY THE PATHWAY

Length of Stay

Our pathway length of stay is set at six days postoperatively, with day of discharge expected by day 6. Before pathway implementation, the average range of length of stay after pulmonary resection was seven to eight days, with one to two days spent in the intensive care unit. Following implementation of our pathway, the median length of stay post pulmonary resection was five days, as shown in Figure 7–3. This represented a significant improvement for our population, since, in addition, our readmission rate remained minimal. The trending of data related to length of stay greater than six days demonstrated that patients exceeding their length of stay generally fell into two categories: those who developed pulmonary complications and those who required a chest tube for a longer time than expected because of persistent air leaks or large amounts of pleural drainage (>200 cc per 24-hour period). Overall, implementing the pulmonary resection pathway decreased the length of stay; however, this may be due to a selection bias.

Utilization of Resources

Review of the relevant literature suggests that costs at M.D. Anderson, an academic comprehensive cancer center, are higher than at nonacademic settings.[16] Using our pathway to determine a baseline of cost information, as seen in Figure 7–4, we found greater resource utilization than expected in the categories of diagnostic tests, medications, and surgery. Further evaluation of the process revealed higher resource utilization in postoperative medications, respiratory therapy, and diagnostic X-rays. Being able to identify these variances from the expected allows opportunity for improvement. An area being addressed currently is working toward a charge based upon the level of respiratory therapy that a patient requires. This continues to be an ongoing process, and with the implementation of a policy in which the respiratory therapist is unit based on the inpatient unit, we will be analyzing resource utilization and quality patient outcomes.

Patient Satisfaction

Patient and family satisfaction outcomes are examined through our institutional patient satisfaction surveys. Our measured outcome is patients' and family members' level of satisfaction with the care that they received from all members of the interdisciplinary team. Our results are ongoing; generally, our patients are satisfied with a shorter

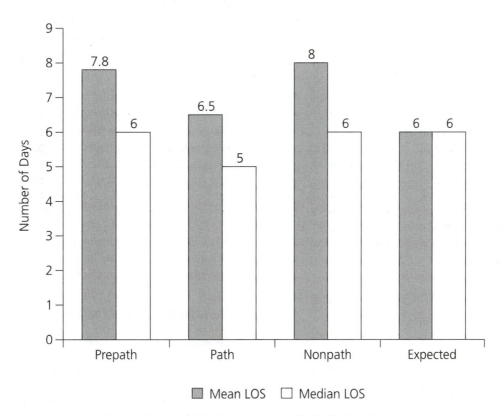

Figure 7–3 Thoracic Surgery: Actual versus Expected LOS. *Source:* Copyright © UTMDA Cancer Center.

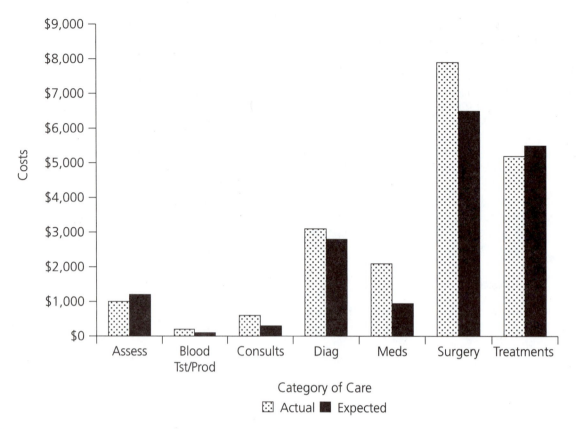

Figure 7–4 Thoracic Surgery: Lobectomy/Wedge Actual versus Expected Resource Utilization. *Source:* Copyright © UTMDA Cancer Center.

hospital stay, provided that they receive quality care. Questionnaires from our acute pain service team reflect that patients were very satisfied with the use of an epidural catheter for pain control and the follow-up by the pain service. There are future plans to obtain patient satisfaction information in relation to the pathway process. This information would provide opportunities for improvement using the patients' perspectives. Another area requiring further evaluation is health care team members' satisfaction with using this model to deliver care. This process should take place in the winter of 1997.

OUTCOMES DRIVEN BY DISEASE MANAGEMENT

Readmission Outcomes

Before pathway implementation, our readmission rate for pulmonary resection was 5 percent. After implementation of our pathway, our readmission rate has dropped to 3 percent, as shown in Figure 7–5. This figure reflects the readmissions of six patients since February 1996 who were readmitted after a pulmonary resection. Of those six patients, two were readmitted with a spontaneous

pneumothorax, one with respiratory distress secondary to emphysema, one with an intestinal obstruction, one with pain secondary to severe headaches, and one with candida vulvovaginitis. Our readmission rate remains low for pathway patients, with few readmissions directly related to their actual surgery. After an in-depth clinical review of readmission rates of pathway patients, the working group concluded that clinical reasons for readmission were not influenced by the utilization of any pathway interventions. There was no need to make any major changes to the pathway, but, some of the expected outcomes required modification, and this will be an ongoing process.

MEASUREMENT OF ALTERATIONS

Alteration Documentation

Pathway variances have been defined as deviations from the plan of care and outcomes outlined on the collaborative pathway that may alter length of stay, utilization of resources, or anticipated day of discharge.[17] Throughout the country, institutions' classification of variances vary widely, with hospitals using anything from

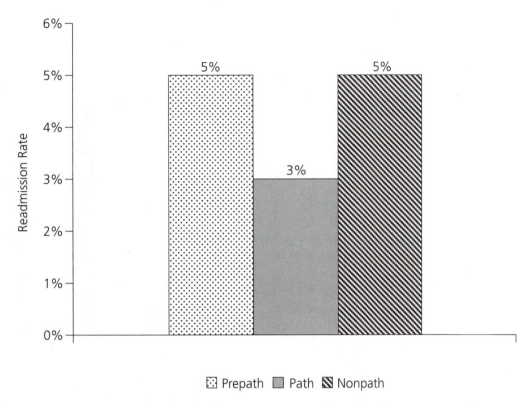

Figure 7–5 Thoracic Surgery: Lobectomy/Wedge Readmission Rate. *Source:* Copyright © UTMDA Cancer Center.

rigid categories for system breakdown to more generic categories based on patients' comorbidity and complications. Just as there are several ways to classify variances, there are several methods to document the variances. At M.D. Anderson Cancer Center, we chose to document patient-centered variances, utilizing the term *alteration,* as shown on our Alterations to Plan of Care Form (Exhibit 7–3). Documentation is performed by the corresponding discipline; nursing may document an alteration of inadequate pain control and the clinical dietitian may document an alteration of inadequate nutritional intake. Using patient-centered alterations allows the plan of care to be individualized, depending on the clinical necessity.

Alteration Collection and Analysis

As a pilot, alterations were collected each day for a one-month period for the purpose of identifying an appropriate tool for collecting alterations and identifying the most common alterations seen in our population of pulmonary resection patients. The clinical nurse specialist responsible for managing the pathway patients performed this collection as a pilot. Our disease management program performed the data analysis. There was trending of alterations correlated to the length of stay.

We identified the three most common alterations within our patient population as pulmonary complications, nausea/vomiting, and pain. With these data, we were able to identify clinical indicators that we believe are critical and reflect quality and influence length of stay. A second pilot is in progress to determine if these indicators prove to be indicative of quality or affect overall outcomes. This is an effort to capture the process of variance analysis. The challenge is chart audits, and if an automated system were in place, this process would not be as cumbersome.

Quality Clinical Indicators

Tracking clinical indicators encourages development of timely alternate plans of care if patients are not meeting expected daily clinical outcomes. Trending these indicators over time assists the health care team to understand shared achievements and opportunities for improvement. For the pulmonary resection patient, we chose the following three quality clinical indicators: control of pain, absence of nausea/vomiting, and ambulatory status. It is our opinion that these three quality clinical indicators need to be achieved; if they are managed successfully, patients will most likely be able to participate in their pulmonary plan of care, resulting in decreased likelihood of developing pulmonary complications. We need to be able to document not only whether patients reached their expected length of stay but, more important, whether the care they received and the outcomes they achieved were of a level of quality based on both the standards of the health care team and patients' and families' expectations. A tool was developed for tracking these three indicators whereby the case manager will record daily whether these three outcomes are met; this information is obtained from the Outcomes Documentation Record and in collaboration with the clinical nurse specialist and nurse. At the end of a one-month period, an interdisciplinary group consisting of the case managers, the disease management representative, the nurse manager, and the clinical nurse specialist will be presented with the results to determine the outcomes and effectiveness of this process.

LESSONS LEARNED WITH PATHWAY USE

Delivery of patient care through the use of a collaborative pathway means that (1) the pathway provides a systematic plan of care for all care providers, (2) patients receive consistent preoperative and postoperative information with regard to plan of care, (3) patients are more able to participate with breathing exercises, and (4) patients experience less pain. The use of pathways encourages nurses to act independently and to use critical thinking in their judgments. For example, before the pathway, nurses were required to obtain a specific order before beginning any therapy for a patient who developed respiratory compromise with wheezing. Since implementation of the pathway process, the nurse may initiate InspirEase treatments with Albuterol and titrate the oxygen as needed; the use of the pathway allows the nurses autonomy in their decision making. Another example is that if an air leak from the chest tube is noted, the nurse will automatically reconnect the patient to suction. The nurse is not using time locating a physician before initiating an action that the nurse knows is the appropriate action to take and will affect the patient's outcome.

Initially, with implementation of the collaborative pathway, a major challenge was a change in the format of documentation. Before implementation of pathways, documentation was straightforward and uniform for all patients. With pathways, a separate patient outcomes record and alteration record is required. An interdisciplinary documentation committee was formed to begin the process of addressing documentation issues such as double documentation. Institutional efforts were made to streamline documentation, such as incorporating the patient outcomes document and complementing it with

a shift to shift assessment form. In doing this, we eliminated a cumbersome nursing flowsheet. At present, this change has been supported by the institution and the health care members. Double documentation is an important issue, but it does take some time to determine which forms to eliminate and which forms to combine; during this time, double documentation will be a challenge with an advantageous outcome. There are plans to combine all documentation in preparation for the computer-based patient record, which will incorporate online documentation and order entry.

CONCLUSION

The development of the pulmonary resection pathway, with subsequent implementation, evaluation, and identification of areas of improvement in our delivery of care, has been demonstrated to be a collaborative and cooperative effort between disciplines. Our goal to provide the highest quality of patient care while attempting to utilize resources appropriately proved to be a challenge.

Several factors are essential for the success of pathways. All disciplines involved must be equally committed as well as taking ownership of the process. There must be ongoing monitoring and evaluation of both the clinical and financial outcomes; without this, there is no clear way to determine the success of the pathway process. The ability to make and accept change must play a role in all disciplines involved, from the patient to the health care team.

With the pathway process, this provided us with a baseline. Once a baseline is established, there can be further focus on the areas requiring a change. We have been able to identify areas that need change. We are working to improve the process by which we deliver pulmonary care, the charges generated within the operating room, and utilization of medications. Our length of stay and past charges associated with return stays in the intensive care unit have decreased significantly. Our goal for the future is to continue to maintain and improve the quality of care that we deliver while managing resource utilization associated with a pulmonary resection; plans for the future are to develop a pathway for an outpatient pleural effusion clinic. For all successful processes, there must be a source of energy, endurance, collegiality, assertiveness, innovativeness, risk taking, and dedication. These major characteristics may be present in one or several of the health care team members. It is these individuals—physicians, clinical nurse specialists, staff nurses, clinic nurses, and respiratory therapists—that make this process successful. The process must be based on a collaborative and cooperative model for the delivery of health care.

REFERENCES

1. U.S. Department of Health and Human Services. *Using Clinical Practice Guidelines to Evaluate Quality of Care.* (AHCPR Publishing No. 95-0045). Washington, DC: Public Health Service Agency for Health Care Policy and Research; 1995.

2. Gould B, Schaeffer L. Cancer care and cost: the Blue Cross of California approach. *Cancer Care and Cost: DRG's and Beyond.* Ann Arbor, MI: Health Administration Press Perspectives; 1989:181–190.

3. Morris M, Jameson S, Murdock S, Hohn D. Development of an outcomes management program at an academic medical center. *Best Pract Benchmark Health.* 1996;1(3):1–8.

4. Ginsberg R, Roth J, Fergusson M. Lung cancer surgical practice guidelines. *Oncology.* 1997;1(6):889–895.

5. Patton M, Katterhagen J. Critical pathways in oncology: aligning resource expenditures with clinical outcomes. *J Oncol Manage.* 1994;3(4):16–21.

6. Katterhagen J. Critical pathways in oncology: balancing the interests of hospitals and the physician. *J Oncol Manage.* 1993:22–26.

7. Doenges M, Moorhouse M, Geissler A. *Nursing Care Plans. Guidelines for Planning and Documenting Patient Care.* 3rd ed. Philadelphia: FA Davis; 1993.

8. Bliss-Holtz J. Computerized support for case management ISAACC. *Comput Nurs.* 1995;13(6):289–294.

9. Hammond W, Hales J, Loback D, Stramber M. Integration of a computer-based patient record system into the primary care setting. *Comp Nurs.* 1995;15(2):61–67.

10. Hamric A, Spross J. *The Clinical Nurse Specialist in Theory and Practice.* 2nd ed. Philadelphia: WB Saunders; 1989.

11. Nugent K. The clinical nurse specialist as case manager in a collaborative practice model: bridging the gap between quality and cost of care. *Clin Nurs Special.* 1992;6(2):106–111.

12. Dewing K. Case management information needs: effective tools in the management of patient care and cost containment. *Nurs Care Manage.* 1997;2(4):168–173.

13. Luquire R. Focusing on outcomes. *RN.* 1994;57(5):57–60.

14. Franklin M. Creating a managed-care product for cancer services. *J Oncol Manage.* 1994;3(1):19–26.

15. Roth J, Ruckdeschel J, Weisenburger T. *Thoracic Oncology.* 2nd ed. Philadelphia: WB Saunders; 1995.

16. Morris M, Levenback C, Burke T, De Jesus Y, Lucas K, Gershenson D. An outcomes management program in gynecologic oncology. *Obstet Gynecol.* 1997;89(4):485–492.

17. Schriefer J. Managing critical pathway variances. *Quality Manage Health Care.* 1995;3(2):30–42.

■ Part III ■
Neonatal/Pediatrics

■ 8 ■

Enhancing Outcomes in Pediatrics: The Development and Implementation of a Pneumonia Protocol

Nancy Shendell-Falik, Katherine B. Soriano, Maureen Bueno, and Barbara J. Maggio

Providing high-quality, cost-effective patient care is becoming increasingly mandatory in today's health care arena. There is no end in sight to the pressure to reduce cost and improve quality. "With increased penetration of managed care, healthcare institutions compete on the basis of quality as well as cost, emphasizing the importance of linking the care planning process with good clinical and fiscal outcomes."[1(p.16)] In response to this demand, many institutions have developed protocols or pathways that define expected outcomes within specific time frames for a particular case type. Robert Wood Johnson University Hospital (RWJUH) in central New Jersey has been a leader in this endeavor. This chapter will focus on the OPTIMedicine[SM] Optimal Practice Taskforce Initiatives in Medicine Pediatric Pneumonia Task Force that was selected at RWJUH related to address variations in patient management and to reduce length of stay.

RWJUH is a 452-bed academic medical center in New Brunswick, New Jersey. It is the core teaching hospital of the University of Medicine and Dentistry of New Jersey-Robert Wood Johnson Medical School and the hub of the Robert Wood Johnson Health System, an integrated health care network. The Children's Hospital is a 70-bed hospital within a hospital and accounts for approximately 15 to 20 percent of the hospital volume and activity.

The mission of RWJUH is expressed in the acronym *CORE,* which stands for *Care, Outreach, Research,* and *Education.* This fourfold mission affects all initiatives and creates a drive to set a standard of excellence in quality patient care, outreach to the community, state-of-the-art research, and education for all professional and paraprofessional staff.

The population of children served is tertiary, diverse, and highly acute in nature. Every medical and surgical subspecialty is represented except for invasive cardiology and transplantation. Pediatric patients are cared for in one of seven distinct units: pediatric, adolescent, pediatric hematology-oncology, pediatric intermediate care, pediatric intensive care, and pediatric medical and surgical same-day units. The philosophy of the Children's Hospital is to provide family-centered atraumatic care that is multidisciplinary in approach. This is achieved through the concerted efforts of all individuals who interact with children and their families.

PATIENT CARE DELIVERY MODEL

The Children's Hospital at RWJUH implemented the professionally advanced care team, PRO-ACT®, in 1993. This patient care delivery model consists of case management within a restructured work environment. The case management model evolved over time from a unit-based system to a service-based orientation. This transformation promoted continuity of care, enhanced efficiency, and improved satisfaction of patients and providers. This was achieved by a single individual following a caseload of patients from admission through discharge.

In 1995, RWJUH commenced a three-part operational enhancement to the present model. The first part, a nonsalary cost reduction endeavor, promoted savings in supply costs and standardized equipment used for patient care. The second initiative was an expansion of ProACT® into ProACT®II™.[2–8] ProACT®II™ is a restructured patient care delivery system that enhanced professional and ancillary roles.

The ProACT®II™ model, specific to the pediatric population, involved restructuring the roles of the professional nurse and the creation of the pediatric technician role. To provide more comprehensive care, the professional nurse provides the respiratory treatments for the children on the inpatient units. This ensures a more holistic approach to care in which the nurse is better able to continually assess and reassess the patient as well as plan the care for the day in an organized fashion, taking into consideration the child's unique needs and schedule.

Additionally, the case manager role was broadened to include functions of discharge planning, quality, outcomes monitoring and tracking, and utilization management. The impetus to develop this role further was the reality that a focus on outcomes was a necessity in the competitive healthcare environment. The title of *outcomes manager* was adopted to reflect comprehensively the scope of the role.

Outcomes managers are integrally involved in the third initiative of operational enhancement, which is the development and implementation of practice protocols, guidelines, algorithms, and standardized orders. A new program entitled OPTIMedicine℠ (Optimal Practice Task-Force Initiatives in Medicine) was developed to promote physician-led intern disciplinary teams. The major purposes of the program are (1) to develop and implement practice protocols to guide medical, nursing, and ancillary staff in the provision of care; (2) to promote state-of-the-art medical practice; (3) to decrease variation in practice; and (4) to maintain the highest standards of individualized patient care.

The protocols, entitled OPTIMaps, provide a multidisciplinary plan of care for a specific patient type that identifies targeted outcomes to be achieved within designated time periods. "Protocols provide a method to decrease variation and manage patients through the health care continuum by ensuring quality and cost-effectiveness."[9(p.138)] The first step in protocol development is the review of length-of-stay (LOS) reports in comparison to those of other institutions and analysis of risk/severity-adjusted data. These results lay the groundwork for focusing the work of the multidisciplinary task force to look at a specific disease process—in this case, pediatric pneumonia.

Institutionally, an outcomes steering committee with membership from hospital administration, medicine, nursing, outcomes, pharmacy, quality, and information systems reviews benchmark and comparative data and commissions each task force. The criteria evaluated in selecting OPTIMedicine Task Forces to be developed include diagnoses that are high volume, high cost, or high risk or that show significant variations in physician management.

CLINICAL PROTOCOL DEVELOPMENT

The strong multidisciplinary nature of protocol development cannot be overestimated. The first step is achieving membership that is truly reflective of all the subspecialties involved in the care of the child and family

with pneumonia. The protocol development committee is chaired by designated community and faculty physicians, in conjunction with the administrator for clinical outcomes, who drives the process. "The management role in this process is fundamental to promoting an outcomes orientation with staff members, establishing project expectations and facilitating the collection of quality data."[10(p.38)] The physicians and administrator work collaboratively to facilitate and coordinate meetings and keep the group focused on the overall goals.

Due to the nature of pediatric pneumonia, physician membership was fourfold, including representatives from pediatric pulmonary and infectious diseases as well as community pediatricians and emergency department physicians. Nursing leadership included members from various important factions: nursing administration, clinical nurse specialists, outcomes management, and representatives in the clinical arena as well as the home care department. The pediatric pharmacist played a critical role in lending expertise on medication types and dosages consistent with state-of-the-art treatment recommendations. The respiratory therapists (RTs) shared their expertise regarding oxygen therapy and assessment. The pediatric nutritionist stressed the importance of meeting the child's nutritional needs during this hospitalization. The child life department provided consultation on the importance of achieving atraumatic care in a developmentally appropriate way. "Building a shared vision by organizational members requires an alliance among key stakeholders involved at all levels of the organization."[11(p.31)]

Once membership is confirmed and assembled, the real work of the committee begins. The first step is identifying the goals of the work group and delegating responsibilities to appropriate parties. An extensive literature review was conducted to ensure that current management regimens were fully considered. This yielded data from many disciplines, including pharmacy, medicine, and nursing. Additionally, several nationally recognized children's centers were consulted to gain knowledge regarding pneumonia treatment nationwide. Clinical protocols/pathways from these institutions were also reviewed. Categories of care were identified, and a grid was developed to plan care over a four-day proposed LOS. On the basis of the scientific literature and a thorough investigation of state-of-the art protocols, the OPTIMedicine pediatric pneumonia task force developed its OPTIMap. Several categories were chosen and are listed on the vertical axis of the OPTIMap: test, consults, respiratory status, cardiovascular status, activity, nutrition, fluid status, nursing intervention, medications, teaching, and discharge plan. The horizontal axis reflects a projected LOS of four days (Exhibit 8–A–1 in Appendix 8–A).

One of the major challenges in developing standardized treatment protocols for pediatric bacterial pneumonia was achieving consensus on the specific medications to be prescribed. A subcommittee consisting of two representatives from pharmacy and the physicians from pediatric pulmonary and infectious diseases met to determine an optimal course of therapy. This group worked closely from the scientific literature review related to medication administration in order to develop "best practice" pharmaceutical management. Recommendations from this group were then presented to the entire task force for approval. Additionally, approval was sought from the hospital pharmacy and therapeutics committee to promote an institutional standard of care.

CLINICAL PROTOCOL IMPLEMENTATION

When a child is admitted with a particular diagnosis for which an OPTIMap exists, the physician and/or nurse initiates the specific OPTIMap. It is the goal of RWJUH to develop OPTIMaps for the majority of diagnoses for which a child is admitted. OPTIMaps are available on every patient care unit and become permanent chart documents when activated.

OPTIMaps exist for both bacterial and viral pneumonia. Much of the treatment plan depends on the clinical presentation of the child as the clinician decides on a working diagnosis. Since obtaining sputum cultures is often difficult with the pediatric population, the assessment of the patient is key in determining which OPTIMap to select.

There are many similar treatment interventions for the patient with either bacterial or viral pneumonia, as indicated on the OPTIMaps. In this chapter, the discussion focuses on the process used for patients with simple bacterial pneumonia who meet certain inclusion and exclusion criteria. For example, children with underlying conditions such as human immunodeficiency virus or cerebral palsy or those that present in status asthmaticus are excluded from this OPTIMap, since their underlying condition will affect the treatment plan. Other examples of patients who would not be eligible are those with pleural effusion, history of prematurity, or hemodynamic instability.

The pneumonia OPTIMap is initiated as soon as a child with a suspected admitting diagnosis of pneumonia enters the emergency department (ED) or is admitted to an inpatient unit. It is critical that the plan be implemented at that time, since the OPTIMap describes the course of hospitalization from admission, regardless of hospital location, through day 4. This entails the admitting physician's conducting a thorough history and physical exam and then determining the course of appropriate medical management.

The OPTIMap also includes a multidisciplinary plan of care for the practitioner's use (Exhibit 8–A–1 in Appendix 8–A). Several diagnoses and outcomes can be modified to meet the patients' and families' unique needs. In this way, a standardized plan of care can easily be adapted for all types of patients. Responsible disciplines to achieve expected outcomes include nursing, respiratory, pharmacy, child life, social work, and outcomes management. Working together, these subspecialties can meet the needs of the patient and family in an organized, coordinated fashion.

It is the responsibility of the ED nurse to initiate the written pediatric pneumonia OPTIMap. In this fashion, the professional nurse is ensuring and documenting that management initiatives are underway in accordance with the overall plan. This is accomplished by the professional registered nurse's (RN) documenting when the interventions are completed. Once the child is transferred to the inpatient setting, accompanied by the OPTIMap, the RN continues the plan where the ED left off in relation to patient management. If the child is in the ED for too short a period of time, the OPTIMap may be started on the inpatient unit.

INTERDISCIPLINARY TEAM RESPONSIBILITIES

Management of the pathway is multidisciplinary, with various subspecialties playing an important role. Overall management and documentation on the protocol is done by the RN, who is also responsible for briefly describing any variances from the plan of care. For example, if the child is still hospitalized on day 4 of admission, the RN must specify the reason.

Any member of the team can contribute to variance documentation. It is often the outcomes manager who identifies situations that negatively affect the patient's LOS. For example, if the child develops a reaction to the prescribed antibiotic, the medication will need to be changed, thereby affecting the LOS. The outcomes manager's responsibilities involve ensuring that every eligible patient is placed on the correct OPTIMap. Since his or her functions include aspects of discharge planning and utilization review, he or she can encourage activities that promote a safe, timely discharge and can help to coordinate any unique discharge needs.

The physician's responsibilities include ongoing assessment and evaluation of the medical status, as well as close monitoring of the types and frequency of medications. Additionally, the physician, in conjunction with the health team, is charged with determining whether the clinical course stays consistent with the selected OPTIMap. The physician then enters the orders into the computerized clinical information system under the pathway that is specific to pediatric pneumonia. These orders are consistent with the OPTIMap. The multidisciplinary team worked diligently to ensure that the order sets for the Pneumonia OPTIMap were both comprehensive and concise to facilitate use by the physicians. Exhibit 8–1 is an example of a computer screen that the physician uses to enter orders.

Exhibit 8–1 Computer Screen Used by Physicians To Enter Orders into the Clinical Information System under the Pediatric Pneumonia Pathway

---

PEDIATRIC PNEUMONIA OPTIMAP

ACTIVITY:

    BEDREST. . . . .   ★★
    OOB . . . .   ★★
    ─

    CONTACT ISOLATION . . . .   ★★
    RESPIRATORY ISOLATION . . . .   ★★

    STRICT I & O . . . .   ★★
    DAILY WEIGHTS . . . .   ★★

PULSE OXIMETRY:

    CONTINUOUS . . . .   ★★
    SPOT CHECK Q__H

    CARDIAC MONITOR . . . .   ★★

                                        *NEXT
                                        *BACK   *INDEX

        RETURN   MASTER   REVIEW
ERR   TYPE   RETRIEVE

PEDIATRIC PNEUMONIA OPTIMAP
BACTERIAL

        5–18 YRS—DOSING BASED ON RENAL FX WNL
        CEFUROXIME (150MG/KG/DAY DIV Q8H TO
                MAX OF 4.5G/D)

CEFUROXIME INJ _____MG, IVPB, NOW & THEN Q8H

OR PENICILLIN (150, 000-400,000U/KG/D
DIV Q4-6H TO MAX OF 24MU/D)

PENICILLIN _____ MILLION UNITS, IVPB,
NOW & THEN Q H
(CONSIDER HIGH DOSE FOR S PNEUMONIA)

*OR CEFTRIAXONE
+/– ERYTHROMYCIN *OR CLARITHROMYCIN
*IF NEEDED, ALBUTEROL

                        *PHARMACY   *INDEX

        RETURN   MASTER   REVIEW
ERR   TYPE   RETRIEVE

Source: Copyright © Robert Wood Johnson University Hospital.

Physician order sets were developed to expedite basic care to OPTIMap patients. Activity, diet, vital signs, and cardiorespiratory monitoring orders are initiated together with the appropriate OPTIMap. The complete computer pathway has a wide variety of options, including choices in oxygen therapy, diagnostic tests, and an array of medications specific to pneumonia treatment. The medication choices are a comprehensive representation of those that would be indicated for patients with varying degrees of respiratory sequelae. The physician can then "customize" the orders to meet the unique and changing needs of the patient and family.

While RTs have limited involvement in executing the OPTIMap, it is critical that they work closely with the physician and nursing staff to continually assess the need for changes in treatment. For example, collaborative decision making occurs when the child's changing status requires respiratory treatments to be increased or decreased depending on the patient's current condition. The RT plays an important role in monitoring and documenting the respiratory status of the patient.

The pediatric social worker is notified when needed to assist with any discharge plans or problems identified upon admission or throughout the hospital course that may be directly or indirectly related to the child's pneumonia. For example, a child who has pneumonia in addition to asthma and who suffers from frequent asthma attacks attributed to warm humid weather but who does not have air conditioning at home will need social work intervention. The social worker can assist the family in utilizing resources either to obtain an air conditioner or possibly to move to an environment that is more conducive to the child's recovery.

The pediatric clinical nurse specialist (CNS) has an important role, functioning as a consultant to the multidisciplinary team. The CNS assists with continual staff education regarding protocol development as well as helping to ensure that children are placed on the correct OPTIMap. The CNS is also instrumental in providing consultation on difficult cases where variances would potentially be the greatest.

Protocols are revised to ensure that they remain state of the art and reflect treatment advances, changes in pharmacological management, and implementation of new technology. These protocols are reviewed at least annually.

OUTCOMES

Several key outcome categories are used to measure the effectiveness of OPTIMedicine[SM] task forces, including LOS, resource utilization, cost, quality, and functional status or quality of life. Each task force identifies specific measures within one or more categories according to the objectives of the task force. The OPTIMedicine[SM] pediatric bacterial pneumonia task force, for example, selected LOS as a measure, since the OPTIMap and physician orders were designed to streamline care and reduce LOS. The objective was also to ensure that patient admissions were appropriate: that is, that individuals who could be safely managed in the community were not admitted. The task force realized that inappropriate admissions would make it difficult to reduce inpatient LOS. Other measures for the pneumonia task force included total and ancillary (e.g., pharmacy, respiratory, laboratory) charges, utilization, complications, disposition outcome, and readmissions.

A pre ($N = 56$) and post implementation ($N = 51$) design was used to evaluate the work of the task force. Data on LOS, charges, and quality measures were ob-

tained from the clinical/financial information system. Actual results are depicted in Table 8–1.

CONCLUSION

The experience in implementing the Pneumonia OPTIMaps has demonstrated a number of challenges and lessons for the health care team. First, due to the large number of practitioners and the varied presenting patient symptomatology, these protocols had a slow implementation phase. When the OPTIMaps went "live," massive inservice education was completed by all members of the health team. It is recognized that educating and reinforcing appropriate use of the Pneumonia OPTIMaps is an ongoing process and essential for success. Direct physician mailings, educational sessions, and resident review are now planned on a regular basis to ensure use by all disciplines. Additionally, orientation of new nursing staff and case managers involves comprehensive review and education of protocol selection, implementation, and variance documentation. Second, it is important to note that since medicine is not an exact science, treating the clinical presentation of the patient drives the system, not the confirmed diagnosis. Working with the physicians toward a commitment to pediatric clinical management improvement continues to create opportunities to optimize the quality and efficiency of pediatric care.

"In today's healthcare environment, institutions are striving to streamline processes, reduce costs of healthcare, and establish best practice patterns while maintaining and improving the quality of care provided."[12(p.160)] It is clear that the development and implementation of protocols can achieve both outcomes of enhanced quality and cost-effectiveness. To this end, RWJUH has realized these goals. Interdisciplinary teams have already completed additional pediatric OPTIMaps, including those for pediatric asthma, head trauma, and splenectomy. Again, the desired goal is to create proto-

Table 8–1 Clinical and Financial Outcomes

Outcomes Measures	Results
Length of stay	Decrease 15%*
Total charges	Decrease 31%*
Ancillary charges	Decrease 31%
Room and board charge	Decrease 30%*
Complications	No statistically significant change
Disposition	No statistically significant change
Readmission	No statistically significant change

*Statistically significant at $p < 0.05$.
Source: Copyright © Robert Wood Johnson University Hospital.

cols for the majority of pediatric patients cared for at the Children's Hospital of RWJUH. This vision will guide the activities of many multidisciplinary teams over the coming months and years.

REFERENCES

1. Ireson CL. Critical pathways: effectiveness in achieving patient outcomes. *J Nurs Adm.* 1997;27(6):16–23.

2. Crabtree-Tonges M. Redesigning hospital nursing practice: the professionally advanced care team (ProACT™) model, part 1. *J Nurs Adm.* 1989;19(7):31–38.

3. Crabtree-Tonges M. Redesigning hospital nursing practice: the professionally advanced care team (ProACT™) model, part 2. *J Nurs Adm.* 1989;19(9):19–22.

4. Luckenbill-Brett JL, Crabtree-Tonges M. Restructured patient care delivery: evaluation of the ProACT™ model. *Nurs Econ.* 1990;8(1):36–44.

5. Ritter J, Crabtree-Tonges M. Work redesign in high-intensity environments: ProACT for critical care. *J Nurs Adm.* 1991;21(12):26–35.

6. Crabtree-Tonges M. Work designs: sociotechnical systems for patient care delivery. *Nurs Manage.* 1992;23(1):27–32.

7. Shendell-Falik N. Perinatal ProACT™: Work redesign and nursing care management. *J Perinat Neonat Nurs.* 1995;8(4):1–12.

8. Shendell-Falik N. ProACT™ for pediatrics: work redesign and nursing case management. In: Flarey DL, ed. *Redesigning Nursing Care Delivery: Transforming our Future.* Philadelphia: JB Lippincott; 1995:162–172.

9. Shendell-Falik N, Soriano KB. Outcomes assessment through protocols. In: Flarey DL, Blancett S, eds. *Handbook of Nursing Case Management: Health Care Delivery in a World of Managed Care.* Gaithersburg, MD: Aspen Publishers; 1996:136–169.

10. Crist L. Outcomes system implementation for subacute care. *Nurs Case Manage.* 1997;2(1):33–21.

11. Kohles MK. The strengthening hospital nursing program: restructuring for a patient-centered health care delivery system. In: Flarey DL, ed. *Redesigning Nursing Care Delivery: Transforming Our Future.* Philadelphia: JB Lippincott; 1995:27–34.

12. Cole L, Lasker-Hertz S, Grady G, Clark M, Houston S. Structured care methodologies: tools for standardization and outcome measurement. *Nurs Case Manage.* 1996;1(4):160–172.

■ Appendix 8–A ■
OPTIMap with Plan of Care

Exhibit 8–A–1 OPTIMap with Plan of Care for Pediatric Pneumonia

Uncomplicated Pediatric Bacterial Pneumonia MULTIDISCIPLINARY PLAN OF CARE				
INITIATED (Date/Initials)	DIAGNOSES	DATE DISCUSSED W/PT. AND/OR FAMILY	EXPECTED OUTCOMES	ACHIEVED OUTCOMES (Date/Initials)
	Alteration in Management of Health Related to knowledge deficit		(Nursing/Respiratory) Patient/family will verbalize signs and symptoms of respiratory distress. (Nursing) Patient/family will be involved in daily care. (Nursing/Respiratory) Patient/family will verbalize understanding of disease process, treatment and outcome. (Nursing/Pharmacy/Nutrition) Patient/family will verbalize understanding of medications, nutrition and rest.	
	Alteration in Oxygenation/Ineffective Airway Clearance Related to ___respiratory insufficiency ___increased secretions		(Respiratory/Nursing) Patient will maintain a patent airway.	
	Alteration in Oxygenation/Impaired Gas Exchange Related to ___atelectasis ___impaired peripheral circulation		(Respiratory/Nursing) Patient will maintain adequate gas exchange as evidenced by arterial blood gases within normal limits, O_2 saturation within normal limits, absence of retractions, dyspnea, nasal flaring, and use of accessory muscles.	

continues

Exhibit 8–A–1 continued

INITIATED (Date/Initials)	DIAGNOSES	DATE DISCUSSED W/PT. AND/OR FAMILY	EXPECTED OUTCOMES	ACHIEVED OUTCOMES (Date/Initials)
	Alteration in Oxygenation/Decreased Cardiac Output Related to ___dysrhythmias ___tachycardia		(Nursing) Patient will demonstrate evidence of brisk capillary refill, palpable peripheral pulses, and pink color.	
	Alteration in Fluid Volume Deficit Related to ___decreased oral intake ___insensible fluid losses		(Nursing) Patient will maintain adequate fluid balance as evidenced by moist mucous membranes, normal skin turgor, soft/flat fontanel (infants), adequate urine output for age (2cc/kg/hr for infants, 1cc/kg/hr for child, and 0.5/kg/hr for adolescent), and serum electrolytes within normal limits.	
	Alteration in Mobility/Activity Related to ___disease process ___impaired O_2		(Nursing) Patient will receive adequate rest periods and increase activity as tolerated.	
	Alteration in Protection Related to ___fever		(Pharmacy/Nursing) Patient will receive IV antibiotics as per pharmacy protocol. (Nursing/Pharmacy) Patient will receive antipyretics as indicated to help maintain normal body temperature.	
	Alteration in Growth and Development Related to ___hospitalization		(Child Life/Nursing) Patient will receive developmental appropriate stimulation/ activities. (Child Life/Nursing) Family will demonstrate interaction appropriate to child's level of development.	
	Alteration in Coping Related to ___hospitalization		(Nursing/Social Work) Patient/family will verbalize fears/concerns regarding diagnosis, hospitalization, and family obligations. (Nursing/Social Work) Patient/family will demonstrate decreased anxiety and effective coping techniques. (Social Work/Outcomes Manager) Patient/ family will verbalize financial and/or insurance concerns.	

Initials	Signatures	Initials	Signatures
_____	_____	_____	_____
_____	_____	_____	_____
_____	_____	_____	_____
_____	_____	_____	_____

continues

Exhibit 8-A-1 continued

Categories	DAY #1 1–12 HOURS	Initials	DAY #1 13–24 HOURS	Initials	DAY #2	Initials	DAY #3	Initials	DAY #4 DISCHARGE	Initials
Tests	Chest X-ray (PA-LAT), CBC with Diff, Chem Profile, Lytes as indicated, Blood Cultures, ABGs as indicated, Sputum Culture if obtainable, for viral/bacteria C&S, UA		Repeat ABGs as indicated		Repeat chest X-ray if condition not improving, Consider CT Scan depending on chest X-ray, Check results of cultures		Check results of cultures and sensitivities.		Check results of cultures and sensitivities.	
Consults	Pulmonary if indicated Infectious Disease if indicated		Pulmonary if indicated Infectious Disease if indicated							
Respiratory Status	Pulse Oximetry continuous if: 1) SaO2 < 92%; 2) <1 year O2 via NC, mask, hood if SaO2 <93%		Pulse Oximetry continuous if: 1) SaO2 <92%; 2) <1 year pulse oximetry spot check q shift O2 via NC, mask, hood if SaO2 <93%		Attempt to wean O2, keep SaO2 >93%, decrease 0.5 L/2 hr		Continue weaning until no longer needed			
C-V Status	Cardiac respiratory monitoring for: 1) <1 yr age 2) >1 yr age if indicated		Discontinue cardiac respiratory monitor if no cardiac events.		Discontinue cardiac respiratory monitoring if no cardiac events.					
Activity	Bedrest. Child Life intervention. Isolation if required.		Bedrest. Child Life intervention. Isolation if required.		Out of bed. Child Life intervention. Isolation if required.		Out of bed. Isolation if required. Child Life intervention.		Out of bed. Child Life intervention.	
Nutrition	Nothing by mouth (depend on respiratory status) Clear fluids (if respiratory status stable)		Clear fluids if respiratory status stable		Advance to age appropriate diet if stable. If <1 year with RR >60, start NGT feeds.		Age appropriate diet. If <1 year, attempt PO feeds. If unable to PO feed, consider NGT feeds.		Advance to age appropriate diet.	
Fluid Status	Admission weight IV fluids—D5 1/2-1/4 NS add 20 MEq KCL after first void		Daily weight. Continue IV fluids if increased RR or inadequate po intake.		Daily weight		Daily weight. Discontinue IV by stages.		Daily weight	
Nursing Interventions	History and assessment: • respiratory status • cardiovascular status • level of consciousness Vital signs and blood pressure q 4 hours—I&O, Orient unit/room, SaO2 and cardiac respiratory monitoring as needed, Suctioning as needed, Chest physiotherapy as needed, Emotional support/OM initial assessment		Assess for any changes in status. SaO2 and cardiac respiratory monitoring as needed, Suctioning as needed, Chest physiotherapy as needed, I&O, Emotional support		Continue assessment and need for: • suctioning • chest physiotherapy • SaO2 cardiac respiratory monitor Begin weaning O2 as indicated. Offer emotional family support.		Continue assessment need for wean: • suctioning • chest physiotherapy • O2 Discontinue cardiac respiratory monitor. Prepare family for discharge.		Assess need for: • suctioning • chest physiotherapy Discontinue O2 Prepare family for discharge.	
Medications	Antibiotics as per pharmacy and therapeutic guidelines. Tylenol based on fever curve. Restart appropriate maintenance meds.		Antibiotics as per pharmacy and therapeutic guidelines. Tylenol based on fever curve.		Antibiotics based on culture and sensitivity data signs and symptoms, pharmacy and therapeutic guidelines. Tylenol based on fever curve. Restart appropriate maintenance meds.		Continue antibiotics. Change to PO or discontinue based on culture results and/or signs and symptoms.		PO antibiotics if indicated	
Teaching	Encourage parental participation. Allow opportunities for parents to ask questions.		Encourage parental participation. Allow opportunities for parents to ask questions.		Encourage parental participation in respiratory care. Reinforce basic handwashing techniques.		Instruct family regarding medications and expected outcomes.		Reinforce information regarding medications, respiratory assessment, and emergency care.	
Discharge Plan	Evaluate and notify Social Services if appropriate.						Assess need for home care, make necessary arrangements. Confirm follow-up appointment at discharge.		Confirm plans with family and Outcomes Manager.	

Note: ABGs, arterial blood gases; C&S, culture and sensitivity; CBC, complete blood count; I&O, intake and output; KCL, potassium chloride; NC, nasal cannula; NGT, nasogastric tube; OM, outcomes manager; PA-LAT, posterior/anterior/lateral; PO, oral; RR, respiratory rate; SaO2, oxygen saturation.

■ 9 ■

Neonatal ECMO: A Pathway to Care

Sharon W. Lake

Nursing case management and clinical pathways began at the University of Kentucky Hospital in 1990 with an initiative started by the Director of Nursing. Over the course of the next five years, pathways evolved from a case management–driven effort to a hospitalwide project with an organized multidisciplinary internal support structure for the development, implementation, and evaluation of clinical pathways. To avoid duplication of documentation, it was decided to use the pathways not only as a guide for care but also as the system of documentation for the hospital. The format for the document was standardized as well as the process by which pathways are reviewed and approved

The clinical process and outcomes management (CPOM) committee oversees clinical pathway activities within the organization. It is responsible for approving clinical pathway development and reviewing related outcomes. Decisions as to pathway format, documentation issues, pathway management, and the system for data collection are made by this group. This chapter discusses the process of developing and evaluating a pathway for neonates requiring extracorporal membrane oxygenation (ECMO). This is a heart/lung bypass therapy for such conditions as sepsis, persistent pulmonary hypertension, diaphragmatic hernia, and some cardiac anomalies.

PATHWAY BEGINNINGS

Patient care is provided using the primary nursing and medical models. The nursing case management model at the University of Kentucky is an advanced-practice nursing model (master's or equivalent). The goals of case management are to coordinate patient care, promote fiscal responsibility, and optimize patient/family outcomes. The case manager collaborates with multidisciplinary groups in the health care system to achieve these outcomes for a group of patients across the continuum of care. Case managers are hospital based and assigned by physician service. Intervention for a patient group occurs independent of the patient's location within the hospital.

The University of Kentucky Children's Hospital is located within the University of Kentucky Hospital, Chandler Medical Center. The neonatal intensive care unit (NICU) is a 50-bed level III nursery that serves as the referral center for central and eastern Kentucky. The unit averages 900 admissions per year. Early efforts using multidisciplinary clinical pathways, which were based on diagnosis, proved to be difficult to manage in a diverse gestational-aged population. From that experience, it was determined that the pathway system for neonates would work better if it was based on gestational age.

Beginning in October 1994, a multidisciplinary NICU clinical group management team (CGMT) was formed to develop and implement clinical pathways (Figure 9–1). Disciplines and departments represented were physicians, nursing, pharmacy, social services, dietetics, respiratory therapy, developmental specialty, data management and analysis, and quality assurance/utilization management. The team met weekly to identify current practice using chart review and clinical, financial, and resource utilization data. Twelve gestational age–based pathways were developed with the intention that all NICU infants would be managed on a clinical pathway. Protocols for specific diagnosis would be used to modify the pathways and meet the needs of the individual patient. Because patients placed on ECMO would require severe modification to the gestational age pathway, the team decided to develop a pathway specifically for neonatal ECMO (Exhibit 9–A–1 in Appendix 9–A). Upon admission, the infant would be placed on a gestational age

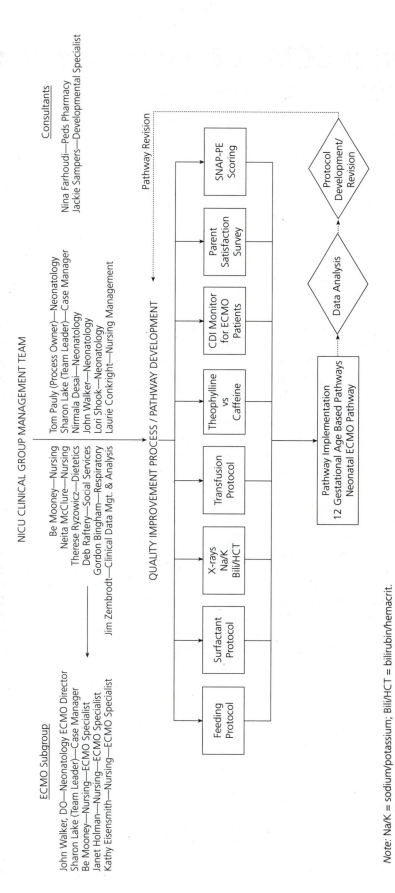

Note: Na/K = sodium/potassium; Bili/HCT = bilirubin/hemacrit.

Figure 9–1 Neonatal Flow Chart. *Source:* Copyright © University of Kentucky Children's Hospital.

pathway and transferred to the neonatal ECMO pathway when the precannulation workup was initiated. The multidisciplinary group continues to meet on a biweekly basis to analyze data and problem-solve issues for the neonatal population. The NICU CGMT reports data outcomes and team activities to the CPOM committee.

NEONATAL ECMO PATHWAY DEVELOPMENT

Since this was the 13th pathway to be written, the development process was well defined and flowed easily. To develop the previous 12 pathways, the entire multidisciplinary group met, with each discipline providing input as to the care of the identified patient population. However, because of the specific nature of this topic, a subgroup of experts was assembled to develop the ECMO pathway. Three nurses, all ECMO specialists, and the case manager met to review the protocols and standardized order set for ECMO patients. Data related to length of stay, number of days on ECMO, and resource utilization were also reviewed The next step was brainstorming the most sensible layout of the information. Since placing an infant on an ECMO pathway requires transfer from the gestational age pathway to the ECMO pathway, it was decided to then manage the infant on the ECMO pathway through to discharge. The group elected to arrange the information in phases, since this provides the best flexibility for length of time on ECMO and overall length of stay.

The next step was defining activities, interventions, outcomes, and progression criteria from the standardized physician order set, ECMO protocol, and nursing protocols. We started with a blank pathway at the beginning of the day and ended with a finished first draft. The draft was then typed, formatted, and presented to the physician director of the ECMO program and the remainder of the ECMO specialist team. Revisions were made, and the second draft was presented to the NICU CGMT. The pathway was then piloted on the next five ECMO patients and again revised The pathway was taken back to the NICU CGMT, and critical indicators were selected for data analysis.

Complementary Pathway Components

Two other components, the patient/family pathway and the parent satisfaction survey, were developed to enhance the clinical pathway system. The set of 12 patient/family pathways directly reflect the gestational age clinical pathway. They are written in lay terms and highlight the significant events and outcomes expected to be achieved in each time interval. The patient/family path-

way is given to the parent(s) upon admission with verbal explanation as to its intent and use. The ECMO specialist team had developed a comprehensive packet of information explaining ECMO: the purpose, expectations, and definitions of the common terminology related to this procedure. It was decided to continue to use the information packet rather than writing an ECMO patient/family pathway at this time.

The patient satisfaction survey is included as a part of the infant's discharge packet of information. It is given to each infant's parent(s) by the nurse one to two days before discharge. Verbal instructions for completing the survey and leaving it in the survey collection box are given. The survey is collected from the survey box by a staff member from the patient representative office, who tabulates the data and passes them to the resource center for data management and analysis to create in report format. The survey data are reviewed quarterly along with the documentation, clinical, and financial data from the pathways.

Pathway Format and Content

The pathway (Exhibit 9–A–1 in Appendix 9–A) is formatted according to the standard developed for the University of Kentucky Children's Hospital. The pathway is a two-sided document. At the top of the front page in the left corner are the diagnosis, patient problem list, and exclusion criteria. The patient problem list is not intended to be exclusive and should be modified to meet the needs of the patient. The clinical team decided that all neonatal ECMO patients would be managed on the clinical pathway; therefore, "none" is listed in the exclusion criteria box. Along the left margin are categories for care. The categories for growth and development and oxygenation/respiratory management are the only difference in format between the adult and pediatric pathways. Across the top of the pathway are listed the time intervals—in this case, phases of care.

Within each block, activities and interventions are written in regular type. Patient outcomes are written in bold type. Each patient outcome is assigned to one or more disciplines for evaluation. The responsible discipline is identified by the abbreviation in parentheses. To conserve space and avoid the description of activities or interventions in minute detail, many interventions were written as "per protocol."* The asterisk indicates that there is a written protocol that can be referred to if the user is uncertain of that information. The NICU was fortunate to have already developed a protocol book containing dozens of protocols related to nursing care. Likewise, as the ECMO program was being developed, a standardized physician's order set and written protocols outlining the care of this patient population had been developed. Other protocols for transfusions, feedings, and surfactant administration were written as a result of the pathway development process (Figure 9–1).

Pathway Movement through the System

When a patient arrives on the unit, the admitting physician resident evaluates the gestational age of the infant using a newborn maturity rating and classification assessment. The physician resident then writes the order to place the infant on the specified pathway. The pathway and all related documents (Table 9–1) are pulled from the NICU storage area and addressographed by the patient clerical assistant. Each clinical pathway has a code that the patient clerical assistant enters into the clinical data/order entry system, so that all hospital census reports and work reports are coded with the pathway number. The clinical pathway and all related documents are kept at the patient's bedside for the length of his or her hospitalization. When an infant is identified as a probable ECMO candidate, the nurse alerts the patient clerical assistant to pull the ECMO pathway and physician order set. The patient clerical assistant then updates the patient's clinical information to reflect that this patient has been removed from the gestational age pathway and placed on the ECMO pathway. Upon discharge, the pathway system becomes part of the permanent medical record. The department of quality assurance/utilization management evaluates all pathways, extracting data on the basis of the identified critical indicators, and enters the information into a database. The Resource Center for Clinical Data Management and Analysis provides the NICU CGMT with quarterly reports containing information determined by the team. On the basis of the data received, the team determines any modifications needed for the pathway.

Pathway Documentation

All disciplines are responsible for reviewing the clinical pathway for their patients. The pathway is used not only as a guide for care, but as a multidisciplinary documentation tool. Nursing is responsible for modifying the pathway to meet the needs of the patient. Activities, interventions and outcomes are dated and initialed on the blank line when completed or met. If an activity or an intervention is unmet, it is circled by the nurse. Any unmet outcome is circled by the responsible discipline, and narrative documentation with a plan and subsequent outcome is required. The reason for the exception is coded by the individual documenting the exception. Narrative documentation is required only for exceptions in patient outcomes. Narrative documentation for exceptions in activities or interventions is done at the discretion of the nurse. Event notes describing occurrences not related to patient outcomes are documented by the nurse as well. All disciplines, excluding physicians, document exceptions on the clinical pathway documentation record. This provides a running narrative of the identified patient exceptions with the plan and outcome evaluated.

We attempted to get physicians to document in the same place and manner as all other disciplines, but we were not successful. Because the University of Kentucky is a teaching institution, certain requirements on the physician resident staff resulted in double documentation. Currently, the physicians are required to document on the pathway itself, but their narrative documentation continues to be done in the history and physical/consults section in the primary medical record.

OUTCOMES MEASUREMENT

The process of developing a clinical pathway is a valuable one. It requires the group to look critically at patient

---

Table 9–1 Components and Purpose of the Clinical Pathway System

Component	Purpose
Clinical Pathway	To outline multidisciplinary care against a timeline
Clinical Pathway Documentation Record	To provide narrative documentation of pathway exceptions
Clinical Pathway Continuation Record	To extend pathway vertically or horizontally
Signature Record	To record name, department, and initials of individuals who have documented
Patient/Family Clinical Pathway	To provide anticipatory guidance for parents
Parent Satisfaction Survey	To evaluate parent satisfaction with care received

Source: Copyright © University of Kentucky Children's Hospital.

care and evaluate the rationale for providing care in that manner. "Because we have always done it that way" is no longer an acceptable answer. Retrospective analysis of medical records and utilization of resources can yield surprising results. Likewise, multidisciplinary involvement allows everyone a comprehensive look at what is provided for the patient. Everyone contributes his or her piece of the puzzle with the result being a complete and clear picture. The process allows each discipline an opportunity to hear and see the contributions of the others. It also sets the stage for ownership for that piece of care. Without ownership, there is little driving force to participate in the remainder of the process.

As the team leader for the multidisciplinary group, the case manager is responsible for facilitating the dissemination of outcomes data to all levels of the multidisciplinary group. Variance data related to critical indicators for each pathway, resource utilization, average cost, average length of stay, and parent satisfaction surveys are pulled from many sources, and a consolidated report is presented to the multidisciplinary team on a quarterly basis by the Resource Center for Data Management and Analysis. The team evaluates the data and analyzes the variances. This may result in the need for additional investigation if the variance falls outside what is considered acceptable by the group. We have not yet set specified standards of deviation for this pathway.

If the variance is appropriate for care, then pathway adjustment is needed. If it is not, then a strategy to solve the issue is in order. Additional individuals with expert knowledge may be requested to join the team for a period of time as necessary to problem-solve issues. It is important to note that variance data rely to a great extent on how well individuals are following and recording information. If documentation is poor, data are unreliable. Operating within a paper system can also complicate data collection. Over time, the critical indicators are anticipated to change as the team evolves and becomes more experienced with each pathway. Pathway modification is expected to occur every three to six months until stable, and then yearly.

Because there is individual patient variance within each patient group on a particular pathway, the NICU team wanted to stratify the data further. The desire to compare outcomes and analyze data for "like" patients led to the search for an acuity scoring system for neonates. Two systems were tried, and one was implemented. It provides each patient with an objective acuity score that improves not only data comparison within our institution but comparison with national data as well.

Patient Outcomes

Before the development of clinical pathways, the plan for patient care changed each month as a new attending physician arrived on service. There were also variances in nursing practice patterns related to care issues. Parents never knew what to expect when they were meeting a new nurse or when a new month began. Would they be allowed to hold their baby outside the isolette today or not? Would the discharge plan now be delayed until their baby weighed 2,200 grams? These variances were also very frustrating for the health care staff. The process of developing pathways significantly narrowed the variances in practice patterns of all disciplines involved. This has resulted in better patient care. Anecdotally, we know that physicians, nurses, and, most of all, parents are much happier having a consistent plan of care.

When one is dealing with a small patient population, the process of data collection and analysis is lengthened. At this time, the neonatal ECMO pathway is in final form, and we are ready to begin official data collection with the next ECMO patient. Therefore, outcomes directly related to the use of the pathway have not yet been identified. However, in developing the ECMO pathway, we were able to identify areas in need of improvement before the actual implementation of the pathway. For example, a seemingly large number of arterial blood gases were being drawn per patient per day. Evaluation of that issue resulted in a capital equipment request for an upgraded piece of equipment that will allow ECMO patients to be monitored more efficiently and will simultaneously require significantly fewer blood gases (Figure 9–2). The savings, both clinical and financial, are projected at this time because we have not yet had the opportunity to use this technology.

Another area of clinical change is in the use of head ultrasounds. During pathway development, it was identified that a head ultrasound was obtained each day that the patient was on ECMO. Following discussion by the group, it was decided that a head ultrasound was necessary for three consecutive days and then only as indicated. This change in practice will reduce the number of procedures that the severely stressed infant undergoes. Both of these practice changes yield not only a clinical benefit for the patient but financial savings as well.

Critical indicators for the neonatal ECMO population are identified in Exhibit 9–1. This is a list of activities and outcomes identified by the NICU CGMT as critical to the management of the ECMO patient. Data collection on these indicators will also begin with the next ECMO pa-

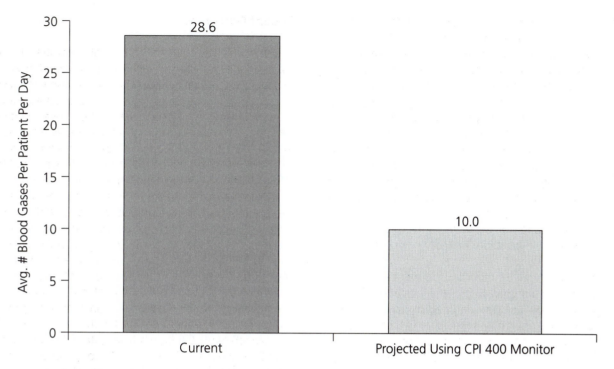

Projected Clinical Savings:
- 3.36 cc of blood saved per patient per day on ECMO
- Decreased number of interruptions to circuit
- Decreased risk of infection

Projected Financial Savings:
- $173.30 per patient per day on ECMO

Figure 9–2 Arterial Blood Gas Utilization in Neonatal ECMO Population. *Source:* Copyright © University of Kentucky Children's Hospital.

---

tient. This information is reviewed on a quarterly basis and modified as determined by the team.

Nursing Staff Outcomes

The most significant outcome for the staff was the emotional response to change. For the department of nursing, this was significant change. Using the pathway as a documentation tool did not fit at all with the documentation system in use. Nursing was already documenting a large amount of information in various places. Adding the pathway only compounded the issue of duplicate documentation. Therefore, it was decided to develop a documentation system around the clinical pathways and eliminate duplicate documentation. The new system supported documentation by exception and allowed for multidisciplinary involvement.

There were two distinct responses from the nursing staff. One camp thought that pathways would provide a better avenue for patient care and were eager to make the change. The other camp consisted of those whose motto was "If I ignore it long enough, it will go away." Anticipating this type of response and allowing time for the emotional letting go of the old was key for those individuals implementing this new system for the documentation and evaluation of care. Significant work in educating, auditing, and providing feedback has been necessary to make the change.

Exhibit 9–1 Critical Indicators for Neonatal ECMO Pathway

Precannulation Phase
 Endotracheal tube stinted at adapter
 Phenobarbitol load administered
 Phenobarbitol maintenance ordered
 ECMO booklet given to parents

Cannulation Phase
 Chest X-ray and echocardiogram confirmation of appropri-
   ate cannula placement (MD)
 Vecuronium given
 Fentanyl/morphine given
 Heparin given

Maintenance Phase
 </= grade I intraventricular hemorrhage per head ultrasound
   (MD)
 Total parenteral nutrition/lipids within 24 hours of beginning
   ECMO
 Range of motion Q 8 hours (after 24 hours on ECMO)
 Measure cannula length with dressing changes
 < 3 cc/hour blood loss around cannula site (NUR/SPECIALIST)

Decannulation Phase
 End-tidal CO_2 monitoring
 Endotracheal tube stinted at adapter

Post-ECMO Phase
 CT scan
 Electroencephalogram
 Ophthalmology exam completed (MD/NUR)
 MEDIC alert identification for venous/arterial ECMO
 Called follow-up physician (MD)

Source: Copyright © University of Kentucky Children's Hospital.

Ancillary Department Outcomes

The departments of dietetics, social services, and respiratory therapy have individuals who are responsible solely for the NICU. These persons were involved with clinical pathways from the beginning. They all have outcomes for which they are responsible, as outlined on the pathway. They have supported the process and participate in the documentation and evaluation process for clinical pathways.

Patient clerical assistants are responsible for assembling the appropriate pieces of the pathway system for each patient. They also enter the pathway code into the clinical data system. Daily monitoring by the case manager has been necessary to ensure that the pathway code has been entered into the system and that the corresponding pathway and appropriate paper components are at the patient's bedside. Patient clerical services is re-

sponsible for ordering and stocking of pathway forms. The patient services coordinator is responsible for managing the portion of materials that require duplicating.

Physician Staff Outcomes

Since the University of Kentucky hospital is a teaching institution, we have several layers of physician care providers (attendings, fellows, residents, and interns) who rotate in and out of the NICU. Some of the attendings were initially concerned that pathways were "cookbook medicine" that deadened the critical thinking skills they were trying to foster among the resident staff. This argument faded once they became more involved and educated in the process of the benefits of clinical pathways. They now are involved, to varying degrees, in the monitoring and evaluation of the pathway data.

Because pathways have not yet been developed for other areas of pediatrics, many of the resident and intern staff have never actually seen a clinical pathway until they are on service in the NICU. This requires monthly education about pathways and the documentation process by the case manager. Their efforts in following and documenting on the pathway vary by individual. We have not developed any consistency from this practice group. Once pathways are more prevalent in other areas of the hospital, perhaps there will be improved compliance from the physician group. Pathways are not used in physician rounds or specifically in reports from nurses. However, nurses and the case manager are quick to point out what needs to be accomplished for that patient "according to the pathway."

Parent Satisfaction Outcomes

Parent satisfaction had never been formally evaluated in the NICU. As part of the quality improvement process and analysis, the multidisciplinary team implemented the survey in May 1997. The survey is anonymous and therefore represents the NICU as a whole. The return rate for the period represented in Figure 9–3 was 33 percent. These data will be analyzed on a quarterly basis.

System Outcomes

Managing pathways using a paper system has been difficult. The hospital is currently evaluating automated documentation systems. The desired state is to be able to manipulate pathways and protocols easily to individualize and manage patient care efficiently. An automated system would also facilitate data collection and reporting.

CONCLUSION

Future pathway plans for the NICU multidisciplinary team include (1) converting a protocol for caring for infants born with neural tube defects into the pathway format and (2) developing a pathway for infants born with abdominal wall defects. A family pathway for each will also be developed. These paths can then be meshed with the gestational age pathways once our system is automated. The team will continue to evaluate data and issues surrounding the neonatal population.

Clinical pathways can be an excellent method of managing care for an identified patient population. The process of critically evaluating practice that occurs in the development of pathways is invaluable. It provides a means of comparing the perceived practice of providing care to the actual processes of caring for patients.

Pathways are a great teaching tool for experienced as well as new care providers. They allow every discipline a look at the comprehensive care for a patient population versus only one discipline's contribution. They are a reminder for all care providers to view the patient as a whole being with multiple care issues. As with all changes, implementation of clinical pathways creates an initial emotional response of fear and reluctance. Given time and patience, these issues resolve, and individuals are able to appreciate the true value of the tool.

Implemented on May 15, 1997
Data Collected on May 15, 1997, to June 30, 1997
Return Rate: 33% (N = 27)
Overall Score for Categories:

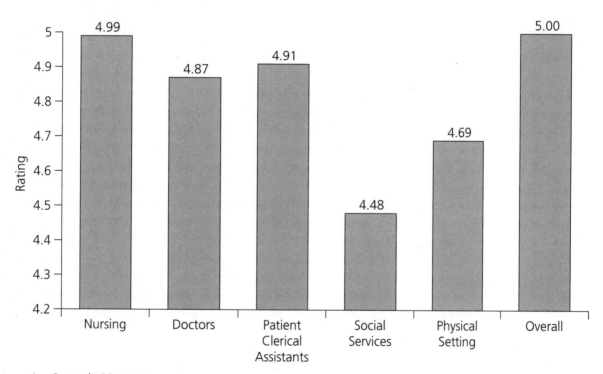

1 = Strongly Disagree
2 = Disagree
3 = Not Sure
4 = Agree
5 = Strongly Agree

Figure 9–3 Parent Satisfaction Survey. *Source:* Copyright © University of Kentucky Children's Hospital.

■ Appendix 9–A ■
Clinical Pathway for Neonatal ECMO

Exhibit 9–A–1 Clinical Pathway for Neonatal ECMO

UNIVERSITY OF KENTUCKY CHILDREN'S HOSPITAL
CHANDLER MEDICAL CENTER
LEXINGTON, KENTUCKY

"Clinical Pathways are guidelines for consideration which may be modified according to the individual patient's needs"

Patient Name:
Medical Record #:
Date of Birth:

Clinical Pathway

Diagnosis Neonatal ECMO	Admission Date	Primary Nurse	Attending Physician

Patient problem list

1. Oxygenation
2. Perfusion
3. Potential/actual insufficient fluid/calorie intake
4. Potential for infection
5. Thermoregulation
6. Potential for skin breakdown
7. Potential for hemorrhage
8. Potential for developmental delay
9. Parental knowledge deficit

Exclusion criteria: None

Parameters

Date:

Interval	Precannulation	Cannulation	Maintenance	Decannulation	Post ECMO
Assessment	Vital signs per protocol* Assess breath sounds, chest movement & respiratory effort Q 4 **Progression criteria:** **Initiation of surgical cannulation (NUR)**	Vital signs per protocol* **Progression criteria:** **CXR & ECHO confirmation of appropriate cannula placement (MD)**	Vital signs per protocol* Assess breath sounds, chest movement & respiratory effort Q 4 Neuro checks per protocol* Right upper extremity restrained per protocol* Assess and maintain skin integrity Physical exam WNL (MD) **Progression criteria:** **Trial off pump meets specific parameters (MD)**	Vital signs per protocol* **Progression criteria:** **Cannula removed (NUR)**	Vital signs per protocol* Assess breath sounds, chest movement & respiratory effort Q 4 Assess readiness to nipple feed Assess for feeding intolerance Physical exam WNL (MD)
Growth & Development Dry Weight: ____	HC: WT: Weight per protocol* HC per protocol* Minimal stimulation per protocol*		Weight per protocol* HC per protocol* Maintain in supported flexion Minimal stimulation per protocol* Comforted by containment (NUR) < grade I IVH per hus (MD)		Weight per protocol* HC per protocol* Stimulation as tolerated Encourage nonnutritive sucking Weaned to crib per protocol* (NUR) Without stiffness, arching, tremors, or tongue extensions with care or holding (NUR) Extremities rest in flexion & movement is symmetrical (NUR) Free of sustained ankle clonus or asymmetry in ankle clonus (MD/NUR) Visually tracks an object or face 90 degrees using eyes and head (NUR) Localizes to voice to the right and left (NUR) Can be comforted (NUR)

continues

Exhibit 9–A–1 continued

Date: Interval	Precannulation	Cannulation	Maintenance	Decannulation	Post ECMO
Tests/Labs	Pre ECMO labs per protocol* Stat CXR Stat HUS Stat ECHO	Stat CXR ECHO	ABGs per protocol* A9 Q 8 hours Hemogram with differential Q AM Blood culture Q AM Total & direct bili Q AM PT/PTT & fibrinogen Q AM ACTs per protocol* CXR or babygram Q AM Hus Q day × 3 days then as ordered Echocardiogram as ordered MG, triglyceride & pre albumin Q week Hematest all stool and urine Monitor glucose, specific gravity and labstix per protocol*		Immediately post decannulation: Hemogram with differential ABG Platelets Q 12 × 3 (beginning 12 hours after hemogram) 24 hours post decannulation: A7 × 1 Ionized Ca × 1 NA/K Q day D/C TPN labs when off TPN D/C NA/K when at full feeds CT scan EEG Neonatal metabolic screen **ALGO passed: (NUR)** **Right** **Left**
Consults	Cardiology stat Pediatric surgery Radiology				Consider developmental specialist Males: consider OB for circumcision Ophthalmology **Ophthalmology exam completed (MD/NUR)**
Oxygenation/ Respiratory Management	Pre & post pulse oximetry ET tube stinted at adapter	Pre & post pulse oximetry Ambu bag/nonventilator O_2 removed from operative area Initiate bypass per protocol*	Pre & post pulse oximetry Pulmonary toilet per protocol* Pump management per protocol* **Cannula secure & in place (NUR/SPEC)**	Pulse oximetry ET tube stinted at adapter Ambu bag/nonventilator O_2 removed from operative area End tidal CO_2 monitoring Sigh breath given as cannula removed Advanced to standard ventilator settings	Pulmonary toilet per protocol* Ventilator wean per protocol* ETT at gumline: **Saturations range 88–94% while on O_2 (NUR)**
Nutrition/ Fluid Balance	NPO Strict I & O NG to straight drain (do not change) Change arterial line fluids to 0.45 NACL with heparin one unit/cc Place hep lock with flushes per ECMO protocol*		NPO Strict I & O TPN/IL within 24 hours of beginning ECMO AC per protocol* **Prealbumin WNL (Dietetics)** **UOP > 1 cc/kg/hr (MD/NUR)** **Growth grid updated biweekly (MD)**		Feedings per protocol* D/C IL when at 2/3 calories from feeds D/C TPN when at minimum 130 cc/kg/day from feeds **UOP > 2 cc/kg/hr while on IVF (MD/NUR)** **Then > 6 saturated diapers Q day (NUR)** **Stools at least Q 3 days (NUR)** **At full feeds (MD/Dietetics)** **PO intake 150 cc/kg/day (Dietetics)** **PO feed Q feed by discharge (NUR)** **Growth grid updated biweekly (MD)**

continues

Exhibit 9–A–1 continued

Date: Interval	Precannulation	Cannulation	Maintenance	Decannulation	Post ECMO
Activity	Place on ECMO bed with X-ray film plate in place		Reposition infant Q 4 hrs (with specialist) ROM Q 8 hours (after 24 hours on ECMO)		Bathe Q 3 days May be held when off ventilator
Treatments	Place urinary bladder catheter **ECMO pump primed (Specialist)**	**ECMO circuit secured to bed (Specialist)**	Skin care per protocol* Dressing changes per protocol* after 24 hours on ECMO Measure cannula length with dressing changes Urinary bladder catheter care per protocol* **ECMO circuit secured to bed (Specialist)** **Signs placed at infant's bedside per protocol (NUR/Specialist)** **<3 cc/hour blood loss around cannula site (NUR/Specialist)** **Absence of skin breakdown (NUR)**		Males: consider circumcision with care per protocol*
Medications Drips Blood Products	Vasopressors as ordered Phenobarbital: Load administered Maintenance ordered	Vecuronium Fentanyl/morphine Heparin	Antibiotics: Ampicillin Gentamicin D/C antibiotics if precannulation blood culture is negative Consider drug levels if antibiotics continue past 3 days Phenobarbital Heparin infusion per protocol* Transfuse blood products per ECMO protocol* Consider D/C vasopressors Analgesic drip as ordered **Blood cultures negative (MD)** **Sedative effect allows for:** **Responsiveness without agitation (NUR)** **Adequate pump function (Specialist)**	Heparin infusion per protocol* Fentanyl/morphine Vecuronium	Phenobarbital Consider D/C phenobarbital if EEG WNL Wean analgesic drip to D/C Hepatitis B vaccine **Consented for hepatitis B vaccine (MD)**
Teaching	ECMO booklet given to parents Encourage verbalization of concerns & fears Prepare parents regarding visual appearance of infant on ECMO		Orient parents to ECMO pump Encourage parents to interact with infant as tolerated Encourage verbalization of concerns & fears Parents taught: Environmental safety issues Plan of care reviewed with parents Q visit		Initiate acute phase teaching sheet Assess for reverse transport Parents taught: Diapering Temperature taking Feeding Elimination Skin care Sleep patterns Infant safety issues Signs of illness How to dress infant Need for pediatrician

continues

Exhibit 9–A–1 continued

Date:					
Interval	Precannulation	Cannulation	Maintenance	Decannulation	Post ECMO

					Bath __
					Use of bulb syringe __
					Formula preparation __
					Feeding schedule __
					Feeding amount __
					Importance of:
					  Follow-up care __
					  Car seat __
					  Immunizations __
					  Medic alert identification for VA ECMO __
					Males: circumcision or foreskin care __
					Plan of care reviewed with parents Q visit __

Discharge Planning			Determine family learning needs, support structure & support __
			Parents will visit Q O day (Soc Svcs) __
			Parents will call between visits (Soc Svcs) __
			Parents will interact with infant (NUR) __
			Parents will express feelings about infant (NUR) __
			Parents will verbalize knowledge of: Environmental safety issues (NUR) __
					Assess home environment and supply readiness __
					Follow-up appointments as ordered __
					Review follow-up appointments __
					Consider referral to:
					  Home health __
					  DME __
					  WIC __
					Verify have car seat __
					Parents will visit weekly (Soc Svcs) __
					Parents will call between visits (Soc Svcs) __
					Parents will self initiate:
					  Change diaper (NUR) __
					  Take temperature (NUR) __
					  Demonstrate:
					    Skin care (NUR) __
					    Use of bulb syringe (NUR) __
					    Bath (NUR) __
					    Feed infant (NUR) __
					    Hold infant (NUR) __
					Parents will verbalize knowledge of:
					  Formula preparation (NUR) __
					  Feeding schedule (NUR) __
					  Feeding amount (NUR) __
					  Signs of illness (NUR) __
					  How to dress infant (NUR) __
					Discharge documentation completed (NUR) __
					Discharge physical complete 24 hours prior to D/C (MD) __
					Discharge planning sheets completed (MD) __
					Parents will identify pediatrician (MD) __
					Called F/U MD (MD) __

Note: HC = head circumference; wt = weight; CXR = chest X-ray; hus = head ultrasound; echo = echocardiogram; NPO = nothing per oral; I&O = input/output; NG = nasogastric; I&H = intraventricular hemmorhage; ABGs = arterial blood gases; PT = protine; PTT = prothromybin time; ACT = actual clotting time; mg = magnesium; TPN = total parenteral nutrition; IL = intra lipids; AC = abdominal circumference; UOP = urine output; A7 = astra 7; Ca = calcium; d/c = discontinue; EEG = electroencephalogram; IVF = intravenous fluids; ROM = range of motion; DME = durable medical equipment; WIC = Women, Infants and Children program; F/U = follow up; MD = physician.

Source: Copyright © University of Kentucky Children's Hospital.

Starting Small: Outcomes Management Design for the Extremely Low– Birth Weight Infant

Anne Milkowski

Outcomes management (OM) is a multidisciplinary process designed to improve the quality of health care, decrease fragmentation, enhance patient outcomes, and constrain costs.[1] Patient outcomes and OM are receiving more attention today from purchasers, payers, institutions, patients, politicians, and accrediting agencies than in years past.[2] Historically, outcomes analysis concentrated on cost, length of stay (LOS), and reimbursement because of skyrocketing costs for health care, pressure from external sources to decrease costs, and easy access to financial data.[3] Although financial outcomes are well defined and easy to quantify, the impact of specialized treatment and traditional care practice on long-term outcomes is not well defined.[4] However, it is the impact on long-term outcomes that is most significant for patients.

The past decade has allowed for improved survival of extremely low–birth weight (ELBW) infants. Since the 1980s, the survival rate of infants with birth weights less than 1,000 grams has increased from 20 to 30 percent to nearly 50 to 60 percent.[5–8] However, the improvements in morbidity rates have not matched improvements in survival rates. Extreme prematurity continues to be associated with severe morbidity, including cerebral palsy, mental retardation, and sensorineural impairments at a rate approaching 30 to 35 percent.[9] Coupled with other comorbidities such as bronchopulmonary dysplasia (BPD) and poor growth patterns, these suboptimal outcomes have lifelong implications for infants and their families.

Extremely tiny patients such as these use more resources and have longer LOSs than their counterparts born closer to full-term gestation. Reported median LOSs range from 100 to 118 days.[8] The expected geometric mean LOS for this population (DRG 386) is only 17.9 days.[8] This discrepancy in LOS and the associated reimbursement forces institutions to stringently control costs associated with neonatal intensive care.

It is not well established, whether a specific approach to ELBW patient management consistently yields optimal outcomes with the least complications. A meta-analysis of ELBW medical outcomes revealed that surfactant therapy improves survival and increases the incidence of BPD, retinopathy of prematurity, and gastrointestinal reflux.[10] Traditional patient care practices have not resulted in improved patient outcomes. New patient care practices, including modified ventilation strategies, altered dosing regimens for corticosteroid therapy, and developmentally supportive care, may provide a lower incidence of long-term sequelae and decrease the LOS associated with extreme premature birth.

Despite advances in technology to improve the survival rates for the ELBW population, there has been a limited effort to study the overall impact of neonatal intensive care unit (NICU) care has on long-term outcomes. Most published studies focus on a narrow range of neurodevelopmental and cognitive outcomes, failing to address other areas of morbidity.[11] Recognizing the need to study and optimize the outcomes of our tiniest patients, the NICU at All Children's Hospital embarked on an outcomes management program. The goals of the program are to study the outcomes of current practice, identify care practices associated with the best outcomes, and modify our approach to optimize outcomes for this population. The expected results are improved morbidity; decreased physical, emotional, and financial burdens to the patient and family; and cost reduction.

I thank Denise Maguire, RNC, MS, for her invaluable assistance in preparing this chapter. Also, thank you to the members of the All Children's Hospital NICU Collaborative Practice Team for their dedication and support.

THEORETICAL FRAMEWORK

Outcomes Management Quality Model

The OM program at All Children's Hospital is based upon the outcomes management quality model (Figure 10–1). The model was selected for its emphasis on development of "best practice" through research within a continuous quality improvement context. The model is approached through four phases.

Phase 1 of the OM quality model is characterized by the identification of the study population and members of the collaborative practice team (CPT). Populations for OM are selected from those that are high volume, high risk, high cost, or problem prone.[12] The selection of the population to be studied may originate from the organizational strategic plan, mission statement, or quality initiatives. Once the population is selected, identification of the CPT members begins. Members of the CPT should include clinicians from all disciplines involved in the care of the selected population, a data/quality process facilita-

tor, and an information systems member. Every effort should be made to keep the CPT to a manageable number of participants. More than 10 to 15 members decreases group cohesiveness and narrows opportunities to meet together regularly. The success of the team is dependent on the strength and commitment of the individuals involved and their ability to work as a group toward a common goal. Members chosen for their expertise and willingness to accept accountability as the voice of their discipline will strengthen the CPT.

Once formed, the CPT will need focused education about OM and current information about the selected population. Their first task is to establish the goals and direction of the team, which they do by identifying the long-term outcomes, intermediate outcomes, and barriers to achievement of intermediate outcomes. Long-term outcomes are those that are measured over time at specific points after the completion of an episode of care.[13] Examples of neonatal long-term outcomes include length of time on home supplemental oxygen, serial developmental scores, readmission rates, and the development

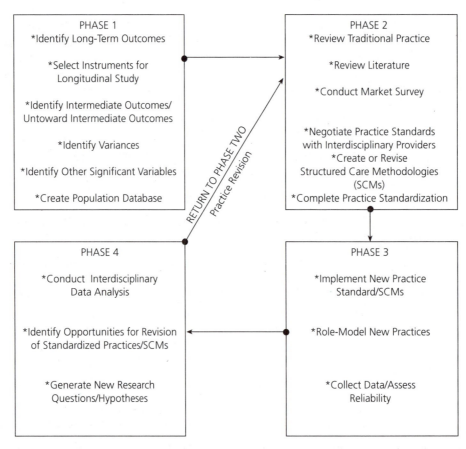

Figure 10–1 Outcomes Management Quality Model. *Source:* Copyright © Health Outcomes Institute, Inc.

of chronic illnesses. Intermediate outcomes are those outcomes that can be measured at the completion of an episode of care. In the ELBW population, intermediate outcomes may include growth parameters adjusted for gestational age, hematocrit at discharge, absence of intraventricular hemorrhage, and absence of retinopathy of prematurity. Untoward intermediate outcomes are those outcomes that prevent the attainment of intermediate outcomes. Examples of untoward intermediate outcomes for the ELBW population include pneumothorax, necrotizing enterocolitis, and anemia of prematurity.

The first phase of the OM quality model is also characterized by a process of descriptive research that describes the population, the current care practices, and outcomes for this population following an episode of care. A data collection tool must be designed and links to databases established. Data collection is a monumental task that often requires tedious medical record review, as well as an intimate knowledge of the population. The data collection tool must be constantly revised in the early stages of research as consistently missing pieces of information are identified. The CPT identifies possible relationships between and among the variables that will help to answer questions such as "Is there a relationship between . . . ?"

Since this is research, it is recommended that members of the CPT discuss their intentions with the hospital's institutional review board to gain their recommendations concerning protection of human subjects.

Phase 2 is characterized by the creation of structured care methodologies and interventions for testing with the population. Structured care methodologies include pathways, protocols, algorithms, guidelines, and order sets (Table 10–1). Each of these structured care methodologies differs in degree of specificity and flexibility. Structured care methodologies are often interrelated to complement each other.[4] For example, the care path for the ELBW infant may have a suctioning protocol, a fluid and nutrition protocol, and a skin care protocol as components, each of which has a complementary order set.

When designing a structured care methodology, the primary objective should be outcome attainment based on the data that were obtained during phase 1 and not merely a cost reduction. While it is important to establish structured care methodologies, they should be regarded as "works in progress." They are evolutionary tools, not finished products.[4] Following the design of structured care methodologies, implementation and testing of the tools begins.

Table 10–1 Structured Care Methodologies

Structured Care Methodologies	Characteristics
Critical Pathways	Represent a sequential, interdisciplinary, minimal practice standard for a specific population Abbreviated format, broad perspective Phase or episode driven Should include a demographic database to facilitate descriptive study Must be complemented by an outcomes measurement worksheet
Algorithms	Binary decision tree that guides stepwise assessment and interventions Intense specificity Useful in the management of high-risk subgroups within the cohort. Complement broad critical pathway minimal practice standard
Protocols	Prescribe therapeutic interventions for a specific clinical problem within a subgroup of patients Multifaceted; may drive specific practices within numerous disciplines Broader specificity than an algorithm but more specific than a critical pathway Broad research-based practice recommendations
Guidelines	Practice resource May or may not have been tested in clinical practice No mechanism for ensuring practice implementation
Order sets	Preprinted provider orders Expedite order process Complement and increase compliance with existing practice standards May include orders for the initiation of protocols within subgroups of patients

Phase 3 of the OM quality model is the implementation of the structured care methodologies and interventions designed during phase 2. During this phase, multidisciplinary staff education occurs for those who will use the structured care methodologies. During this phase, data collection concentrates on whether the changes in practice were beneficial to the population.

Phase 4 is characterized by data analysis. Relationships between and among variables are explored in search of care practices that support optimal outcomes. Practices that are associated with poor outcomes are also identified. The analysis generates new questions and identifies new interventions to test. Structured care methodologies are revised to reflect the knowledge gained from the first phase of research, optimizing patient care outcomes. The process continues to cycle through phases 2, 3, and 4 as the new interventions are tested.

Change Theory

Change theory plays a large role in the OM process. Traditional practices are questioned, historical boundaries are blurred, and new methods for practice are introduced. Change produces periods of unrest and discomfort, often followed by a period of apathy. Finally, acceptance of the new way occurs. Change has been described to occur in three stages. The first stage is *unfreezing,* when the pressure to change is greater than the pressure not to change. The second stage is *moving,* when the actual change occurs. In the final stage of *refreezing,* the change is accepted and incorporated into the system.

Organizations considering an OM program should thoroughly appreciate the impact of change process on the organization. Increasing knowledge about the change process and the expected new behaviors enables individuals to adjust better to the change. Change can be a slow and time-consuming process, and attitudes toward change vary. Capitalize on the enthusiasm of those in favor of the change to drive the program.[4] People who are against the change often need additional proof that the change works. There are always some individuals who remain adamantly against the change, despite apparent proof.

ELEMENTS OF AN OUTCOMES MANAGEMENT PROGRAM

Necessary elements of a successful OM program include administrative support, information systems support, research and data analysis support, and strong staff education. The first priority is to establish commitment, leadership, and support from the organization's adminis-

tration and medical staff. Administrative support is essential to the success of the OM program. Support may include altering job responsibilities so that individuals on a CPT are available to concentrate on the OM program. Support may also be energizing the outcomes program through public recognition of the program in the institution's publications and meetings. Medical support is established by including key physicians in the planning and implementation of CPTs, and the identification of the study populations. Physicians often have the power and authority to ensure CPT success or failure, and individuals who can influence their peers will be helpful in achieving program goals.

An accessible information system is critical to a successful OM program. The information system stores data useful to the outcomes manager. An important organizationwide assessment of all data currently collected related to outcomes and performance should occur before the establishment of the CPT.[12] This kind of inventory is valuable to prevent duplication of collected data. Medical information systems that have the ability to download multiple types of information and generate automatic reports eliminate paper data collection tools and promote a more efficient collection. However, if the information system is not developed to this extent, data must be collected and entered manually into a computer program for analysis. Future developments in medical information systems such as a clinical data repository will enable the CPT to access data from various systems within and beyond the institution's walls.

Data analysis and research experts are vital supports to the CPT. Persons knowledgeable in applied statistics and research methodologies enable the teams to concentrate on clinical issues. If individuals with expertise in these areas are not members of the staff, consultation is highly recommended.

CPTs need education about the OM process, team building, group dynamics, statistical analysis, and research methodology. This education must occur early in the process and be the focus of a continuing education plan. Education for staff not directly involved in the CPTs is equally important so that they are better prepared to participate in and support the OM process.

ALL CHILDREN'S HOSPITAL NEONATAL INTENSIVE CARE UNIT EXPERIENCE

All Children's Hospital chose the NICU as a pilot group for the OM program because a strong interdisciplinary team existed, the patient population was managed throughout their hospitalization in the NICU, and an advanced-practice nurse was willing and available to act as outcomes manager. The NICU CPT selected the extremely low–birth weight infant population because it fit the high-cost, problem-prone, and high-risk categories. The NICU team admits 50 to 70 patients every year with birth weights less than 1,000 grams. The average length of stay for patients discharged during 1996 was 94 days, with a range of 53 to 232 days. Like other NICUs in the country, this NICU is using many combinations of new care practices with this population. However, research to support any one combination as "best practice" is lacking.

Many disciplines in the NICU have a direct impact on the outcome of care delivered to the extremely low–birth weight infant. It is critical that each of these disciplines is represented on the CPT (Exhibit 10–1).[13] The NICU CPT is a diverse group with a history of working with each other individually but not as a team. Because they form a new group, development follows the typical pattern of forming, storming, norming, and performing.[14] Following an education session about outcomes and OM, the NICU CPT was established. The *forming* stage is characterized by the team getting to know each other better. Even though members of the team knew each other, they learned and shared each other's skills, talents, and resources at the first meeting. This provided an informative venue to share accomplishments and goals with colleagues for the first time.

The next stage, *storming,* is characterized by conflict. The presence of conflict suggests a willingness within the group to allow free and open discussion of ideas. Perceived territorial threats or verbalization of uncommon goals may lead to conflict. Although some members of the team may be very uncomfortable with conflict, conflict is necessary for change to occur. Members who are willing to discuss issues on which they do not agree help

change to occur by putting the issues on the table. Members who are unwilling to voice their disagreements become roadblocks to success. Too much conflict is just as unproductive as too little conflict. The challenge for the leader is to manage the conflict so that it moves the group in the direction of their goals.

Norming, the next stage, is where true collaboration occurs. The members gain a healthy respect for their differences and become willing to trust that territories will expand rather than shrink when they work together. Collaboration is a skill that must be learned, and permission must be given for the CPT to collaborate. Permission is initially given by hospital administration, but it must be operationalized and lived by persons with authority. *Performing* is the final stage, in which meeting targeted goals is the priority, and pride within the team is evident.

ELBW OUTCOMES

The NICU CPT determined their current end point in the episode of care as discharge from the NICU. The expected outcomes for the ELBW population at discharge are that infants are on full enteral feeds, are consistently gaining weight, and are without major sequelae secondary to treatment. Some of the untoward intermediate outcomes being tracked are pulmonary complications, gastrointestinal complications, neurological complications, hematologic complications, and sepsis (Table 10–2).

One of the greatest challenges that the NICU CPT faces is collection of baseline information and establishing the database. The tool designed by the NICU CPT includes variables such as demographics, LOS, total ventilator days, average daily weight gain, average calories per kilogram per day, discharge medications, information about the infant's birth, psychosocial data, and types of complications experienced. Relationships between and among these variables will help the NICU CPT to understand the impact of the therapies and interventions on intermediate and long-term outcomes. Several problems have been identified with data collection. The very large volume of information being collected on each infant has required many revisions to the data collection tool to streamline the process. Completing data collection on each patient required much more time than was initially anticipated. Although data collection still takes time, loading the data collection tool onto a laptop computer has eliminated duplication of work. Finally, many of the hospital's databases were not integrated into a compatible information system, and tighter links with information systems to address integration issues are being made.

Developing and implementing structured care methodologies is also a focus of the NICU CPT. An earlier version of the care path was changed from a day-by-day approach to a phase approach (Exhibit 10–A–1 in Appen-

Exhibit 10–1 Neonatal Intensive Care Unit Collaborative Practice Team Members

Neonatologist	Quality Process Facilitator
Department Director	Staff Nurse
Social Worker	Physical Therapist
Utilization Review Nurse	Clinical Educator
Dietitian	Discharge Nurse Specialist
Speech Therapist	Information Systems
Respiratory Therapist	Representative
Neonatal Nurse Practitioner	Pharmacist
Audiologist	Early Intervention Program
Occupational Therapist	Representative
Lactation Consultant	Developmental Specialist
Neonatal Clinical Nurse	
Specialist	

Table 10–2 Selected Untoward Intermediate Outcomes Tracked by the Neonatal Intensive Care Unit Collaborative Practice Team of All Children's Hospital

System	Complications Tracked
Pulmonary	Apnea of prematurity, chronic lung disease, respiratory arrest, pneumonia, pulmonary hemorrhage
Cardiac	Cardiac arrest, persistent patent ductus arteriosus, congestive heart failure, endocarditis
Neurologic	Interventricular hemorrhage, seizures, meningitis, shunt placement, shunt infection
Renal	Nephrocalcinosis, urinary tract infection, acute tubular necrosis, dialysis
Hematologic	Anemia of prematurity, thrombocytopenia
Sensory	Retinopathy of prematurity, hearing loss

dix 10–A). As the pathway is implemented, it is expected that data collection and the analysis of the data by the NICU CPT members will alter the path. The first path revision is planned at the end of six months.

The next challenge for the NICU CPT is the establishment of the longitudinal arm of the project. This aspect of the project involves descriptive research of the long-term health and functional status of ELBW patients. The development of chronic conditions, number of rehospitalizations, and growth parameters are some of the variables under consideration. As the CPT begins to analyze data beyond our walls, the lifelong impact of the ELBW NICU admission will become more clear.

CONCLUSION

As the health care environment continues to evolve, new definitions of quality and value begin to emerge. We can anticipate that some of those definitions will be linked to long-term outcomes of patient management practices. Health care professionals must begin to analyze systematically the outcomes of their care and implement measures to continually improve those outcomes. Linking the results of an episode of care to the overall health and function of the individual enables health care providers to determine what is "best practice" and its impact on the individual and society. In the case of extremely low–birth weight infants, the impact of their stay in the NICU may affect their entire lifetime.

The initial steps to create an OM program for the extremely low–birth weight infant include the formation of an interdisciplinary CPT, the development of structured care methodologies, and descriptive research to determine "best practice" for the population. As the program grows and expands, there will be more opportunities for enhanced collaborative practice and clinical research. And what started small will grow tall.

REFERENCES

1. Moss MT, O'Connor S. Outcomes management in perioperative services. *Nurs Econ.* 1993;11(6):364–369.

2. Harris MR, Warren JJ. Patient outcomes: assessment issues for the CNS. *Clin Nurs Spec.* 1995;9(2):82–86.

3. Nadzam DM. Nurses and the measurement of health care: an overview. In: *Nursing Practice and Outcomes Measurement.* Oakbrook Terrace, IL: Joint Commission on Accreditation of Healthcare Organizations; 1997:1–15.

4. Wojner AW. Outcomes management: from theory to practice. *Crit Care Nurs Q.* 1997;19(4):1–15.

5. Godson E. The micropremie: infants with birth weight less than 800 grams. *Inf Young Child.* 1996;8(3):1–10.

6. Hack M, Taylor HG, Klein N, Eiben R, Schatschneider C, and Mercuri-Minich N. School age outcomes in children with birth weights under 750 grams. *N Engl J Med.* 1994;331(12):753–759.

7. Roth J, Resnick MB, Ariet M, et al. Changes in survival patterns of very low birth weight infants from 1980–1993. *Arch Pedia Adoles Med.* 1995;149:1311–1317.

8. Tyson JE, Younes N, Verter J, Wright LL. Viability, morbidity, and resource use among newborns of 501–800-g birth weight. *JAMA.* 1996;276(20):1645–1651.

9. Hack M, Friedman H, Fanaroff AA. Outcomes of extremely low birth weight infants. *Pediatrics.* 1996;98(5):931–936.

10. Dusick AM. Medical outcomes in preterm infants. *Sem Perinat.* 1997;21(3):164–177.

11. McCormick MC. The outcomes of very low birth weight infants: are we asking the right questions? *Pediatrics.* 1996;99(6):869–876.

12. Hoesing H, Karnegis J. Nursing and patient care processes: interdisciplinary care outcomes management. In: *Nursing Practice and Outcomes Measurement.* Oakbrook Terrace, IL: Joint Commission on Accreditation of Healthcare Organizations; 1997:35–62.

13. Wojner AW. Outcomes management: an interdisciplinary search for best practice. *AACN Clin Iss.* 1996;7(1):133–145.

14. Tuckman BE. Developmental sequence in small groups. *Psycho Bull.* 1965;63:384–399.

■ Appendix 10–A ■

Care Path for Infants Weighing Less Than 1,000 Grams at Birth

Exhibit 10–A–1 Care Path for Infants < 1,000 Grams at Birth

	Phase I: Acute Stabilization	Phase II: Stabilization and Maintenance	Phase III: Growth	Phase IV: Preparation for Discharge
Assessment Evaluation	• Strict I & O • Daily weight • Skin Assessment • Remain at bedside after interventions to assess infant response	• Evaluate skin for maturity and consider discontinuing humidity • Evaluate need for long-term IV access • Assess iron intake needs	• Evaluate frequency of Apnea and bradycardia incidents	• Evaluate need for home monitoring • Evaluate need for home oxygen therapy
Tests	• Consider Echo to assess for PDA • Astra 8 q 12 then PRN • Cultures as ordered CBC w/Diff. • RPR IMS @ 72 hrs of age • Newborn Antibody Screen • Newborn Urine Drug Screen • CUS by day 7	• Astra 8 or TPN profiles • IMS # 2 at 3 weeks of age • HCT • Renal US after 4 weeks of diuretic use • Consider need for follow-up CUS	• Consider need for follow-up CUS • Consider need for follow-up Renal US • Consider eye exam at 6 weeks of age	• Hearing exam • Consider need for f/u eye exams • Consider need for f/CUS • Consider need for f/u Renal US • Consider need for f/u hearing tests
Treatments	• Assisted Ventilation, wean per RCP • Initiate Suctioning Protocol • Initiate Skin Care Protocol	• Assisted Ventilation, wean per RCP • Suctioning protocol while intubated		
Medication IVs	• Consider Surfactant Protocol • Consider use of Indocin • Consider use of Sedation • Consider use of Antibiotics • Consider use of Inotropes • Consider use of Aquaphor	• Evaluate need for antibiotics • Consider starting steroids • Consider starting bronchodilators • Consider use of diuretics • Consider initiating use of Caffeine • Consider use of Epogen • Consider Iron supplements	• Evaluate Bronchodilator use • Evaluate use of Xanthines • Evaluate use of diuretics • Consider immunizations @ 60 days of life • Evaluate need for Epogen	
Activity Safety	• Initiate Developmental Sheet • Minimize noise and activity at the bedside • Shield eyes from bright lighting • Place infant on sheepskin or gel mattress • Move to incubator	• Encourage skin to skin contact between parents & infant • Dress infant in clothing • Update Developmental Sheets Weekly	• Update Developmental Sheets Weekly	• Update Developmental Sheets Weekly • Wean to open crib
Diet Nutrition	• NPO • Start IVF D5W 120–150 ml/kg/day • Adjust fluids & electrolytes based on Astra 8 • Consider initiating FEN Protocol	• Initiate Feeding Protocol	• Evaluate readiness for bolus feedings • Evaluate readiness for breast-feeding or nipple feeding	• Advance to ad lib feeds • Change to discharge formula • Assess need for continued fortification of breast milk
Consults	• Social Services	• BEST as appropriate	• Pulmonology Consult prn • Developmental Consult prn • OT/PT Consult prn • Speech Consult prn	

continues

Patient Family Education **Discharge Planning**	• Orient family to NICU environment, equipment & visitation policy • Educate parents on blood transfusion policy, obtain consents • Inform family of status, condition, & prognosis of their infant • Demonstrate & explain care giving procedures during parental visits • Initiate teaching tool w/primary care givers	• Encourage parental participation in infant's care • Begin assessment of family learning needs • Identify enhancers & detractors to learning • Bi-weekly multidisciplinary discharge planning rounds • Consider family & care givers conference • Encourage parental participation in parent support group	• Bi-weekly multidisciplinary discharge planning rounds	• Bi-weekly multidisciplinary discharge planning rounds • Evaluate need for CPR training • Evaluate need for parental rooming in • Instruct parents in the preparation and administration of D/C medications • Give D/C prescriptions to parents • Obtain name of pediatrician or clinic • WIC referral

Note: CBC, complete blood count; CPR, cardiopulmonary resuscitation; CUS, cranial ultrasound; *d/c*, discharge; FEN, fluid, electrolyte and nutrition; *f/u*, follow-up; HCT, hematocrit; IMS, infant metabolic screening; I&O, intake and output; IVF, intravenous fluids; NPO, nothing by mouth; OT, occupational therapist; PDA, patent ductus arteriosus; PT, physical therapist; RCP, respiratory care plan; TPN, total parenteral nutrition; w/, with; WIC, Women, Infants and Children.

■ 11 ■

Measuring Outcomes for Spleen and Liver Trauma in Children

Anita Gottlieb and Patti Higginbotham

Most would agree that if you haven't measured something you can't really say much about how well you are doing.
—Steven B. Kritchevsky, *Joint Commission Journal on Quality Improvement*

Improving patient outcomes has always been the focus for health care providers. In today's managed-care environment, achieving positive outcomes while decreasing cost has become a major goal. To achieve this, direct providers are developing strategies for cost containment at the clinical care level. Arkansas Children's Hospital, a private nonprofit organization, has begun to use clinical pathways to address both quality and cost issues. Since trauma cases represent an emergent, nonelective, and usually high-cost diagnosis, blunt abdominal trauma was chosen as one of the diagnoses for pathway development. Approximately 30 percent of pediatric and adolescent trauma cases involve splenic or hepatic injury. After severity and hemodynamic stability of the patient have been assessed, nonoperative management has proven to be both safe and effective in the majority of cases. Due to the increased risk of sepsis and death after splenectomy in children, the nonoperative approach is extremely important in pediatrics. Many cases of spleen and liver trauma are accompanied by other severe injuries that further affect the overall morbidity and mortality of the patient. According to Coburn et al., in cases of multiple injury in pediatric patients (including those with other injuries requiring surgical intervention), nonsurgical management of abdominal trauma is associated with a positive outcome.[1] Although the potential risks of nonopera-

tive management of splenic injury include delayed hemorrhage, abscess formation, and peritonitis, these complications rarely occur in children.[2–5]

Shock resuscitation must be the initial priority in treatment of all pediatric trauma victims, followed by evaluation of injuries, diagnosis, and intervention. If hemodynamic stability can be achieved and the diagnosis can be established, most children and adolescents with blunt abdominal trauma can be safely treated without a laparotomy. Although today's radiological diagnostic capabilities have become increasingly useful in directing therapy, the literature stresses that the physiologic condition of the patient must be the primary determinant of treatment.[6] With nonoperative therapy, the following are avoided: postoperative pain, anesthetic complications, wound healing, and the expense of surgery. Additionally, nonoperative treatment may prevent the removal of salvageable organs. Regardless of the method of treatment, abdominal trauma requires rapid evaluation and intensive monitoring initially for the best outcome to be achieved.[7–9] A review of the literature provides much information regarding the nonsurgical management of the child or adolescent with blunt abdominal trauma and focuses primarily on acute care. After stabilization of the patient, standards of care are minimally addressed.

Blunt abdominal trauma was the fourth pathway to be developed and implemented at Arkansas Children's Hospital. After the successful implementation of the first three, the surgical service identified blunt abdominal trauma as the next diagnosis that could benefit from pathway development. Although current practice at our institution used nonsurgical management, the surgical and nursing staff wanted to standardize the treatment plan, evaluate outcomes, compare diagnostic findings, and monitor cost. Although this was not a high-volume diagnosis, a compari-

We wish to recognize and thank Raye West for her technical support in the preparation of this chapter.

son of the number of patients seen at Arkansas Children's Hospital with that of the Pediatric Health Informations Systems database showed that a representative number were being treated at our facility. Since Arkansas Children's Hospital is the only pediatric hospital in the state, the majority of the state's pediatric blunt abdominal trauma cases are treated in this facility.

DEVELOPMENT OF THE BLUNT ABDOMINAL TRAUMA PATHWAY

From lessons learned on previous pathways, Arkansas Children's Hospital has a process for clinical pathway development. The steps were appointment of the pathway team, review of related literature, obtaining and reviewing the hospital's historical data, comparing cost and outcome data with other pediatric hospitals, setting timelines and goals for the pathway, identifying outcome measures, establishing mechanisms for evaluation, educating staff, implementing the pathway, and monitoring outcomes on an ongoing basis. Pathway development, implementation, and evaluation are overseen by the institution's pathway committee.

DEVELOPMENT OF A PATHWAY TEAM

The pathway development team was composed of representatives from the surgical staff, medical-surgical nursing, cost accounting, nursing education, laboratory, quality improvement, and nutrition. The team leader was a surgical staff physician who had expressed an interest in development of the pathway. It has been noted that the hospitals that have achieved the greatest success with clinical pathway implementation have used a physician "champion" to gain support from the medical staff.[10] We were fortunate to have had the support of the physicians and physician residents for this project. This does not lessen the importance of the other team members and their equally important contributions to development and implementation of the pathway. Each step of the process was facilitated by various team members

as their specialty or discipline was included. The team used the problem-solving methodology FOCUS-PDCA for pathway development, as shown in Exhibit 11–1. To keep the team meeting time to a minimum, much of the actual drafting of the pathway and data collection was completed before the team meetings.

HISTORICAL DATA ON SPLEEN AND LIVER TRAUMA

The first step was to obtain historical data for spleen and liver trauma. The ICD-9 codes being used for this diagnosis were provided by the medical records department. The next step was to obtain data through cost accounting. The data retrieved and assessed were a two-year historical profile of the length of stay (LOS) and cost associated with this diagnosis (see Table 11–1).

The cost data revealed a decrease in total charges for spleen and liver trauma from 1995 to 1996 in nonsurgical cases. This probably demonstrated that some practice changes had occurred before pathway development. The LOS for nonsurgical patients had decreased an average of 2.48 days while there was an increase in LOS for surgical patients.

DEVELOPING DESIRED OUTCOMES AND OUTCOME MEASUREMENTS

The objectives of the pathway included reducing LOS and costs and improving or maintaining the quality of care. It was determined at the first pathway team meeting that an algorithm as well as the standardized pathway format would be used. Review of the literature provided support for the methods of care outlined in the pathway draft. The laboratory representative provided the information necessary to make determinations related to number and frequency of hemocrits necessary for evaluation of hemodynamic stability. The goal was to keep venipuncture to a minimum to decrease patient discomfort and anxiety. After reviewing the first draft, the medical and nursing representatives agreed that it would be helpful to divide the pathway into two separate documents: "Stable

---

Exhibit 11–1 FOCUS-PDCA Problem-Solving Methodology

F	Find an opportunity for improvement		P	Plan the improvement
O	Organize a team		D	Do the improvement
C	Clarify the current process		C	Check/evaluate
U	Understand variation in the process		A	Act to hold the gains
S	Select a solution			

Source: Copyright © Arkansas Children's Hospital.

Table 11–1 Length of Stay for Spleen and Liver Trauma Patients, Historical Data

Year	Type	Number of Cases	Average Length of Stay
1995	Surgical	3	10.33
	Nonsurgical	8	5.63
1996	Surgical	2	13.50
	Nonsurgical	13	3.15

Spleen and Liver Trauma Patients" (see Exhibit 11–2 and Figure 11–1) and "Unstable Spleen and Liver Trauma Patients" (see Exhibit 11–3 and Figure 11–2).

The pathway was not used as a form of documentation and did not become a part of the patient's medical record. The surgical staff physicians and physician residents were responsible for writing the order for the patient to be placed on the pathway. At the recommendation of the risk management and legal affairs office, each pathway has the following disclaimer at the top: "This clinical pathway is provided as a general guideline for use by the physician and families planning the treatment of patients/families. It is not intended to and does not establish a standard of care. Each patient is individualized according to their specific needs." Color-coded laminated copies of all the hospital's pathways are posted on each nursing unit. Color-coding the pathways was also helpful in familiarizing the staff with them and in making it easier to locate specific pathways on the units. This pathway was introduced to the nursing staff by the nursing representatives on the team. The team leader was responsible for education of the medical staff.

Exhibit 11–2 Pathway for Stable Spleen and Liver Trauma Patients

---

ARKANSAS CHILDREN'S HOSPITAL
CLINICAL PATHWAY
SPLEEN AND LIVER TRAUMA

Patient Population: Spleen and Liver Trauma Patients
ICD 9 Code: 865.09, 864.09, 864.05
LOS: 2 Days

NOT FOR DOCUMENTATION
FINAL 11/11/96

> This clinical pathway is provided as a general guideline for use by physicians and families in planning care and treatment of patients/families. It is not intended to and does not establish a standard of care. Each patient's care is individualized according to their specific needs.

STABLE

	Day 1/Admission	Day 2	Day 3
Lab/Test	Admission: CBC, Amylase, SGOT, SGPT, UA, Type and Screen Hct at 6 and 12 hrs post admission Transfuse if Hct <21% then Hct q 12 × 2	Hct at 42 hours post admission	Hct if transfused
Medications/ Pain Control	Morphine .05–.1 mg/kg (IV) q 2 hrs prn Pain at Admission	Tylenol PO per wt q 4 hrs prn pain ——————————➤	
Nutrition/Hydration	NPO D$_5$ 1/2 NS IV with 20 meq. KCL	Regular diet as tolerated ————— Heplok IV with adequate intake ————➤	
Assessment/ Monitoring	Vital signs q 2 hrs × 4 I & O, Observe for Tachycardia, Hypotension, abd pain or distention	Vital Signs q 4 hrs ——————————➤	
Treatment	Bedrest	Out of Bed ——————————————➤	
Education/ Discharge Planning	Introduction of Pathway	Sign/Symptoms of Abdominal Pain	DC Instructions re: Activity Limitations Surgery Clinic—4 wks Appt. Made a X of DC
Expected Outcome	Hct Stable	Hct Stable, without abdominal pain Tolerating diet	Discharge Asymptomatic and Tolerating Diet

Note: For abbreviations, refer to reference key at the bottom of Figure 11–1.
Source: Copyright © Arkansas Children's Hospital.

Patient Population: Emergency Department/Inpatient
Inpatient: Acute Spleen and/or Liver Trauma
Stability:
0–1 years old, systolic ≥ 70mg/hg
1–13 years old, systolic ≥ 80mg/hg
14–18 years old, systolic ≥ 90mg/hg

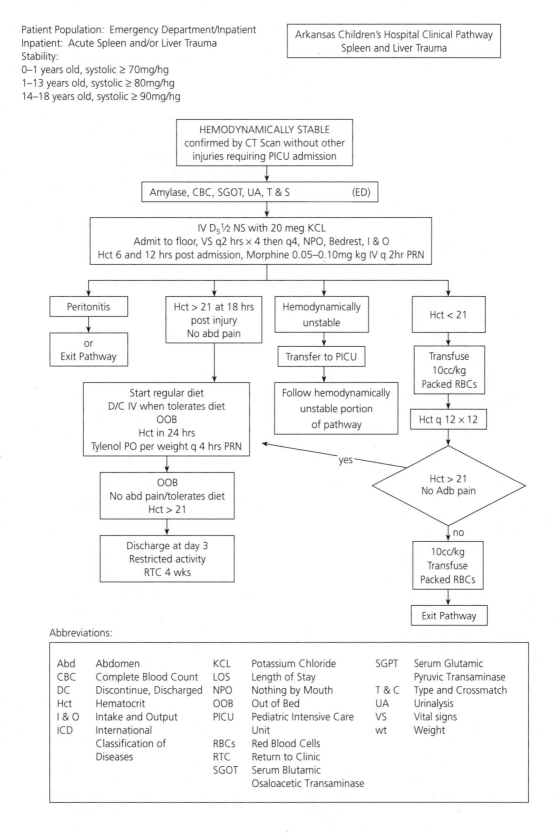

Figure 11–1 Algorithm for Stable Spleen and Liver Trauma Patients. *Source:* Copyright © Arkansas Children's Hospital.

Exhibit 11–3 Pathway for Unstable Spleen and Liver Trauma Patients

ARKANSAS CHILDREN'S HOSPITAL
CLINICAL PATHWAY
SPLEEN AND LIVER TRAUMA

Patient Population: Spleen and Liver Trauma Patients
ICD 9 Code: 865.09, 864.09, 864.05
LOS: 3 Days

NOT FOR DOCUMENTATION
FINAL 11/11/96

This clinical pathway is provided as a general guideline for use by physicians and families in planning care and treatment of patients/families. It is not intended to and does not establish a standard of care. Each patient's care is individualized according to their specific needs.

UNSTABLE

	Day 1/Admission	Day 2	Day 3	Day 4
Lab/Test	CBC, Amylase, SGOT, SGPT, UA, T&C Infuse up to 40 cc/kg. Ringers lactate in less than or equal 2 hrs. Hct q 6 hrs × 3 (if Hct stable and >21 @ 18 hrs) Transfuse up to 20 cc/kg. PRBCs in less than or equal to 2 hrs as indicated by Hct	Hct × 2 Transfuse if Hct <21 with 10 cc/kg PRBCs then Hct q 12 hrs × 2	Hct	Hct
Medications/ Pain Control	Morphine .05–0.1 mg/kg (IV) q 2 hrs prn pain	Morphine .05–0.1 mg/kg (IV) q 2 hrs prn pain	Tylenol PO per wt q 4 hrs prn pain	————▶
Nutrition/Hydration	NPO ———— D$_5$½ NS IV with 20 meq. KCL	————————▶	Regular diet as tolerated/Heplok	Regular diet
Assessment/ Monitoring	Vital signs q 2 hrs × 4 I & O, Observe for tachycardia, hypotension, abd pain or distention	Vital signs q 4 hrs	Vital signs q 4 hrs	————▶
Treatment	Bedrest/admit to PICU	Transfer to floor on bedrest If transfused cont. In PICU	OOB (transfer to floor if transfused)	————▶
Education/ Discharge Planning	Introduction of pathway and PICU	Activity limitations Precautions		DC instructions re: Activity limitations Surgery clinic—4 wks Appt. made a X of DC
Other	Admit to PICU			Discharge home
Expected Outcome	Hct stable	No return to PICU, Hct stable	No abd pain, tolerating diet well	Hct stable, no abd pain, tolerating diet

Note: For abbreviations, refer to reference key at the bottom of Figure 11–1.
Source: Copyright © Arkansas Children's Hospital.

DEVELOPING OUTCOMES ASSESSMENT

The cost of care in using both stable and unstable blunt abdominal trauma pathways was summarized and compared to historical cost data. Projected cost of care was less for both stable patients and unstable patients after reviewing the first six months of the project. To assess outcomes adequately, specific objectives and measurements were defined, as shown in Table 11–2.

ASSESSMENT OF THE PATHWAY

After the quality measures were identified and before pathway implementation, a data collection sheet was developed to ensure that each case was retrospectively reviewed. All patients with spleen and liver trauma were placed on the pathway. It was the responsibility of the surgical staff physician to notify the pathway coordinator of all admissions for blunt abdominal trauma. Notifica-

Patient Population: Emergency Department/Inpatient
Inpatient: Acute Spleen and/or Liver Trauma
Stability:
0–1 years old, systolic ≥ 70mg/hg
1–13 years old, systolic ≥ 80mg/hg
14–18 years old, systolic ≥ 90mg/hg

Arkansas Children's Hospital Clinical Pathway
Spleen and Liver Trauma

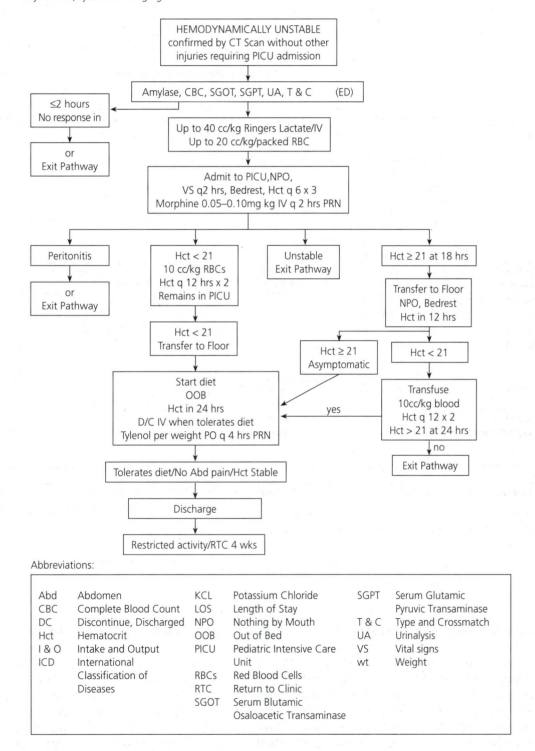

Abbreviations:

Abd	Abdomen	KCL	Potassium Chloride	SGPT	Serum Glutamic
CBC	Complete Blood Count	LOS	Length of Stay		Pyruvic Transaminase
DC	Discontinue, Discharged	NPO	Nothing by Mouth	T & C	Type and Crossmatch
Hct	Hematocrit	OOB	Out of Bed	UA	Urinalysis
I & O	Intake and Output	PICU	Pediatric Intensive Care	VS	Vital signs
ICD	International		Unit	wt	Weight
	Classification of	RBCs	Red Blood Cells		
	Diseases	RTC	Return to Clinic		
		SGOT	Serum Blutamic		
			Osaloacetic Transaminase		

Figure 11–2 Algorithm for Unstable Spleen and Liver Patients. *Source:* Copyright © Arkansas Children's Hospital.

Table 11–2 Pathway Objectives and Measurements

Objectives	Measure
Improve/maintain quality	Cases requiring operative exploration
	Patient on pathway remains on pathway
	No unplanned readmissions or returns to pediatric intensive care unit
Reduce length of stay	Hemodynamically stable: 2 days (72 hours)
	Hemodynamically unstable: 3 days (120 hours)
Reduce cost	Total cost at or less than projected
	Medications altered from pathway
Laboratory	Laboratory altered from pathway
	Radiology altered from pathway

tion by the physician was feasible due to the small volume of cases involved. Since blood transfusions were used in patients unable to achieve hemodynamic stability, it was necessary to incorporate a mechanism to evaluate the administration of blood and blood products. The blood utilization review process, presented in Table 11–3, was approved by the Arkansas Children's Hospital transfusion committee for evaluation of transfusions received by blunt abdominal trauma pathway patients.

INITIAL OUTCOME ASSESSMENT

The initial assessment of the blunt abdominal trauma pathway was conducted after the pathway had been used in practice for a six-month period. The preliminary review included all patients admitted between November 1996 and April 1997 with the diagnosis of blunt abdominal trauma confirmed by CT scan. The majority of the cases were the result of motor vehicle accidents. The most frequent comorbidities were fractures of the head or extremities. Although all the cases were treated nonsurgically, one did require a percutaneous drainage. Only three patients received transfusions, and 70 percent were discharged within three days or less of admission. Thirty percent, or three patients, had extended LOSs. Two of the extended LOSs were due to comorbidities. Only one extended LOS was directly related to the blunt abdominal trauma. The order for the pathway was not consistently written, but all patients were cared for according to the objectives of the pathway. Our initial pathway assessment results are shown in Table 11–4.

One patient required admission to the pediatric intensive care unit for reasons unrelated to the abdominal trauma. One patient returned to the emergency department two days post discharge and was readmitted for hematuria. This resulted in a one-day LOS for observation only. Another finding was that documentation on dis-

charge planning and patient/family education was minimal. In the majority of cases, the laboratory tests were done in accordance with the pathway, with the exception of one patient who was receiving additional unrelated lab and two patients who did not have a urinalysis. Criteria for being placed on the unstable pathway versus the stable pathway were related to ability to achieve and maintain hemodynamic stability. Only three of the patients met the criteria for the unstable pathway. All three responded well and were discharged within the projected three-day LOS.

Information on cost for the first six months of the pathway was obtained from the cost accounting department. The findings reflected cost of care to be 14 percent less than projected by the pathway for stable patients and 24 percent less than projected for unstable patients.

The initial data review identified the following areas for improvement: (1) written orders for placement on pathway, (2) documentation of patient/family education, (3) development of a patient education pamphlet describing home care for the spleen and liver trauma, and (4) consistency in ordering lab and medication per pathway. The responsibility for follow-up on the areas of improvement was assigned to designated team members. For example, the physician team leader was responsible for improving the mechanism for documentation of orders. The data will be reviewed again in six months, and improvements will be made as indicated.

CONCLUSION

Development of the spleen and liver pathway was a successful attempt to improve care for children and adolescents presenting with blunt abdominal trauma. Although this was not a high-volume diagnosis, it was associated with high risk for the patient and met Arkansas

Table 11–3 Process of Blood Utilization Review

Process	Measure
Ordering	Transfuse with packed red blood cells if hematocrit < 21% (hemodynamically stable patients)
Distribution, handling, and dispensing	Product transfused is the product ordered (amount, preparation, and product)
Administering	Consent was obtained
Monitoring blood and blood components	Hemodynamically stable:
Effects on patient	0–1 year old, systolic > 70 mm Hg
	1–13 years old, systolic > 80 mm Hg
	14–18 years old, systolic > 90 mm Hg
	Hematocrit > 21 after transfusion(s)
	Hemodynamically unstable

Table 11–4 Initial Pathway Analysis

Objective	Measure
Improve/maintain quality	Cases requiring operative exploration • 1 CT guided pericutaneous drainage
10 cases admitted 11/96 through 4/97 Additional codes 864.03, 864.01, V71.8	Patient on pathway remains on pathway • All 10 remained on pathway (100%) • Documented order/reference to pathway 4/10 referred to in orders (40%) No unplanned readmissions or returns to pediatric intensive care unit • 1 patient readmitted for hematuria (10%) (1 admission at 31 days—related) (10%) • 1 patient initially admitted to pediatric intensive care unit/no transfers or returns (10%) (due to comorbidity) • No readmit or transfer to pediatric intensive care unit
Reduce length of stay (Historical length of stay: nonsurgical, 3.15)	Hemodynamically stable: 2 days (72 hours) 4 patients had a 2-day length of stay (40%) Hemodynamically unstable: 3 days (120 hours) 3 patients had a 3-day length of stay (30%) (Range was 1 to 15 days with 4, 13, and 15 days for 3 patients)
Reduce cost	Total cost at or less than projected For patients with a 2-day or less LOS (stable) was 14% less than projected For patients with a 3-day LOS (unstable) was 27% less than projected
Laboratory	Lab altered from pathway • Additional lab ordered/not related Radiology altered from pathway • 8/10 had CAT scan confirmation at Arkansas Children's Hospital, and 2 had prior to transfer (80%; 20%)

Children's Hospital's criteria for pathway development. Surgical versus nonsurgical management of blunt abdominal trauma is well documented in the literature, but standards of medical and nursing management for this diagnosis are not as well documented. The importance of adapting plans of care that fit the diagnosis cannot be overstressed. Protocols and pathways are appropriate methods for providing patterns of care while incorporating ways of monitoring outcomes. As demonstrated by our development, implementation, and evaluation of the blunt abdominal trauma pathway, using these tools in practice can help ensure that we are reviewing patient outcomes and striving to improve care. When compared to historical data, LOS had not shown a significant decrease with the pathway implementation, but the average cost per patient was less than projected. In addition, the need for improving patient/family discharge information was identified and addressed through the pathway team.

In the past, we believed that we were doing a good job, but in today's health care industry, it is important to provide facts, figures, and other data to demonstrate outcomes. Using the quality improvement process for pathway development and continually monitoring patients on the pathway is one way to successfully move into the 21st century in health care.

REFERENCES

1. Coburn MC, Pfeifer J, DeLuca FG. Nonoperative management of splenic and hepatic trauma in the multiply injured pediatric and adolescent patient. *Arch Surg.* 1995;130:332–338.

2. Pranikoff T, Hirschl RB, Schlesinger AE, Polley TZ, Coran AG. Resolution of splenic injury after nonoperative management. *J Pedi Surg.* 1994;29(10):1366–1369.

3. Witte CL, Esser MJ, Rappaport WD. Updating the management of salvageable splenic injury. *Ann Surg.* 1992;215(3):261–265.

4. Velanovich V, Tapper D. Decision analysis in children with blunt splenic trauma: the effects of observation, splenorrhaphy, or splenectomy on quality-adjusted life expectancy. *J Pedi Surg.* 1993;28(2):179–185.

5. Morse MA, Garcia VF. Selective nonoperative management of pediatric blunt splenic trauma: risk for missed associated injuries. *J Pedi Surg.* 1994;29(1):23–27.

6. Ruess L, Sivit CJ, Eichelberger MR, Taylor GA, Bond SJ. Blunt hepatic and splenic trauma in children: correlation of a ct injury severity scale with clinical outcome. *Pedi Radiol.* 1995;25:321–325.

7. Eichelberger MR. *Pediatric Trauma: Prevention, Acute Care, Rehabilitation.* St. Louis, MO: Mosby-Year Book; 1993.

8. Buntain WL. *Management of Pediatric Trauma.* Philadelphia: WB Saunders; 1995.

9. Bond SJ, Eichelberger MR, Gotschall CS, Sivit CJ, Randolph JG. Nonoperative management of blunt hepatic and splenic injury in children. *Ann Surg.* 1996;223(3):286–289.

10. Birdsall C, Sperry SP. *Clinical Paths in Medical-Surgical Practice.* St. Louis, MO: Mosby-Year Book; 1997.

■ 12 ■

Bronchiolitis: Improving Outcomes in the Pediatric Population

Maura MacPhee, Lynne Hedrick, and James K. Todd

In 1993, representatives from nursing, medicine, finance, and social work were appointed by the chief operating officer and the medical board to propose strategies for improving the efficacy of our care delivery system. The group's goal was to find ways to deliver quality care in a cost-effective manner within our management philosophy framework of continuous quality improvement (CQI). CQI strives to eliminate duplication of effort, to respond proactively to client needs, and to continuously assess the processes and outcomes of care while developing and testing ways to improve performance. The CQI approach has been able to lower costs, raise morale, achieve excellence, and beat the competition.[1]

To meet its goal, the task force decided to operationalize the CQI framework. The basic tenets of CQI include data collection to document baseline performance, an analysis of similar institutions' clinical performance, and a review of the literature. Members of the group were assigned responsibility for these concurrent activities.

Interdisciplinary focus group discussions were a source of qualitative data. It was clear to the task force that there was great variability among caregivers providing care to clients with similar diagnoses. The medical records and finance departments provided evidence that for the same diagnoses, there was considerable variability in the length of stay (LOS) and resource utilization. These cost-benefit analyses were conducted on the institution's top 10 admitting diagnoses. Cursory chart reviews did not reveal health outcome differences, despite resource utilization variability. The professional literature substantiated similar findings at other institutions.[2]

The task force recommended the use of practice guidelines and clinical care paths to optimize client care and to achieve best practice. Care paths provide a detailed description of ideal care management. Daily interventions by specific care providers are explicitly outlined. They often include definitions of the diagnosis with clear-cut criteria for admission and discharge.[3] Guidelines are narrower in scope with an outline of necessary medical orders.[4] These tools have been shown to contribute positively to the CQI approach.[5]

The task force agreed to serve as facilitators for the development of these guidelines and care paths. Seven care paths were initially created for our most common admitting diagnoses: acute osteomyelitis, bronchiolitis, diarrhea and dehydration, failure to thrive, fever and neutropenia, reactive airway disease, and rule out sepsis.

CARE PATH PROTOCOL

Respiratory illness in the pediatric population is the chief reason for hospital admissions. The top two diagnoses for admission to our institution in 1993 were reactive airway disease and bronchiolitis, respectively. We chose the care path for bronchiolitis because it is currently an active source of best-practice debate among our staff and the community health care providers. Ongoing discussions, for instance, have included the appropriate use of cardiorespiratory monitors, pulse oximetry, and nebulizer treatments (see Exhibit 12–A–1 in Appendix 12–A for bronchiolitis care path). The implementation of our failure-to-thrive care path already has been documented.[6]

The care path format was standardized for all care paths and their guidelines (see Exhibit 12–A–2 in Appendix 12–A for care path template). The first step was to identify key members for the bronchiolitis care path task force. These members included medical representatives from emergency medicine, infectious diseases, and a community physician. Nursing had representatives from the outpatient clinics, the inpatient medical unit, and home care. A pulmonary nurse clinical specialist and a pulmonologist served as consultants. Respiratory care, pharmacy, and finance were also represented in the group. The facilitator was responsible for convening the group, delegating task responsibilities, and monitoring the timeliness of task force activities and outcomes.

A task force meeting was necessary to establish consensus on the format for the care path/guidelines. After the initial meeting, most work was accomplished by individuals and/or small groups who focused on their assigned tasks. The facilitator collated and circulated completed assignments to all members according to a predetermined timeline. The facilitator solicited reviews and editions from task force members on a series of drafts. A second group meeting was held to approve the final draft. The facilitator was also responsible for ensuring consistency and format standardization for the final drafts of the care path and guidelines.

CASE MANAGEMENT

When the bronchiolitis care path was completed in 1994, our hospital was just beginning to establish a case management department. Utilization review was performed by the finance department, and there was a clear separation between the clinical and business perspectives of client care. With the advent of case management in 1995, the care path became an important tool for realizing the potential of the CQI approach.

At The Children's Hospital of Denver (TCH), nurse case managers are responsible for specific client care units. These nurses are clinical experts with on-the-job training about financial case management. The case managers are primarily responsible for monitoring the achievement of client care outcomes within the specified resource and LOS parameters of the care paths. Positive and negative variances are tracked by the case managers. Tahan and Cesta provide an excellent discussion of case management variance analysis.[7] (See Exhibit 12–1 for a copy of the TCH variance tracking record.) The nurse case managers also are responsible for identifying the need for other unit-based care paths and instigating and coordinating their development. Education is another critical component of the case managers' job, since members of the health care team, community care providers, third-party payers, and families require explanations about the purpose and function of the care paths.

The nurse case managers meet monthly with their respective unit's nursing and medical directors. These meetings are designed to review the performance of the care paths and to elicit further ideas for CQI. In addition, the nurse case managers meet weekly as a group with

the head of the finance department and a member from the quality assurance committee to update each other's awareness of pertinent clinical and managed-care issues. Variances in particular are a common source of clinical and finance-based discussion.

The routine of the nurse case managers is similar to the systematic process described by Hurt.[8] The nurse case managers obtain shift reports and ensure that applicable care paths have been appended to the new admission charts by the unit secretary. The care paths are referenced by all care providers and are utilized during daily client care rounds. These rounds include the physicians, case managers, and pharmacy. During respiratory season, respiratory therapists also do rounds with the medical staff. "Working" rounds are used to summarize client status and to plan for discharge needs. After rounds, variance issues are documented and discussed between the case managers and individual care providers.

OUTCOMES

Several changes have occurred since the initial implementation of the care paths and guidelines. The care path for bronchiolitis consists of four pages in addition to a page of order guidelines. Feedback from health care staff indicated that the care path is too unwieldy in its present state. The hospital is changing over to a computerized system that will include clinical databases, patient charges and other financial information, and a decision support system. The decision support system will consist of the care paths in a condensed, user-friendly format. Computer "linkage" of client data will allow case management to track variances in a number of different ways. The case managers will be able to correlate variances with client clinical information as well as charges for tests, procedures, medications, and so forth. This system was expected to be in place by the end of 1997.

Physician orders also will be computerized and will be formatted to correspond with the care path categories. (See Exhibit 12–2 for the draft of the guidelines for bronchiolitis.) If a physician orders tests and procedures outside the care path parameter, the computer will highlight these order variances and prompt the physician to consider the necessity for additional tests.

TCH is a teaching hospital, and for educative purposes, clinical tool boxes will be added to the computerized system so that physicians can access information screens that delineate important features of the care paths and guidelines. For bronchiolitis, a clinical tool box will contain information about respiratory distress severity scoring, evaluation of client responses to medication and chest physiotherapy (CPT), weaning of oxygen, and weaning from nebulized bronchodilators. In addition,

Exhibit 12–1 Case Management Patient Tracking Form

Variance

1. ❑ Treatment w/o clinical criteria:
   - a. ❑ Pharmaceuticals
   - b. ❑ Imaging study
   - c. ❑ Lab tests
   - d. ❑ Therapy
   - e. ❑ Redundant testing

2. ❑ Delay of Discharge due to:
   - a. ❑ Transportation
   - b. ❑ Suitable site to be discharged to
   - c. ❑ Pharmacy delay
   - d. ❑ Home health care delay
   - e. ❑ Family preparedness
   - f. ❑ MD delay in orders
   - g. ❑ DME delay
   - h. ❑ Lack of competent caregiver/therapist

3. ❑ Discharge criteria not defined within 24 hours of admit

4. ❑ Re-admit within 48 hours

5. ❑ Inappropriate Admit:
   - a. ❑ Does not meet medical necessity criteria
   - b. ❑ Could have been treated in alternate setting

6. ❑ Financial Variance:
   - a. ❑ Self-pay
   - b. ❑ Correct insurance not identified at admission—delay in notification

7. ❑ Quality of Care concern—incident report filed

8. ❑ Financial Planning/Self-pay: _____

9. ❑ Teaching/Physician Rounds: _____

10. ❑ Staff Education (Nursing, etc): _____

11. ❑ HHC Referral: _____

12. ❑ Care Conference: _____

13. ❑ Payor Contact: _____

EXEMPLARS:

_____

_____

_____

_____

Source: Copyright © The Children's Hospital.

each care path and guideline will be accompanied by professional references with abstracts that document the efficacy of our care path recommendations.

Controversy over bronchiolitis management has focused on the use of nebulized medications, such as albuterol, the use of cultures and antigen detection tests for respiratory syncytial virus (RSV), and discharge criteria for infants requiring home oxygen. During the past respiratory season, a cursory survey of bronchiolitis clients by case management revealed that we were using bronchodilators on almost all our clients. The research literature indicates that only one third of bronchiolitis infants respond favorably to nebulized medications.[9–11] We found that RSV tests were frequently ordered in cases where no change in care was likely to result. Clients on oxygen were kept in the hospital awaiting pulse oximetry tests despite clear signs of pulmonary function improvement and clinical stability. Finally, many clients were maintained on continuous pulse oximeters as well as cardiorespiratory monitors with no indication of the benefits

Exhibit 12–2 Draft of Guidelines for Bronchiolitis

DIAGNOSIS: BRONCHIOLITIS
MONITORING:
1. TPR on admit and q 4h until normal parameters; then q 8h.
2. Notify house officer (HO) if temp > _____, HR > _____, or < _____, RR > _____, or < _____.
3. Weight upon admission.
4. CR monitor if child < 2 mo old or on home monitoring.

DIAGNOSTICS:
1. Oximetry: Baseline oximetry in room air. Repeat q 8h × 24h; then q 24h.
2. If wheezing: Trial bronchodilator therapy (per Respiratory Care Order Protocol):
                     Albuterol inhalation solution 2.5mg in 2 ml NS via NEB × 2 treatments.
                         —Positive response (decrease wheezing, increased pO_2); continue therapy q 4h and evaluate in 24 hr.
                         —No positive response: D/C nebulized therapy; contact HO
3. Cultures: If indicated; child < 3 months of age: RSV? chlamydia ? pertussis?
4. Chest X-ray: If indicated × 1.
5. Urine specific gravity: If clinically dehydrated.

FLUIDS, ELECTROLYTES, NUTRITION:
1. IV fluids only if not taking adequate fluids PO or dehydrated:
   Bolus = _____ ml of _____ solution over _____ min.
   Maintenance = _____ ml/hr of_____ solution.
2. Oral fluids: _____ .
3. Regular diet for age.

MEDICATIONS:
1. Oxygen per protocol (Respiratory Care Orders).
2. Acetaminophen 10 to 15 mg/Kg q 4–6h PO or pr prn temp > 38.5
3. Ibuprofen 10 mg/Kg q 6–8h prn.
4. Albuterol inhalation solution:
   < than 1 yr. of age: 1.25mg with 2 ml NS q 4h (0.1–0.15 mg/kg/dose q 4–6h prn).
   > than 1 yr. of age: 2.5 mg with 2 ml NS q 4h (0.1–0.15 mg/kg/dose q 4–6 h prn).
5. Other considerations: Terbutaline, Intal and Atrovent.

TREATMENTS:

CONSULTS:

OTHER:

DISCHARGE CRITERIA:
1. Improved work of breathing and respiratory rate (Refer to Respiratory Distress Score under development).
2. Improved oxygenation or referral for Home Care Therapy.
3. Adequate oral intake.
4. Parent education completed.
5. Follow-up with P.C.P arranged.

Source: Copyright © The Children's Hospital.

of double monitoring. Case management recorded these as variances, although due to the newness of the program, a systematic tally of these variances was not achieved. Our new computer system will record objective measurement elements on bronchiolitis clients without complicating conditions.

For the upcoming respiratory system, the measurable outcomes will be (1) percentage with RSV/viral tests, (2) percentage fulfilling admission criteria, (3) percentage treated with bronchodilators or CPT without documentation of benefit, (4) average LOS, and (5) percentage receiv-

ing necessary monitoring via cardiorespiratory monitors and/or pulse oximetry monitors. Another measurable outcome may include an evaluation of a teaching program that we have instituted to encourage smoking cessation. We will record the percentage of infants from homes and/or day care with smokers and compare this percentage with the percentage of families who receive smoking abatement instruction from the health care staff.

Although many care path variances are conceivable, case management will initially track only five or six outcome criteria on a day-to-day basis. These outcomes cover the most important aspects of care path management, such as adherence to admission and discharge criteria and therapies. Charges will also be tracked daily by case management, particularly with regard to additional expenditures from unnecessary tests and procedures. Outcomes variances and financial variances will be discussed with the health care staff during rounds to correct problems as they occur. The case management variance record will be computerized, and monthly summary reports from each unit will be printed out and shared with health care staff. There is no level of acceptability for established outcomes. It is hoped that by addressing variances on a daily basis, the case managers will eliminate negative variance trends and potentially improve standards of performance.

Our hospital has instituted many changes simultaneously. In addition to the relatively new case management model, the conversion to a computerized system has required considerable staff adjustment. The medical director has established daily rounds guidelines to ensure regularly scheduled times for the case managers to discuss variance issues with the interns, residents, and attendings. For the nursing staff, a new section is being added to the nursing Kardexes to highlight day-to-day variance issues as well as discharge planning needs. Since computer monitoring of variances will be relatively new to the staff, the case managers will initially track and record variances of concern on the nursing Kardexes, but eventually, through regular inservicing, the nurses and other health care professionals, such as respiratory therapists, will actively participate in identifying care variances and implementing relevant plans of care.

When the case managers give client reviews and updates to third-party payers, established care paths are frequently faxed to support decisions on admission, therapies, discharge criteria, and any home care needs or concerns. It is becoming more commonplace for third-party payers to ask for care paths or other evidence for the hospital's plan of care. It has also been helpful for case managers to acknowledge any/all variances with an explanation of the cause for the variances. When complicating factors arise, communications between case management and third-party payers is improved by tracking progress on the care paths.

In the past, care paths have been the domain of the health care professionals. A health care professional, usually the intern, is responsible for discussing client status on a daily basis with the client/family and with discussing disposition to home as needed. Client satisfaction surveys have shown that some clients and their families are dissatisfied with how information is shared. TCH uses Press-Ganey surveys, which are distributed, collected, and collated through our volunteer office. Results of the surveys are shared with staff on a monthly basis by the medical and nursing directors. The surveys have indicated a minority of individuals who feel either overwhelmed by too much information or underinformed of their plans of care. Communication with clients, families, and community care providers is a primary concern for the hospital's quality improvement committee, and the committee is presently debating the best way to use the care paths as a primary information source and teaching tool. The medical unit is in the process of completing a communications survey with community health care providers. The surveys, therefore, will provide us with a better idea of where our strengths and weaknesses are so that we can determine the best way to systematize our communications. The care paths will no doubt play an important role in improving communications between and among families, staff, and community health care providers. Since we are moving away from our original paper versions of the care path, we will have to create a "hard copy" care path for sharing with clients, families, and our third-party payers.

CONCLUSION

Care of our patients with bronchiolitis has become well planned, coordinated, and streamlined. The variations in practice seem to have diminished, and patient care appears to have become efficient while continuing to be as or more effective.

Our subjective experience has been a positive one with regard to the case management model, our care paths, and the new computer system. These are critical components of the CQI approach established by our hospital. There is commitment from top management, and there has been extensive education and training throughout the hospital ranks to ensure consistent standards of care.

Objective evaluation will be the responsibility of the unit-based case managers working with the information systems staff and the quality improvement committee. We have begun the evaluation process with a few key outcomes that we can measure daily. A finite number of critical outcomes based on the care paths will allow us to track and to report the impact of our care to each other, to the community, and to our clients and their families.

REFERENCES

1. Walton M. *The Deming Management Model.* New York: Putnam Publishing, New York, 1986.
2. Zander K. Nursing case management: strategic management of cost and quality outcomes. *J Nurs Adm.* 1988;18:23–30.

3. Marr J, Reid B. Implementing managed care and case management: the neuroscience experiment. *J Neurosc Nurs.* 1992;24:281–285.

4. Puma J, Schiedermayer D. *Managed Care.* New York: McGraw-Hill; 1996.

5. Dawson P. On the scene: managed care at Johns Hopkins hospital. *Nurs Admin Q.* 1992;17:54–79.

6. Macphee M, Hoffenberg E. Nursing case management for children with failure to thrive. *J Pediat Health Care.* 1996;10:63–74.

7. Tahan H, Cesta T. Evaluating the effectiveness of case management plans. *J Nurs Adm.* 1995;25:58–63.

8. Hurt L. Care management: providing a connecting link. *Nurs Manage.* 1995;26:27–33.

9. Chowdury D, al-Howasi M, Khalil M, al-Frayh AS, Chowdury S, Ramia S. The role of bronchiodilators in the management of bronchiolitis: a clinical trial. *Ann Trop Paediatr.* 1995;15(1):77–84.

10. Gadomski AM, Lichenstein R, Horton L, King J, Keane V, Permutt T. Efficacy of albuterol in the management of bronchiolitis. *Pediatrics.* 1994 June;93(6 Pt 1):907–912.

11. Reijonen T, Korppi M, Pitkakangas S, Tenhola S, Remes K. The clinical efficacy of nebulized racemic epinephrine and albuterol in acute bronchiolitis. *Arch Pediatr Adolesc Med.* 1995 June;149(6):686–692.

■ Appendix 12–A ■

Clinical Pathway and Template for Bronchiolitis

Exhibit 12–A–1 Clinical Care Path for Bronchiolitis

DEFINITION OF DX:
Previously normal child < 2 years old with lower respiratory tract infection, typified by tachypnea, T < 39 C, with or without wheezing, retractions, grunting, or chest crackles. Viral cult. not necessary for cohorting purposes.

DISCHARGE CRITERIA:
Improved respiratory status as evidenced by improving work of breathing, respiratory rate and oxygenation (if measured). Adequate PO. Parent education complete re: disease and follow-up. Adequate follow-up arranged. Visiting nurse referral, if indicated.

PRIMARY CARE PHYSICIAN: _____

PRIMARY ATTENDING: _____

PRIMARY RESIDENT: _____

PRIMARY NURSE: _____

	DAY 1 (1st 24 hr.)	DAY 2 (2nd 24 hr.)
Monitoring	• TPR and BP on admit to the ER and unit • TPR q 4 hr. • Weight on admit • Notify HO if TPR outside normal parameters for age and/or ordered parameters • C/R monitor × 24 hr. when a baby presents with a hx of apnea, severe respiratory distress or cyanosis. • For children < 2 months or home monitoring, continue to monitor. • Strict I & O × 24 hr. • Urine specific gravity bid if clinically presenting with dehydration and not taking adequate fluids PO feeds. Continue specific gravity if PO intake < maintenance.	• PR q 4 hr. • T q 8 hr. • Discontinue strict I & O if taking maintenance PO. • Discontinue C/R monitor if no apnea and taking adequate PO. • For children < 2 months or home monitoring, continue to monitor.
Diagnostics	• RA pulse ox. on admit to the ER—graph SaO_2 to monitor trends • O_2 saturation q 8 hr. × 24 hr. • If wheezing or respiratory distress, trial bronchodilator treatment with 0.5 ml Albuterol in 2 ml normal saline via nebulizer. Monitor breath sounds, P, R, respiratory effort and SaO_2 (if available) at beginning of tx and 5–15 minutes after tx, noting effect. If improvement, continue nebulizer Tx q 4 hr, using appropriate dose (see Medications). • Follow Albuterol regime under medications	• O_2 saturation qd while at rest.

continues

138

Exhibit 12–A–1 continued

	DAY 1 (1st 24 hr.)	DAY 2 (2nd 24 hr.)
	Discontinue if no continued benefit × 2 Txs • Suction for severe upper airway congestion • If indicated by patient history, RSV, chlamydia and/or Pertussis cults. for infants < 3 mos of age. No routine cult. necessary for infants > 3 mos of age, depending on symptoms Chest X-ray indicated for infants presenting with a asymmetrical chest exam or history suggestive for a focal pneumonia or foreign body.	
FEN	Diet appropriate for age. • If patient presenting with apnea and/or not taking adequate PO feedings or clinically dehydrated, begin IV with D_5 0.2 NS 20 meq/Kcl/L IV + PO = maintenance plus rehydration.	• Encourage PO feedings as tolerated • Discontinue IV when PO feedings equal maintenance, dehydration resolved and no apnea × 24 hrs.
Medications	• O_2—Administer when resting SaO_2 < 90% • Acetaminophen for fever and irritability 15 mg/kg PO or p.r. q 4–q 6 hr. PRN. • Ibuprofen 10 mg/kg q 6 hr. PRN for fever not resolved by Acetaminophen (do not give acetaminophen and Ibuprofen together) • Albuterol nebulizer q 4–6 hr. x 24 hr if clinically improved after initial challenge < 1 yr: 0.25 ml Albuterol with 2 ml NS > 1 yr: 0.5 ml Albuterol with 2 ml NS	• Discontinue O_2 if resting RA SaO_2 90% or greater and if clinically improving and taking all fluids PO. • Decrease Albuterol to q 8 hr. if clinically beneficial to infant, or discontinue Albuterol if no clinical improvement with 2 consecutive treatments.
Activity	Activity as tolerated appropriate for age • Encourage parent involvement • Isolation: As per protocol	• Continue as in Day 1.
Treatments	• None	• None
Nursing Assessment	• Physical Assessment q 4 hr. with TPR • Respiratory status and activity level • Nutrition/hydration • Elimination • Learning needs of family • Coping of parents & child with hospitalization	• Continue as in Day 1.

Education/ Prevention	Begin teaching:
	  handwashing
	  unit & facility orientation
	  team members
	  cardiac monitor
	  IV
	  O_2
	  infant care
	  illness recognition—assessment of respiratory status
	  nutrition
	  community resources
	  adjustment to ill child (wheezing without respiratory distress may be acceptable)
	  nebulizer tx if responds to challenge and requires q 4 hr. nebs
	  No smoking in home or car (provide smoking cessation information)
	  Parents to call primary physician for follow-up questions
	  Medication education
		If clinically improving and Albuterol treatment cannot be decreased to q 8 hr., begin R.T. home teaching for nebulizer tx
		• Order home equipment
		• Begin home O_2 teaching if patient clinically improving and RA SaO_2 < 90%
		• Continue respiratory assessment teaching, including parameters for which child should be seen by health care provider

continues

Exhibit 12–A–1 continued

	DAY 1 (1st 24 hr.)	DAY 2 (2nd 24 hr.)
Psychsocial	• Family assessment • Provide emotional and resource support • Identify family needs	• Arrange for visiting Home Care referral for parents who have difficulty coping and/or inability to understand/assess child's care needs.
Communication	• Daily communication with parents and PCP • Communicate with referring MD	• Continue as in Day 1 • Communicate with follow-up physician
Discharge Planning	• Consult with Discharge Planning Coordinator, as needed • Identify primary pediatrician • Identify community resource • Consult with PCP regarding Home Care referral plan • Immunization Status	• Arrange for visiting Home Care referral (see above "Psychosocial") • Order appropriate home equipment • Schedule physician/clinic appointment within 24–48 hr. of discharge if discharged on O_2 and/or nebulizer • Immunization *not* recommended • HO to dictate D/C summary

Abbreviations:

< = less than	hx = history	mg = milligram
> = more than	I&O = intake and output	kg = killigram
T = temperature	PO = oral	RSV = respiratory syncytial virus
C = centigrade	RA = room air	nebs = nebulizer(s)
TPR = temperature, pulse, respirations	SaO$_2$ = oxygen saturation	PCP = primary care provider
P = pulse	O$_2$ = oxygen	MD = physician
R = respirations	ml = milliliter	RT = respiratory therapist
BP = blood pressure	qd = every day	d/c = discharge
ER = emergency room	pulse ox = pulse oximeter	prn = as needed
q = every	tx = treatment	pr = per rectum
hr = hour	cults. = cultures	yr = year
HO = house officer	IV = intravenous	NS = normal saline
C/R = cardio-respiratory	D$_5$0.2NS20meqKCl/L = Dextrose 5% solution with ¼ normal saline and 20 milliequivalents of potassium chloride in one liter solution	

Source: Copyright © The Children's Hospital.

Exhibit 12–A–2 Clinical Care Path Template

CLINICAL CARE PATH FOR:

DEFINITION OF DX:

DISCHARGE CRITERIA:

PRIMARY CARE PHYSICIAN: _____

PRIMARY RESIDENT: _____

PRIMARY NURSE: _____

	DAY 1 (1st 24 hr)	DAY 2 (2nd 24 hr)	DAY 3 (3rd 24 hr)	DAY 4 (4th 24 hr)
Monitoring				
Diagnostics				
Fluid/Electrolytes/Nutrition (FEN)				
Medications				
Activity				
Treatments				
Consults				
Nursing Assessment				
Education/Prevention				
Psychosocial				
Communication				
Discharge Planning				

Source: Copyright © The Children's Hospital.

◾ 13 ◾

Positive Outcomes from a
Pediatric Appendicitis Pathway

Patti Higginbotham and Anita Gottlieb

One of the most frequently performed surgeries in the United States is appendectomy for acute appendicitis. Although much has been written on appendicitis since it was first described over a century ago, the etiology and epidemiology remain poorly understood.[1,2] Even with today's technological advancements, appendicitis continues to be difficult to diagnose. The importance of immediate and appropriate diagnosis cannot be overstressed. Further, the misdiagnosis of pediatric appendicitis is associated with increased rates of perforation, abscess formation, wound infection, and death.[3,4] According to a study by Addiss et al., the following epidemiological patterns have been noted: acute appendicitis occurs most frequently in the 10- to 19-year-old age group, is more common during the summer, and has a higher incidence in males than in females, in whites than in nonwhites, and in certain regions, particularly the upper Midwest.[2] Much of the research on appendicitis has focused on the importance of diagnosis and timely treatment, and morbidity and mortality have been shown to relate to the perforation of acute appendicitis. Diagnostic accuracy has been reported to be the most difficult to attain in women of childbearing age and children.[5] The literature supports the performance of surgical intervention even at the expense of increasing the overall appendectomy rate. Many variables, such as the patient's age, childbearing status, pregnancy, immunocompromised state, and other patient characteristics may influence the overall patient outcome.[6] This chapter presents an overview of the approach used by Arkansas Children's Hospital and the results achieved.

STRATEGIC PATHWAY DEVELOPMENT

In today's managed-care environment, cost reduction of common medical and surgical procedures has become a challenge for health care providers. The question is, "How do we do things better and more economically while continuing to provide high-quality services to patients and families?" The standardization of care and development of clinical pathways has been one method that many health care institutions have used to address the issue. As health care providers are being asked to provide data related to high-volume and high-cost diagnoses and procedures, the need to evaluate outcomes in a more scientific manner has become an expectation of managed-care companies and other payers. At Arkansas Children's Hospital, clinical pathways have been developed to reduce cost, to shorten length of stay, and to improve or maintain overall quality for specific diagnoses. Since appendicitis is a high-volume diagnosis for the surgery service and since there was physician support for an appendicitis pathway, it was chosen as the first pathway to be developed and implemented in our institution.

PATHWAY OR PROTOCOL DEVELOPMENT

Clinical pathways are among the approaches incorporated into the hospital's strategic plan for improving quality. Development of clinical pathways is also consistent with the goals of quality improvement at Arkansas Children's Hospital—to provide quality care and services in a cost-effective, efficient manner. The FOCUS-PDCA method of problem solving is used to present the development, implementation, and evaluation of the appendicitis pathway:

F Find an opportunity for improvement
O Organize a team
C Clarify the current process
U Understand variation in the process
S Select a solution
P Plan the improvement
D Do the improvement
C Check/evaluate
A Act to hold the gains[7]

A flowchart (Figure 13–1) is used to describe the process for pathway development at Arkansas Children's Hospital.

Find an Opportunity To Improve

The surgery service in a 265-bed pediatric university-affiliated hospital identified the opportunity to reduce length of stay and cost while improving or maintaining quality of care for patients with appendicitis. For the calendar year 1994, appendicitis was one of the five highest volume diagnoses for the surgery service.

Organize a Team That Knows the Process

The team was organized with representatives from nursing service, utilization management, pharmacy,

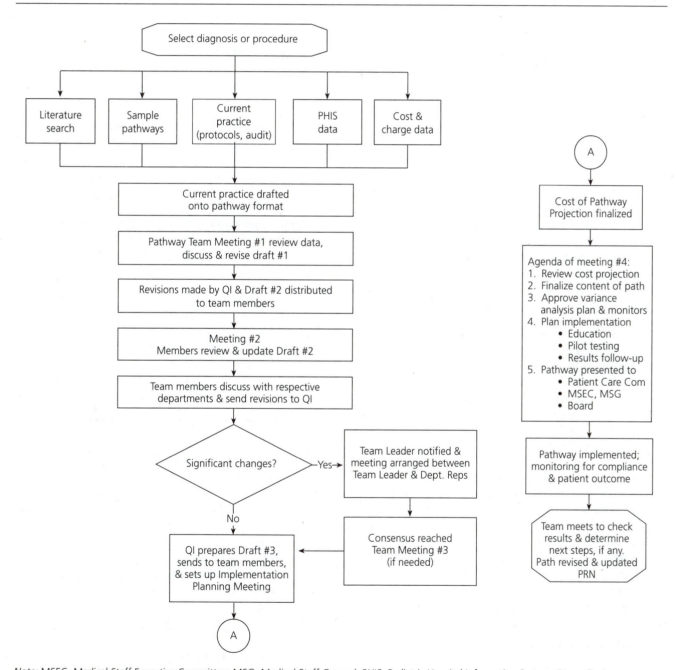

Note: MSEC, Medical Staff Executive Committee; MSG, Medical Staff General; PHIS, Pediatric Hospital Information System; QI, quality improvement.

Figure 13–1 Process for Pathway Development

quality improvement, financial service, nursing education, and surgery service. The team leader was the chief of the surgery service, and the facilitator was the director of the quality improvement department. As the pathway process continued, other surgeons and surgery residents joined the team.

The pathway process, member responsibilities, and team goals were discussed at the first team meeting. Key responsibilities for team members included developing the pathway, serving as a two-way conduit for staff members who would be using the pathway but were not on the team, developing controls and quality improvement plans, designing and monitoring implementation, education, and evaluating the effectiveness of the pathway at intervals post implementation.

Clarify the Process

Historically, data on appendicitis were available to the pathway team from a variety of internal and external sources. Practice information regarding charge and length of stay was compared with other pediatric institutions using the Pediatric Hospital Information System (PHIS) database. This database is composed of medical record abstracts and financial data submitted by 20 participating children's hospitals. Initial review of data for appendicitis identified that the best method for evaluating performance and for development of an appendicitis pathway would be to use ICD-9 codes rather than Health Care Financing Administration or APR diagnosis-related groups. The PHIS data identified Arkansas Children's Hospital as a top performer for length of stay for acute appendicitis with peritonitis (ICD-9: 540.0, 540.1) and below the mean length of stay for acute appendicitis (ICD-9: 540.9). (See Figure 13–2.)

The surgery service began to make changes in practice in 1994, using a research protocol, but believed that additional improvements could be achieved by expanding the protocol into a multidisciplinary pathway. In clarifying the current process, the team determined that the research protocol for perforated appendicitis could reasonably be expanded into a clinical pathway. The team also agreed that the pathway should include appendicitis without peritonitis. Review of financial data demonstrated that early changes were being seen with length of stay and costs, and these changes were attributed to the research protocol.

Understand Variation

To identify "best practice," examples of appendicitis pathways were obtained from other pediatric hospitals, and a literature search was completed. The example pathways were consistent with the current management of patients at Arkansas Children's Hospital. Comparative data on outcomes, as well as on the management for care, showed that Arkansas Children's Hospital outcomes were comparable. On the basis of review of these findings and the PHIS results, Arkansas Children's Hospital's management of appendicitis was deemed "best practice." Therefore, it seemed reasonable to draft the clinical research protocol into the pathway format as a first step in pathway development.

Select a Solution

The process for development of clinical pathways has been defined on the basis of experience with the appendicitis pathway, as well as two other pathways that were being developed at the same time. The use of a flowchart (Figure 13–1) assisted teams to understand what would be expected to occur during pathway development. The "appy team" required four team meetings over a period of six months. Other meetings between a few team members and hospital staff and/or medical staff were needed to gain additional input and clarification. The goal was to have a clinical pathway that reflected best practice and to gain buy-in from those who would be required to use the pathway most frequently.

An appendicitis pathway was developed for inpatients with acute appendicitis with generalized or localized peritonitis (ICD-9: 540.0, 540.1) and acute appendicitis without peritonitis (ICD-9: 540.9). The pathway for appendicitis with peritonitis was for five days; the pathway for patients without peritonitis was for two days. The complete pathway is presented in Exhibit 13–A–1 in Appendix 13–A.

Plan

Several activities were included in planning for implementation of the pathway. These included quality improvement plans, pilot testing, and education. These activities took place during the last team meeting. To facilitate planning, team members brought a draft plan to the meeting: for example, the quality improvement team member developed plans based on pathways, the nursing manager and educator had plans for education and implementation on the pilot unit, and so on.

Quality improvement plans included compliance with key indicators from the pathway, projected costs and lengths of stay, and met regulatory requirements for required review functions (i.e., Joint Commission on Accreditation of Healthcare Organizations and health department). Measures for quality, length of stay, and costs were minimum components of a quality improvement plan for pathways, as shown in Table 13–1.

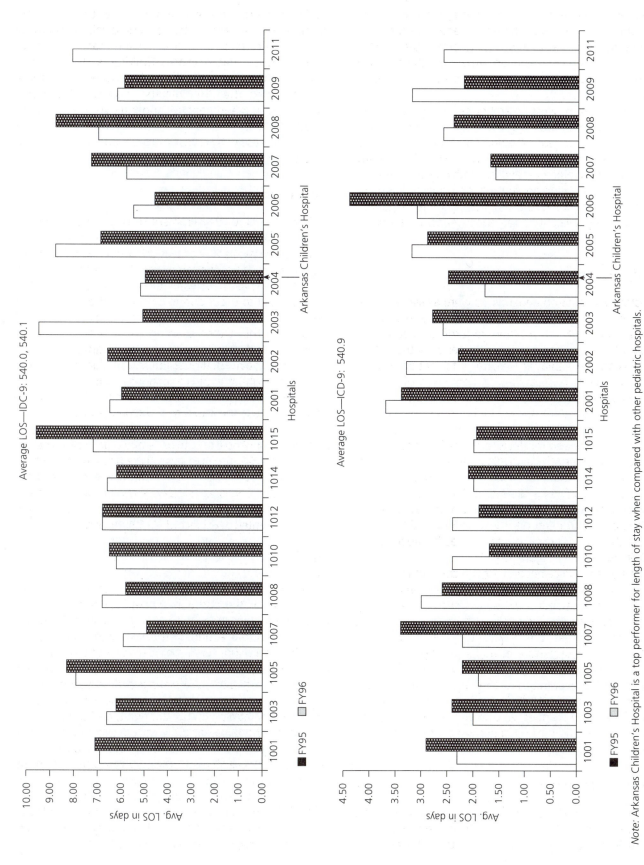

Note: Arkansas Children's Hospital is a top performer for length of stay when compared with other pediatric hospitals.

Figure 13–2 Length-of-Stay (LOS) Comparisons with Other Pediatric Hospitals

Table 13–1 Quality Improvement Plan for Appendicitis Pathway

Objective	Measure
IMPROVE/MAINTAIN QUALITY 3 rates: Overall rates, 540.9, and 540.0 + 540.1 Denominator: Total appendectomies performed	Rates of postoperative complications for: abscess wound infection obstruction Readmission within 14 days to treat postop complications (rate)
REDUCE LENGTH OF STAY Overall LOS, LOS for 540.9, and LOS for 540.0 and 540.1	Patients will be discharged by: postop day 4 for 540.0 and 540.1 postop day 2 for 540.9 Average length of stay compares favorably with other children's hospitals (PHIS data annual comparison)
REDUCE COST Per diem and total costs Laboratory costs Radiology costs	Decrease total cost for appendicitis Reduce laboratory and radiology costs during preop (Comparison of average costs/charges for ICD-9 codes 540.0 and 540.1 and for 540.9 over time; prepath years of 1992–1994)

Note: PHIS, Pediatric Hospital Information System.

Some of the indicators of interest to the team were translated into review of operative and other procedures. Data collection and findings summary were assigned to the quality improvement office. These areas of interest are reported in Table 13–2.

After discussion, the team agreed that the pathway would be implemented on the surgical unit in October

Table 13–2 Review of Surgical and Other Procedures

Surgical and Other Procedures	Expected Outcome
Selecting Appropriate Procedures	
Pre- and postoperative diagnoses consistent	100%
Pathology confirm appendicitis *or* indications are met	100%
Preparing the Patient for the Procedure	
Preoperative lab per pathway	100%
Preoperative radiology per pathway	100%
Premedication per pathway	100%
Performing the Procedure and Monitoring the Patient	
Intraoperative cultures	0%
Intraoperative complications	0%
Providing Postprocedure Care	
Patient afebrile, and WBC normal by Postop day 3	100%
Length of stay per pathway	100%
Readmission within 14 days	0%
Postoperative complications	0%

Note: WBC, white blood cell count.

1995 for a three-month pilot test. All patients with the diagnosis of appendicitis intraoperatively would be placed on the pathway, and a written physician's order would not be required for the patient to be placed on the path. The operative notes were used to identify the presence or absence of peritonitis as observed intraoperatively. This approach was also discussed with medical records coding staff to ensure that records would be appropriately coded after discharge.

Two key groups needed to be well informed about the pathway: surgeons and nurses. Education of surgeons was simplified by the fact that surgeons and surgery residents had participated in the development of the pathway as team members or by review and approval of the pathway. Arkansas Children's Hospital is a teaching facility with most of the surgeons on staff serving as faculty for University of Arkansas Medical Sciences. Nursing education was conducted by the nurse manager and nurse educator.

Do

The appendicitis pathway was implemented October 1995 as a pilot test on the nursing unit to which most appendicitis patients were admitted. The pathway was not designed to be used for documentation of care; rather, it was a document to guide patient care. Its value as a guide became apparent almost immediately as nurses began discussing the course of patient care with surgeons when variances from the pathway occurred.

Check

The pathway was piloted for three months and reviewed for process and outcome. Findings at the end of three months identified some variances from the pathway. The more significant results were after one year of implementation. These results will be discussed later in this chapter. After the initial pilot, team members reviewed the findings from 30 patients: 12 with acute appendicitis and 18 having appendicitis with peritonitis. Of these 30, 23 (77 percent) had remained on the pathway throughout their hospitalization. A few variations were discovered, and action plans were developed, as follows:

- More lab work and X-rays were ordered. Most of the additional lab tests were for liver function tests or urine pregnancy tests. The chief of surgery (team leader) would discuss these findings with the director of the emergency department.
- Gentamicin levels were ordered on all 18 patients who received gentamicin. A gentamicin level was indicated for only one of the patients. The team reviewed initial doses and results of levels. About 60

percent had doses ordered per the pathway (2.5 mg/kg every 8 hours). Doses were adjusted based on gentamicin levels for six of seven patients. The surgeons and surgery residents agreed that the pathway doses and use of gentamicin levels continued to be appropriate management. The pharmacist agreed to monitor this within the Therapeutic Drug Monitoring Program.

- Intra-abdominal cultures were done on five patients. All were performed by one surgeon. The chief of surgery would discuss this finding with the surgeon.
- Documentation of discharge education was incomplete and inconsistent. The compliance score was 47 percent. The nurse manager would be responsible for reviewing the discharge instruction sheets and educating nursing staff on the components for discharge instructions within the pathway.

Outcomes

After the initial review, the team recommended continuation of the pathway and monitoring of patient outcomes. Subsequent reviews were conducted annually, and results from October 1995 through March 1997 were compiled. Results demonstrated that the percentage of patients remaining on the pathway through hospitalization had decreased. Patients with acute appendicitis without peritonitis were more likely to remain on the pathway throughout their hospitalizations, as demonstrated by 94 percent for appendicitis without peritonitis compared with 67.5 percent for appendicitis with peritonitis.

The projected length of stay and cost for appendicitis without peritonitis were met for pathway patients (Table 13–3). Average length of stay for patients on the pathway was below projections after implementation of the pathway. The projected cost for patients on the pathway was 35 percent lower than historic costs. Average cost for all patients with acute appendicitis (ICD-9) met the projection. Patients on the pathway had costs 40 percent less than historic, thus exceeding expectations.

Reductions in length of stay for appendicitis with peritonitis began in 1994 when the research protocol was implemented. A four-day length of stay for this group of appendicitis patients represented a stretch target. Projected costs were 48 percent of historic costs. While improvements were achieved, projections were not met. On the basis of the overall results, the team examined more details of the data. As Figure 13–3 demonstrates, at least two different patient populations were included. The most homogeneous group was patients with length of stay of three to five days. These were the pathway patients. The second group of patients had length of stay of eight to twelve days. Review of these patients revealed that these patients had complications: namely, abscess, ileus, or other complicating illnesses, such as cystic fibrosis. The team decided to develop a third branch of the pathway for eight days for patients who develop postoperative complications.

The team questioned the effectiveness of the pathway for appendicitis with peritonitis when compared with that of other pediatric hospitals. The data reported at team meetings were data that had been validated and corrected (i.e., for coding errors). The data in the PHIS database would not have been adjusted in any way. When data were reviewed, Arkansas Children's Hospital was identified as the benchmark performer for length of stay in fiscal year 1995. For fiscal year 1996, Arkansas Children's Hospital had the third shortest length of stay (Table 13–4). When the data for both years were combined, Arkansas Children's Hospital had a lower average length of stay and standard deviation, which was indicative of the effectiveness of the pathway to reduce variation.

An important element for evaluating the effectiveness of the pathway was to ensure that infectious complications were at or below historic levels. As shown in Table 13–5, there was no statistically significant difference in infectious complications for patients with appendicitis

Table 13–3 Length-of-Stay and Cost Results Compared to Historic and Projected

	Length of Stay	Cost as % of Historic
Historic	7 days	100%
Projected	4 days	48%
All patients after pathway implemented	5.68 days	64%
Pathway patients	5.37 days	60%

Table 13–4 Length-of-Stay Comparisons

	Acute Appendicitis	Appendicitis with Peritonitis
Average Length of Stay:		
Historic	2.6 days	7 days
Projected	2.0 days	4 days
PHIS Comparative Data for		
Fiscal Years 1995 and	2.4 ± 0.5	6.71 ± 1.0
1996 $p < .05$	(all PHIS hospitals)	(all PHIS hospitals)
	2.04 ± 0.2 (ACH)	5.19 ± 0.83 (ACH)

Note: ACH, Arkansas Children's Hospital; PHIS, Pediatric Hospital Information System.

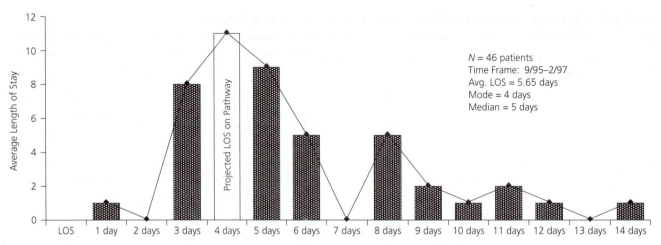

The length of stay for most patients was as expected. A second group of patients could be identified with this tool. Those were patients who developed postoperative complications necessitating additional hospital days.

Figure 13–3 Histogram of Length of Stay (LOS) for Patients with Appendicitis with Peritonitis (IDC-9: 540.0 + 540.1).

with peritonitis. Other improvements attributed to the pathway include the following:

- Gentamicin levels were ordered more appropriately. During the pilot test, 18 gentamicin levels were done, and of these, 1 level had been ordered by the indicators defined in the pathway. In the last review, 8 gentamicin levels had been done, and of these, 6 levels had been ordered in compliance with the pathway.
- Documentation of discharge instructions improved from 47 percent compliance to 100 percent.

Results of clinical pathways can be used to meet another regulatory requirement: physician-specific data for reappointment to the medical staff. Performance data for appendicitis with peritonitis were evaluated by use of box plots to show variation among surgeons and to compare performance with departmental results. Data analysis revealed that differences in performance by surgeon were not significant (Exhibit 13–1).

Table 13–5 Complications Following Appendicitis with Peritonitis (ICD-9: 540.0 + 540.1)

p = ns	No. of Patients	Wound Infection	Intra-abdominal Abscess	Total
1985–1991				
Prepathway	68	8 (11.8%)	9 (13.2%)	17 (25%)
Postpathway	42	1 (2.4%)	7 (17.1%)	8 (20%)

Act (To Hold the Gains)

To "hold the gains," the pathway team meets to formally review the annual report and determine if changes need to be made to the pathway to maintain or improve the gains. Findings are shared through various medical staff committees, including the surgical affairs committee and the patient care committee. The pathway team has recommended extending the pathway to eight days to include care of patients with perforation requiring an extended length of stay.

CONCLUSION

A reduction in hospital length of stay and cost in patients placed on a clinical pathway for appendicitis has been achieved since implementation of an appendicitis pathway. The most significant changes in cost and length of stay resulted from the implementation of the protocol before pathway development. Implementation of the pathway supported the initial improvements. Evidence from evaluation of appendicitis patient data since implementation confirms the effectiveness of the pathway. The support of the surgery staff and having a physician team leader to introduce the pathway positively affected acceptance of practice changes. Nursing staff found the pathways helpful in planning care and discussing variances with physicians. Discussions at present include consideration of a patient version of the pathway and use of pathways for order entry and documentation. Positive results from this first pathway set the tone for the future of pathways at Arkansas Children's Hospital.

Exhibit 13–1 Box Plot of Length of Stay (LOS) for Appendicitis with Peritonitis by Surgeon

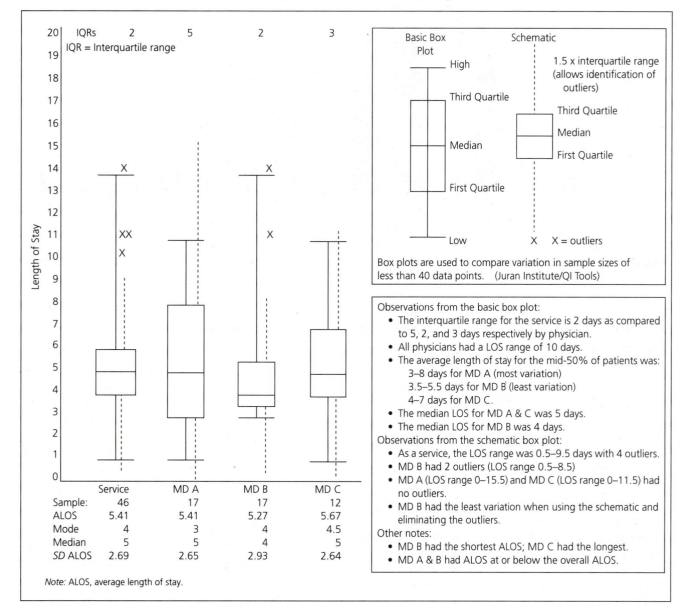

Observations from the basic box plot:
- The interquartile range for the service is 2 days as compared to 5, 2, and 3 days respectively by physician.
- All physicians had a LOS range of 10 days.
- The average length of stay for the mid-50% of patients was:
  3–8 days for MD A (most variation)
  3.5–5.5 days for MD B (least variation)
  4–7 days for MD C.
- The median LOS for MD A & C was 5 days.
- The median LOS for MD B was 4 days.

Observations from the schematic box plot:
- As a service, the LOS range was 0.5–9.5 days with 4 outliers.
- MD B had 2 outliers (LOS range 0.5–8.5)
- MD A (LOS range 0–15.5) and MD C (LOS range 0–11.5) had no outliers.
- MD B had the least variation when using the schematic and eliminating the outliers.

Other notes:
- MD B had the shortest ALOS; MD C had the longest.
- MD A & B had ALOS at or below the overall ALOS.

Box plots are used to compare variation in sample sizes of less than 40 data points.   (Juran Institute/QI Tools)

	Service	MD A	MD B	MD C
Sample:	46	17	17	12
ALOS	5.41	5.41	5.27	5.67
Mode	4	3	4	4.5
Median	5	5	4	5
SD ALOS	2.69	2.65	2.93	2.64

Note: ALOS, average length of stay.

REFERENCES

1. Dolgin SE, Beck AR, Tartter PI. The risk of perforation when children with possible appendicitis are observed in the hospital. *Surg, Gynecol Obstet*. 1992;175:320–324.

2. Addiss DG, Shaffer N, Fowler BS, Tauxe RV. The epidemiology of appendicitis and appendectomy in the United States. *Am J Epidemiol*. 1990;132:910–925.

3. Rothrock SG, Skeoch G, Rush JJ, Johnson NE. Clinical features of misdiagnosed appendicitis in children. *Ann Emerg Med*. 1991;20:45–50.

4. Maxwell JM, Ragland JJ. Appendicitis: improvements in diagnosis and treatment. *American Surg*. 1991;57:282–285.

5. Bond GR, Tully SB, Chan LS, Bradley RL. Use of the MANTRELS score in childhood appendicitis: a prospective study of 187 children with abdominal pain. *Ann Emerg Med*. 1990;19:1014–1018.

6. Velanovich V, Satava R. Balancing the normal appendectomy rate with the perforated appendicitis rate: implications for quality assurance. *American Surg*. 1992;58:264–269.

7. Joint Commission on Accreditation of Healthcare Organizations. *Monitoring, Assessment, and Improvement: An Introduction to Quality Improvement in Health Care*. Chicago, IL: Joint Commission; 1991: 56–57.

■ Appendix 13–A ■

Clinical Pathway for Pediatric Appendicitis

Exhibit 13–A–1 Clinical Pathway for Pediatric Appendicitis

Arkansas Children's Hospital
Clinical Pathway
APPENDICITIS PATHWAY

Patient Population: Emergency Dept/Outpatient: Abdominal pain, possible appendicitis
Inpatient: (A) Acute appendicitis with generalized or localized peritonitis (ICD-9: 540.0, 540.1)
 (B) Acute appendicitis without peritonitis (ICD-9: 540.9)

This clinical pathway is provided as a general guideline for use by physicians and families in planning care and treatment of patients and their families. It is not intended to be and does not establish a standard of care. Each patient's care is individualized according to specific needs.

	Pre-op Dx: Appendicitis	Day of Surgery	Post Op Day 1	Post Op Day 2 (End of Pathway for appendicitis without peritonitis)	Post Op Day 3	Post Op Day 4 (End of Pathway for appendicitis with peritonitis)	Day 5
Lab/Tests	CBC w/diff; UA c spec. gravity; *If dehydrated:* ED 7; *If diffusely tender abd:* Amylase			*If still NPO, repeat ED7*	CBC w/diff; *IF febrile (38°) OR WBC elevated OR not tolerating diet, obtain gent blood levels*		*IF febrile (38°) and WBC > 14,000, CT of abd.*
Radiology	*Acute abd. series only if requested by senior surgery resident*						Abscess drained
Treatments	Insert IV	Wound care; Infection protection ——→					
Medication	(A) Ampicillin[1] (25 mg/kg q6h) Gentamicin (2.5 mg/kg q8h) Clindamycin (10 mg/kg q8h); (B) Unasyn[1] x 3 doses (25 mg/kg q6h)	Discontinue Unasyn after third dose ——→			*DC IV antibiotics when afebrile WBC normal, & tolerating diet. Change to oral antibiotics* Bactrim 5 mg/kg q12h ——→		
Pain Management	Comfort measures ——→	Morphine IV; Retrograde c caudal (PCA if no caudal); Positioning ——→			Tylenol or Codeine ——→		
Nutrition/Hydration	NPO	(A) IV fluids; Fluid management; (B) DC IV after antibiotics; Liquids; Advance diet	DC IV when tolerating diet; Clear liquids ——→	Diet as tolerated ——→			

continues

Exhibit 13–A–1 continued

	Pre-op Dx: Appendicitis	Day of Surgery	Post Op Day 1	Post Op Day 2 (End of Pathway for appendicitis without peritonitis)	Post Op Day 3	Post Op Day 4 (End of Pathway for appendicitis with peritonitis)	Day 5
Activity	Ad lib	(A) Up at bedside (B) Begin ambulation	Begin ambulation Ambulate	Ambulate			
Assessment & Monitoring	Physical assessment Vital signs monitoring Fluid monitoring	Respiratory monitoring Pain monitoring					
Education & DC Plan	Consent	Involve patient/family in care	Discharge Planning			Appointment for follow-up	
Other	If obstructed, NG tube If dehydrated, IV (Ringers 10 cc/kg bolus, may repeat) and/or foley	Intra-operative protocol[2]	Remove NG when the following met: BS return, drainage less, abd not bloated, pt. hungry Remove foley				
Expected Outcome	No other X-rays or lab	No intra-operative complications Compliance with intra-op protocol[2]		(B) Discharge	(A) Afebrile Normal WBC	(A) Discharge	Discharge by Day 7

Note: BS, bowel sounds; CBC, complete blood count; CT, computed tomography; DC, discontinue, discharge; IV, intravenous; NG, nasogastric (tube); NPO, nothing by mouth; PCA, patient-controlled analgesia; UA, urinalysis; WBC, white blood cell count.

■ Part IV ■
General Surgery

■ 14 ■

Managing Outcomes for Breast Procedures

Susan Mott Coles

With the advent of managed care throughout the United States,[1] Saint Thomas Hospital began to explore tools that would result in a smooth transition into this new care delivery environment. While managed care and capitation had not yet begun to affect Tennessee in a significant way, the feeling seemed to be "Forewarned is forearmed." This chapter discusses how, in the spring of 1994, a group of nurses working on the general surgery unit opened discussions with the physicians in the Division of General Surgery about developing a case management model, along with critical pathways, for certain procedures. Two physicians verbalized significant interest in the development of critical pathways for patients undergoing surgical procedures for breast cancer.

HISTORICAL PERSPECTIVE

The Beginning

Since its inception, case management has focused on those cases that were considered outliers, or variant.[2] These have included (1) patients who had chronic repeat admissions, (2) high-volume population/diagnosis-related groups (DRGs), (3) notoriously expensive cases, (4) high-risk diagnoses or procedures, and/or (5) patient populations with high-risk socioeconomic factors. In actuality, breast *surgery* by itself fits almost none of these criteria. A diagnosis of breast cancer would become a potentially high-risk diagnosis/high-expense case type only after it crossed from the surgical oncology realm into that of medical oncology. Nonetheless, we felt that breast procedures were a good starting point, primarily because there was physician support associated with the project. The pathway developers felt that physician support would lend more credence to the model in general

and aid in future case management projects through positive feedback from these physicians to their colleagues. The other factors supporting our choice were the statistics related to the incidence of breast cancer. It was estimated that there would be 182,000 cases of female breast cancer in the United States in 1995, and 46,000 deaths. Because the incidence and mortality for male breast cancer were estimated to be 1,400 and 240, respectively, the critical pathway was directed to the female breast cancer population only.[3]

All available information for breast surgery DRGs was examined for fiscal year (FY) 1994. When the data were evaluated, the group detected two primary DRGs into which this particular patient population fell. These were DRG 257 (Breast Surgery with Comorbid Conditions) and DRG 258 (Breast Surgery without Comorbid Conditions), composing 92 percent of the total breast surgery population. After much discussion about these data, the first breast surgery critical pathway was implemented in July 1994 (the beginning of FY95). The pathway itself was designed to assess, and then address, various physiologic and psychosocial patient needs (Exhibit 14–1).

The First Pathways

The patient was initiated on the pathway at the time and location of admission. The steps contained within the pathway were the preoperative date, the day of surgery, and two other days on the medical/surgical unit until discharge. Other than a specific step for both the preoperative time and the operative day, the other two steps were not restricted by a set period of time. Areas of care included tests, diet, medication/intravenous (IV) fluids, vital signs/monitoring, pulmonary care, treatments, activity, discharge planning/teaching, and outcomes.

Exhibit 14–1 Front of Breast Critical Pathway Form

SAINT THOMAS HOSPITAL
Nashville, Tennessee
Admit Date: _____
Physician: _____
Discharge Date: _____

(addressograph)

BREAST SURGERY

STEP	PRE-OP	OPERATIVE DAY	STEP 1	STEP 2/DC DAY
Usual Location/ Bed	Early Morning Admission or Medical/Surgical unit	Medical/Surgical	Medical/Surgical	Medical/Surgical
Tests	Lab work as ordered Chest X-ray if ordered Electrocardiogram			
Medication/ IV Fluid	HS sedation if ordered	IV antibiotics on call & Q 8H x 24 H →→→→→→→→→→ IV fluids as ordered post-op Antiemetics IM/IV PRN →→→→→→→→ Pain meds as ordered PRN →→→→→→→→	Discontinue IV fluids Oral pain meds PRN →→→→→→→→→→→→ →→→→→→→→→→→→ →→→→→→→→→→→→	Oral pain meds PRN Discharge prescriptions & instructions documented
Vital Signs	Routine V/S	Routine Post-op V/S →→→→→→→→	→→→→→→→→→→→→	Discharge vital signs documented →→→→→→→→→→→→
Pulmonary Care		Turn, cough, deep-breathe Q 2H →→→	→→→→→→→→→→→→	→→→→→→→→→→→→
Treatments		Post-op dressing changes as ordered →→→→→→→→→→→→→→ Drain care Q 4–8H as ordered →→→→→ I/O Q shift →→→→→→→→→→→→→→→→ Affected arm up on pillow →→→→→→ Affected arm exercises as ordered	→→→→→→→→→→→→ →→→→→→→→→→→→ →→→→→→→→→→→→ →→→→→→→→→→→→ →→→→→→→→→→→→	Appearance of wound documented
Activity	As tolerated	Out of bed to chair in PM Bathroom privileges	Progressive ambulation Bath with assistance	Progressive ambulation & document tolerance self care
DC Planning/ Teaching	Call Social Services if problems anticipated on discharge. Pre-/Post-op teaching (Saint Thomas Education for Patients, teaching booklets, turn/cough/deep-breathe (TCDB), exercises). Review Pathway. Notify Reach to Recovery	Drain care teaching Wound care Arm exercises	Anticipate discharge in AM Arrange transportation. Drain care teaching (no driving or showering while drain is in, care of drain). Wound care & signs and symptoms of wound infections	Drain management Wound care Discharge instructions
Outcomes	Patient/Family verbalize understanding of pre- & post-op instructions Permit signed	Use of affected arm for basic activities of daily living (ADLs) Tolerating oral fluids	Perform return demonstration of emptying drain & wound care. Incision intact with no signs of infection. Reach to Recovery visit Use of affected arm for ADLs	Incisions intact without purulent drainage Verbalizes understanding of discharge instructions

Note: DC, discharge; HS, hour of sleep; I/O, intake and output; IM, intramuscular(ly); IV, intravenous(ly); V/S, vital signs; PRN, as needed.

Source: Copyright © Saint Thomas Health Services, Center for Clinical Evaluation.

Most of the instructions within the pathway covered basic nursing care, such as "turn, cough, deep-breathe every two hours," and did not address more specific criteria. Listed in the category entitled "tests" were instructions to perform diagnostic workup activities if/as ordered rather than instructions to perform specific diagnostics if certain predetermined criteria were met. The positive aspects seen with this approach were:

- Each physician could customize his or her care to meet each specific patient's needs.
- The lack of what physicians frequently considered "directives" about how they could manage their patients' care eventually would encourage more physicians to participate actively in this model.
- This lack of specificity decreased or prevented the possibility of overlooking some vital medical condition due to rote practice.

Unfortunately, this same generic approach that could create desirable outcomes could also have negative aspects:

- The perpetuation of historical interactions between the nursing and medical staffs resulted in significant reluctance on the part of the nurses to request orders for blood work, tests, and referrals that had not been already ordered.
- A lack of standardized protocols created the potential for vital patient or family caregiver needs, such as reordering home medications, making necessary referrals (i.e., Reach to Recovery), or following preexisting medical conditions such as diabetes or arthritis, to be overlooked on occasion.

The pathway indicated the patient's location at time of admission, usually the early morning admission unit, and any changes in location throughout the entire hospital stay. At the inception of the breast critical pathway, the beginning of FY95, the average length of stay (ALOS) in DRG 258 was 2.69 days. The staff responsible for developing the breast critical pathway felt that the ALOS, through utilization of the case management model, would be no more than two days. Data analysis done at the end of FY95 found that the ALOS had decreased to 2.48 days. The group viewed this as the beginning of the downward trend toward their desired goal.

In addition to reducing the patient's total hospitalization to two days or less, the goal of our case management model was to reduce patient charges and the utilization of supplies. For example, after three years of utilizing the case management model for breast surgery patients, data showed that the average overall patient charges for FY97 had decreased by just over $800 from previous years. The group felt that this new practice model would have its greatest hospitalwide impact in the

financial arena. There were noticeable improvements in the areas of IV therapy, blood bank, laboratory, anesthesia, and pharmacy. IV therapy, defined as the amount of IV fluids used, demonstrated an 8 percent savings from one year to the next. The physicians recognized that performing a type and screen was just as effective as a type and crossmatch of two or more units of packed red blood cells. This resulted in a patient savings approaching 53 percent with no untoward patient incidents or complications. Savings in the anesthesia field were also significant, with a 20 percent savings recognized in one year. The two physicians employing the practice model determined that in the absence of significant medical problems, the patient needed only minimal blood work checked, including electrolytes, hematocrit/hemaglobin, and baseline clotting studies. If already performed within 30 days of the time of biopsy, tests with normal results were not repeated. Decreasing the amount of testing done resulted in a savings of close to 13 percent in laboratory charges. These same physicians noted that antibiotics in the absence of frank infection could—and should—be limited to one preoperative prophylactic dose, resulting in a 12 percent decrease in pharmacy charges.

The next area in which the group desired to have significant impact was that of outcomes. Again, with this tool, the outcomes tracked were somewhat generic in content and/or criteria. With the exception of the patient's ability to perform return demonstration of surgical drain care, the others were relatively subjective. Statements such as "tolerating oral fluids" had the potential for many interpretations. Was the patient considered to be tolerating fluids after just 8 ounces, or did this mean drinking ad lib and in large quantities? At what point did the patient cross that fine line between barely meeting the outcome and becoming a variant patient? According to Kenneth K. Boggs, tracking outcome variances presents the greatest challenge to a case management practice model. Case managers often become so entrenched in gathering the outcome data that they overlook the outcome itself.[4] Nevertheless, the group did try to track patient outcomes through the utilization of variance codes. The outcomes not met were categorized into one of four variance classifications:

1. *Patient/Family*—patient condition, refusal of treatment or procedure, or "other," a rather global and nebulous term requiring further explanation
2. *Provider*—medication error, delay in provision of services or treatment, or "other"
3. *System*—unavailable bed, equipment, supply, or service, or "other"
4. *Community*—nursing home or rehabilitation facility bed needed/placement pending, or "other"

When an outcome could not be met, the nurse caring for the patient was to document this variance information on the reverse side of the critical pathway. The nurse was to record the date and time of the variance, at what step the patient was at the time of the variance, the variance code, the reason for the variance, and any action taken (Exhibit 14–2). If, for example, a patient was unable to care for her surgical drains due to a preexisting medical condition and she had no other caregiver, this indicated a patient/family variance, "other," in step 1. The action taken should have been to secure orders for home health to assist the patient in caring for her drains.

First Data Returns

By May 1995, the breast critical pathway model had been in operation for 10 months. In an attempt to determine the strengths and weaknesses of the current tool, the clinical nurse specialist (CNS), in the dual role of case manager (CM), arranged a meeting with the two principal physician participants. A review of data revealed that between November 1994 and May 1995, only 17 critical pathways had been returned for extrapolation of data. Because this was such a new practice model to Saint Thomas, the information gleaned from review was often incomplete and/or incorrect. Only five of the documents had the admission/preoperative date noted. Ten pathways lacked dates for the different steps within the pathway, so there was no way to know the patients' actual LOS without doing a closed chart review. Only 11 pathways indicated the date of surgery, and only 13 were stamped with the patient's name and medical record number, making the other four impossible to track. Most variance codes were incorrect, and only a few had explanations as to why the variance had occurred.

Recognizing that documentation on a piece of paper was only a part of the puzzle about the effectiveness of the model, the CNS/CM proposed conducting a survey

Exhibit 14–2 Back of Breast Critical Pathway Form

Date	Time	Step	Variance Code	Reason for Variance	Action Taken/Initials

CASE MANAGEMENT VARIANCE DATA

Variance Codes
A. Patient/Family
1. Patient condition
2. Refuses treatment/procedure
3. Other (explain)

B. Provider
1. Medication error
2. Delay in providing services or treatment
3. Other (explain)

C. System
1. Bed unavailable
2. Equipment/supply unavailable
3. Service unavailable

D. Community
1. Nursing home bed needed awaiting placement
2. Rehabilitation bed needed awaiting placement
3. Other (explain)

Name/Initials   Name/Initials

Source: Copyright © Saint Thomas Health Services, Center for Clinical Evaluation.

among the patients who had been involved in this pilot project. Both physicians were agreeable. It was decided that the patients would receive a simple survey (Exhibit 14–3) at the time of their first or second postoperative office visit. The CNS/CM would then be able to determine how these first patients in the new case management model perceived their care and their hospital experience. During the months of May and June, 12 patients had some type of breast surgery; because 2 of these had simultaneous reconstructive procedures, they were not included in the survey results. Survey results indicated that the majority (90 percent) of patients felt positive about the care and education they received during their hospitalization.

Logistical Problems

Looking at the overall management of care process, we felt that it was time to move forward with this model. We began approaching other physicians to enlist their participation in using the breast critical pathway for their

breast surgery patients. One physician agreed to allow the pathway to be used for his patients. Although this was a start in the quest to integrate case management into all of general surgery, we knew that we would have to overcome many more obstacles.

Because the ALOS had been decreased to two days or less, one aspect of care that was becoming neglected due to lack of time was the psychosocial realm of the patient's care. Previously, patients having mastectomies or other breast procedures stayed in the hospital for three or more days. With extremely rare exceptions, such as a patient suffering with some form of dementia, breast cancer patients were seen by a volunteer from the Reach to Recovery organization, a breast cancer support group connected with the American Cancer Society. Because these volunteers were all breast cancer/surgery survivors, they could offer the patient education and support on a personal level, complementing the clinically focused education given by physicians and nurses. With the patients now leaving only a day or two after surgery, there was often not time to contact a volunteer and schedule a hospital visit. This created an en-

Exhibit 14–3 Breast Surgery Patient Survey

FOR SAINT THOMAS HOSPITAL PATIENTS ONLY

We are interested in checking our teaching methods. Please help us by filling out the questionnaire below. Information is confidential, and signature is not required. Thank you for your time.

1. Were you given a patient teaching sheet for breast surgery?   Yes ❑   No ❑

2. Did you still have a drain in place when you were discharged?   Yes ❑   No ❑

   If yes, were you able to care for it?   Yes ❑   No ❑

3. Did someone from Reach to Recovery contact you?   Yes ❑   No ❑

4. Are you doing your exercises?   Yes ❑   No ❑

5. Did every nurse give you the same basic information about your surgery?   Yes ❑   No ❑

6. Is this your first office visit since your surgery?   Yes ❑   No ❑

   If no, how many times have you seen the doctor since surgery, *NOT COUNTING* today?   _____

Additional comments welcome:

_____

_____

_____

_____

Source: Copyright © Saint Thomas Health Services, Center for Clinical Evaluation.

tirely new set of problems. It was permissible for the volunteer to schedule a visit with the patient in her home. However, if the patient lived in a neighborhood with a high crime rate, frequently there was not a volunteer who would agree to make a home visit. Additionally, some patients were opposed to allowing a stranger to enter their homes, viewing this as an invasion of privacy. Some patients even considered this to be a breach of medical confidentiality when the contact came outside the clinical setting. Finally, a significant number of our patients lived in small rural Tennessee counties where there were generally few or no volunteers to whom patients could be referred.

The solution seemed simple: involve the CNS/CM from the beginning point of the diagnostic workup through postmastectomy care. The group expressed a desire to aim revisions of the current pathway to include mammography, preadmission testing (PAT), outpatient biopsy, the surgical oncology setting, and then the medical oncology setting if necessary. We began the process of making revisions to include these services, as well as to create more measurable outcomes. However, because of the unwieldy logistics of this process (i.e., physician support, collaboration among the departments, continuity of the document utilization through the various services), we were unable to fulfill this goal.

INSTITUTIONAL APPROACH

At about this same time, Saint Thomas initiated the process of developing a case management model that would encompass all patients on all services. Up to this point, various CNSs and CMs had been facilitating groups in creating critical pathways specific to their own patient populations. We recognized that while each pathway and practice model had its own inherent good qualities, our overall process, structure, and outcome standards were less than optimal. Very few, if any, pathways addressed the same generic needs or outcomes. Designs differed greatly. Interpretation of the development, implementation, and management processes also differed among CMs, CNSs, and the various diagnostic-related services. Recognizing the need to standardize the case management model as a whole, a task force began working to develop a multidisciplinary case management model and the tools associated with this type of practice model. This task force included CNSs, CMs, staff nurses, and representatives from other patient-related disciplines within the organization. Our goal was to implement case management and critical pathways, either diagnosis/disease specific or generic, for every patient within the system.

We also enlisted the support and assistance of all the medical directors within the hospital in making the new documents a permanent part of the patients' medical records. The hope was that this would give the documents themselves increased credibility, promoting better utilization. But this change created another obstacle. Without some type of entry in the document by nursing or another discipline, there was nothing to transform the generic pathway to a patient-specific pathway. Thus, our group began to develop the pathway into a documentation tool, redesigning the entire documentation system at Saint Thomas Hospital. The idea behind the redesign was to direct the focus onto outcomes rather than tasks. We established as our goal, using clinical standards as a practice framework,[5] reporting outcome variances needing to be addressed from the CareMap® rather than tasks completed or to be performed.

For the redesign of the breast critical pathway, now referred to as CareMap®, the general surgery task force examined the number of patients in the primary breast surgery DRGs. Because issues relating to the care of patients undergoing breast reconstruction procedures were often radically different from breast surgery without reconstruction, we elected to exclude these procedures from this CareMap®. (The CNS for the Division of Plastic Surgery was charged with the task of developing a document for patients having breast surgery with reconstructive procedures.) Because the total number of patients undergoing breast surgery was small (260 for FY94–95) in comparison with other patient populations in general surgery, the group decided to examine other homogeneous patient populations. They found that the number of patients undergoing laparoscopic cholecystectomy or appendectomy procedures during the same time period, though not as small (831 total), had the same approximate ALOS. Therefore, in an attempt to streamline the transition into total case management, the CareMap® developers clustered these three patient populations onto one document, entitled Breast Surgery, Laparoscopic Cholecystectomy/Appendectomy CareMap®.

In an attempt to develop a totally multidisciplinary document, the group approached the PAT nurse manager, the nurse manager and CNS for the intraoperative and postanesthesia care units, and representatives from non-nursing disciplines to elicit their assistance. The goal of the development team was to create a document that could be initiated in the preoperative setting, be used through the actual surgical procedure (Exhibit 14–A–1 in Appendix 14–A), and then address postoperative care and outcomes until discharge (Exhibit 14–A–2 in Appendix 14–A). There would be specific, measurable outcomes to be addressed before the patient could move from a particular step. Outcomes not met would result in a variance. Some method would need to be developed allowing the nurse to indicate the specific reason or reasons that an outcome was not met, rather than the more global reasons used with the original pathway. In addition, the nurse would be instructed to document other information, such as certain tasks being performed or the reporting of certain data to the physicians in the body of the CareMap®.

Other disciplines, such as pastoral care and the discharge planning nurse/social worker, would date and initial as indicated when they interacted with the patient or family. The date and initials would direct anyone having interactions with the patient to check the progress notes section of the chart for further information and/or recommendations made by other disciplines. Sections would be available to address individual patient needs/outcomes not included in the standardized document.

Finally, an outcomes tracking tool was developed (Exhibit 14–A–3 in Appendix 14–A). Because each outcome was a potential variance, it was important to know why each variance occurred and how often any one variance did occur. If it was found that one particular variance was occurring on a repeated basis, the original outcome would need review. The standards might need adjusting upward or downward. The process for recording a variance would be that when a patient was unable to meet an outcome, the nurse would indicate "no" and then check any reason(s) applicable on the outcome tracking tool. The reason might be that the patient's temperature was 100.8°, which would be over the desired outcome of a maximum temperature of 99.8°. Since this was primarily a task-oriented behavior (take the temperature, record, note if over acceptable limits, document variance reason), the nurse was then to document in the progress notes what action was taken to change the patient's status from unacceptable to acceptable. The action might be nothing other than calling the physician about the temperature elevation (Figure 14–1).

After their development, the new CareMaps® were initiated in May 1996. The breast surgery patients were grouped, as planned, with laparoscopic cholecystectomy and appendectomy patients on one document (Exhibit 14–A–4 in Appendix 14–A). Hospital personnel involved in direct or indirect patient care were given extensive education about the process of case management and the purpose/use of the CareMap®.

Within the first six months, flaws in the process became glaringly evident. Even though breast surgeries and laparoscopic procedures had similar ALOSs, the numerous differences in the scope of care given precluded the successful clustering of these patient populations into

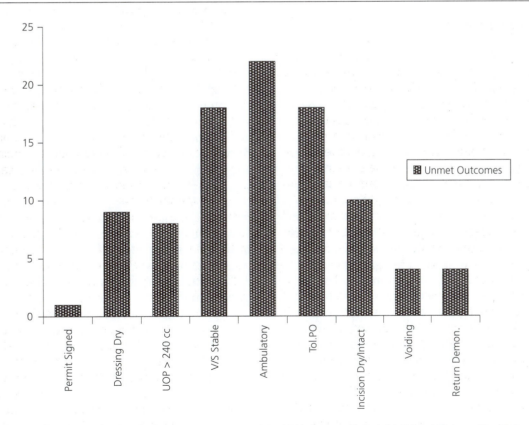

Figure 14–1 Breast/Laparascopic Procedures, Unmet Outcomes (*N* = 191). *Source:* Copyright © Saint Thomas Health Services, Center for Clinical Evaluation.

one practice pattern. Also, attempts to incorporate task-oriented documentation into all the CareMaps® were noted to be virtually impossible for the clerical and nursing staff to manage. Physician orders changed sometimes two or three times in one day, necessitating changes in the notations made in order-related sections. A morning order of "nothing by mouth" could be changed to clear liquids by noon and to soft diet by supper. Because we set a goal of having the majority of documentation on one form, we had discarded the patient care profile, or kardex, used by the clerical associate to transcribe orders other than medications and then to erase and change orders when necessary. With this new document being a permanent part of the medical record, all entries had to be made in ink. This resulted in extremely illegible documents and frequently overlooked orders (especially lab work) and caused great confusion about which orders were pertinent and which were old.

The documentation task force reconvened to begin work on the first revisions of the CareMaps®. The first order of business was seen as developing a worksheet that would *not* be a permanent part of the medical record but would be a worksheet on which orders could be recorded in pencil (Exhibits 14–A–5 and 14–A–6 in Appendix 14–A). The next task facing the group was to further standardize the documents, transferring the majority of task-related documentation to the daily flowsheet, graphic sheet, and/or progress notes. The only documentation within the body of the CareMap® would be the initiation of referrals to the multidisciplinary team, the acknowledgment of these referrals by the appropriate team member(s), and the status of patient outcomes.

With the Breast/Laparoscopic Cholecystectomy/Appendectomy CareMap®, the staff from general surgery acknowledged the failure of having the three procedures clustered. The only true similarity was the short LOS. Otherwise, the educational process and the desired/expected discharge outcomes were different. Therefore, these procedures were separated, creating two CareMaps®, one for laparoscopic procedures and one for breast procedures without reconstruction. Outcomes were decreased in number and focused more on the patient's ability to resume self-care, manage drains, and have psychosocial needs met. Reach to Recovery referrals became part of the preoperative outcome criteria. Another significant preoperative assessment criterion became that of patient risk for deep-vein thrombosis/pulmonary embolus (DVT/PE). The goal of this assessment was to decrease the morbidity and mortality of DVT/PE through the prophylactic use of low-dose heparin, antiembolitic stockings, and/or lower extremity sequential compression devices/foot pumps. Once the revisions were complete, these newest documents became operational in February 1997 (Exhibit

14–A–7 in Appendix 14–A). Since the revisions, the CareMaps® have been viewed as more user friendly. Questions relating to outcomes are more objective, requiring less subjective interpretation on the part of the nursing staff. Other disciplines have the opportunity to have more ownership in the roles they play in the patient's care.

One vital area of any institution in an outcomes-based practice model is the database department. Recording outcomes is meaningless without someone to track the outcome variances and supply data to the CNS about his or her patient population's outcome status. During the early days of case management, the critical pathway outcomes had no tracking mechanism attached to them. The only way to know how many and what kind of variances occurred was to collect and tabulate the data manually. This meant reviewing every critical pathway and the associated chart when variances were noted. Because the process for critical pathway retrieval was marginal at best, it was rare to have a critical pathway returned to the CNS for review. When the documents *were* returned, it was often impossible to track the patients to whom they belonged due to missing demographic data. Also, because there was no yes/no notation within the body of the pathway, the only way to know if there had indeed been a variance was if the nurse had completed the reverse side of the pathway. This was not seen as a high priority of the nursing staff, who knew that the documents were not part of the permanent medical record, so documentation was poor. When the documentation and case management task force began redesigning the critical pathways into the CareMap® format, at the top of our resources "wish list" was strong financial and outcomes database support. It would be much simpler to extrapolate variance data from the CareMap® with the outcomes more clearly defined and the creation of an outcomes tracking tool. Personnel working in the database would review each CareMap® for variances, create graphs of these variances, and report back to the CNS and/or the CM. Additionally, the financial support team could give us information relating to improvements or regressions in ALOS and utilization of supplies, providing us with support for this practice model from a fiscal standpoint.

Tracking outcomes, however, proved to be a monumental task. The staff in the database department asked if each CNS and CM could select one or two salient outcomes to track, allowing them to execute the data collection with a better return time than previously possible. By selecting a few of the more outstanding outcomes, the CNS or CM would know much more quickly if current practice was appropriate or if changes needed to be made to address patient needs better. The outcomes chosen could be those that had the highest morbidity/ mortality potential, were the most difficult to achieve, or addressed a service-specific quality indicator.

During the past year, surgical services had been examining the quality indicator of pain management and pain control. Because we had recognized that an uncomfortable patient is often a noncompliant patient, we had begun educating the staff on the guidelines for acute pain management.[6] We had been tracking the patient's self-report of pain relief through the use of the Agency for Health Care Policy and Research (AHCPR) 0-to-10 scale. With the creation of the first CareMap®, we decided that we would also begin examining the types of analgesia being used throughout the general surgery/surgical oncology setting. To achieve this, we included as an outcome "pain less than 5 on the 0–10 scale." In the first draft document, the outcomes tracking tool gave as the reason for a variance simply that the patient reported a pain level of 5 or greater. Recognizing that this was much too global, we revised it in the second document, giving us information of pain reports between 5 and 7, and pain reports of 8 or above. This change allowed us to have a more objective, rather than subjective, goal. We would then have concrete objective evidence to report to the physician when asking for analgesia changes, either increases or decreases. We also felt it would be beneficial to the study as a whole to know on a day-by-day basis

what types of analgesia were being used in our area. We were not so much concerned with specific drugs as with routes.

We attempted to have this information recorded within the body of the CareMap®. As with other documentation data requested, the nurses often overlooked this entry because of illegible or confusing entries. With the reformatting, this query was moved into the "Outcomes" section, directly below the statement about the patient's rating the pain at or below 5. We wrote this statement in such a manner to allow the nurses to denote original analgesia orders as well as changes, such as from IV to oral (PO), or from patient-controlled analgesic to IV (Exhibit 14–A–8 in Appendix 14–A). We found that as a rule, patients undergoing breast procedures without reconstruction generally will report pain in the range of 5 or less within the first 24 hours postoperatively (Figure 14–2). Also, their need for IV or intramuscular (IM) analgesia will be less, with their achieving pain control with PO narcotics within that same time period. Because of the pioneering work with the AHCPR pain management guidelines within general surgery/surgical oncology, the rest of the hospital adopted this method of pain assessment. Additionally, when the graphic flowsheet was revised in May 1996, a horizontal column was added to allow the recording of the patient's pain report as frequently as every four hours.

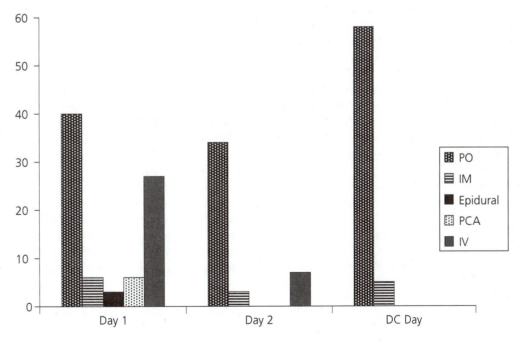

Note: DC, discharge; IM, intramuscular; IV, intravenous; PCA, patient-controlled analgesia; PO, oral.

Figure 14–2 Analgesia Usage. *Source:* Copyright © Saint Thomas Health Services, Center for Clinical Evaluation.

Another outcome monitored was the patient's "discharge disposition." This indicated the type of care the patient would require after she left Saint Thomas. The categories were (1) self-care, (2) home health, (3) hospice, (4) unskilled nursing facility, and (5) skilled nursing facility. Of 202 patients, 166 were discharged as self-care patients; 15 had either home health or hospice care arranged; and 1 was admitted to an unskilled nursing facility. Unfortunately, 10 documents had nothing recorded about the discharge disposition, so our statistics were incomplete by just under 20 percent. While this is a very small percentage, our goal is to know this information for all our patients. By tracking what services are used most frequently by what patient populations, we can begin the process of planning for our patients earlier in their hospital stay.

PERSPECTIVES ON THE FUTURE

Our goal is to have this outcomes-based practice model functioning in an even more multidisciplinary mode than at present. Currently, the model is primarily a nursing tool, with varying degrees of input from other disciplines. To begin working toward achieving this goal, we have attempted to conduct multidisciplinary rounds twice weekly. By discussing the patient's case with nursing, social services, pulmonary, rehabilitation, pastoral care, and a nutritionist, we can better plan how to meet the patient's needs within the walls of Saint Thomas and then extend our care outside the walls. While the process has been a desirable one, the outcome has not been as good as we had hoped. The fact that everyone is extremely overcommitted in many respects has caused us to review the process. Our current plan has been to have our rounds once weekly, late in the week to ensure that as many potential problems/needs as possible are addressed before the weekend arrives and outside resources are limited (i.e., home health agencies or companies that supply medical equipment are not open to take new patients).

We do not currently have CareMaps® that can be shared with the patient. This makes it somewhat difficult to review the plan of care with the patient using our medically directed format and be certain that every nurse is consistent in his or her teaching. Creating a companion CareMap® for patients would allow them to follow their progress on a daily basis. These CareMaps® for patients have been incorporated into use at other institutions with great success, with patients reporting increased understanding of what is expected of them, thus increasing compliance.[7]

Because we see the importance of the multidisciplinary approach to patient education and care, we have created a preoperative video. For the past two years, a task force with representatives from all patient-related departments has been working to design a short (13-minute) patient education video. The original idea was to create a video that would be generic in content and focus, yet would include references to all areas of Saint Thomas Hospital with which a patient might have direct contact. Starting with a very standard "Welcome to Saint Thomas" statement, this video traces the route of a patient from his first contact with our institution until discharge. Two of our staff members played the roles of Mr. Patient and his wife. The video shows him traveling from the PAT center through the admission process, a surgical procedure, and the recovery period and finally out of the hospital on his way home. By viewing this film, real patients can have an idea of the various people they may meet and services they may receive at Saint Thomas Hospital. Each departmental representative was charged with developing the script for his or her area so that what was said in the actual production would be an accurate reflection of that department's potential interaction with the patient and family. Our goal is to have these videos placed in the physicians' offices to be given to the patient when a hospital admission reservation is made.

One very much desired goal among the nursing staff is that of increasing collaborative practice between nursing and medicine. While many physicians are open to this type of discussion regarding patient care issues, others are more reluctant to have a truly collaborative practice with nursing. It is our hope that we can demonstrate, through increased patient satisfaction in all phases of care, how developing these alliances will benefit all concerned.

One position that we wish to see added to our services is that of a cancer care coordinator. This person would contact the patient preoperatively to discuss fears and concerns and explore needs. She or he could begin the education process for the patient and family, providing information regarding adjuvant therapy if it was planned or strongly anticipated postoperatively, or making referrals to appropriate resources and support groups. In the case of the breast cancer patients, this would mean that the Reach to Recovery volunteer could make contact with the patient preoperatively, rather than waiting until the person was groggy from sedation or anesthesia. With planned ostomy or laryngectomy patients, a visit with an ostomate or a laryngectomy patient who had recovered to lead a full life might give hope where there had been none. These are just some of the ways in which such a program could benefit surgical patients.

While we have developed workable tools, we are continually striving to improve our process through staff education and through refining our practice model. Change is slow and often meets with resistance. Statements of

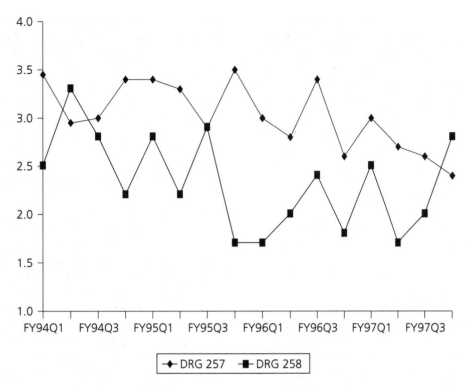

Figure 14–3 Average Length of Stay for Breast Surgery Patients by Quarter for DRGs 257 and 258. *Source:* Copyright © Saint Thomas Health Services, Center for Clinical Evaluation.

"We've never done it this way before!" have been heard on more than one occasion. The first few months of CareMap® use were extremely frustrating. Nurses could not remember many of the hows and whys of the documentation process. The first data returned indicated that only 191 CareMaps® had any variances; there were, however, over 800 total patients having breast or laparoscopic procedures during the time frame examined. Education in why we are using this model and our own outcomes is vital to gain staff compliance.

One method we use to disseminate education to the staff on a daily basis is a tool that has affectionately been named "potty notes." The quality resources department has installed heavy clear plastic jackets on the stall doors of all staff restrooms. Bullet-point information is placed in these on a weekly basis. The information is either a "Don't forget to chart . . . ," a new policy, a new documentation form being initiated into the system, or a note related to the standards of the Joint Commission on Accreditation of Healthcare Organizations addressed by our professional practice model. Random verbal quizzing has demonstrated an increase in knowledge. Periodic chart audits, along with one-on-one instructional sessions between the CNS and the staff nurse, have resulted in improvements in the overall process of documentation and understanding. Dissemi-

nation of data related to how outcomes-based practice has had an impact on our institution overall has been positively received by the staff. They were impressed to learn that between FY94 and FY97, the ALOS in DRG 257 had decreased from 3.21 to 2.73 days. The ALOS in DRG 258 had decreased from 2.69 to 2.20 days during the same time period (Figure 14–3).

Outcomes-based practice has gone through many changes at Saint Thomas Hospital. The most positive of these is the increase in patient-centered rather than task-centered practice. Our goals include providing both the care and the service required to meet patient outcomes, improve patient satisfaction, and increase collaborative practice across the scope of services. The accomplishment of these goals will make outcomes-based practice a meaningful endeavor for our practice within surgical services and throughout Saint Thomas Health Care System as a whole.

REFERENCES

1. Wojner AW. Outcomes management: from theory to practice. *Crit Care Nurs Q.* 1997;19:1–15.
2. Girard N. The case management model of patient care delivery. *Assoc Oper Room Nurs J.* 1994;60:403–415.

3. Wingo PA, Tong T, Bolden S. Cancer statistics, 1995. *CA-A Cancer J Clinic*. 1995;45:8–30.

4. Keep focus on the 'big picture' when tackling variance analysis. *Hosp Case Manage*. 1993;1:150–152.

5. Woodyard LW, Sheetz JE. Critical pathway patient outcomes: the missing standard. *J Nurs Care Qual*. 1993;8:51–57.

6. Acute Pain Management Guideline Panel. *Acute pain management: operative or medical procedures and trauma. Clinical practice guideline*. AHCPR Publication No. 92-0032. Rockville, MD: Agency for Health Care Policy and Research, Public Health Service, U.S. Department of Health and Human Services. February, 1992.

7. To involve patients in their care, give them user-friendly critical paths. *Patient Educ Manage*. 1995;2:105–116.

■ Appendix 14–A ■
CareMaps® and Worksheets

Exhibit 14–A–1 Pre-/Intraop Day, Original CareMap®

	CareMap® for Breast Surgery, Laparoscopic Cholecystectomy/Appendectomy		Addressograph	
Time Interval	**Pre-Op**		**Intra-Operative/PACU**	
Date	(PAT)_____ (Admission Day)_____			
Nutrition	NPO after 12 MN			
Activity	Ad lib			
IVs Medications	Pre-ops per Anesthesia Pre-op Antibiotics _____ SQ Heparin _____			
Assessment	Report abnormal assessment data Identify allergies Assess pt/family readiness to learn YES ☐ NO ☐	Date Time Init		
Treatments	Hibiclens: scrub_____ shower_____ ☐☐ TEDs_____ SCDs_____ Other preps: Specify_____ ☐☐			
Diagnostics	HCT_____ EKG_____ Copies of previous tests on chart YES ☐ NO ☐	Date Time Init		
Teaching DC Planning	Pre-op teaching ☐☐ STEPS ☐☐ Notify Reach to Recovery ☐☐ DC needs assessment ☐☐			
Psychosocial	Pastoral Care ☐☐ Notify CNS/Case Manager ☐☐			
Individual Patient Needs				
Individual Patient Outcomes		Date Time Init		Date Time Init
Outcomes	**Pt/family verbalize understanding of:** **1. Pre-op teaching/STEPS ☐ YES ☐ NO** **2. Scrub to be done ☐ YES ☐ NO** **3. Permit signed ☐ YES ☐ NO**		**4. Pt transferred to PACU** **☐ YES ☐ NO** **5. Pt discharged from PACU ☐ YES ☐ NO**	

Treatments documented on the CareMap® do not need to be duplicated on the Flowsheet.

Source: Copyright © Saint Thomas Health Services, Center for Clinical Evaluation.

Exhibit 14–A–2 Discharge Day, Original CareMap®

	CareMap® for Breast Surgery, Laparoscopic Cholecystectomy/Appendectomy		*Addressograph*	
Time Interval	**Continuation Day**		**Discharge Day**	
Date				
Nutrition			Diet:_____	
Activity			Ad lib	
IVs Medications			**22. Analgesia**—document on outcomes page	
Assessment			V/S Q_____ I/O Q 8 H	
Treatments				
Diagnostics				
Teaching DC Planning			Review MD's discharge instructions ☐☐ Observe pt/family in drain care ☐☐	
Psychosocial				
Individual Patient Needs				
Individual Patient Outcomes				
Outcomes		Date Time Init	23. **Pt./family verbalize understanding of MD's instructions for diet, activity, meds, wound care, return appointment** ☐ **YES** ☐ **NO** 24. **Return demonstration of drain care** ☐ **YES** ☐ **NO** 25. **Incision clean, dry, well approximated** ☐ **YES** ☐ **NO** 26. **V/S stable; Temp. < 99.8** ☐ **YES** ☐ **NO** 27. **Ambulating independently** ☐ **YES** ☐ **NO** 28. **Pain < 5 on 0–10 scale** ☐ **YES** ☐ **NO** 29. **Voiding w/o difficulty** ☐ **YES** ☐ **NO** 30. **Discharge disposition—document on outcomes page** ☐ **YES** ☐ **NO** 31. **Pt. discharged** ☐ **YES** ☐ **NO**	Date Time Init

Treatments documented on the CareMap® do not need to be duplicated on the Flowsheet.
Note: DC, discharge; I/O, intake and output; V/S, vital signs.

Source: Copyright © Saint Thomas Health Services, Center for Clinical Evaluation.

Exhibit 14–A–3 Outcomes Tracking Tool, Original CareMap®

Outcomes Tracking Tool—CareMap® for Breast Surgery/Laparoscopic Cholecystectomy/Appendectomy. If outcome unmet, check reason(s); if reason not listed, document in Progress Notes. Actions taken as a result of unmet outcomes also are charted in the Progress Notes.   *Addressograph*

1. Pt./family verbalized understanding of Pre-op teaching/STEPS
❏ Pt. too sick to comprehend
❏ No family available
❏ Pt. confused
❏ Language barrier

2. Pt./family verbalized understanding of Scrub to be done
❏ Pt. too sick to perform
❏ No family available
❏ Not ordered

3. Permit signed
❏ Pt. too sick to sign
❏ Pt. refuses to sign
❏ Pt. confused
❏ No family available

4. Pt. transferred to PACU
❏ Condition unstable at time of transfer

5. Pt. discharged from PACU
❏ Requires return to OR

6. Analgesia
❏ IV
❏ PCA
❏ Epidural
❏ IM
❏ PO

7. Dressing dry/intact
❏ Old drainage in dressing
❏ Dressing saturated with bloody drainage

8. UOP > 240 cc/8H
❏ UOP < 240 cc/H

9. Pain < 5 on 0–10 scale
❏ Pain > 5 on 0–10 scale

10. V/S stable; Temp. max. 99.8
❏ V/S > ordered parameters
❏ V/S < ordered parameters
❏ Temp. > 99.8

11. Amb. length of hall BID
❏ Pt. c/o fatigue
❏ Pt. c/o SOB
❏ Pt. c/o > pain
❏ Unable to ambulate
❏ Syncopal episode

12. Drains patent
❏ Drains clogged
❏ Suction not patent

13. Tolerating PO intake with w/o c/o n/v
❏ C/O N/V with PO intake
❏ Patient still NPO

14. Analgesia
❏ IV
❏ IM
❏ PO

15. Incision clean, dry, well-approximated
❏ Dressings not removed
❏ Incision red
❏ Drainage noted
❏ Wound open

16. Voiding w/o difficulty
❏ Foley in place
❏ Unable to void

17. Pain < 5 on 0–10 scale
❏ Pain > 5 on 0–10 scale

18. V/S stable; Temp. max. 99.8
❏ V/S > ordered parameters
❏ V/S < ordered parameters
❏ Temp. > 99.8

19. Amb. length of hall at least TID
❏ Pt. c/o fatigue
❏ Pt. c/o SOB
❏ Pt. c/o > pain
❏ Unable to ambulate
❏ Syncopal episode

20. Drains patent
❏ Drains clogged
❏ Suction not patent

21. Return demonstration of drain care
❏ Unable to comprehend/perform
❏ No family present
❏ Pt. confused

22. Analgesia
❏ IM
❏ PO

23. Pt./family verbalize understanding of MD's instructions for diet, activity, meds, wound care, return appointment
❏ Pt. unable to comprehend
❏ Language barrier
❏ Pt. confused
❏ No family present @ time of DC

24. Return demonstration of drain care
❏ Unable to comprehend/perform
❏ No family present
❏ Pt. confused

25. Incision clean, dry, well-approxi- mated
❏ Dressing not removed
❏ Incision red
❏ Drainage noted
❏ Wound open

26. V/S stable; Temp. max. 99.8
❏ V/S > ordered parameters
❏ V/S < ordered parameters
❏ Temp. > 99.8

27. Ambulating independently
❏ Pt. c/o fatigue
❏ Pt. c/o SOB
❏ Pt. c/o > pain
❏ Unable to ambulate
❏ Syncopal episode

28. Pain < 5 on 0–10 scale
❏ Pain > 5 on 0–10 scale

29. Voiding w/o difficulty
❏ Foley in place
❏ Pt. unable to void

30. Discharge Disposition
❏ Self care
❏ Home Health/Hospice
❏ Rehab facility
❏ Skilled nursing facility
❏ Unskilled nursing facility
❏ Home medical equipment

31. Pt. discharged
❏ No order
❏ No support system
❏ Awaiting nursing home placement (bed)
❏ Awaiting nursing home placement (financial)

Note: BID, twice a day; C/O, compliments of; DC, discharge; IM, intramuscular; IV, intravenous; N/V nausea/vomiting; OR, operating room; PACU, postanesthesia care unit; PO, oral; SOB, shortness of breath; TID, three times a day; UOP, urinary output.

Source: Copyright © Saint Thomas Health Services, Center for Clinical Evaluation.

Exhibit 14–A–4 Front, Original CareMap®

Allergies:		Addressograph

CareMap®:

Breast Surgery, Laparoscopic Cholecystectomy/Appendectomy

Inclusion:

Admission Date/Diagnosis:

Pertinent Medical History:

Advance Directives:
Living Will   Yes or No
Organ Donor   Yes or No
DNR   Yes or No
Durable Power of Attorney   Yes or No

Room Numbers/Transfers:

#_____   #_____   #_____   #_____
Date   Date   Date   Date

Concurrent Medical Conditions:
ICD on   Yes   No   NA
Permanent Pacemaker   Yes   No   NA
Pregnant   Yes   No   NA
Diabetic   Yes   No   NA
   On Metformin/Glucophage?   Yes   No   NA

Primary Physicians:

Consulting Physicians:

Procedures: Date:
_____   _____
_____   _____
_____   _____
_____   _____
_____   _____

Procedures: Date:
_____   _____
_____   _____
_____   _____
_____   _____
_____   _____

Special Needs:
_____
_____
_____
_____
_____
_____
_____
_____
_____

Isolation: No_____ Yes_____
                     Type: _____
CAPS: No_____ Yes_____
                     Date started: _____
                     Date ended: _____
Contact Person:
Name: _____
_____
Phone: _____
_____

Printed Name	Initial	Printed Name	Initial	Printed Name	Initial

Source: Copyright © Saint Thomas Health Services, Center for Clinical Evaluation.

Exhibit 14–A–5 Front of New Patient Care Worksheet

Allergies:	*Addressograph*
Advance Directives:	**Admission Date/Diagnosis:**
Durable Power of Attorney Yes No	
Living Will Yes No	**Date** **Procedure**

Allergies:

Advance Directives:
Durable Power of Attorney	Yes	No
Living Will	Yes	No
Organ Donor	Yes	No
DNR_____	Yes	No

Concurrent Medical Conditions:
ICD on	Yes	No	NA
Permanent Pacemaker	Yes	No	NA
Pregnant	Yes	No	NA
Diabetic	Yes	No	NA
On Metformin/Glucophage?	Yes	No	NA

Room Numbers/Transfers

#_____   #_____   #_____   #_____
Date   Date   Date   Date

#_____   #_____   #_____   #_____
Date   Date   Date   Date

Primary Physician:

Consulting Physicians:

Case Manager:

Miscellaneous: _____
_____
Wears glasses	❑ _____
Hard of hearing	❑ _____
Uses cane	❑ _____
Uses walker	❑ _____
	_____

Started on _____ CareMap®; Date _____
Transferred to _____ CareMap®; Date _____
Transferred to _____ CareMap®; Date _____
Transferred to _____ CareMap®; Date _____

Isolation: YES ❑ NO ❑
Type _____
_____
Date Started_____ Date Ended _____

CAPS: YES ❑ NO ❑
Date Started_____ Date Ended _____

Addressograph

Admission Date/Diagnosis:

Date   **Procedure**
_____
_____
_____
_____
_____
_____
_____
_____

Pertinent Medical History:

SPECIAL ORDERS:

Level 1 Extubation: Yes ❑ No ❑

PCA/Epidural Information: _____
_____
_____
_____

Other: _____
_____
_____
_____

Restraints on Yes ❑ No ❑
If restraints ordered, check patient every hour and prn
Obtain re-order for restraints every 24 hours

Contact Person: _____
Name: _____
_____
_____
Phone: _____

This worksheet is not a part of the permanent medical record and will be discarded upon discharge.

Note: DNR, do not resuscitate; ICD, Internal cardiac defibrilator.

Source: Copyright © Saint Thomas Health Services, Center for Clinical Evaluation.

Exhibit 14–A–6 Back of New Patient Care Worksheet

INTRAVENOUS THERAPY			
	Addressograph		
	Date Ordered	**DAILY TESTS**	To Be Done
Venous Access Devices:	Date Ordered	**LAB TESTS**	To Be Done
Drsg △ DUE:			
Tubing △ DUE:			
Site △ DUE:			
FOLEY: YES ❑ DATE INSERTED			
DRAINAGE DEVICES:			
TEMPORARY PACEMAKER SETTINGS:			
Mode Rate	Date Ordered	**DIAGNOSTIC TESTS**	To Be Done
VENTILATOR SETTINGS:			
Rate AC ❑ IMV ❑			
FiO_2 TV			
PEEP PS			
OXYGEN/RESPIRATORY ORDERS:			
	Date Ordered	**ORDERED TREATMENTS**	To Be Done
VITAL SIGNS FREQUENCY:			
Neuro ✓ 's Circ ✓ 's			
ACTIVITY:		**DAILY WEIGHT: YES ❑ DATE ORDERED**	
	Date Ordered	**DIET/TUBE FEEDINGS:**	
		Fluid Restriction:	
		7–3 3–11 11–7	

This worksheet is not a part of the permanent medical record and will be discarded upon discharge.

Source: Copyright © Saint Thomas Health Services, Center for Clinical Evaluation.

Exhibit 14–A–7 Front of New CareMap® for Breast Procedures

CareMap®: Breast Procedures	Addressograph

Inclusion: **Any patient having mastectomy, lumpectomy, wedge resection, or axillary node dissection. DOES NOT include patients who also have reconstructive procedures.**

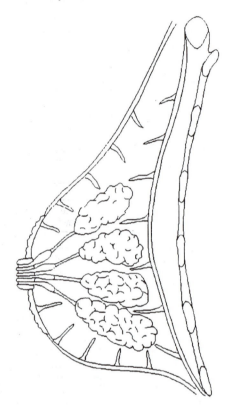

Printed Name	Initial	Printed Name	Initial	Printed Name	Initial

All signatures are those individuals recording information in the CareMap®.

Exhibit 14–A–8 New CareMap® for Breast Procedures, Postop Period

CareMap®: Breast Procedures	Addressograph

Time Interval	POST-OP
Nutrition	As ordered
Activity	As ordered—document on Flowsheet
IVs Medications	As ordered Evaluate for food/drug interactions
Assessment	Vital signs every 4–8 hours Pain assessment every 4–8 hours I/O every 8 hours or as ordered Dressing assessment—document on Daily Flow Sheet House Officer parameters
Treatments	TEDs if ordered SCDs if ordered TCDB every 2 hours and PRN I/S every 1–2 hours and PRN Drain care
Diagnostics	As ordered
Teaching	Evaluate learning needs; document teaching on Education Record Review post-op teaching Begin drain care instructions
Discharge Planning	Evaluate potential Discharge Planning needs Social Services Consult Indicated: **YES** ☐ **NO** ☐ Requested: **Date**_____ **Initials**_____ ☐ **MD deferred**
Psychosocial	Reach to Recovery Consult request verified: **YES** ☐ **NO** ☐ **Initials**_____ Pastoral Care Consult Indicated: **YES** ☐ **NO** ☐ Requested: **Date**_____ **Initials**_____ Consult Initiated: **Date**_____ **Consultant's Initials**_____
Individual Patient Needs	
Individual Patient Outcomes	

Outcomes		Date	Time	Initial
Complete by 12 midnight of Operative Date: **5. Dressings clean, dry** YES ☐ NO ☐	5			
6. UOP within ordered parameters, or minimum of 30 cc/H YES ☐ NO ☐	6			
7. Patient rates pain < 5 on 0–10 scale YES ☐ NO ☐ **ANALGESIA (ROUTE):**_____ **CHANGED TO (IF APPLICABLE):**_____	7			
8. Hemodynamically stable YES ☐ NO ☐	8			
9. Drain(s) patent YES ☐ NO ☐	9			
10. Tolerating PO intake YES ☐ NO ☐	10			

Treatments documented on the CareMap® do not need to be duplicated on the Flowsheet.

Source: Copyright © Saint Thomas Health Services, Center for Clinical Evaluation.

■ 15 ■

Radical Prostatectomy: Process and Outcomes

Roxelyn G. Baumgartner, Nancy Wells, and Michael Koch

Critical or collaborative pathways are a central focus in the collaborative care process. The pathways define interdisciplinary plans of care for high-volume and/or high-cost diagnosis-related groups (DRGs). The pathway is structured to meet specific patient outcomes or goals. It typically includes therapies, diagnostic and laboratory tests, diet, activity progression, and medications during an episode of illness. Therapies, which are dependent upon the patient's needs, may include medical, respiratory, physical, occupational, and nursing therapies. Collaborative pathways provide a structure for communication and coordination of care among health care providers from multiple disciplines. This chapter describes the development and use of a collaborative pathway for a high-volume DRG—radical retropubic prostatectomy—and demonstrates how monitoring outcomes can contribute to improved patient care.

The delivery of high-quality, cost-effective patient care is a primary goal of all health care institutions. At Vanderbilt University Medical Center, we selected the collaborative care process, which follows a continuous quality improvement model,[1,2] to examine and refine our patient care delivery processes. The collaborative care process begins with development of a standard plan of care, the collaborative pathway, with clearly defined patient goals and measurable outcomes (Figure 15–1). Data are gathered on process and outcome indicators, including financial, quality, and patient satisfaction outcomes. The improvement process consists of analyzing process and outcome indicators, comparing results to internal and external stan-

dards, and modifying the plan of care to meet patient goals consistently. Results of this process are evolving high-quality, cost-effective care. The development, data gathering, and analysis of collaborative pathways are completed by interdisciplinary teams composed of caregivers who are directly involved in care delivery for the case type or DRG. The collaborative pathway, then, is a central focus in the collaborative care process.

Collaborative pathways have evolved in our institution over the past eight years. The first pathways were developed by clinical nurse specialists (CNSs) in collaboration with staff nurses who cared for that particular patient population. These pathways were based on "usual care." Physicians' involvement in developing collaborative pathways followed four years later, at the time when the role of CNS was modified to CNS/case manager.[3] This physician involvement proved to be difficult at times but rewarding as physicians began to examine variations among their practices in an attempt to develop a standardized plan of care for specific case types. One powerful method used to examine "usual care" involved the review of patient bills. This highlighted chargeable therapies ordered for patients and provided guidance in determining a reasonable patient trajectory for specific DRGs. Since 1993, collaborative pathways have been developed and modified by interdisciplinary teams, with primary roles for the physician and an advanced-practice nurse in a case manager role. While many inpatient units continue to use paper documentation, there are some inpatient units where documentation is automated so that variances in the pathway can be monitored electronically.

The radical prostatectomy (DRG 334 and 335) is a high-volume procedure in our institution. Standardization of care for this DRG was relatively easy because of the homogeneous patient population and clearly defined

The initial work on the collaborative pathways was supported through a Robert Wood Johnson/Pew Charitable Trust grant awarded to Judy Spinella, MS, RN, MBA. We thank Sue Erickson, RN, MPH, for her ongoing support of the case management program and Beth Hodge, MS, RNCS, for her contribution to the original radical prostatectomy pathway.

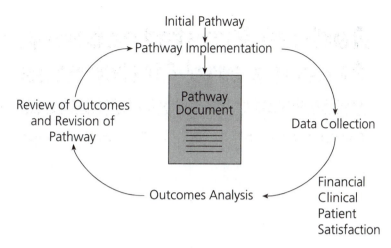

Figure 15–1 Collaborative Care Process

surgical outcomes.[4,5] This collaborative pathway was developed with high physician involvement. Because of the work by the department of urology and the surgeons who perform this operation, an extensive database is available to track patient outcomes. These data are used by all care providers involved with radical prostatectomy patients to modify practice to achieve better patient outcomes. Additionally, patients are currently admitted to a unit that has computerized documentation, which provides electronic variance tracking for this DRG.

RADICAL PROSTATECTOMY PATHWAY

Our pathways are designed to meet intermediate (daily) and terminal (discharge) goals. The plan of care is developed to address 10 broad areas: laboratory tests, diagnostic tests, treatments, activity, diet, consults, equipment and supplies, medications, teaching and discharge planning, and patient flow. To develop the radical prostatectomy pathway, the physicians reviewed the hospital records and charges of 87 patients. This review revealed wide variation in multiple aspects of patient care, such as length of stay (LOS), surgical time, blood loss during surgery, and use of ancillary services.[6] Standards of care were then identified from the literature, and the pathway was developed on the basis of data in the literature rather than our "usual" care. Interventions for which there was no literature to support their use were eliminated or modified by consensus among attending physicians who performed the procedure. Charges for therapies and services also were considered in pathway development.[4] For example, reviewing laboratory charges for preoperative blood donation and blood administration during surgery suggested that a substantial cost saving could be achieved if autologous

blood donation was eliminated. The available medical literature has never demonstrated that donation of autologous blood before surgery reduces the use of "banked" blood during surgery. In fact, there is evidence that it does not.[7–9] This process was repeated for changes made in each intervention used in the pathway. The major changes in clinical care are displayed in Exhibit 15–1.

Collaborative pathways are developed by interdisciplinary teams. The primary disciplines involved in the development of the radical prostatectomy pathway were physicians (urology surgeons), a CNS/case manager, and staff nurses. Because of the nature of the disease and expected recovery, social work, physical therapy, occupational therapy, speech therapy, and respiratory therapy had limited involvement in the development of the radical prostatectomy pathway.

The current collaborative pathway for radical prostatectomy defines goals and interventions from the preoperative clinic visit through hospitalization to the two-week postdischarge clinic visit. Our LOS was five to seven days before introduction of the pathway, which was consistent with the literature on this procedure.[10–12] The

Exhibit 15–1 Major Changes in Clinical Practice of Radical Prostatectomy

- Standardization of surgical carts
- Use of single-dose preoperative antibiotic
- Reduction in preoperative autologous blood donation
- Elimination of postoperative epidural analgesia
- Early postoperative feeding

Source: Data from Koch MO, Smith JA Jr. Clinical outcomes associated with the implementation of a cost-efficient programme for radical retropubic prostatectomy. *Br J Urol.* 1995;76:23–33.

planned LOS defined by the pathway for this procedure is three days. Because of the short LOS, adequate patient education is essential for a smooth and uneventful recovery from surgery. Patient teaching is started during the preoperative clinic visit and includes discussion of the content of the collaborative pathway. A patient information brochure, which also includes content from the collaborative pathway, is provided for the patient and family members to review at home during the preoperative period. This is done to prepare the patient and his or her family members for a three-day hospitalization and the care that will be necessary after discharge. Four goals must be met for discharge: (1) presence of bowel sounds and flatus, (2) toleration of regular food, (3) toleration of analgesics by mouth, and (4) completion of discharge teaching (Exhibit 15–2). Discharge teaching begins on the first postoperative day and involves early identification of complications, activity restrictions, diet progression, and care of the foley catheter, since patients are discharged with a foley catheter in place. A home care instruction sheet is provided before discharge to assist in transition to care at home.

CARE DELIVERY MODEL

Patient care in our academic medical center is currently organized into Patient Care Centers, which include staff from inpatient and outpatient settings. Thus, all nursing staff working with the radical prostatectomy patients have one administrative group (nurse, physician, and business manager). Care delivery on the inpatient units employs a team or modular approach, with licensed nurses paired with nonlicensed personnel providing direct care to a group of patients. Licensed providers from other disciplines are included on the team when patients cared for on the unit require them. For example, respiratory therapists are assigned to the medical intensive care unit, and physical therapists to orthopaedics.

Our case management model employs primarily advanced-practice nurses as case managers. These advanced-practice nurses are Master's prepared and may also have a nurse practitioner license. Because of their clinical expertise, case managers are responsible for coordination of care to a specific patient population that is defined by the collaborative pathway. This coordination entails identifying and addressing patient and system problems that disrupt the efficient flow of patients through our complex academic institution. While patient care is the primary focus, collecting data for evaluation of the collaborative pathway is part of the case manager role.[3] Case managers work collaboratively with social work and utilization review to manage their patient population. The case management team works closely with the medical team and nursing staff. While the case manager is responsible for coordinating care, staff nurses remain the primary care providers, covering both clinical

care and patient/family education. In most instances, the case managers move between the inpatient and outpatient settings, thus providing a link between settings to ensure continuity of care.

Radical prostatectomy patients were admitted to a general surgery unit during the development and early implementation of the radical prostatectomy pathway. Ten months after implementation, urology service patients were moved to a 31-bed combined orthopaedics/urology unit, where they have stayed for the past three years. During development of collaborative pathways, unit staffing was 100 percent registered nurse. Skill mix on the unit where radical prostatectomy patients are currently cared for has ranged from 53 percent registered nurses in 1994 to 45 percent in 1996.

OUTCOME MEASUREMENT

Outcomes of radical prostatectomy are categorized into three broad areas: financial, quality, and satisfaction. Financial data include LOS, cost, and hospital charges. Quality data are data obtained by monitoring aspects of the process that were initially modified on the basis of patient charges and literature review. These data include surgical time, estimated blood loss during surgery, blood transfusions, and antibiotic use. Additionally, minor and major complications are obtained from the time of surgery through 30 days after surgery. Readmission rates also are monitored for the first 30 days after hospitalization. In addition to specific patient outcomes, variances or deviations from the pathway are obtained from the computerized patient record. Patient satisfaction is routinely collected six weeks postoperatively through the use of a mailed, anonymous Patient Satisfaction Questionnaire.

Table 15–1 displays the type of data, source(s) of retrieval and storage, and schedule of review by the interdisciplinary team. Data that are readily available, such as financial data from information systems and antibiotic use from pharmacy, are reviewed continuously by the surgeons and the CNS/case manager. The quarterly department of urology case management meeting, attended by all disciplines involved in care of urology service patients, is used to systematically review financial, quality, and satisfaction outcomes for radical prostatectomy. The interdisciplinary group identifies deviations from the pathway, identifies trends found in the data, and plans modifications to the plan of care that will result in improved outcomes for radical prostatectomy patients. Thus, these department case management meetings provide the structure for systematic review of data and continuous improvement in patient care and outcomes.

OUTCOME TRACKING

The financial outcomes and the majority of quality outcomes were obtained retrospectively for 125 patients undergoing radical prostatectomy before July 1993,

Exhibit 15–2 Collaborative Pathway for Radical Retropubic Prostatectomy

DRG Number: _____

ELOS: _____ 2–3 days _____

	Pre-Op (Outpatient)	Day of Surgery (OR → RW)	Post Op: Day One (RW)	Post Op: Day Two (RW → Home)	Post Op: Day Three (RW → Home)	F/U: 14 Days (Clinic)
Goals:	Pre-op testing completed WNL Patient/family teaching completed Consent signed		Bowel Sounds present, no flatus Drain Output <150cc/day UOP > 150cc over 4 hours	Bowel Sounds present, +/– flatus Drain output <100cc/day Possible D/C if tolerate diet Remove drain if <60cc and D/C planned today	Bowel Sounds present, passing flatus Follow up appt. w/MD scheduled D/C drain if <60cc/24* Discharge home with Foley catheter D/C teaching completed	Pt to bring Depends undergarment to clinic visit
Labs:	SMA 18 CBC w/plt T&S		Hct			
Tests	H&P CXR EKG					
Treatments	Consent signed	VS q 4* x 24, then q 8* ─────────────▶ I&O q 4 hrs ─────────▶ Foley catheter ────────────────────────────▶ IMED Pump ─────────────▶ JP drain ─────────▶ PAS stockings Pulm. toilet: TCDB and IS q 1 hour WA		D/C Possible D/C drain ─────▶	D/C drain	Foley remains in for 2 weeks after discharge
Activity	Ad Lib	May be OOB tonight	Ambulate in hall TID ──────────────────────────▶			
Diet	Clear liquid diet for supper, then NPO at 12 MN, night before surgery	NPO No ice chips	Full liquid evening meals	Regular diet	Regular diet	
Consults	Anesthesia Case Manager					
Equipment & Supplies		OR Supplies Anesthesia Supplies IMED/IV tubing JP drain ──────────────────────▶ PAS stockings Foley/Urine Bag PCA pump ──────▶ D/C		Possible D/C drain ─────▶	D/C drain	
Meds/IV	2 Bisacodyl Tabs PO in AM the day before surgery	PREOP: • Cefazolin IV 1 hr prior to surgery POSTOP: • IV (D5 1/2 NS with 20 mg KCL at 150cc/hr) • Ketorolac 30 mg IV in RR then 15 mg IV q 6° x 36° ─────▶ • PCA pain med ─────▶ D/C PCA • Promethazine IV Q 6 HR prn nausea • Tylenol gr x prn T > 101°	IV Fluid (decrease rate 100cc/hr) Oxycodone and Tylenol prn pain	Heplock IV Dulcolax supp. if pt. hasn't passed flatus ─────▶ D/C	D/C Heplock	
				Home care instructions May shower No driving No lifting >5–10 lbs	Wound care: Keep clean & dry except when showering MOM for constipation Call for fever >101, wound erythema or increased tenderness, nausea/vomiting Percocet PRN pain Normal for some leakage around Foley	
Teaching/ D/C Plan	Procedure Plan of Care VUMC Orientation	Procedure Post-op Care RW routines	Begin Home Care Instructions	Instruct pt in use of leg bag Complete home care instructions and give pt HCI sheet		
Patient Flow	Complete tests and labs	Admit EMA Usual surgery time 1.5–2 hours Average blood loss: 550 cc General Anesthesia Lower abd. midline incision			Schedule F/U appt w/attending M.D. in 14 days Pt to bring urine protective pad (Depends under-garment) to F/U appt.	

Note: CBC, complete blood count; CXR, chest X-ray; D/C, discontinue, discharge; D5½ NS, 5½% dextrose in normal saline; EKG, electrocardiogram; EMA, early morning admission; F/U, follow-up; Hct, hematocrit; H&P, history and physical; HCI, home care instruction; I&O, intake and output; IS, incentive spirometer; KCL, potassium chloride; IMED, infusion; IV, intravenous; JP, Jackson Pratt; MN, midnight; MOM, milk of magnesia; NPO, nothing by mouth; OOB, out of bed; OR, operating room; PAS, pneumatic antishock; PCA, patient-controlled analgesia; PO, oral; PRN, as needed (for); plt, platelets; RR, recovery room; RW, round wing; SMA, blood chemistries; T&S, type & screen; TCDB, turn, cough, deep-breathe; TID, three times a day; UOP, urinary output; VS, vital signs; VUMC, Vanderbilt University Medical Center; WNL, within normal limits.

Source: Copyright © Vanderbilt University Medical Center.

Table 15–1 Outcomes Monitored for Radical Prostatectomy

Outcome	Definition	Source	Storage	Review
LOS	Days hospitalized	Financial CM record[a]	Dept.[b]	C[c]
Cost	Ratio of charges in $	Financial	Dept.	C
Charges	In $	Financial	Dept.	C
Surgical time	Minutes from incision to close	OR CM record	Dept.	Q[d]
Estimated blood loss	ml of blood lost during surgery	OR CM record	Dept.	Q
Blood transfusions	No. of units of blood used during hospital stay; autologous and autogenic	Patient record CM record	Dept.	Q
Antibiotic use	No. of doses of antibiotics during hospital stay	Pharmacy CM record	Pharmacy	C
Minor complications	Complications that extend LOS and/or require intervention	Patient record	Dept.	Q
Major complications	Life-threatening events or events that require emergency surgery	Patient record	Dept.	Q
Readmission rate	No. of times hospitalized in first 30 days after discharge	CM record	Dept.	Q
Variances	Deviation from pathway goals	Information management	Same	Q
Patient satisfaction	10-item questionnaire	Anonymous mailed survey	Dept.	Q

[a]CNS/case manager tracking record.
[b]Department of urology database.
[c]Continuous review.
[d]Quarterly review.

when the pathway was implemented. Since implementation, data have been gathered prospectively for 322 consecutive patients. Table 15–2 displays the demographic and medical characteristics for patients cared for before and after pathway implementation. In the aggregate, patients having radical prostatectomy after implementation are older and have lower preoperative prostate-specific antigen than patients operated on before implementation. Clinical stage of disease and surgical risk (American Society of Anesthesiology grade) are similar for patients before and after implementation. These data indicate some change in the patient population before and after implementation of the radical prostatectomy pathway.

FINANCIAL OUTCOMES

LOS was 5.7 ($SD = 2.1$) days before the pathway was implemented in July 1993. The pathway defines a three-day LOS, which is similar to the mean LOS since imple-

mentation ($M = 3.3$, $SD = 1.5$). This represents a 42 percent decrease in LOS over this four-year period. Cost and hospital charges have shown similar reductions from before to after implementation (Table 15–3). Average hospital charges were reduced by $4,847 per case; with a concomitant reduction in cost of $3,292, representing reductions of 39 percent for costs and 37 percent for charges. In addition to reductions in all three financial outcomes, implementation of the pathway has led to a more consistent pattern across patients (Figure 15–2). This pattern reflects the reduction in variation in clinical care that patients received once the pathway was implemented.

One method of continuous improvement is comparison of outcomes to a "gold standard" or benchmark. Financial outcomes are compared to an external standard provided by the University Health Systems Consortium (UHC). These comparisons reveal that aggregate data from 1996 compared favorably with the benchmark provided by UHC (Table 15–4).

Table 15–2 Patient Population before and after Implementation of Radical Prostatectomy Pathway

Variable	Before Implementation (N = 125)	After Implementation (N = 322)	p
Age in years (SD)	64 (6.9)	62 (1.5)	< .01
PSA[a] (SD)	17.2 (30.8)	11.1 (10.2)	< .01
ASA[b] (SD)	2.0 (0)	2.1 (0.38)	ns
Clinical stage (%)			ns
T1 a–c	71%	67%	
T2	22%	26%	
T3	2%	2%	

[a]PSA, prostate-specific antigen.
[b]ASA, American Society of Anesthesiology.

QUALITY OUTCOMES

Preoperative and Operative Indicators

Quality measures of the surgical experience show reductions similar to those of financial outcomes. Mean surgical time and estimated blood loss declined from before to after implementation (Table 15–5). Only 7 percent of the first 132 patients undergoing radical prostatectomy after pathway implementation donated blood preoperatively. Of the 93 percent of patients who chose not to donate blood preoperatively, as recommended by the pathway, 2.4 percent required transfusions.[8] With fur-

Table 15–3 Changes in Financial Outcomes from before to after Implementation of Radical Prostatectomy Pathway

Variable	Before Implementation Mean (SD)	After Implementation Mean (SD)	p
LOS in days	5.7 (2.1)	3.3 (1.5)	< .001
Cost in $	8,423 (2,894)	5,131 (1,398)	< .001
Charges in $	13,197 (4,364)	8,350 (2,289)	< .001

ther experience using the surgical approach and postoperative plan of care, less than 1 percent of patients have required blood transfusions. This reduction in use of blood transfusions also is reflected in reduced variation in estimated blood loss during surgery (Figure 15–3) and lower surgical times (Table 15–5) from before to after implementation. All of these changes translated into lower operative, anesthesia, and laboratory charges for patients managed on the radical prostatectomy pathway.

Postoperative Indicators

The quality measures for postoperative care, antibiotic use and complication rates, have demonstrated that the pathway produces beneficial patient outcomes following radical prostatectomy (Table 15–5). Readmission rates, which have been tracked only since implementation, have been low. For example, there was only one readmission in the last five months, which was unavoidable because of a urinary tract infection.

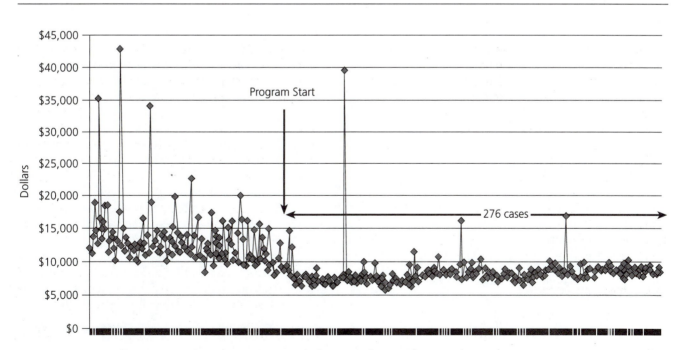

Figure 15–2 Financial Outcomes of Consecutive Patients before and after Implementation

Table 15–4 Comparison of Financial Outcomes to University Health Systems Consortium (UHC) Outcomes

Variable	Institutional Performance		UHC Best Performance	
	With CC	Without CC	With CC	Without CC
LOS in days, 1996	3.21	3.03	4.36	3.61
Cost in $, 1996	6,320	5,850	7,890	6,400
Cost in $, 1995	5,940	5,490	8,510	7,070

CC = comorbid condition.

Variance Analysis

Failure to meet intermediate and discharge goals is recorded on the computerized patient record. The pathway is included in the computerized patient record, and documentation is by exception.[13] Sources of variance are categorized to assist in modifying either the pathway or the care delivery system. The first level identifies the source of variance as (1) patient/family, (2) provider, (3) system, (4) community, or (5) not applicable. Within each category, a more specific reason for the variance is documented. For example, patient/family variances may be related to physiological or psychosocial problems that preclude meeting intermediate goals, poor pain control, noncompliance with treatment recommendations, or un-

availability of patient/family for teaching. Provider variances are associated with change in plan of care and lack of communication among care providers working on the unit, whereas system variances are associated with problems in coordination of required diagnostic tests and therapies provided by other departments. The major source of community variance is the lack of beds for placement at a lower level of care, which is rarely a problem with the radical prostatectomy population.

Figure 15–4 displays the variances for the last quarter of 1996 and the first two quarters of 1997. Few variances occur with the radical prostatectomy population (2.7 to 4 percent), and when they occur, they are usually failure to meet the intermediate goals. Sources of variance in this pathway that may be modified include (1) patient/family factors (e.g., teaching not complete, patient noncompliant with plan); (2) provider factors (e.g., orders not written, plan of care changed without communication to other staff); and (3) system factors (no extended care facility bed available). Variances related to the specific clinical care changes in pain management and diet progression (Exhibit 15–1) support the appropriateness of the plan of care. No variances have occurred because of poor pain control. In addition, few variances have occurred because of early feeding. The lack of variances suggests that the changes in pain control measures and diet progression made during the initial pathway development are appropriate interventions for this procedure.

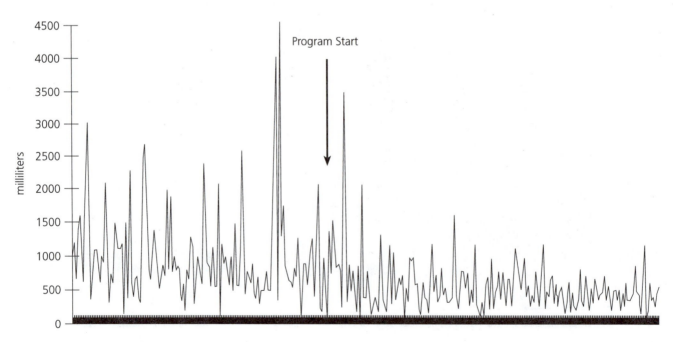

Figure 15–3 Estimated Blood Loss of Consecutive Patients before and after Implementation

Table 15–5 Quality Outcomes before and after Implementation of Radical Prostatectomy Pathway

Variable	Before Implementation	After Implementation	p
Surgical time in min. (*SD*)	198 (67)	161 (57)	< .001
Estimated blood loss in ml (*SD*)	1,023 (726)	576 (415)	< .001
No. of blood transfusions (*SD*)	1.81 (1.6)	0.28 (1.1)	< .001
1-dose antibiotic %	Not available	100%	—
Minor complications %	16.8	10.5	trend *p* = .07
Major complications %	3.2	2.2	ns
Readmission rate %	Not available	.04%[a]	—

[a]First five months of 1997.

PATIENT SATISFACTION OUTCOMES

A 10-item Patient Satisfaction Questionnaire is mailed to each patient six weeks after surgery. Because of the substantial reduction in LOS, the questionnaire focuses on the preparation for and comfort with discharge. Additional satisfaction items address preparation for hospitalization and overall satisfaction with hospitalization. Satisfaction items use a binary (yes/no) response format, with open-ended questions to probe for additional information on problems encountered with hospitalization or discharge. Patient satisfaction data are reported quarterly and are reviewed at the department level.

Of the 143 questionnaires distributed, 123 were completed and returned, representing an 86 percent response rate. The overall satisfaction with hospitalization was high (87 percent). Almost all patients who returned the questionnaire (97 percent) felt adequately prepared for discharge. Ninety-five percent received at least one follow-up telephone call, and 92 percent felt that this telephone call was helpful. Approximately one-third of the respondents experienced some problem after discharge. Incontinence was one frequent problem experienced after discharge and identified from the Patient Satisfaction Questionnaire. Standard care includes discussion of incontinence at multiple points during the episode of illness. Incontinence is addressed (1) by the physician and clinic nurse during the preoperative visit, (2) in the patient information brochure given to each patient before surgery, (3) in postoperative teaching provided by the CNS/case manager, and (4) in the home care instruction sheets given to patients before discharge. While incontinence is typically seen on a temporary basis after radical prostatectomy, our patient responses emphasize the importance of adequate education about this issue. To provide better understanding of incontinence after surgery, the patient information brochure was modified to state clearly that incontinence occurs in all patients and that the duration ranges from two weeks to three months. In addition, increased emphasis was placed on this topic during the postoperative discharge teaching.

The information obtained about patient satisfaction also is systematically reviewed and revised. After we reviewed current data at the quarterly department meet-

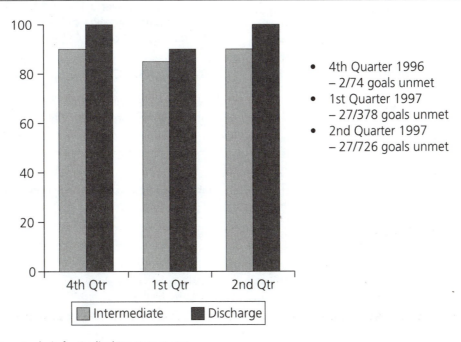

- 4th Quarter 1996
  – 2/74 goals unmet
- 1st Quarter 1997
  – 27/378 goals unmet
- 2nd Quarter 1997
  – 27/726 goals unmet

Figure 15–4 Variance Analysis for Radical Prostatectomy

ing, we revised the Patient Satisfaction Questionnaire to include items on pain management and knowledge of illness trajectory. More general items that did not contribute useful information were deleted from the questionnaire. This illustrates how outcome data can be used to improve care delivery.

PROVIDER ACCEPTANCE OF THE COLLABORATIVE PATH

Anecdotal evidence suggests that the collaborative path for radical prostatectomy has been incorporated by the staff who care for these patients. The attending physicians who developed the pathway refer to it in their progress notes and use it as the basis for discussion during patient rounds. As residents rotate through the urology service, they are initially oriented to the pathway, and they receive a laminated pocket card with the radical prostatectomy path for easy referral. Daily orders are standardized and available from a computerized order entry program, which reduces variation in how patient care is managed. The nursing staff are familiar with the radical prostatectomy pathway. All staff document in a computerized patient record, which is structured so that the only charting necessary is of the exceptions or variances.[13] In addition, the intershift report is structured from the pathway to ensure that the plan of care is followed.

The pathway also provides a basis for patient teaching during the preoperative and postoperative periods. During their preoperative clinic visit, the attending physician and clinic nurse review the plan for hospitalization and early discharge, using the patient information brochure. The discharge plan is reviewed again after surgery using standardized home care instruction sheets, with emphasis on discharge with a foley catheter in place and incontinence once the foley is removed.

INTERDISCIPLINARY COLLABORATION

Collaborative pathways provide a structure for interdisciplinary collaboration.[2,14] Development of the pathway requires examination of patient care delivery practices of all disciplines involved with the case type and fosters communication and negotiation among disciplines. Use of the pathway ensures coordination of care focused on attainment of patient goals within a specific time frame. Consistent with the anecdotal evidence presented, the pathway also fosters communication among staff. To elucidate the impact of collaborative pathways on interdisciplinary collaboration, a general measure of collaboration was obtained from licensed staff on selected units at the beginning of collaborative care implementation and 16 months later.

Collaborative Practice Scale

Licensed staff were surveyed using the modified Collaborative Practice Scale.[14-16] The modified CPS-M is a 10-item scale with a 4-point response format. The total scale score reflects the degree to which providers communicate, coordinate, and give and receive input from other providers when making decisions about patient care. A five-item scale tapping perceived physician involvement in collaborative care was administered with the CPS-M.[14,16] Data were collected early in the implementation of collaborative paths (1993) and 16 months later (1995).

Licensed staff from the unit that currently cares for radical prostatectomy patients completed the survey in 1993 ($N = 21$) and 1995 ($N = 34$). Thus, the preimplementation data were collected before the radical prostatectomy patients were admitted to this unit. This sample included nurses (57 percent in 1993; 59 percent in 1995), physicians (5 percent in 1993; 27 percent in 1995), and other licensed providers (38 percent in 1993, 14 percent in 1995) such as therapists, social workers, and pharmacists. Collectively, the staff from this unit reported relatively high levels of collaboration that remained stable over time (Table 15–6). In addition, perceived physician involvement was moderate and rose from 1993 to 1995 as licensed staff incorporated the collaborative pathways into their care delivery routine. These findings suggest that licensed providers believe that they do collaborate to provide care to their patients and perceive that physicians are involved in collaborative practice. Although the level of perceived physician involvement rose over a 16-month period, the implementation of pathways did not lead to an increase in collaboration measured by the CPS-M.

STAFF SATISFACTION

Staff satisfaction data are routinely obtained in our institution every three years. Satisfaction items reflect mul-

Table 15–6 Interdisciplinary Collaboration in Early and Late Phases of Pathway Implementation

Scale	Mean (SD)		p
	1993 (n = 21)	1995 (n = 34)	
Collaborative Practice Survey—Modified	3.18[a] (0.56)	3.12 (0.57)	ns
Perceived physician involvement	2.40 (0.68)	2.82 (0.62)	ns

[a]Possible range, 1 to 4; higher scores reflect greater collaboration/involvement.

tiple aspects of work, including teamwork, professional identity, patient care, work environment, and pay/benefits.[17] A 4-point response format is used to measure both importance of and satisfaction with each item on the survey. These data are collapsed into six multi-item scales; the two scales most affected by collaborative pathways are teamwork and patient care.

Satisfaction data are displayed from staff working on the unit that currently cares for radical prostatectomy patients in 1993 (*N* = 22) and 1996 (*N* = 28). At the first data collection point, radical prostatectomy patients were not cared for on the unit. The composition of staff includes nurses (54 percent in 1993; 61 percent in 1996), other licensed personnel (30 percent in 1993; 7 percent in 1996), and nonlicensed personnel (10 percent in 1993; 32 percent in 1996). Physicians are not included in this satisfaction survey; therefore, no data are available on physician satisfaction. Within the institution, staff satisfaction with patient care improved significantly from 1993 to 1996, with no significant change in satisfaction with teamwork over the same time period. The changes on the unit where radical prostatectomy patients are hospitalized are similar to those seen for the institution, but the changes from 1993 to 1996 are not statistically significant for patient care or teamwork scales (Table 15–7).

CONCLUSION

The radical prostatectomy pathway and the work on measuring and using outcomes to improve patient care are viewed positively within the organization. Implementation of the pathway has demonstrated beneficial financial outcomes that have resulted in substantial cost savings to the institution and payers. In addition, the financial outcomes compared favorably with internal and external benchmarks. These findings are consistent with previous research demonstrating reduced LOS and cost savings with the implementation of collaborative pathways and case management.[18,19] While many work redesign and quality improvement initiatives occurred in our institution during the implementation, none of these innovations could be clearly related to the reduction in LOS, cost, and charges. These financial indicators for radical prostatectomy, however, clearly began a downward trend shortly after the implementation of the pathway and reductions in variation that have been maintained for the last two years. These findings suggest that financial outcomes for radical prostatectomy are likely to be related to collaborative pathway implementation.

Reducing variation in practice has resulted in reduced variance in operative and postoperative quality indicators. The quality of care provided also is reflected in the high levels of satisfaction reported by patients. Though

Table 15–7 Staff Satisfaction in Early and Late Phases of Pathway Implementation

Satisfaction Scale	Mean (SD)		p
	1993	*1996*	
Teamwork	2.87[a] (0.64)	2.91 (0.63)	ns
Patient care	3.29 (0.45)	3.32 (0.42)	ns

[a]Possible range, 1 to 4; higher scores reflect greater collaboration/involvement.

our patient satisfaction data are limited to the period after implementation, Goode[2] reported a significant increase in overall patient satisfaction from before to after implementation of pathways and case management in women undergoing Caesarean section.

Consistent with the collaborative care process, analysis of process and outcome indicators has guided change in clinical care. Examination of variances assists in the identification of patient, provider, and system problems that may interfere with attaining patient goals. These data are used to modify aspects of the care delivery process to better meet patient goals. The use of patient satisfaction data to modify the timing and content of education that patients and their family members receive is a clear example of the continuous improvement process.

Four years after implementation, the radical prostatectomy pathway has been integrated into the care delivery model by all staff working with this DRG. The lack of change in provider collaboration and satisfaction is generally consistent with previous research[2] in which general measures of collaboration and satisfaction have been used. However, staff satisfaction was increased over a shorter time (9 to 18 months) with the implementation of case management and pathways.[20]

In conclusion, the radical prostatectomy pathway provides an example of how an interdisciplinary plan of care used within collaborative care process enhances patient outcomes. Continuous improvements are made through systematic collection and analysis of financial, quality, and patient satisfaction data. The work done on radical prostatectomy has become an internal benchmark for successful implementation of collaborative pathways.

REFERENCES

1. Donabedian A. The role of outcomes in quality assessment and assurance. *QRB.* 1992;18:356–380.
2. Goode C. Impact of CareMap and case management on patient satisfaction, staff satisfaction, collaboration, and autonomy. *Nurs Econ.* 1995;13:337–349.

3. Wells N, Erickson S, Spinella J. Role transition: from clinical nurse specialist to clinical nurse specialist/case manager. *J Nurs Adm.* 1996;26:23–28.

4. Koch MO, Smith JA Jr. Cost containment in urology. *Urology.* 1995;46:14–26.

5. Bennett CL, Buchner DA, Ullman M. Approaches to prostate cancer by managed care organizations. *Urology.* 1997;50:79–86.

6. Koch MO, Smith JA Jr. Clinical outcomes associated with the implementation of a cost-efficient programme for radical retropubic prostatectomy. *Br J Urol.* 1995;76:28–33.

7. Etchason J, Petz L, Keeler E, et al. The cost effectiveness of preoperative autologous blood donation. *N Engl J Med.* 1995;332:719–724.

8. Koch MO, Smith JA Jr. Blood loss during radical retropubic prostatectomy: is preoperative autologous blood donation indicated? *J Urol.* 1996;156:1077–1080.

9. Goh M, Kleer CG, Kielczewski P, Wojno KJ, Kim K, Oesterling JE. Autologous blood donation prior to anatomical radical retropubic prostatectomy: is it necessary? *Urology.* 1997;49:569–574.

10. Licht MR, Klein EA. Early hospital discharge after radical retropubic prostatectomy: impact on cost and complication rate. *Urology.* 1994;44:700–704.

11. Litwin MS, Kahn KL, Retcus N. Why do sicker patients cost more? A charge-based analysis of patients undergoing prostatectomy. *J Urol.* 1993;149:64–88.

12. Toy PT, Mendozzi D, Strauss RG, Sehling LC, Kruskall M, Ahn DK. Efficacy of preoperative donation of blood for autologous use in radical prostatectomy. *Transfusion.* 1993;33:721–724.

13. Ashworth G, Erickson S. Enhancing an outcome-focused collaborative care model with charting by exception. In: Burke L, Murphy J, eds. *Charting by Exception Applications.* Albany, NY: Delmar; 1994.

14. Wells N, Johnson R, Salyer S. Interdisciplinary collaboration. In review.

15. Weiss S, Davis H. Validity and reliability of the Collaborative Practice Scales. *Nurs Res.* 1985;34:399–405.

16. Wells N, Holder G, Dengler S. Staff-nurse managed collaborative care: evaluation on a rehabilitation unit. *Ser Nurs Adm.* 1996;8:187–202.

17. Ames A, Adkins S, Rutledge D, et al. Assessing work retention issues. *J Nurs Adm.* 1992;22:37–41.

18. Blegen M, Retter R, Goode C, Murphy R. Outcomes of hospital-based managed care: a multivariate analysis of cost and quality. *Obstet Gynecol.* 1995;86:809–814.

19. Cohen EL. Nursing case management: does it pay? *J Nurs Adm.* 1991;21:20–25.

20. Cohen EL, Casa TE. *Nursing Case Management: From Concept to Evaluation.* St. Louis, MO: CV Mosby; 1993:156–163.

■ 16 ■

Developing an Outpatient Surgery Clinical Path for Laparoscopic Cholecystectomy

Mae Taylor Moss and Sue Brown

Clinical care pathways are outcome-oriented guidelines for care meant to provide cost-effective systematic care for high-volume predictable cases. These pathways, sometimes called *critical pathways,* delineate patient care, carefully sequencing interventions and standardizing the process, in an effort to maximize quality care while minimizing costs. Provision of care within these planned pathways is then assessed through outcome measures that contribute to continuous quality improvement, creating a system in which care is continuously redefined and refined. In addition, adherence to these pathways strengthens documentation and reduces variances from case to case.

At Saint John's Health Center in Santa Monica, California, the creation of an outpatient surgery clinical pathway for laparoscopic cholecystectomy became both an end and a means. Because 20 million Americans have cholelithiasis and 1 million cases are diagnosed each year,[1] we knew that laparoscopic cholecystectomy fit the profile of a predictable, high-volume procedure suitable for documentation in a clinical pathway. We also knew that such a clinical path would affect a large segment of our patient population and have far-ranging implications. Beyond that, creating a clinical path became a way of achieving the end—improving the perioperative experience—and of finding a means—discovering how to apply our traditional skills in new ways. We knew how to care for patients, but we learned how to make that caring more efficient, more ready for multidisciplinary interventions and cooperation, and more amenable to refinement and improvement.

Our clinical care pathway sought to outline the procedures involved in perioperative care and to define outcome measures that, when assessed, would help shape improvement. Aristotle said, "A whole is that which has beginning, middle, and end," and we believed that each step could tell us something about the next while creating a "whole" perioperative experience that met the high standards of patients and providers alike. This meant that outcome measures had to include not only discharge criteria but intermediate criteria as well[2]; it meant that review of postoperative measures might have roots of success or failure in preoperative preparation.

This chapter describes Saint John's development of its first outpatient clinical path and the path's use as a pilot to create a generic ambulatory clinical path for all outpatient surgery procedures.

PATHWAY

When the California earthquake of 1994 closed Saint John's to inpatients and limited the outpatient procedures that surgeons could perform, we took advantage of this unique break in service. While our architectural infrastructure was being rebuilt, we began to reexamine our organizational infrastructure and to develop the institution's first outpatient clinical path. Our initial task was to choose an inpatient procedure that we believed was on the threshold of becoming an outpatient procedure.

Laparoscopic cholecystecomy was introduced in France in 1987[3-5] and in the United States in mid-1988.[6] Within 12 months of its introduction in the United States, almost 100 percent of surgeons and hospitals had adopted the laparoscopic technique.[7] In 1991, only 3 percent of all cholecystectomies were performed laparoscopically; by 1993, the percentage had grown to 67 percent, and surgical groups were reporting series of as many as 500 patients.[1,8] Today, open procedures constitute only 1 percent of the cholecystecomies performed in our institution. Readily embraced by patients and physicians alike, the laparoscopic cholecystectomy was associated with less postoperative pain, shorter hospitalization, and more rapid return to work. Rightly, we believed that an opportunity existed to convert laparoscopic cholecystectomy to an outpatient procedure, allowing us not only to maintain quality of care but also to improve patient outcome, improve patient satisfaction, and lower costs.

To develop, institute, assess, and refine the laparoscopic cholecystectomy outpatient path, we employed multidisciplinary team management, intercommittee cooperation, data collection, and communication. As a foundation, we gathered together a multidisciplinary team including nurses, anesthesiologists, surgeons performing laparoscopic cholecystectomy, gastroenterologists, and outpatient program administrators. We divided the outpatient surgical process into the obvious three stages—preoperative, intraoperative, and postoperative—but incorporated the patient's preoperative interview in the surgeon's office and the preoperative workup in the first stage in an effort to better reflect the patient's experience and not only ours (Table 16–1). To each of these we assigned a time frame in which we believed the tasks could be completed: the office interview (one hour); the preoperative workup (one-hour process to be completed 48 hours before surgery); the day-of-surgery preoperative process (one to two hours); surgery (two hours); and the postoperative period, including recovery in the postanesthesia care unit (PACU) (one hour) and in the second-stage recovery unit (one to two hours). Overall time requirement was 8 to 10 hours.

Six general outcomes were named: (1) patient is discharged within 23 hours, (2) patient's pain is managed, (3) patient's nausea is managed, (4) no signs of postoperative infection are visible, (5) patient's operative site is intact, and (6) patient is knowledgeable of signs and symptoms necessitating notification of surgeon. In a more specific approach, we described expected outcomes for each of the segments of the patient's experience. Within the path, we named actions for each of the following variables:

- laboratory or diagnostic studies
- case management
- patient education
- prophylaxis
- consultations
- intravenous therapy
- pain management
- activity
- nutrition
- assessment
- consent
- medications
- protocols

Arranging these orders into a chart facilitated tracking variances and provided a guideline for the entire process, streamlining performance by outlining tasks (Table 16–1). Actions in the category of *activity*, for example, were:

- In preoperative assessment before the day of surgery: "Assess home activity level."
- In preoperative assessment on the day of surgery: "Allow patient up ad libitum until IV inserted or preoperative sedative administered."
- In postoperative period: "Take fall/risk precautions. Consider patient's ability to sit up in chair. Ambulate before discharge."

Preoperative

Because of its power to set expectations, the patient-surgeon interview became an area of focus of the preoperative period—a beginning that we believed would affect the "middle" (outpatient surgery) and the "end" (discharge). A poll of the surgeons told us that the interview was restricted to obtaining informed consent, but we believed that this was too limiting. Researchers in England had postulated a relation between extended recoveries (mean time to return to work–3.3 weeks) following laparoscopic cholecystectomy and physicians' advising significant restrictions in the postoperative period. They recommended better communication to adjust prevailing attitudes to align with improved procedures.[9] Therefore, to rescue this lost opportunity in the surgeon's office, we suggested that surgeons view this exchange as an opportunity to create expectations of an *outpatient*—not an *inpatient*—procedure and to educate the patient about the patient flow process (Figure 16–1), all the way to the postoperative period.

Furthermore, extending the reach of perioperative services to the surgeon's office allowed better coordination in admissions, preoperative assessments, and discharge planning. Part of this experience has led to a lighter load of day-of-surgery nursing assessments because making telephone follow-up calls, a practice implemented during the path's first official year of practice, allows incomplete assessments to be completed ahead of time, even if the patient failed to make or keep an appointment for preoperative evaluations.

Standardization of laboratory work was another focus of the analysis of the preoperative period. Preoperative intravenous orders for patients 11 years of age and older were generally standardized and committed to a form. On this form, the specifications for intravenous preparations, including heparin, are generally outlined. To individualize testing, we provided a selection of diagnostic tests in checklist form to allow physicians to base testing on the

Table 16–1 Laparoscopic Cholecystectomy Clinical Pathway

Category	Physician's Office ≤0.5 hr	Preoperative Patient Preparation 1 hr	Preoperative 1–2 hr	Intraoperative 2 hr	Postoperative 1 hr PACU; 2 hr second-stage care
Time frame	≤0.5 hr	1 hr	1–2 hr	2 hr	1 hr PACU; 2 hr second-stage care
Outcome	• Ensure patient understands concept of outpatient procedure	• Ensure patient able to verbalize understanding of surgical procedure. • Ensure patient completes preoperative testing and assessment. • Validate all essential paperwork (history and physical, consent, lab work, and physician's orders) on chart day before surgery.	• Validate all essential paperwork on chart. • Ensure that patient prepared for surgery and has minimal anxiety level.	• Ensure surgery completed without complications.	• Meet discharge goals and discharge patient from outpatient setting.
Lab/diagnostic studies	• Obtain chemistry panel: CBC, platelet count, prothrombin time, partial thromboplastin time.	• Obtain lab/diagnostic studies as ordered.	• Obtain needed lab/diagnostic studies not done in POPP.	• Verify results of lab/ diagnostic studies are on chart.	• Obtain lab/diagnostic studies as ordered.
Case management		• Assess discharge needs. • Review discharge instructions. • Ensure adult available to drive home after discharge.	• Ensure responsible adult available to drive patient home.		• Phone to follow up the day after discharge.

Patient/SO education	• Create expectation of *outpatient—not inpatient*—procedure
	• Educate patient about patient flow process, including postoperative management protocol
		• View video.
		• Review preoperative and discharge instructions.
		• Review clinical pathway with patient and/or significant other.
			• Reinforce previous education.
			• Initiate outpatient postoperative management protocol.
			• Inform/direct family and/or significant other to surgical waiting area.
				• Overview briefly inappropriate phase.	• Instruct patient/family regarding
					  a) wound care
					  b) activity
					  c) signs and symptoms to report to physician
					  d) instructions for follow-up with physician.
					• Give patient: "Ambulatory Surgery Postoperative Instructions."

Prophylaxis				• Ensure sequential TEDS and Foley catheter present.	• Discontinue Foley in PACU if started in OR.
Consultations		• Arrange consultations as needed for comorbidities.	• Ensure anesthetist assesses patient prior to OR and initiates needed treatments, lab work, preop medications.		

continues

Table 16–1 continued

Category	Physician's Office	Preoperative Patient Preparation	Preoperative	Intraoperative	Postoperative
IV therapy			• Initiate IV therapy per anesthesia standing orders unless ordered otherwise.		• Discharge when able to tolerate fluids, ambulate, and void
Pain management		• Discuss availability of postop analgesia.	• Review availability of postop analgesia.	• Administer Toradol at end of case.	• Administer IM/PO analgesia. • Review discharge medication. • Arrange for prescription to be filled if requested and approved.
Activity		• Assess home activity level.	• Allow patient up ad libitum until IV inserted or preop sedative administered.		• Take fall/risk precautions. • Ambulate prior to discharge.
Nutrition		• Give instructions for NPO status.	• Determine patient NPO as instructed.		• Allow liquids postop as ordered.
Assessment		• Assess for help at home after surgery. • Initiate "Ambulatory Services Nursing Flowsheet."	• Perform admission physical assessment • Complete "Ambulatory Services Nursing Flowsheet."	• Assess patient for allergies, loose teeth, contacts, last time anything to eat or drink, hepatitis exposure. • Report intraoperative interventions to PACU.	• Complete physical assessment in PACU per protocol. • Assess per "Outpatient Postop Management Protocol."
Consent	• Ensure that consent forms signed.	• Ensure that consent forms signed.	• Ensure that consent forms are signed if not done in preoperative patient preparation unit.	• Validate consent forms.	
Medications		• Give instructions regarding which medications should be taken day of surgery.	• Administer preop sedatives and other medications as ordered. • Administer antiemetics as indicated. • Administer 1 gm Ancef 1° prior to surgery.	• Administer antibiotics.	• Administer antiemetics as needed.
Protocols			• Follow "Outpatient Postop Management." • Follow "Peripheral IV Therapy Management."	• Follow "Management of the Perioperative Patient." • Follow "Peripheral IV Therapy Management." • Follow "Outpatient Postop Management."	• Follow "Outpatient Postop Management." • Follow "Fall/Injury Risk Prevention." • Follow "Post-Anesthesia Management." • Follow "Peripheral IV Therapy Management."

Note: CBC, complete blood count; IV, intravenous; IM, intramuscular; NPO, nothing by mouth; OR, operating room; PACU, postoperative acute care unit; PO, by mouth; SO, significant other; TEDS, antiembolism stockings.

Source: Copyright © Saint John's Health Center.

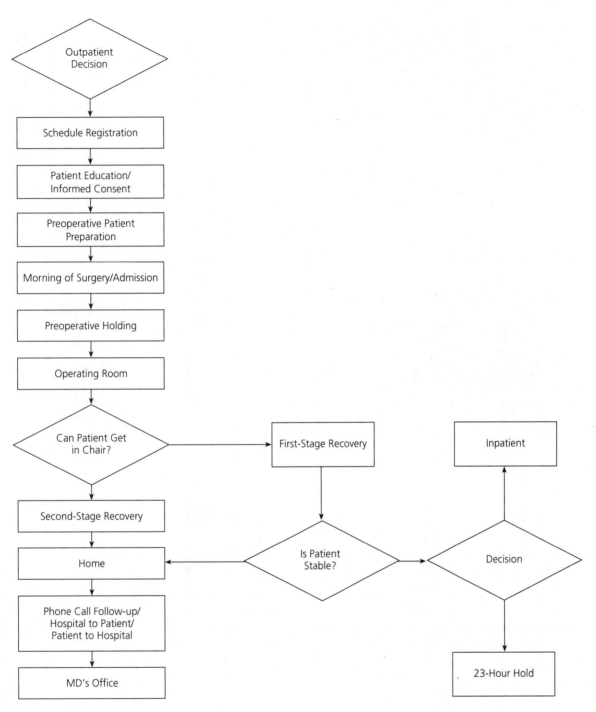

Figure 16–1 The patient flow process, as indicated in this patient flow chart, begins in the surgeon's office with the decision to undergo surgery, and, in this case, with the decision to have outpatient surgery. *Source:* Copyright © Saint John's Health Center.

patient's needs and avoid unnecessary measurements and blood collection. Space is allowed for additional orders, and special pediatric orders are also included.

In conceiving the patient flow process, we were confronted with specific issues. Below we outline a few of these issues:

- Making the outpatient decision: What problems are specific to this patient, and what is his or her ability level? Is the outpatient decision within the surgeon's "comfort level"?

- Educating the patient: Are the instructions consistent?

- Preparing the patient preoperatively: Do the preoperative instructions meet high standards for clarity? Are there specific issues that need addressing before the anesthesia consultation?
- Preparing the patient on the morning of surgery: What last-minute tests are required? Can we get the patient to the right place at the right time? Are the history and physical recorded, and is the consent form signed? Is the surgeon late? Has the anesthesiologist met with the patient? How accurate is the scheduling?

These questions required answers in devising our clinical path and in ensuring smooth patient flow preoperatively.

Intraoperative

After the path was implemented in 1995, the multidisciplinary committee began to look at the intraoperative process. Included within the scope of the multidisciplinary committee's recommendations in developing the path intraoperatively were standardization of surgical equipment and practice and a review of anesthesia practice. Standardization began with a review of equipment and practices of the 12 surgeons performing laparoscopic cholecystectomy. It found, for example, that these surgeons used two types of trocars, five types of catheters, five types of adhesive bandages, and two types of drapes. Nurses would have to query as many as 11 of the 12 physicians about a specific practice element. Despite a limited refractory response to standardization and a claim that efforts were aimed at making "cookie-cutter surgeons," efforts to standardize proceeded, with standardization based on practice, cost, scope of use, and other similar factors. Standardizing practice in elements such as patient positioning and draping paid dividends in shortening operating room turnover and saving surgeons' time by expediting preparation.

Standardization efforts in anesthesia encompassed drug of choice, time of administration, and relation to length of stay (LOS). Cost was a major factor in choosing between two drugs that physicians considered equal in effect: one was 8.9 times the cost of the other. Collaboration between surgeons and anesthesiologists led the surgeons to stop routinely injecting local anesthetic near trocar sites postoperatively. Instead, accepting new understanding of the pain pathways and their neurological and anatomical implications, surgeons began injecting the local anesthetic before trocar insertion to attempt better postoperative pain relief.

A review of charts also revealed that use of ketorolac tromethamine (Toradol) was linked with a noticeable difference in LOS. Though the sample of patients was small

($N = 37$), the mean LOS without it was 2.68 days and with it was 4.42 days. Similarly, ketorolac tromethamine use was lower among those whose LOS was one day or less, prompting its discontinuation as well.

Postoperative

LOS in the PACU and the second-stage recovery unit were originally estimated to be one hour each, but, believing we had been overly optimistic, we revised our estimate by adding a second hour to the PACU and two hours to the second-stage recovery unit. Though now average LOS in both units combined is 2.5 hours, at the outset we still had many patients relegated to 23-hour hold, which often destined them to be hospitalized overnight, and many were remaining overnight because of pain, failure to void, and/or nausea.

Pain and emesis are common postoperatively, posing substantial distress to outpatients as well as inpatients.[10] They have been called "the big little problem" following ambulatory surgery[11] and are often controlling factors in discharge decisions.[12] Health care providers must meet the challenge of controlling pain without increasing postoperative nausea and vomiting. Improved understanding of acute pain mechanisms, special consideration of patients at high risk for postoperative emesis, employment of minimally effective doses of anesthetics, broader use of nonopioids, and the promise of newer anesthetics are all strategies for better ensuring the expected outcome of safe discharge within 23 hours of outpatient surgery.[12]

At Saint John's, PACU nurses, working with the anesthesiologists, linked drugs with LOS, as mentioned above, and were able to devise guidelines associated with shorter stays. At Saint John's, having pain and nausea does not prevent hospital discharge: control, not absence, of pain and nausea is the criterion. Using preoperative injection, attentive postoperative control of pain, and preoperative and postoperative teaching concerning signs and symptoms to report to the physician, we are able to discharge most patients on a day-surgery basis. Telephone follow-up (see below) at 24 hours postsurgery has not indicated any adverse developments.

CARE MANAGEMENT

Ultimately, anyone who cares for a patient within the scope of clinical practice is responsible for implementing, maintaining, and ensuring the viability of the clinical path. Primarily, in our institution, the task falls to the nurses in preoperative patient preparation units and the admission/outpatient ambulatory unit. Guidelines and records of the clinical path are kept on both nursing units charged with the care of the patients undergoing laparoscopic cholecystectomy. The registered nurses on staff implement the path, educate the patient about it, and help identify variances from it. Colleagues in other disciplines support these nurses in establishing and maintaining the pathway's procedures by actively reviewing presentations they make about it. Graphs of significant quality indicators are posted on bulletin boards within the working area as visual reminders of the pathway's aims. In our setting, the role of the case manager is being redefined, and no clinical nurse specialist works within these two groups. All seven nurses in the PACU and the 15 nurses in second-stage recovery bear data-gathering responsibilities for pathway evaluation. They submit their findings to the case manager, who collates the information and submits it to the director of perioperative services, who in turn reports the findings to the care providers working within the pathway and to the clinical practice improvement department.

OUTCOMES

The rapid and broad acceptance of laparoscopic cholecystectomy by patients, physicians, and hospital administrators made prospective, randomized studies comparing its outcome with that of open traditional cholecystectomy or minicholecystectomy difficult. As a technological advance, minimal-access surgery, with its shorter inpatient stay, reduced pain, and more rapid return to work, could hardly be refused. Therefore, it is particularly suited to an outcomes research approach.

Outcomes Measurement

The outcomes of the patients in the outpatient laparoscopic cholecystectomy path are routinely measured by the nursing staff. On each patient, nurses also collect clinical path–specific variances, including

- incomplete patient preparation (unfinished lab work, medical history or physical examination incomplete)
- conversion to open procedure
- intraoperative complications
- change in admission status and reason for it
- LOS greater than two hours in PACU
- LOS greater than three hours in second-stage recovery phase
- uncontrolled pain or nausea

On a monthly basis, nursing staff complete a form that focuses particularly on LOS and reasons for extending stay. Physicians had identified time of surgery as a factor affecting same-day discharge, so this evaluation also tracked that along with surgeon, patient's age, and sta-

tus affecting LOS. These findings are given to the director of perioperative and women's services, who generates a report for tracking and for presentation to the clinical practice improvement resource committee.

Length of Stay

In 1995, a series of 37 patients who underwent laparoscopic cholecystectomy had a mean LOS of 3.2 days (range, 1 to 12 days; *SD,* 3.12 days). By the next year, the LOS was almost halved. Of 52 patients undergoing surgery in that series, mean LOS was 1.8 days (range, 0 to 4 days; *SD,* 1.2). Each series was drawn from a six-week period within the year.

In late 1996 and early 1997, our efforts focused on ensuring that patients who were admitted as outpatients retained that status and were discharged as outpatients. In a comparison of two quarters, the percentage of patients admitted as laparoscopic cholecystectomy outpatients who were discharged on schedule in the same status fell from 43 percent to 23 percent. Those patients whose status changed were neither standard outpatients nor true inpatients. Unfortunately, we found that specific reasons for failure to discharge at outpatient status were obscured by physicians' orders to put patients on 23-hour hold (an admit-for-observation order) and a failure to record in the chart the reason for the change in status. Holding these patients impeded patient flow through the system, complicated provision of adequate staffing, and jeopardized our efforts to clearly establish laparoscopic cholecystectomy as an outpatient procedure. Not knowing why they were held hampered the hospital's efforts to provide efficient, appropriate care. Furthermore, we wanted to determine that surgeons were committed to the outpatient concept and were not, as some researchers had found, committed to the concept but still believing that the patient's postoperative period would be characterized by significant restrictions.

To bring about change, the physician who was head of the clinical practice improvement department in charge of laparoscopic cholecystectomy clinical practice sent out a letter to surgeons to request attention to this problem. The letter brought results. In the following quarter, the percentage of standard outpatient discharges almost doubled.

Cost

Mean cost to the hospital of providing laparoscopic surgery and perioperative care was $4,643 (range, $2,000–$10,000; *SD,* $1,732) for 40 patients at Saint John's in 1995. Costs of other equipment or materials used intraoperatively were compared; savings were found when comparisons yielded equal but cheaper alternative goods, and savings multiplied when surgeons could agree to consolidate preferences. Some identified savings opportunities reached 75 percent on some small products, but savings up to 25 percent were also possible on such big-ticket items as trocars and catheters for cholangiography.

Satisfaction Outcomes

Perioperative services measures patient satisfaction through a customer satisfaction survey conducted by the hospital and by a telephone assessment conducted 48 hours after discharge by perioperative services (Figures 16–2 and 16–3). Eighty-six percent of patients are followed up, giving the hospital vital information about patients' perceptions of their hospital experience and significant feedback about how patients are doing. We have been able to incorporate findings—including the fact that many of these patients report feeling great—in preoperative teaching. In studies of surgical outpatients, a significant outcome has been the ability to track patients' concerns by offering to refer them for solutions (Figure 16–2). On a hospitalwide survey employing a 0-to-5 scale, customer satisfaction has been high on pre- and postprocedure surgical patient preparation (Figure 16–3). Components of the operative process for which scores rose included preprocedure teaching, anesthesia infor-

mation, and overall satisfaction. All measures were ranked above 4 during the four quarters of 1996, and improvement was shown between the third and fourth quarters on seven of eight measures, including overall satisfaction.

While the laparoscopic cholecystectomy outpatient program does not poll physicians and surgeons treating these patients specifically regarding their satisfaction with the path, the hospital performs an annual physician satisfaction survey. Reports of the staff's satisfaction or dissatisfaction with the program remain anecdotal.

MOVING ON TO A BROADER APPLICATION

In efforts to broaden applicability of this pilot clinical path, Saint John's is developing an ambulatory clinical path for use with all outpatient surgery cases. Like the laparoscopic cholecystectomy path, the overall outpatient surgical path includes the visit in the physician's office, the preoperative preparation visit to the hospital, and the preoperative procedures on the day of surgery, all within the preoperative phase. The intraoperative phase is different for each procedure, so it cannot be delineated in the overall path as it can be for the laparoscopic cholecystectomy path. That piece of the puzzle must be supplied for each procedure. Finally, whereas the postoperative assessment and discharge are condensed into the postoperative phase in the laparoscopic cholecystectomy

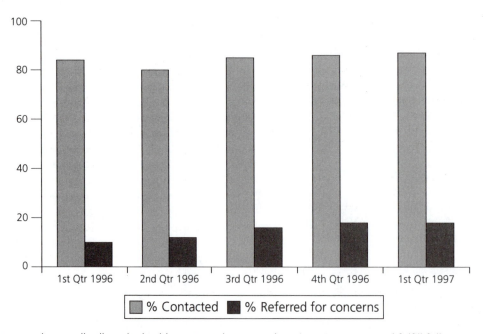

Figure 16–2 Follow-up phone calls allow the health center to better track patients' concerns and fulfill follow-up responsibilities by making referrals if necessary.

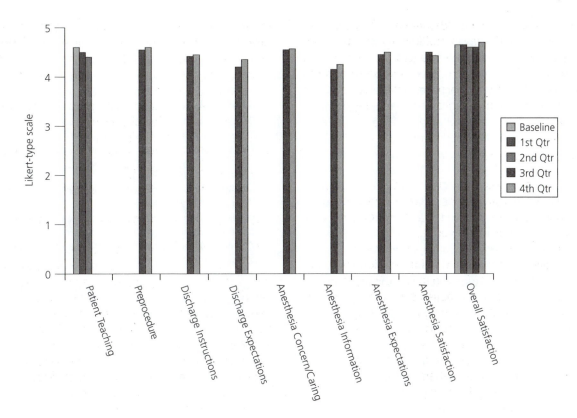

Figure 16–3 Customer satisfaction continues to climb, as recorded in this hospitalwide survey, including issues of preoperative and postoperative preparation.

path, they are two separate entities in the ambulatory outpatient path.

Committing the steps of each phase to a one- or two-page checklist streamlines routines, focuses essential information for each phase in the chart, and provides a record showing initials of the health care provider who made the determination. Below, we summarize briefly the elements of each phase.

Preoperative

At the office visit, the availability of standard laboratory test results and the necessity of other consultations are determined, the surgery is scheduled, and the patient's precertification is obtained from the insurance company. Patient instruction is initiated, and the first steps are taken to ensure that logistical failings do not transform an outpatient procedure into an inpatient procedure. (Is transportation ensured? Does patient need hospitality services?) Further, patient flow delays are prevented by reminding patients about preoperative intake prohibitions (concerning milk, solids, clear liquids, and blood thinners, including aspirin and nonsteroidal anti-

inflammatory drugs), having the physician dictate the history and physical, and obtaining signatures giving permission to operate.

At the hospital's preadmission evaluation, the interview and examination focus narrows to the patient's physical cardiovascular condition and mental alertness, and the patient instruction becomes more specific. Prohibitions for day of surgery are repeated, and the anesthesia interview is held. Any contraindications to surgery, missing paperwork, or abnormal test results are matched with actions meant to ensure resolution before the day of surgery. By the time the preoperative clinical path is followed on the day of surgery, little remains to be done. In fact, at this point the path is in its simplest form—a confirming of facts previously gathered, a recording of vital signs, a reminding about anesthesia initiation, and a checklist of presurgery tasks (jewelry and dentures removed, for example).

Postoperative

Assessments dominate the postoperative portion of the ambulatory clinical path. Beside vital signs, nurses record the patient's pain medication administration, ability to tolerate activity, nausea or vomiting, and comprehension of instruction. Also on this form, the registered nurse records the follow-up phone call placed within 48 hours of discharge. The discharge record itself guides the health care provider in helping the patient gauge what is an appropriate activity level, when to call the doctor, and how to take postoperative medications.

CONCLUSION

To improve patients' perioperative experience and to streamline surgical services and make them more cost-effective, we have worked to improve the beginning, middle, and end—the preoperative, operative, and postoperative components—of the ambulatory surgical process at Saint John's, working first to delimit the laparoscopic cholecystectomy outpatient surgery process and then to create a clinical path applicable to all types of outpatient surgery. Consequently, outcomes measures indicate a perceived improvement in almost all aspects of the path, and the broader application of the elements of this approach is expected to further streamline a systematic delivery of care that enhances, rather than compromising, the safety and quality of the perioperative experience at Saint John's.

REFERENCES

1. Mushinski M. Average charges for cholecystectomy open and laparoscopic procedures, 1994. *Stat Bull.* 1995;5:21–30.
2. Little AB, Whipple TW. Clinical pathway implementation in the acute care hospital setting. *J Nurs Care Qual.* 1996;11:54–61.
3. Dubois F, Icard P, Berthelot G, Levard H. Coelioscopic cholecystectomy: preliminary report of 36 cases. *Ann Surg.* 1990;211:60–62.
4. Perissat J, Collet D, Belliard R. Gallstones: laparoscopic treatment—cholecystectomy, cholecystostomy, and lithotripsy: our own technique. *Surg Endosc.* 1990;4:1–5.
5. Mouret P. How I developed laparoscopic cholecystectomy. *Ann Acad Med (Sing).* 1996;25:744–747.
6. Reddick EJ, Olsen DO. Laparoscopic laser cholecystectomy: a comparison with mini-lap cholecystecomy. *Surg Endosc.* 1989;3:131–133.
7. Chernew M, Fendrick AM, Hirth RA. Managed care and medical technology: implications for cost growth. *Health Aff.* 1997;16:196–206.
8. Rubio PA. Laparoscopic cholecystectomy: experience in 500 consecutive cases. *Int Surg.* 1993;78:277–279.
9. McLauchlan GJ, Macintyre IM. Return to work after laparoscopic cholecystectomy. *Br J Surg.* 1995;82:239–241.
10. Carroll NV, Miederhoff P, Cox FM, Hirsch JD. Postoperative nausea and vomiting after discharge from outpatient surgery centers. *Anesth Analg.* 1995;80:903–909.
11. Kapur PA. The big "little" problem. *Anesth Analg.* 1991;73:243–245.
12. White P. Management of postoperative pain and emesis. *Can J Anaesth.* 1995;42:1053–1055.

Getting a Head Start:
Improving Outcomes with a
Craniotomy for Tumor Care Pathway

Cathy A. Campbell, Lisa R. Cohen, Sharon E. Harpootlian, and Susan J. Ashcraft

As health care expenditures continue to rise, hospital-based providers of health care are called upon to control costs and improve quality of care. A clinical or care pathway is one strategy that minimizes variation and promotes best practice while positively affecting patient satisfaction. At Henry Ford Hospital in Detroit, Michigan, a large urban hospital, an initiative for multidisciplinary teams to develop care pathways for specific patient populations is underway. This chapter discusses the development, implementation, and outcome evaluation of a care pathway for patients undergoing a craniotomy for tumor.

INITIAL EVALUATION

Patients undergoing a craniotomy for a brain tumor were identified as a population for care pathway management as a result of discussions by members of the department of neurological surgery, nursing, and the utilization care team department. The discussions identified the desire on the part of the clinicians to provide patient-centered care rather than care that is organized to serve the clinicians and the departments. It was felt that to achieve patient-centered care, there would need to be an emphasis on maintenance and improvement of quality of care, with measurable clinical outcomes. Collaboration among health care providers and an orientation toward the patient as customer were also viewed as important.

Other criteria recognized as indicators that this population might benefit from care path management included the presence of a high-volume population with high costs and variations in care within a fairly predictable range. Since becoming a member of the National Cancer Institute-funded consortium, New Applications in Brain Tumor Therapy, and since expanding our local referral base, we have seen an increased number of brain tumor patients in our clinic. Consequently, more surgical procedures have been performed each year. Review of data from previous years found that there were 112 procedures for brain tumor in 1993, 131 in 1994, 134 in 1995, and 146 in 1996.

The data also demonstrated that there were variations in care not based on the patient's condition. These variances included the utilization of the different levels of nursing care (intensive care, intermediate, and general practice units) and the plan of care and length of stay (LOS). Inconsistencies in the utilization of nursing services and the plan of care resulted in longer LOSs with a fluctuating range. For example, in 1995, the average LOS was 6.86 days, with a range of 1 to 22 days, for this group of patients.

Other factors identified by nurses that supported the development of a care pathway for management of these patients were the high levels of anxiety and fear related to the diagnosis and surgery, which led to increased needs for emotional support. In addition, knowledge deficits and teaching needs, requirements for careful discharge planning, and the potential for serious complications were recognized as important considerations in the care of these patients. As a result of the data review, preliminary investigations, and discussions, it was determined that an opportunity for improvement existed, and the decision was made to develop a care path management strategy for patients having a craniotomy for brain tumor.

FOUNDATIONS FOR CARE PATHWAY DEVELOPMENT

Once the decision was made to proceed with the project, preliminary meetings with project champions were held. The project champions were instrumental in planning and implementing the care pathway and in generating acceptance of the plan by other team members. The champions included a neurosurgeon and a nurse associate from the neuro-oncology program and a member of the utilization care team department. This small group met before the formation of the multidisciplinary team to lay the foundation for the project. At these early meetings, discussions centered on potential goals for the project, expected patient outcomes, care elements, and boundaries for patient inclusion in preparation for the initial meeting of the entire team.

It was important at the beginning of the process to identify and include representatives from various disciplines who were directly involved in caring for the selected patient population. This ensured that all concerns were addressed and facilitated successful implementation of the care pathway. Multidisciplinary team members key to the development of the pathway included neurosurgeons, neurosurgical nurse associates, physician assistants, representatives from the utilization care team department, neuroscience clinical nurse specialists, pharmacists, nurse administrative managers, anesthesiologists, physical therapists, and social workers.

Early meetings by the multidisciplinary team focused on determining pathway goals. Emphasis of these goals was on quality improvements as well as cost reductions. Goals formulated by the team included the following:

- Maintain the quality of care.
- Increase the consistency of care.
- Maintain patient and staff satisfaction.
- Decrease LOS.
- Decrease cost.

Once these goals were set, the team's work focused on defining the boundaries for the patient population and targeted LOS.

In defining the boundaries for inclusion in the care pathway, several options were discussed. Care for the patient undergoing a craniotomy for brain tumor can be organized by tumor characteristics (i.e., histology or location) or type of admission (i.e., scheduled or emergent). Other clinical pathways and guidelines have been developed within these limitations.[1-3] However, the team decided to put all patients having a craniotomy for tumor, regardless of tumor characteristics or type of admission, on the pathway to allow data gathering and monitoring of the entire group. This included patients with newly diagnosed tumors, recurrent tumors, benign or malignant tumors, and primary or metastatic tumors, as well as patients receiving chemotherapy wafer placement during surgery. Patients with presurgical neurological deficits and significant past medical histories would also be placed on the pathway.

Team discussions regarding LOS included the experience of the team members with this patient population. Before planning for the care pathway began, several team members from the neuro-oncology program had recognized the need to begin decreasing LOS. In select cases in which the patient's condition was stable and there was good family support, the patient was discharged to home earlier than common practice. These cases were presented to the team in support of the target of a two-day LOS. Several team members expressed concern about reducing LOS this aggressively, identifying a number of issues to be addressed before continuing with the development of the care pathway. These issues were postoperative peak swelling time, the education and discharge needs of the patient and family, and the availability of staff for family assistance and support after discharge.

Cerebral edema occurs postoperatively, reaching a peak some time between 24 and 72 hours.[4,5] Because this is a very well-known postoperative event, patients are routinely medicated with high-dose steroids. The steroids are gradually tapered as the cerebral edema resolves. In planning for earlier discharges, plans were made to adjust the tapering schedule to prevent the patient from becoming symptomatic after discharge. In addition, to provide the follow-up thought necessary by the team, a process was developed for a neuroscience registered nurse (RN) to contact the patient by telephone for the first two days after discharge. This has proven to be a vital component of the care pathway. The design and implementation of the postdischarge telephone calls will be discussed in more detail in the following section.

To address the concern that less time would be available for patient education and discharge planning as a result of a shortened LOS, a recently piloted program in the department of neurosurgery was accepted by the team to assist with the implementation of the care pathway. This program, "Your Next Step Visit," was created by the nurse associates for electively scheduled surgery patients due to an increase in same-day surgery admissions, resulting in reduced opportunities for patient education. The focus of this visit is to prepare the patient and family for the surgery, hospital stay, and discharge from the hospital. Figure 17–1 diagrams the events that occur with this meeting. In cases where distance, patient illness, or other circumstances make it too difficult to return for this meeting, a telephone session is substituted for a face-to-face meeting.

The components of the visit include information related to the surgery and disease process. Written materials are provided and reviewed with the patient and family. Information about hospital routines, the nursing units, expected LOS, nursing care, and patient activities that are geared toward the prevention of complications

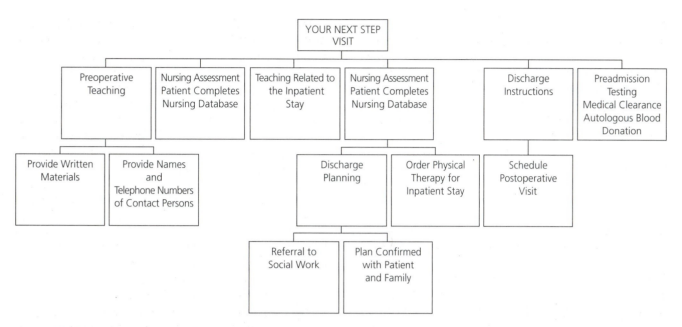

Figure 17–1 Your Next Step Visit. This appointment with the nurse associate prepares the patient and family for the surgery, the hospital stay, and discharge from the hospital. *Source:* Copyright © Henry Ford Health System.

and restoration of health is provided. In addition, information on who the members of the patient's health care team are, who the patient's contact persons are, and how to access assistance is discussed. Business cards and telephone numbers are provided that represent availability 24 hours a day, seven days a week.

Another element of the "Your Next Step Visit" is the nursing assessment for discharge needs. The patient completes a nursing database, and this is followed by an interview with the patient and family to identify patient resources and potential risks related to discharge from the hospital following the surgery. The nurse associate assists the patient to recognize how much assistance will potentially be needed and to develop a plan that will allow for that level of assistance.

For those patients who cannot name a person who can assist them at home following discharge, possible sources are explored. An agreement is made with the patient to follow up on the identified resources, and he or she is requested to call the nurse associate to confirm that arrangements have been made that will provide the necessary assistance. If indicated, a referral to a social worker is made to ensure that every patient has a potential discharge plan in place before admission to the hospital. If different care needs are present at the time of discharge, the plan is adjusted. This process is aided by the preadmission assessment that is completed. The nurse associate is familiar with the patient resources and shares this information with other team members in developing an alternative plan.

Discharge instructions are also reviewed with the patient and family. Included are diet, activity, care of incision, medications, pain management, signs and symptoms of infection, potential complications, and follow-up care with the physician. At this time, the patient is scheduled for a routine return appointment with his or her physician to ensure adequate postdischarge follow-up.

Another factor in the scheduling of the "Your Next Step Visit" is the need to coordinate it with other appointments required before admission so as to reduce the number of days on which the patient must return to the hospital. These other appointments can include preadmission laboratory tests, medical clearance, autologous blood donation, and preoperative radiological studies. At the completion of the "Your Next Step Visit," the patient and family are encouraged to call the nurse associate with any questions and concerns that they have.

CARE PATHWAY DESIGN

Once these supporting processes were in place, the team began to work on the content of the pathway. The pathway itself was developed by the multidisciplinary team over several months. Before the formation of this team, a template for care pathways was designed by the utilization care team department in conjunction with other care pathway teams. Care elements on the template are listed vertically, and the timeline is listed horizontally. Care elements, or categories of activities, include patient outcomes, assessment/evaluation, tests and diagnostics, treatments and interventions, medications and IVs, elimination, nutrition, activity/safety, patient education, discharge planning, and consults. Current practice patterns were discussed in relationship to pathway goals, and consensus was reached regarding best practice under each of the care element categories. Concurrently, patient outcomes were determined for each day of the pathway (Exhibit 17–1).

Key process variables (KPVs) are patient care activities that the team feels are significant in helping patients to achieve desired outcomes. Completion of these key process activities (variables) is documented on a data collection tool called a KPV. Key activities identified by the team at the time of pathway development included pain management, time surgery ended, time of admission to the unit, progression of activity (from dangling on the day of surgery to ambulation on postoperative day [POD] 1), development of a new neurological deficit, development of a new comorbidity, or instability of an existing comorbidity. The data collection tool is revised by the team as needed on the basis of analysis of data. A more detailed discussion is presented later in this chapter in the section, "Outcomes Routinely Measured."

To facilitate continuity of care among all practitioners, a set of preprinted orders was prepared to accompany the care pathway. This set consists of three separate order sheets: postoperative, POD 1, and POD 2 orders. The preprinted order sheets reflect the activities listed on the care pathway, although physicians may individualize the orders as indicated to ensure that each patient receives optimal care.

As discussed earlier in this chapter, appropriate education and support provided to our patients and families were identified as vital components to the successful implementation of the care pathway. The care pathway was adapted to supplement patient education during the "Your Next Step Visit" with the nurse associate. This version of the care pathway was written in appropriate language for patient use and put into a booklet format (Exhibit 17–2). The focus on expected outcomes, interventions, and anticipated length of hospitalization meant that the patient and family were more prepared for the surgery, had realistic expectations of their recovery, and could anticipate their discharge care needs.

To address the concerns of the team related to the two-day LOS, a plan to provide assistance and support and a vehicle to assess the patients' clinical status were developed. Telephone follow-up calls were to be made on postdischarge day 1 and day 2 by an RN. A form was devised with specific questions to ask, with space for the nurse to record the patients' responses and status. During the preoperative visit, the content of the postdischarge telephone calls would be reviewed so that

Exhibit 17–1 Excerpt of Henry Ford Hospital Craniotomy-for-Tumor Care Pathway

---

Day 2 (POD 1) ICU to Neuro Floor
DATE: _____

Outcomes
   Transfer to neuro floor
   Ambulate out of room w/assistance or at baseline activity
      level
   Absence of unanticipated neuro instability
   Absence of new comorbidity
   Tolerate PO
   Acceptable pain management

Assessment
   Neuro exam
   RN daily assessment
   VS q8 hrs
   Neuro checks 2hrs x 12 hrs. then q4hrs x 12 hrs. as specified
      by physician
   Basic neuro check inc. LOC, pupils, motor

Tests and Diagnostics
   MRI if not yet completed
   Consider anticonvulsant level
   Consider chemstix bid

Treatments
   Triflow q 1–2 hrs
   D/C sequential compression device
   D/C dressing
   Saline lock IV as PO tolerated

Medications
   Dexamethasone
   Oral antacid
   D/C IV Cimetidine
   Anticonvulsant
   Acetaminophen with codeine
   Consider stool softener—esp if getting codeine

Diet and Nutrition
   Advance as tolerated

Elimination
   Foley catheter—d/c prior to
   Transfer to neuro floor

Activity
   Ambulate in hallways, or baseline activity level—goal 100' &
      stairs prn
   Self care in AM—or at baseline

Education
   Pain management
   Discharge instructions
   Care of incision
   Patient pathway

Discharge Planning
   Implement discharge plan
   Social work to see if indicated

Note: BID, twice daily; D/C, discontinue; ICU, intensive care unit; IV, intravenous; LOC, level of consciousness; MRI, magnetic resonance imagery; PO, oral (intake); POD, postoperative day; prn, as needed; qs, quantity sufficient; VS, vital signs.

Source: Copyright © Henry Ford Health System.

Exhibit 17–2 Excerpt from the Craniotomy-for-Tumor Patient Pathway

Day 1 In the Intensive Care Unit	
General	You will stay in the Intensive Care Unit until your health care team feels that you are stable and feeling well enough to be transferred to the General Practice Unit. You will probably stay overnight in the Intensive Care Unit and then go to the General Practice Unit.
Nursing Care in the ICU	You will have a dressing covering your head like a turban to protect your incision. It fits snugly.
	Special socks are put on your legs when you are in the Operating Room. These socks are covered with another wrap called a "Sequential Compression" boot. They help prevent blood clots in your legs while you are in bed.
	Your pulse, blood pressure, temperature, and neurological status will be checked hourly. Your nurse will wake you hourly. She/he will ask you what your name is, where you are, what the date is, and other simple questions. This may be annoying as you begin to feel better, but it is an important part of your care.
	Your nurse will show you how to cough, deep breathe, and use a breathing device. This helps prevent pneumonia.
	The IV fluids, arterial line, and foley catheter will be continued.
Medications	There will be medication for pain ordered for you as needed. To receive this medication, tell your nurse of any headache or other discomfort you are experiencing.
	Other medications you may receive include an anticonvulsant (to prevent seizures), dexamethasone (to reduce swelling of the brain), and an antacid. The antacid helps to balance stomach acidity caused by dexamethasone.
	You will continue to take any medications you usually take unless your physician decides they should not be taken during this time.
	If you have nausea (sick to your stomach), let your nurse know so you can be given some medication for this.
Activity/Safety	The nurse will help you sit at the side of the bed. You may be assisted to get out of bed and sit in a chair and then walk a short distance.
	The care team will help you and remind you to start doing the simple exercises you learned in the clinic before surgery. Like the special socks, these exercises help prevent blood clots in your legs.
Diet	After surgery you will start with ice chips and then clear liquids (soup, broth, tea, gelatin, clear juices). As your digestive system begins to return to normal, you will start regular food.
Things You Should Know	Intensive Care Unit visiting hours are 12 noon to 8 pm. Your family can see you briefly each hour based on how you are feeling. There is a lounge outside the Intensive Care Unit where they can wait. They must use the phone in the lounge to call the unit before entering.

Source: Copyright © Henry Ford Health System.

the patient and family would be familiar with the process. The questions would also be included in the patient pathway booklet (Exhibit 17–3).

To offer daily coverage and ensure that calls were made on weekends and holidays, collaboration between the nurse associates, clinical nurse specialist, and the nursing staff on the neuroscience general practice unit was essential. A detailed flowchart outlined the process, as well as the actions to be taken if there was a change in the patients' condition or if problems were reported (Figure 17–2). After the call, the completed form would be reviewed by the neurosurgeon and nurse associate, and subsequent action would be taken as indicated.

After the care pathway was completed and approved by the team, a nursing care guideline was written by the neuroscience clinical nurse specialists. Care pathways are used in our institution as a guideline for the care of the patient, not as a documentation tool. Documentation of nursing care is accomplished using the Nursing Information System. Standards of care, or nursing care guidelines, are the foundation for this documentation system. Nursing care guidelines identify the critical elements of nursing performance (process) and the expected outcomes of the care received (patient outcomes) for a given patient population, nursing diagnosis, or collaborative problem. The format for care guidelines is based on the

Exhibit 17–3 Excerpt from the Craniotomy-for-Tumor Patient Pathway: Questions Asked during the Telephone Follow-up Call

ACTIVITY:	Length of time out of bed
	Length of time sitting up in a chair
	Walking—distance and frequency
NUTRITION:	How many meals have you eaten in the last 24 hours?
	Have you had nausea and/or vomiting?
ELIMINATION:	Are you having any difficulties passing urine?
	Increased frequency or pain or burning?
	Have you had a normal bowel movement since surgery?
	Since you have been home?
INCISION:	Is there more swelling than when you were discharged from the hospital?
	Is there any drainage? What does it look like?
	How are you caring for the incision?
RESPIRATORY:	Do you have a sore throat? fever?
	Are you using your triflow? How often?
PAIN:	Are you having pain/headaches?
	Rate pain on scale of 1–10 now.
	Rate the worst pain you have had since discharge.
	Location and duration of pain:
	Time of day pain occurred:
	Was it related to activity?
MEDICATIONS:	Decadron: dose, frequency, taper schedule
	Anticonvulsant: type, dose, frequency
	Pain medication: type, dose, frequency
	Is it controlling your pain?
	Other medications you are taking:
	Any medication side effects:
	rash, GI upset, hiccups, insomnia, imbalance, other
NEURO STATUS:	Any changes since discharge in level of consciousness, confusion, weakness of arms or legs, other?
	When is your next clinic appointment?

Source: Copyright © Henry Ford Health System.

same 11 categories or care elements used in the care pathways. The nursing care guideline is structured on phases of care mirroring the care pathway that was developed for the patient undergoing a craniotomy for tumor. The nursing care guideline reflects two phases of care: intensive care unit (ICU) and general practice unit to discharge (Exhibit 17–4). The nursing care guideline was approved via the practice wedge of the nursing shared governance process. The nursing care guideline allows nurses to document the patient's responses to nursing care on the basis of expected outcomes.

IMPLEMENTATION

Before putting the care pathway into practice, all health care team members were educated about the care path and their individual role in carrying out the process.

The nursing staff on the three neuroscience units (neurosurgical ICU, neuro intermediate care unit, and neuroscience general practice unit) as well as the surgical/trauma ICU were previously inserviced on the development of care pathways in our institution and had experience with several care paths already in use. Additional education was given regarding the specifics of the craniotomy-for-tumor care pathway, the preprinted physician order sheets, and how to document on the KPV form as well as the accompanying nursing care guideline. The entire staff of the department of neurological surgery, staff and resident physicians, nurses, secretaries, and others also attended informational sessions in preparation for the implementation of the care pathway.

In July 1996, as planned by the team, all patients who had a craniotomy for tumor were placed on the care pathway. Communication of this was continued throughout the patient's hospital stay through various processes. Care path folders were available in the operating room. The contents of the care pathway folders included a copy of the care pathway; preprinted order sheets for day of surgery, POD 1, and POD 2; the KPV sheet; the telephone follow-up form; and the patient and family educational brochure. Additional folders were available in the ICU if the patient was transferred from the postanesthesia care unit without a folder. The ICU nurse was alerted that the patient was on the care pathway because this information was included on the operating room (OR) schedule. The care pathway folder was coupled with the medical record throughout the patient's hospitalization and kept at the patient's bedside for use as a communication tool during shift-to-shift nursing report. The pathway could also be used during morning multidisciplinary rounds.

All members of the health care team are accountable for using and following the care pathway. The physician completes and signs the preprinted post-op orders, and this starts the inpatient portion of the care pathway. To advance the patient on the pathway each day, the physician must write "Advance on pathway" and sign the appropriate order sheet.

Documentation of care activities and outcomes occurs on the data collection tool, the KPV form. This form is used for quality assurance purposes and is not part of the patient's medical record. The RN(s) caring for the patient completes the KPV form by 11:00 PM every day, explaining any variances. A variance, or unexpected outcome, is designated as any response that is shaded on the KPV form. This would require the RN to explain the situation and circumstances surrounding the variance. Expected outcome responses are not shaded on the form and do not require additional documentation. Expected outcomes to be evaluated include

Craniotomy-for-Tumor Clinical Pathway
Postdischarge Follow-up Phone Call Process

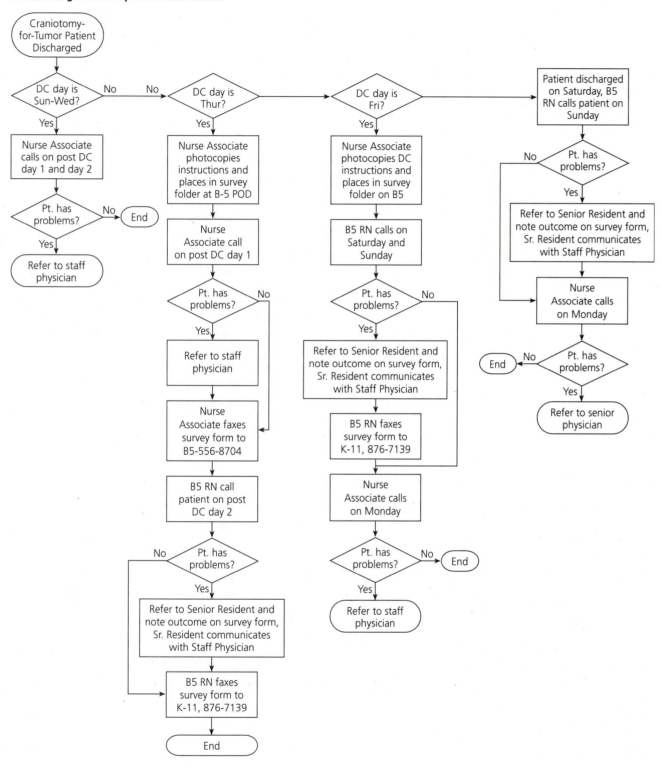

Note: DC, discharge; POD, postoperative day.

Figure 17–2 Flowchart of Telephone Call Follow-Up Process. *Source:* Copyright © Henry Ford Health System.

Exhibit 17–4 The Nursing Care Guideline for Patients Having a Craniotomy for Tumor

NURSING CARE GUIDELINE

Patient Population: Craniotomy for Tumor

Phases Definition:

Phase 1: ICU phase

Phase 2: GPU phase

PHASE 1

Outcome:

Upon transfer to the GPU, the patient:

A. Maintains pre-op neurological status for 24 hours post-op.

B. Maintains Glasgow Coma Score > 12.

C. Tolerates dangling for 15 minutes.

D. Vital Signs within normal limits.

The patient/family verbalizes understanding of the ICU routine and plan for transfer to the GPU.

PROCESS:

Assessment/Evaluation:

1. Assess heart rate, blood pressure, and respiratory rate q 1 hour.
2. Assess neurological status q 1 hour: levels of consciousness, pupils, extraocular movements, additional cranial nerves as indicated, and motor strength.
3. Assess pain per care guideline: Pain Management, Adult.
4. Assess dressing/incision q 8 hours.
5. Assess respiratory status q 4 hours.
6. Assess for seizure activity.

Tests & Diagnostics:

7. Obtain serum lytes and hematocrit on admission to ICU.
8. Assess pulse oximetry q 1 hour.
9. Obtain MRI for gliomas and metastases.
10. Chemstick BID.

Treatments & Interventions:

11. Keep head in neutral position.
12. Antiembolic measures per physician prescription.
13. Encourage patient to use triflow q 1 hour.

Medication & IVs:

14. Administer pain medication prn per physician prescription.
15. Administer anticonvulsants per physician prescription.
16. Administer steroids per physician prescription.
17. Administer H2 blocker per physician prescription.
18. On POD 1, change IV to saline lock when tolerating oral fluids/diet.

Elimination:

19. Maintain and assess intake and output.
20. Discontinue indwelling urinary catheter before transfer to GPU.

Nutrition:

21. Clear liquids, advance as tolerated on POD 1.

Activity/Safety:

22. Turn and reposition q 2 hours.
23. Elevate head of bed per physician prescription.
24. Maintain bed in low position with side rails up.
25. Dangle at bedside for 15 minutes, advancing to up in chair if tolerated.

Patient Education:

26. Orient patient/family to ICU environment and routine.
27. Review craniotomy-for-tumor patient pathway with patient/family.

Discharge Planning:

28. Identify discharge needs.

Consults:

29. Physical Therapy for home safety evaluation.

PHASE 2

Outcome:

Prior to discharge, the patient exhibits signs of recovery of functional capabilities as evidenced by:

A. Increased physical endurance.

B. Improvement in and/or adaptation to neurological deficits.

The patient/family verbalizes understanding of GPU routine and discharge plan.

If outcome is achieved, resolve care guideline.

PROCESS:

Assessment/Evaluation:

1. Assess vital signs with neuro checks q 2 hours x 12 hours, q 4 hours x 12 hours, then q 8 hours.
2. Assess pain per care guideline: Pain Management: Adult.
3. Assess dressing/incision q 8 hours.
4. Assess respiratory status q 8 hours.
5. Assess for seizure activity.

Treatments & Interventions:

6. Encourage patient to use triflow q 1–2 hours.
7. Discontinue sequential compression device when ambulating in hallway.

Medications & IVs:

8. Change IV to saline lock when tolerating oral fluids/diet.
9. Administer pain medication prn per physician prescription.
10. Administer anticonvulsants per physician prescription.
11. Administer steroids per physician prescription.
12. Administer oral antacid per physician prescription.
13. Discontinue saline lock prior to discharge.

Elimination:

14. Maintain and assess intake and output.

Nutrition:

15. Advance diet to regular as tolerated.

Activity/Safety:

16. Ambulate in hallway, progress activity to baseline.

Patient Education:

17. Orient patient/family to GPU environment and routine.
18. Review pathway with patient/family.
19. Teach patient/family about activity restrictions and incision care.

Discharge Planning:

20. Review discharge plan with patient/family.

Consults:

21. Consult Physical Therapy for home safety evaluation/rehab evaluation.
22. Consult Social Work for assistance with additional discharge needs/placement.

Note: BID, twice daily; GPU, general practice unit; ICU, intensive care unit; IV, intravenous; MRI, magnetic resonance imaging; POD, postoperative day; prn, as needed.

Source: Copyright © Henry Ford Health System.

- patient location (ICU, neuroscience general practice unit, or other)
- occurrence of new neurological deficit (yes or no)
- activity level (none, dangle, ambulate with assistance, ambulate independently)
- activity tolerated (yes or no)
- diet (NPO, clear liquids, or regular)
- diet tolerated (yes or no)
- voiding independently (yes or no)
- patient able to be discharged (yes or no)

For example, on the OR day, it is expected that the patient will be in the ICU that evening, and the RN will circle that answer on the KPV form. If the patient stays in the ICU for POD 1 or longer, that circled response will be shaded. The RN will document on the KPV form an explanation of the variance—specifically, what kept the patient in the ICU an additional day and prevented transferring to the floor.

The KPV forms are reviewed by the clinical nurse specialist on the unit to ensure that the information documented is accurate and complete. When the patient is discharged, the KPV forms are forwarded to the utilization care team representative for data analysis. This concludes the inpatient portion of the care pathway.

OUTCOMES ROUTINELY MEASURED

Currently, the following outcomes are measured quarterly for all patients started on the care pathway, including those with path detours:

1. LOS—overall and postsurgical, mean and median
2. Percentage of patients discharged by the goal LOS and the goal LOS plus one day
3. Total charges—mean and median
4. All patients readmitted within 30 days of the craniotomy-for-tumor discharge
5. Percentage of patients having adverse care outcomes reflected in the ICD-9 coding, including septicemia, pneumonia, deep-vein thrombosis/pulmonary embolism, ileus, urinary tract infection, and wound disruption or infection.
6. Percentage of patients having none of these adverse care outcomes (since a patient could have several adverse outcomes, this helps to see how much double counting occurred in #5).

Data are collected and maintained by the utilization care team department for all patients meeting the definition of craniotomy for tumor (a series of CPT 4 codes). The data are obtained from a relational database maintained by Henry Ford Health System. In addition to the data elements mentioned above, the data from the KPVs are entered into an access database that was developed by the utilization care team department.

Length of Stay

Although average LOS decreased in the craniotomy-for-tumor patient, it was not as significantly influenced as the team had expected. Examination of the data on all patients who completed the craniotomy-for-tumor pathway revealed no common factor to account for the longer LOS. However, when the data were stratified by scheduled versus emergent admissions, it was evident that patients who were scheduled for surgery and had received the preadmission education and discharge planning experienced a shorter LOS. Postoperative LOS for scheduled patients has decreased 54 percent from 1993 to 1997. The emergent admission patients were placed on the care pathway postoperatively, even though they did not receive the preadmission education and discharge planning. The reported data indicate that the postoperative LOS for emergency patients has decreased 46 percent. Figure 17–3 demonstrates the median postoperative LOS by quarter for scheduled and emergent admissions.

It is interesting to note that in early 1995, before development of the care pathway, the LOS decreased. This is thought to be attributable to the effect of informal discussions and individual efforts to change practice patterns. By the time the path was formally implemented in July 1996, a downward trend in LOS was observed. Figure 17–3 has been annotated to illustrate the evolution of the clinical path, indicating that the decreased LOS occurred before design of the care pathway. This phenomenon occurred with several other pathways developed at Henry Ford Hospital.

Current literature cites LOS for craniotomy for tumor as three to seven days.[1–3,6] However, during the development of the pathway, the team believed that a significant number of patients could be discharged earlier. The goal LOS for our pathway is two days. Review of the data shows that all patients are not discharged on POD 2. Table 17–1 demonstrates the percentage of patients discharged by POD 2 and POD 3. Since the implementation of the pathway, the median LOS for all emergent admissions, including those with pathway detours, is seven days. The median postoperative LOS for all scheduled admissions is four days. The median postoperative LOS for patients that remained on the pathway is three days.

The team found that data from the KPV were not clearly explaining why a patient could not be discharged on POD 2. In April 1997, the KPV form was revised, adding a question to obtain this information. When this item is missing from the form, it is returned to the staff physi-

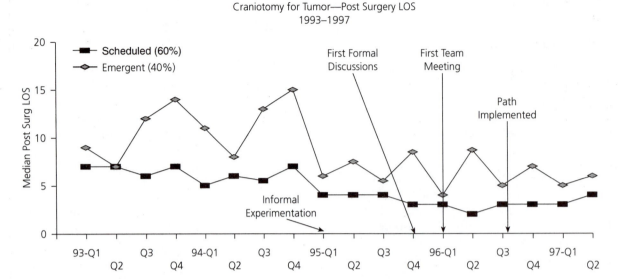

Figure 17–3 Median Postoperative LOS for Elective and Emergent Craniotomy-for-Tumor Patients. *Source:* Copyright © Henry Ford Health System.

cian for completion. Early review of the data shows that in most cases there are clinical reasons for delaying discharge. However, the data also suggest that there is a tendency not to discharge patients on the weekend. The team is exploring ways to manage this problem.

Charges

The team tracks mean and median charges for all scheduled and emergent admissions. The data indicate that charges decreased at the same time LOS decreased—more than a year before the path was implemented. At an upcoming meeting, the team will be reviewing an analysis of 20 patient billing statements. Attention will be focused on how closely the billing statements reflect the activities ordered per the pathway, primarily regarding radiologic studies and laboratory tests.

Table 17–1 Percentage of Patients Discharged on POD 2 and POD 3 by Quarter

Quarter	% Discharged by POD 2	% Discharged by POD 3
Q1—1996	29	44
Q3—1996	46	54
Q4—1996	48	54
Q1—1997	40	63
Q2—1997	38	47

Source: Copyright © Henry Ford Health System.

Readmissions

On a quarterly basis, the team reviews data on patients who are readmitted within 30 days of their discharge. Craniotomy-for-tumor patients tend to be readmitted for various reasons, including chemotherapy treatments. However, since the pathway has been implemented, the readmission rate has not significantly changed. In the year preceding implementation of the care pathway, readmission rates ranged from 9 percent to 34 percent. For the year since launching the care pathway, readmission rates ranged from 6 percent to 26 percent.

Adverse Outcomes of Care

On a quarterly basis, information regarding discharged patients is reviewed for the occurrence of septicemia, deep-vein thrombosis and pulmonary embolism, pneumonia, ileus, urinary tract infections, and wound healing disruption or infection. These items were chosen because our care pathway includes patient care activities aimed at preventing these adverse outcomes. The occurrence of these adverse outcomes in this patient population is uncommon, and no significant increases have been observed since initiating the care pathway. Before implementation of the path, 14 percent of patients experienced an adverse outcome. Since implementation, the average percentage of patients with an adverse outcome is 17 percent.

TEAM MEETINGS: DATA REVIEW

In addition to the data described previously (LOS, charges, readmissions, and adverse outcomes), the team also reviews data from the KPV data collection forms and ad hoc data. For example, at a recent team meeting, the following topics were discussed regarding the data collected:

- care path completion and detours
- LOS data stratified by the day of the week of surgery
- KPV data collection and analysis

These topics are typical of those on the agenda of a quarterly team meeting.

At this meeting, a list of all patients placed on the pathway during the previous quarter was reviewed, with attention to pathway completions and detours. It was noted that 62 percent of the patients completed the pathway as planned, with 38 percent falling off. Factors contributing to these detours were discussed and included evidence of a new neurologic deficit, postoperative complications such as hemodynamic instability, fever and nausea, possible wound infection, and activity intolerance. Analysis of these data to date has not revealed any specific variable that significantly contributes to the detour rate. Therefore, the team decided to continue placing all craniotomy-for-tumor patients on the care pathway.

From review of previous quarterly data, the team identified a group of patients whose LOS was three to five days. Even though these patients had a slight increase in LOS, there was no actual detour off the pathway. Documentation on the KPV form indicated that although patients frequently met the criteria for discharge, they remained hospitalized. For this meeting, LOS data were stratified by the day of the week that the patient had surgery. Analysis of these data revealed that due to the neurosurgeons' operating room schedule, over 60 percent of craniotomies were performed on Thursdays and Fridays. Therefore, on the basis of the pathway, these patients would be scheduled for discharge on the weekend. The team was interested to know if weekend discharges contributed to hesitation on the part of the practitioners as well as the patients and their families to go ahead with discharge.

Concerning the KPV forms, one issue that continues to pose a problem for the team is incomplete data. Missing data rates range from 13 percent to 41 percent for each of the elements collected. The data collection process was designed to have a clinical nurse specialist available to work with the nursing staff on the general practice unit as well as in the ICU to ensure that the data collected were accurate and complete. However, six months after implementation of the care pathway, we lost one clinical nurse specialist, and the position has been vacant for seven months. This has left a large void in our ability to collect data on a daily basis. Other alternatives to ensure completeness of the KPV forms until another clinical nurse specialist could be hired were not explored because the missing data rates were considered to be within an expected range.

MODIFYING THE KPV FORM

As data were reviewed at quarterly team meetings, it was found that the KPV form was not providing the team with adequate information in some areas. For example, in a review of the first-quarter data in 1997, it became apparent that the data collection tool was not revealing why a patient could not be discharged on POD 2. Team members felt that delayed ambulation might have been a contributing factor. As a result, the KPV form was revised, adding three questions about ambulation: who assisted the patient to ambulate (nursing, physical therapy, family), how many times per day the patient ambulated (1–2, 3–4, >4), and the maximum distance attained by the patient (<50 feet, 50 to 100 feet, and >100 feet). Review of second-quarter data demonstrated that patients still hospitalized on POD 3 were ambulating independently more than four times per day and going further than 100 feet, suggesting that delays in discharges were not related to lack of ambulation. When the KPV form was revised, the nurse associates put in place a process to obtain a physical therapy consult for these patients. Review of the second-quarter data showed that patients who were not scheduled for a preoperative clinic visit with the nurse associate missed initiation of a physical therapy consult. Once this problem was identified, steps were taken to remedy this situation.

In addition to adding the above questions, a very direct statement was added to the KPV form asking for the reason the patient was not discharged on POD 2. If the form was returned with these data missing, the staff physician was asked to provide this information. We are interested in correlating these data with the tumor histology and location. This may lead to some refinements of the care pathway.

SATISFACTION SURVEYS

In the fourth quarter after implementation of the care pathway, surveys were distributed to staff nurses, physicians, resident physicians, nurse associates, and patients to identify areas for improvement. The team has found the survey data helpful and plans to send out surveys on a periodic basis.

Patient Survey

Henry Ford Health System uses the Parkside Survey to obtain information about patient satisfaction. These surveys are mailed to 20 percent of patients discharged from each unit, with a return rate of approximately 30 percent. Therefore, at best, a small percentage of craniotomy-for-tumor patients would have an opportunity to receive a Parkside survey. Since Parkside data do not address areas specific to the craniotomy-for-tumor care pathway, we developed our own questionnaire. The survey includes questions about preoperative teaching, preparation for hospital stay and discharge, and problems encountered and support available in the early discharge period. The survey used a Likert scale of 1 to 5, with 1 designating *poor* and 5 designating *outstanding.*

Patients who had a craniotomy for tumor during the first five months of 1997 were mailed a survey that specifically asked questions related to the goals of the care pathway. Of 37 surveys mailed, 12 were returned. The compilation of the surveys showed that:

1. Patients were very pleased with how the preoperative teaching had prepared them for what to expect for their hospitalization, with 82 percent of the patients rating this a 4 or a 5.
2. Patients also indicated that they felt reasonably prepared for discharge. Seventy-five percent of patients felt prepared for discharge, and 50 percent rated their preparation with a 4 or a 5. Some of the comments suggested that the patients did not feel that they had achieved the goals of the care path at the time the physician came to discharge them. Specifically, they felt that the ambulation goals had not yet been achieved—indicating that they were familiar with the content of the care pathway.
3. Several patients indicated that they did not receive follow-up phone calls and felt that a call would have been appreciated.
4. Patients did feel supported by their health care team for issues that occurred after discharge, with 58 percent scoring a 4 or a 5 on the questionnaire.

Nursing Staff Survey

The nursing staff in the neuroscience inpatient units were also surveyed. Thirty-seven nurses responded (i.e., more than a 50 percent return rate). They were polled about education related to the pathway, helpfulness of the pathway in planning patient care, the data collection (KPV) tool, the coordination of care, and physician collaboration. Findings included:

1. Thirty-five percent of the staff reported that they did not receive inservice education on the path, possibly due to turnover. The team plans to add a process to orient new staff members to the care pathway.

2. Comments indicated that the font on the KPV form was too small. Subsequently, it was reformatted with a larger font.
3. Forty-one percent of the staff felt that coordination of care was improved, and 27 percent felt that physician collaboration had also improved.

Comments specifically related to coordination and collaboration identified that the physicians do not consistently write orders to place the patients on the care pathway or to advance the patient on the path from one day to the next. In response to these concerns, the team is now making the signing of orders part of collaborative patient rounds. Nurse satisfaction will continue to be measured on an intermittent, ongoing basis.

Nurse Associate Survey

All of the nurse associates in the department of neurological surgery responded to the questionnaire. This group responded favorably to all aspects of the path, indicating that patient satisfaction was improved and that overall patient care was more efficient.

Physician/Resident Survey

Table 17–2 presents the questions and results of the physician survey. These results generally indicate that the residents are more positive about the path than staff physicians. It is possible that the path made more of an impact on the work that the residents do and less of an impact on the work done by the staff physicians. Therefore, the residents view the path in a more favorable

Table 17–2 Physician Satisfaction Survey

	Staff	Residents
(1 = Not Useful, 4 = Very Useful)		
Overall, has the CFT path been helpful to you?	3.0	3.5
Rate the usefulness of the patient path	3.5	3.7
Rate the usefulness of the clinical path	3.0	3.3
Rate the usefulness of the preprinted orders	3.8	3.9
(1 = Never, 4 = Always)		
Availability of packets in the OR	3.0	3.0
Availability of packets on the unit	3.0	3.0
(% Answering "Yes")		
Has the path reduced patient care variation not based on patient condition?	83%	100%
Has patient care been improved?	50%	90%
(% Positive)		
How have patient outcomes been affected?	17%	70%
How has patient satisfaction been affected?	0%	80%

Note: CFT, craniotomy-for-tumor; OR, operating room.
Source: Copyright © Henry Ford Health System.

light. The team will continue to work toward acceptance of the care pathway from all senior staff physicians in the department.

CONCLUSION

Care pathways are designed to be a multidisciplinary strategy to improve continuity and quality of care and reduce variation and cost while increasing patient satisfaction. At Henry Ford Hospital, this method of care design has provided an opportunity for collaboration and objective examination of current practice. Members of the department of neurological surgery, nursing, and the utilization care team department were brought together to participate in this endeavor.

The primary objective of this multidisciplinary team was to develop a care pathway for the craniotomy-for-tumor patient population. As the team met and discussions occurred, two areas along the continuum of care that could affect the expected outcomes of the pathway were identified. The team modified its focus toward developing these processes before continuing with care pathway development. These processes, the preadmission education and discharge planning visits and the postdischarge telephone calls, were developed and implemented through a collaborative effort of the nurse associates, the clinical nurse specialist, and staff nurses. In retrospect, the importance of the preadmission visits and postdischarge telephone calls was confirmed when the data were stratified by scheduled versus emergent admissions. This demonstrated that hospital stays for scheduled admissions were closer to the target LOS set for the craniotomy-for-tumor care pathway.

The reduction in LOS for patients on the craniotomy-for-tumor care pathway has affected the hospitalization charges. Average charges have decreased 40 percent since 1993. The team recognizes that the reduction in charges is primarily related to fewer hospital days. In future meetings, the team will be examining the utilization of resources during the inpatient hospitalization for opportunities for additional cost savings.

In accordance with the Henry Ford Health System's mission of providing excellent customer service, an important outcome to be monitored over time is patient satisfaction related to the use of the care pathway. More than half of the patients responding to the survey rated their satisfaction high in the following areas: preoperative teaching, discharge planning, and postdischarge

support. Although the results of this survey are favorable, some of the written comments provide valuable information to focus on when planning for improvements. In future questionnaires, areas to be surveyed will include pain management, communication between disciplines, and coordination of care.

The current process for care pathway development is time consuming, paper intensive, and restrictive. Future considerations for care pathway management in our institution include computerization. This will reconcile the current documentation system and allow improved comparative reporting and variance analysis. Computerization will provide the ability to link financial data to patient severity, psychosocial risk factors, and comorbidities. Currently, elements of the care process, specifically nursing interventions, are not calculated in the cost of providing care. Computerization will aid in our ability to make real-time, patient-specific modifications, to capture actual utilization of resources, and to manage patient outcomes proactively.

The development and implementation of care pathways cannot occur without a commitment to collaboration and continuous quality improvement by all team members. By recognizing the importance of maintaining a patient-centered approach, we have taken the steps to meet the demands of the changing health care environment. We will continue to strive for improved quality of care with emphasis on patient satisfaction.

REFERENCES

1. Sarkissian S. Length of hospital stay and contributing variables in supratentorial craniotomy patients with a brain tumor: a pre-care map study. *Axon.* 1994;15:86–89.

2. Tucker SM, Canobbio MM, Paquette EV, Wells MF. *Patient Care Standards: Collaborative Practice Planning Guides.* St Louis, MO: CV Mosby; 1996.

3. Urban NA, Greenlee JM, Krumberger JM, Winkelman C. *Guidelines for Critical Care Nursing.* St Louis, MO: CV Mosby; 1995.

4. Allen MB Jr, Johnston KW. Preoperative evaluation: complications, their prevention and treatment. In: Youmans JR, ed., *Neurological Surgery: A Comprehensive Reference Guide to the Diagnosis and Management of Neurosurgical Problems.* Philadelphia, PA: WB Saunders; 1990:803–900.

5. Bilsky M, Posner JB. Intensive and postoperative care of intracranial tumors. In: Ropper AH, ed. *Neurological and Neurosurgical Intensive Care.* 3rd ed. New York: Raven Press; 1993:309–329.

6. Doyle RL. *Healthcare Management Guidelines. Volume 1: Inpatient and Surgical Care.* New York: Milliman & Robertson; 1995.

■ 18 ■

Clinical Paths for Prostatectomy: Linking Outcomes to Practice through Continuous Quality Improvement

Jan Randall, Christine Roeback, Cary Robertson, and Joyce Arcus

With health care costs at $650 billion a year and rising, government and health care payers are challenging health systems to provide cost-effective care.[1] The task facing health care systems within this managed-care environment is the provision of care with improved financial performance. The survival of quality health care within this environment depends upon effective delivery of care with managed resource utilization and monitored clinical outcomes. The development and implementation of clinical paths has served as one strategy or tool within the Duke University Health System to address these issues. We have defined clinical paths as interdisciplinary treatment plans with specific lengths of stay and essential time-sequenced interventions to achieve defined clinical goals. Clinical paths are unique because they define key events for patients, allowing a clinical continuum in a timely, cost-effective manner.

With five years' experience, we have noted numerous benefits. Because the clinical path provides a written plan describing patient outcomes and essential interventions, it serves to promote the delivery of cost-effective and efficient care. Since the path is a mutually agreed-on plan, it is a tool that promotes continuity of care, regardless of location of patients within the health care system. The clinical path document, which describes a process of care, can also assist staff to understand and improve important patient management. Variance data from clinical paths serve as a basis for practice change and continuous quality improvement. Overall, greatest value comes from interdisciplinary dialogue that leads to improved outcomes for pa-

tient care management. This chapter presents an overview of establishing a collaborative care model for health care delivery through clinical paths. We would also like to share our success of linking clinical paths with quality improvement efforts to change practice.

CLINICAL PATHWAYS AND CARE MANAGEMENT

Since its inception, Duke University Medical Center has maintained a long-term interest in health care excellence for its patient population and improved methodology. In 1991, a major initiative of clinical path development in response to the need for cost containment was launched. Maintaining clinical quality and patient satisfaction in a managed-care environment was a coexistent goal. After selecting a multidisciplinary model for clinical paths, we provided extensive marketing and educational sessions for all involved. A specific office of care management was established to provide continuity for this multidisciplinary endeavor.

We identified certain criteria for clinical path development. Initially, clinical paths focused on high-risk, high-cost, and high-volume patient populations. Additionally, we targeted clinical areas with expressed interest and the presence of a strong physician leader. With consideration of these factors, we chose to initiate some of our first paths within the department of surgery, specifically where populations of elective surgical patients allowed a predictable clinical course with identifiable outcomes.

One of the initial paths developed within our system focused on the care of radical prostatectomy patients. There are several reasons that the prostatectomy patient group is an ideal target population for clinical paths. First, over the past several years, the national rates of radical prostatectomy have increased sixfold.[2] In accordance

We thank Dr. David Paulson, Dr. Peter Kussin, and Kevin Sowers for ongoing support of this initiative and Kathy Morris for editorial support. We also thank the urology staff on the inpatient unit and the urology clinic for support of this project.

with the national trend, the prostatectomy patient population in our health system was already at high volume and was projected to experience a future volume growth. Second, because this patient population carried a substantial component of Medicare fixed reimbursements, we determined that the development of a clinical path would offer financial opportunity for cost savings. Finally, the care for this patient group occurred in a variety of settings along a continuum that often created fragmentation. We realized that patients and clinicians could benefit from a tool to assist coordination of cost-effective care.

On the heels of success with the prostatectomy path, we continued efforts in clinical path development for additional diagnoses. To date, we have 65 paths throughout the health care system. Since the original prostatectomy path was prototypical, it serves as a model for development, implementation, and evaluation for all paths. We would like to share our experience of evolution of outcome measurement, particularly linkage of clinical paths to continuous quality improvement efforts.

PATHWAY: PROSTATECTOMY

In 1992, a multidisciplinary team of clinical staff and physicians within the division of urology identified a need to develop clinical paths for the purpose of promoting seamless care. This team agreed to address system issues that prolonged length of stay and fragmented care. Additionally, they identified the need to meet on an ongoing basis to discuss implementation issues, manage variance data from clinical path data, and revise the paths to meet changing health care environments. This team also recognized an opportunity to change their practice and delivery system on the basis of outcome analysis. This long-term vision included linking clinical paths with ongoing continuous quality improvement efforts.

Developing the Prostatectomy Clinical Path

The initial steps of the process included forming a team and determining key players necessary for successful development of the path. Since clinical paths emphasize multidisciplinary and collaborative practice, all health care providers involved in direct care of the patient populations were invited to participate in the development process. After the physicians were educated about the process, one physician was identified to be the team leader of the project. This physician leader served as the pathway's champion with peers and colleagues. The nurse manager from the inpatient urology unit and physician leader then identified the appropriate support departments and requested representatives for the team. The team included physicians, nurses, personnel and staff from inpatient and outpatient settings (social worker, discharge planner, pharmacy representative, utilization management, and dietitian), and a facilitator from the care management office. The care management facilitator guided the process by gathering benchmark data, assisting with format, and coordinating implementation of the paths.

During the initial phase of the process, the care management facilitator met with a variety of resources for information exchange and goal setting. The facilitator established an initial meeting with the physician leader and nurse manager to discuss goals of the clinical path process, agendas for the meetings, and goals for the patient population. Next, the utilization management coordinator provided information about contractual expectations from third-party payers and managed-care contracts. This information served as the basis for defining a target length of stay, resource utilization, and clinical outcome. Additionally, the facilitator gathered baseline data on the patient population, including average length of stay, total cost per case, payer type, number of cases, and daily utilization of resources during the hospital stay.

After reviewing a variety of information, the care management facilitator developed a working draft before the first team meeting. External benchmarking data, including sample paths from other medical centers, along with target length-of-stay data, were used in developing the draft. Information from chart reviews proved to be helpful in defining current practice patterns, treatments, medications, diagnostic tests, discharge planning, and patient teaching. Additionally, unit-based standards of care, policies, procedures, and protocols for patient care management were incorporated into the path document. These provided an infrastructure for expectations of standards of care and expected outcomes for patient populations. The initial draft of the path reflected relative best-practice standards. These were compared to external benchmarks. The goal was to provide a draft to the team as a starting point for discussion about currently held practice patterns and possible changes to be incorporated in the path document.

Prostatectomy Path Team Meetings

After establishing the team and developing an initial draft, the care management facilitator orchestrated four weekly team meetings. During the initial meeting, a variety of issues were discussed:

- What should be the target length of stay?
- What were the overall objectives or goals for the patient population?
- What were the specific milestones that should be reached to meet a specific target length of stay?
- What variations in care existed, and what was their source?

- What were the issues leading to fragmentation of care and loss of connection for patient and family?

As these issues were discussed through brainstorming techniques, the team began to build consensus about common values and goals for the target patient population. At the end of the first meeting, the team collectively agreed on a system and process to highlight fragmentation of care and identified daily goals for the surgery experience.

Over the next month, the team refined the original draft and addressed a variety of practice issues. Various interventions surrounding current practice were reviewed and challenged by the team. Each intervention, including lab, diagnostic tests, and medications, was reviewed to determine if it was essential to patient care. If a suggestion was made by the team to include a specific item, then the team discussed its relevance and cost. Additionally, the team addressed the most effective approach for patient teaching, with a view of achieving a two-day decrease in length of stay. Partnership with outpatient clinics allowed the team to agree to initiate patient teaching before admission. Teaching protocols were developed to assist unit staff in providing patient teaching more progressively rather than on the discharge day. During weekly meetings, the clinical path team revised documentation to include essential interventions and outcomes for each day of the expected length of stay. Additionally, they developed a patient version of the clinical path that would provide additional information to patients and families about expectations for the hospital stay.

Implementation of the Path

After developing the path, the team proceeded with implementation. A process for written feedback and consensus approval was established by sending the path document to those who would interact with the path. This included all physicians who performed prostatectomies, as well as others in departments that provided services to the patient population. After receiving written feedback, the team reviewed suggestions and revised the path document. A letter was sent to individuals whose feedback was not incorporated in the path. After revisions were made, the team prepared information sessions for all disciplines along the continuum about the path. Education for health care providers included goal setting for multidisciplinary documentation and utilization of paths through daily collaborative discussions of care issues.

Since implementation, the prostatectomy path has undergone several revisions. The original path, written in 1992, reflected a seven-day length of stay (Exhibit 18–A–1 in Appendix 18–A, postoperative day 1 of the original path). Originally, a multidisciplinary format was selected with the hope that all disciplines would use the path as a working document. However, in reality, nursing staff viewed it only as a documentation tool. Over the past year, other disciplines have seen the benefits associated with clinical paths and have moved to make the path more than a nursing tool. We have revised our original format to include patient outcomes charting from a multidisciplinary perspective. We have also streamlined documentation on the path by focusing on outcomes and avoiding written duplication of interventions. The current prostatectomy path, detailing a four-day length of stay, was modified in 1997 to reflect the new format (Exhibit 18–A–2 in Appendix 18–A).

Care Management

With the original and revised clinical path format, a registered nurse (RN) initiates the path and serves as a coordinator of care. Upon admission, the path is reviewed by the respective patient's RN and physician for appropriateness. Although an RN may initiate a path, there must be physician orders accompanying the path. Currently, many of the paths have associated preprinted physician orders. If a physician house officer chooses to write orders separate from the preprinted sets, the attending physician is notified, except in emergent cases, to coordinate the clinical plan.

Clinical paths within our system now consist of two primary components, one that describes a plan of care and one that serves as a tool for documenting clinical outcomes. The admitting RN initially reviews and individualizes the path (plan of care) to meet patient needs. Other disciplines may also adjust the path to reflect daily changes in patient status. Outcomes or interventions may be added to the clinical path or deleted from the path by drawing a single line through the item. Initials and full signatures must accompany any change to the path. All disciplines, including physicians on some services, document outcomes at least once every 24 hours, using clinical path progress notes. If an outcome is not met, then a variance note must be written on the back of the form describing a reason and providing details about action steps to keep the patient on track. The clinical path serves as the daily tool to coordinate care for all disciplines. The paths are used as a focal point for daily rounds by multidisciplinary teams and end-of-shift reporting for nursing staff. Physicians have demonstrated visible support for clinical paths by incorporating them into daily rounds and discussions with patient care teams. Because the clinical path is part of the permanent medical record, it serves as a plan of care for all disciplines.

Case managers use clinical paths in negotiations with third-party payers as part of ongoing certification for hospital stay. These clinical paths provide a clearly written plan describing an expected length of stay and goals for patients. Staff nurses and other disciplines use clinical paths as part of a daily dialogue about care issues and patient progress. Case managers and discharge planners are notified by nursing or physician staff whenever there are deviations from clinical paths resulting in a prolonged length of stay. This creates multidisciplinary dialogue concerning best approaches for keeping patients on the path and meeting target goals.

The clinical path multidisciplinary team reviews ongoing feedback about the effectiveness of the paths. Generally, all paths are revised every six months or sooner if necessary. As paths are evaluated, the clinical path multidisciplinary team simultaneously reviews patient outcomes. These data provide a basis for continuous quality improvement within the practice setting.

Outcomes

Since development at Duke University Medical Center, clinical paths have become more sophisticated and accepted, and variance data have become more realistic. Following the implementation of change within any clinical path, it has been vital to measure and evaluate the effectiveness of change. Was there a difference in the financial and clinical aspects of care? Did patient and/or customer satisfaction improve? How can outcome data be used to determine the need for further changes?

Outcome Measurement from Prostatectomy Clinical Path

As the prostatectomy team started to discuss outcomes, they realized the importance of linking outcomes

to continuous quality improvement efforts. As a result, the prostatectomy team decided to routinely evaluate the following outcomes:

- financial management and resource utilization
- patient satisfaction
- staff satisfaction
- clinical outcomes

The task for the team was to ascertain how it would collect the data and distribute the information. Fortunately, the financial and patient satisfaction data were already being collected by hospital administration. The team realized that staff satisfaction and clinical outcome data would need to be collected at the patient care unit level.

Financial Management and Resource Utilization

Financial and resource utilization data are distributed quarterly from the hospital financial office. These reports reflect cost and length-of-stay trends for clinical path patient populations. The clinical path team and other physicians within each service review these data elements. Clinical path teams compare actual data to goals for length of stay and resource utilization.

As the payer mix for the prostatectomy patient population within our health care system shifted to more managed care, we realized that prospective reimbursement placed a financial risk on the hospital. When the prostatectomy path was developed in 1992, approximately 45 percent of the patient group was covered by fixed reimbursement, either managed care or federal/state programs. These percentages continued to climb over the next few years (Figure 18–1).

At the implementation of the current clinical paths in 1991, it was believed that cost savings could be realized by reducing length of stay and managing resources. As expected, reductions in length of stay were positively as-

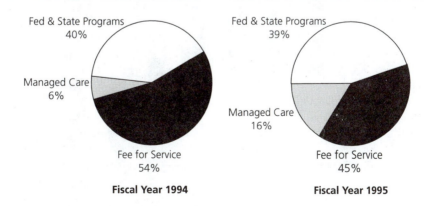

Figure 18–1 Payer Analysis for Radical Prostatectomy (Uncomplicated Cases) at Duke University Medical Center. *Source:* Copyright © Duke University Medical Center.

sociated with reduced hospital cost (Figure 18–2). Net margin for hospitalization, costs, and positive or negative cash flow could be directly correlated to length of stay and patterns of clinical care. An interesting subanalysis of this data set for radical prostatectomy was a comparison between surgical approaches. Whether a retropubic or a perineal approach was used for a prostatectomy determined hospital costs on the basis of operating time, transfusion, and length of stay. This correlated with the positive net margin in favor of radical perineal prostatectomy (Figure 18–3).

Data analysis of this nature strongly demonstrated the utility of clinical path implementation and data acquisition. Partnering with physicians, nursing staff, and hospital administration allowed clarification of issues in relation to hospital cost structure. This also allowed easy identification of areas for improvement and critical review. Data sharing occurred through formats of formalized presentations as well as data sheet distribution.

Patient Satisfaction

Within our health care system, patient satisfaction outcomes are routinely reported at a unit level. These data, collected via written and telephone surveys on a biannual basis, are distributed to each nurse manager and physician leader for review. This feedback is subsequently shared with nursing staff and serves as a cornerstone for operation improvements.

Patients often express a high level of satisfaction with the patient version of the path. Staff nurses review this tool with patients and families before surgery (Exhibit 18–A–3 in Appendix 18–A). As patients become familiar with the path, they often remind staff to ensure that interventions described on the path actually occur.

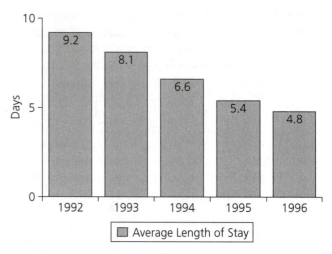

Figure 18–2 Average Length of Stay for Radical Prostatectomy Patients after Pathway Implementation

Because prostatectomy patients are generally located on a specific urology unit, data collection is simplified. Patients on clinical paths are surveyed regarding their satisfaction with care via a patient satisfaction survey. These patient satisfaction surveys are routinely administered hospitalwide. Questions pertinent to satisfaction with clinical paths include those presented in Exhibit 18–1.

When comparing survey results between the dedicated urology unit and the hospital overall, the urology unit mean scores were consistently higher than the overall hospital mean scores. The majority of patients on this specific urology unit are on clinical paths. Even though many variables may have affected patient satisfaction, we feel that this observation reinforces the principle that clinical paths enhance overall patient satisfaction. Pa-

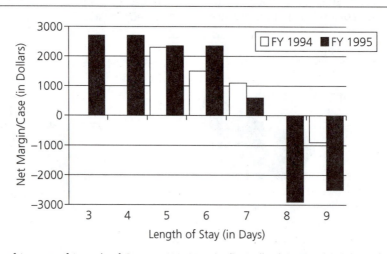

Figure 18–3 Comparison of Impact of Length of Stay on Net Margin (in Dollars) in Fiscal Years 1994 and 1995 of Clinical Path Utilization

Exhibit 18–1 Patient Satisfaction Survey Questions Related to Clinical Path

---

You knew what was going to happen to you because your nurses kept you informed.

Overall, you were satisfied with the quality of care you received from your nurses.

Your doctors explained what was happening to you and why.

All of your doctors worked together as a team.

Your doctors and nurses worked together as a team.

Overall, you were satisfied with the quality of care you received from your doctors.

You were provided with adequate home care instructions.

You felt the Duke staff worked together as a team to coordinate your care during your visit.

Source: Data from Patient Satisfaction Questionnaire, © Healthcare Research Systems.

---

tients respond favorably when their care is coordinated and they are aware of what to expect.

Staff Satisfaction

Besides patient satisfaction data, the prostatectomy team decided to collect and report staff satisfaction at a unit level for this project. Unit staff note a variety of benefits associated with the use of clinical paths. A survey conducted in 1994 of 40 nursing staff on the urology unit revealed that staff identified benefits of using clinical paths in two important areas. On a scale of 1 (*strongly disagree*) to 4 (*strongly agree*), the following statements received an overall response of *strongly agree:*

- Clinical paths benefit the patients.
- Clinical paths improve the way a health care provider performs.

Staff members have identified positive outcomes of using clinical paths in their practice. First, they have seen time savings associated with using clinical paths as documentation tools to meet internal and external documentation requirements. Because the clinical path is part of the permanent medical record, it serves as a standard plan of care for the patient, eliminating the need to develop a plan. The staff need only modify the path for each patient. Staff members have also noted a benefit of using the path as a teaching tool for new team members. The path serves as a vital part of the cross-training process for staff who might be new or unfamiliar with the urology patient population. Specifically, clinical paths are now included in the residency training manuals for the physician urology house staff. The path also serves as a catalyst to promote collaborative discussion about expectations for patient care. The staff has stated that paths provide a mechanism

to assist in meeting the requirements of the Joint Commission on Accreditation of Healthcare Organizations on multidisciplinary planning and coordinating patient needs along the health care continuum.

Physician Satisfaction

Physician acceptance and use of clinical paths in relation to the prostatectomy project diverged on two important points. From a practical standpoint, physicians in training (i.e., house staff) found the clinical paths to be useful and pragmatic. Feedback indicated wide acceptance among the physician house staff, with the clinical paths being a tool for discussion with nursing and ancillary staff concerning critical daily goals for postoperative patient care. The clinical paths are incorporated in the urology house staff training manual and are used for cross-checks on critical daily events during morning and afternoon rounds.

Attending staff and faculty, after an initial period of skepticism, grew to accept the clinical paths as documents worthy of use. This was in specific reference to efforts to standardize antibiotic usage, drain output measurement and removal, advancement of diet, activity levels, and discharge planning. Physician extenders/ancillary staff have also come to rely on the clinical path documents for establishing a framework for communication with patients, as well as physicians in training. Few negative comments were received after the 18-month initiation. This led to greater enthusiasm for additional path development, leading to well over 20 clinical paths for multiple urologic diagnoses.

Clinical Outcomes and Practice Changes

Within our system, clinical outcomes and path variances are monitored by nursing staff daily and reported by the care management office on a quarterly basis. Since the variance tool is not part of the permanent record, the data sheets are collected at the end of each month. Data are entered into a Paradox database, and reports are analyzed and distributed on a quarterly basis from the care management office.

Variances from clinical path outcomes provide the clinician with vital information about health care practices. However, clinicians must be willing to use the information to create change. This information can help clinicians know how much change is necessary by creating a "conscience" of managed care, especially as attention shifts from length of stay and resource utilization to outcome and risk reduction.

As our prostatectomy team met monthly, they progressed along a continuum with their comfort level at addressing variances. As previously mentioned, a generic variance tool was developed to track variances from the path. Originally, staff recorded all deviations or variances, positive or negative, from expected interventions and outcomes on a generic variance form (Exhibit 18–A–4 in Appendix 18–A). As anticipated, an overwhelming amount of data was collated and provided to the team for review.

Initial data revealed that patients were achieving certain outcomes ahead of prediction or expectation on the path. The original path had ambulation out of bed occurring on postoperative day 2. However, variance data revealed that the majority of patients were able to ambulate on the night of surgery or at least by the first postoperative day. Additionally, the majority of patients were able to take clear liquids on the first postoperative day instead of being restricted to nothing by mouth, as described on the path. The team agreed to modify practice to promote patient progression. As the team monitored outcomes, they noted less gut motility dysfunction associated with advancing diet and activity levels of the patients. By advancing the diet and activity progression without untoward effects, the team collaborated with physician attending and house staff to reduce expected length of stay by two days. With this change, the team also identified the need to convert intravenous-preparation antibiotics to oral on postoperative day 2 rather than day 3.

As the team became more comfortable with addressing variance data, they agreed to look beyond the hospital environment. Through partnership with the outpatient clinic, they began to review outcomes related to wound management and home care. Specifically, the team received feedback from the outpatient clinic that patients were returning for follow-up appointments after discharge with dermatitis of the groin area. The clinic staff reported that patients were verbalizing concern about skin irritation from the povidone-iodine sitz baths during home care. After reviewing infection rate data, the team decided to change the home care routine from requiring povidone-iodine sitz baths to requiring warm-water sitz baths and showers. Improvement in dermatitis complaints was realized. There has been no increase in wound infection rates associated with this change in practice. We have seen cost savings from eliminating povidone-iodine utilization. Additionally, prostatectomy patients report a higher level of satisfaction with the elimination of povidone-iodine from the home care routine.

Over the past two years, the team has continued to review variance and outcome data. As a result, there have been numerous practice changes (Exhibit 18–2). With these changes and the revised length of stay, there have been no increases in readmissions for this patient population.

As the team gained experience dealing with the variances, they decided to refine the variance data collection

Exhibit 18–2 Practice Changes with Prostatectomy Path

Lab/Tests
  Minimized labs
Operating Room
  Implemented standardization of surgical case carts
Routine Care
  Began immediate ambulation on POD 1
  Began early feedings (POD 1)
  Targeted POD 3 for discharge
  Reduced length of stay from 9 to 4 days
  Simplified taping procedure for Foley catheters, which saved time for staff and patients
Patient Teaching
  Created video for patient/family teaching
  Initiated patient teaching in clinic prior to admission
  Provided follow-up teaching in clinic with return visits
Pharmacy
  Eliminated povidone-iodine for sitz baths with home care
  Switched antibiotics earlier from IV route to oral (POD 2)

Note: IV, intravenous; POD, postoperative day.

process and to monitor specific indicators that would affect cost, length of stay, and/or quality. Recently, the variance tool was modified to include indicators monitored as part of ongoing quality improvement activities on the patient care unit (Exhibit 18–A–5 in Appendix 18–A).

The team continues to monitor clinical trends and to use the data to enhance overall practice performance and outcomes. Throughout this process, the team reviews system and process issues of care delivery to determine if changes are necessary. Sometimes, patient conditions and complications contribute to unresolved outcomes that may be independent of the health care team. Therefore, it is critical for the team to differentiate those deviations or variances from expected outcomes related to (1) patient issues and (2) nonpatient issues, such as patient complications resulting in discharge delay.

Advantages of and Barriers to the Prostatectomy Path

The overall response to the prostatectomy path has been very positive. Although there was some initial resistance to change, the team kept everyone informed and accepted constructive feedback. This technique promoted a supportive, trusting environment that made the change more acceptable.

The ongoing support of the physicians, nurses, and other disciplines has contributed to the success of this path. As a result, there are many advantages associated with clinical paths. The greatest benefit to the staff was their perception of transformation to a health care team that accepted responsibility for patients. The path enhanced staff accountability by delineating interventions within expected time frames. The clinical path also allowed us to streamline care and cost within the hospital, without cost shifting to the outpatient setting. There was no increase in home health referrals associated with a decreased length of inpatient stay. Additionally, the path has provided a mechanism for a nurse manager to justify nursing hours required to provide patient care for a population with a reduced length of stay and high acuity.

Some barriers have been noted. The overall cost for path development and implementation can be quite high. Team members need time away from their duties for ongoing team meetings. It has been estimated that the cost of pathway development, including labor charges for biweekly meetings, is $12,000 to $15,000 per diagnosis-related group.[3] Also, during the initial phase of path implementation, staff may need time and training for conflict resolution as they manage variances and confront practice issues. Staff may also need extra time to become comfortable with multiple documentation systems, especially if an institution has a separate system for patients who are not on clinical paths.

Future Plans

Since the development of this path, we have converted our utilization management department to a case management department. Case managers are now using clinical paths as a tool to guide patient care management. As part of this process, Duke Hospital is piloting a project in which case managers capture inefficiencies of care. Care inefficiencies are described as deviations from an appropriate plan, resulting in a delay in patient care delivery. These inefficiencies are validated with the respective physician and entered into a database. Reasons are assigned for each delay and collated as systems, process, or patient-related issues.

Data from quarterly reports will be reviewed for trend analysis and shared with physicians and others as part of a discussion for improving our health care system. Case managers will also continue to collect readmission data. Once this system has been implemented, we will move away from daily clinical path variance collection by the nurse. However, staff nurses and other health care members will still be responsible for documenting patient outcomes.

Future plans for our system include moving clinical paths from a paper system to a computer/electronic record system. Currently, it is difficult to move paper with the patient along the continuum. If a patient has several diagnoses or comorbidities associated with his or her diagnosis, then it is difficult to pull together multiple paths on paper to meet the patient's needs. Computerization will facilitate this process and ease the burden of documentation. We are beginning to discuss the need for computerization of paths. Initial discussions will include

placing the paths on the network within our system to allow staff the opportunity to view them from any computer workstation. We are also in preliminary discussions regarding the addition of clinical paths to our commonly performed electronic documentation. Our goal is to continue to refine the paper version to prepare for computerized documentation.

As we move toward acquiring more managed-care contracts, we anticipate that clinical paths will become an integral component of the negotiations. For us, a new method of clinical path development is emerging that will allow us to automate path development and track outcomes.

With a recent software addition to our hospital financial system, we can now determine the actual standard cost for providing care according to the path. This system automates clinical path development by producing histograms of length of stay and daily utilization of resources. This process allows us to determine the best practice internally without the laborious task of chart reviews. Additionally, the program supports the ongoing monitoring of outcomes, such as mortality, readmission to the hospital, and return visits to the clinic. Cost data, along with expected outcomes, will be a vital part of any discussion in future managed-care contract negotiation, particularly when it is necessary to discuss financial and quality performance.

CONCLUSION

Clinical paths optimize resource utilization and facilitate high-level patient care outcomes. It is vital to define a process for the development, implementation, and evaluation of clinical paths within any health care system. We have identified principles for successful clinical path development. More important, we have shared our past experience and future plans for linkage of clinical outcomes to practice changes. The development of clinical paths and the development of quality improvement programs have much in common. The key to success is joining the two for the greater goal of understanding processes important to excellence in patient care delivery.

REFERENCES

1. Lu-Yao GL, McLerran D, Wasson J, Wennberg JE. An assessment of radical prostatectomy time trends, geographic variation and outcomes. *JAMA*. 1993;269:2633–2636.
2. Zander K. Use of variance from clinical paths: coming of age. *Clin Perf Qual Healthcare*. 1997;5:24–30.
3. Williams M. Three alternative methods of developing critical pathways cost and benefits. *Best Pract Benchmarking Healthcare*. 1996;3:126–128.

■ Appendix 18–A ■
Clinical Pathways and Tools

Exhibit 18–A–1 Original Prostatectomy Path

CLINICAL PATH FOR: RADICAL PERINEAL PROSTATECTOMY DRG#/Description _____335_____ ICD-9-CM: _____60.62_____ Projected LOS: _____7 days_____ Date Path Initiated: _____ Date Patient Discharged: _____	 ADDRESSOGRAPH

DATE	POD 1 Place Initial beside each intervention and outcome as completed
PATIENT PROBLEM	**DESIRED OUTCOMES**
1. Knowledge Deficit/anxiety. Related to pre-op surgical procedures, post-op care, home care instructions, and S/S to report to MD	_____;_____;_____ Pt. verbalizes understanding of post-op care
2. Potential alteration in respiratory function related to effects of surgery, immobility, and pain.	_____;_____;_____ Pt. maintains effective airway clearance _____;_____;_____ Pt. performs TCDB independently
3. Acute pain r/t surgical incision.	_____;_____;_____ Pt. verbalizes pain control with IM or PO pain needs
CLINICAL PATH	**POD 1**
1. Consults	—Anesthesiology post-op evaluation
2. Laboratory Tests	Chem CS, ABC
3. Diagnostic Tests	
4. Treatments	___ Remove perineal dressing _____;_____;_____ Apply gauze fluffs w/promise pants, change prn _____;_____;_____ Incentive spirometer q 2h _____;_____;_____ TCDB q 2h _____;_____;_____ Nothing per rectum ___ Perineal penrose out 1st BM _____;_____;_____ Begin sitz baths if necessary: 30cc Betadine/1000cc saline TEDs/SCDs, VS q 8h, I/O q 8h, Foley to SD, Irrigate foley prn, Bowel sounds q shift, Horseshoe cushion
5. Medications	H₂ Blocker prn, Antibiotic IV, Antispasmodics prn, Antiemetics IM/IV prn, IVF, Sleep med prn, Stool softener prn, IM/PO pain meds prn, Toradol IV q 6 hr
6. Diet	Clear liquid breakfast, full lunch & dinner
7. Activity	_____;_____;_____ OOB to chair w/horseshoe cushion
8. Teaching/Education	Medication _____;_____;_____ Diet _____;_____;_____ Activity _____;_____;_____ Treatments _____;_____;_____ Clinical Path _____;_____;_____ Pain control _____;_____;_____ Incision care/wound healing _____;_____;_____ _____;_____;_____ Begin leg bag instructions

Signatures:		

Circle variance & document on variance record.

Note: ABC, automated blood count; BM, bowel movement; CS, chemistry sugar; IM, intramuscular; I/O, intake and output; IV, intravenous; IVF, intravenous fluid; OOB, out of bed; PO, oral; POD, postoperative day; prn, as needed; r/t, related to; SCDs, sequential compression devices; SD, straight drainage; S/S, signs and symptoms; TCDB, turn, cough, deep-breathe; VS, vital signs.

Source: Copyright © Duke University Medical Center.

Exhibit 18–A–2 Revised Prostatectomy Clinical Path

DUKE UNIVERSITY MEDICAL CENTER CLINICAL PATH FOR: RADICAL PERINEAL PROSTATECTOMY

M040-T 6/97

DRG #/Description: 335   ICD-9-CM: 60.62   Date Initiated: _____

Projected LOS: 4 DAYS

Date Discharged:

Date Discontinued from Path: _____   ☐ New Plan of Care M052B

☐ New Path:

Order sets need to accompany the Clinical Path and must be signed prior to implementation. There is room for individualization of patient care within the order sets and clinical path. Pre-printed forms do not preclude clinical judgment.

If modifications are made: cross out text with single line using black ink, initial, date & time; full name and signature at the bottom. Please update Advance Directives as appropriate.

ADDRESSOGRAPH

Advance Directive: ☐ Deferred ☐ Resolved   Date/Time
                    ☐ Waiting for copy ☐ Resolved   Date/Time

	Pre-Admission/Admission	Day of Surgery	Post Op Day 1	Post Op Day 2	Post Op Day 3
Consults	Anesthesia Cardiology prn	Respiratory prn Pharmacy prn	Anesthesia post op assessment Social Worker prn		
Laboratory Tests	PSA, Acid Phosphate ABC, PT/PTT, Chem CS Urinanalysis with Culture Type and Crossmatch	PACU standard	Chem CS, ABC	Chem CS, ABC	Chem CS, ABC
Diagnostic Tests	Bone Scan prn Chest X-ray & EKG				
Treatments	– Mechanical bowel prep – Hibiclens shower @ hs – Order TEDS – History and physical: note allergies	– TEDS/Venodynes – Void on call – If Hct <25, transfuse prn – Bowel sounds q shift – Vital signs q 4 h – I/O q shift. Foley to straight drainage – Nothing per rectum – TCDB/Incentive spirometry q 2 – Reinforce dressing prn – PACU per standard – Do not manipulate foley	– Remove perineal dressing. Apply gauze fluffs with promise pants – TCDB/incentive spirometry every 2 hours – Nothing per rectum – Perineal penrose out 1st BM – Begin sitz baths prn: 30cc Betadine/1000 saline – TEDS/Venodynes – Vital signs every 8 hours – I/O, Foley to Straight drain, irrigate prn – Assess bowel sounds q shift – Horseshoe cushion	– Vital signs every 8 hours – Foley to SD, irrigate prn – Spirometer every 2 h prn – TCDB – Nothing per rectum – Perineal penrose out 1st BM – Promise pants with gauze fluffs to perineal incision – After rectal penrose removed, sitz baths BID with Betadine 30 cc's/1000 water – TEDS – Do not manipulate foley	– Vital signs every 8 hours – Foley to SD irrigate prn – Spirometry every 4 h prn – TCDB prn – Bowel sounds q shift – TEDS – Gauze fluffs and promise pants to perineum prn – D/C Glans stitch – D/C penrose if no BM yet – Nothing per rectum – Sitz bath BID – Cleanse sutures and paint with Betadine swabs after BM – Do not exceed BID sitz baths even with increased BMs
Medications	– Antibiotic bowel prep – Sleep med prn – Insert saline lock – Acetaminophen 60 mg q 4 h prn	– IV antibiotics – Neomycin enema at 6 AM – Sleep meds prn – IV fluids – Torodol IM in OR at closure – Torodol IV q 6 h x 24 hours	– H_2 Blocker prn – IV antibiotics – Antispasmodics, prn – Antiemetics IM/IV, prn – IV fluids, sleep meds, stool softener prn – IM/PO pain meds prn – Torodol IV q 6 h	– IV antibiotics. Change to PO after 48 hours of IV – Antispasmodics, stool softener, antiemetics, sleep meds, prn – IV fluids, PO pain meds – H_2 Blocker, prn – Change to Heparin lock if po intake >800cc's by lunch	– PO antibiotics – Antispasmodics prn – PO pain meds, stool softener prn – Antiemetics prn – Sleep meds prn – Heparin lock
Diet	– Clear liquids. NPO after midnight. – Meds may be taken with sips.	– NPO – Meds with sips	– Clear liquid breakfast, full lunch and dinner	– Regular	– Regular
Activity	– As tolerated	– Bedrest	– OOB to chair with horseshoe cushion	– OOB to hall QID	– Ambulate QID and prn

	Pre-Admission/Admission	Day of Surgery	Post Op Day 1	Post Op Day 2	Post Op Day 3
Teaching	Pre-op teaching Review clinical path—give patient version. Complete pre-op checklist.	– Pain management – TCDB – Post-op expectations	– Medication, diet, activity, treatments, pain management, incision care/wound healing. – Begin leg bag instructions.	– Medications, treatments, pain control, diet, activity, incision care, wound healing – Review sitz bath technique – Begin leg bag instructions – Give patient "Home Foley Cath Care" book and typed RPP discharge sheet.	– Provide M01B – Provide equipment for home use – Review meds, provide leaflets on meds – Provide prescription, return appointment and reasons to call MD
Continuum of Care	– Complete M014 form Pre-admission certification with length of stay. Case manager to assess.	– Assess patient progress toward discharge	– Assess patient progress toward discharge Case Manager to assess if anticipated LOS >3 days or patient falls off path	– Assess need for home health referral	– Return to clinic in 14 days from surgery to have catheter removed

Initials	Signature/Title	ID #		Initials	Signature/Title	ID #

Note: ABC, automated blood count; BID, twice daily; BM, bowel movement; CS, chemistry sugar; D/C, discontinue; EKG, electrocardiogram; Hct, hematocrit; hs, at night; IM, intramuscular; I/O, intake and output; IV, intravenous; LOS, length of stay; NPO, nothing by mouth; OOB, out of bed; OR, operating room; PACU, postanesthesia care unit; PCA, patient-controlled analgesia; PO, oral; prn, as needed; PSA, prostate-specific antigen; PT, prothrombin time; PTT, partial thromboplastin time; QID, four times daily; RPP, radical perineal prostatectomy; SD, straight drainage; TCDB, turn, cough, deep-breathe.

Source: Copyright © Duke University Medical Center.

continues

Exhibit 18-A-2 continued

DUKE UNIVERSITY MEDICAL CENTER CLINICAL PATH FOR: RADICAL PERINEAL PROSTATECTOMY

TODAY'S DATE: _____   ADMISSION/DAY OF SURGERY

Addressograph

MULTIDISCIPLINARY DOCUMENTATION TOOL

	INITIAL/TIME TO INDICATE GOAL/ PROCESS MET, NOT MET *If not applicable, write NA beside goal		
	Met	Met	Not Met
1. PATIENT PROBLEM: Post Op Stability EXPECTED OUTCOMES: Patient will remain stable post operatively.			
• Patient maintains adequate intake and output			
• Vital signs stable			
• Lungs clear			
• Skin integrity intact without reddened areas			
• Incision clean and dry without evidence of infection			
• Abdomen soft and not distended			
• Performing turn, cough, deep breathing exercises and uses incentive spirometer every 2 h with staff reminders			
2. PATIENT PROBLEM: Pain Management EXPECTED OUTCOMES: Patient will have pain managed satisfactorily.			
• Patient verbalizes adequate pain control with epidural/PCA/IV or IM medications.			
• Patient identifies pain control measures.			
•			
3. PATIENT PROBLEM: EXPECTED OUTCOMES:			
•			
•			
•			
4. PATIENT PROBLEM: Patient/family education EXPECTED OUTCOMES: Patient will be able to verbalize preop and postop care procedures, understanding of plan of care.	SEE PATIENT EDUCATION RECORD		

Progress Notes:

Initial	Signature/Title	ID #	Initial	Signature/Title	ID #

DUKE UNIVERSITY MEDICAL CENTER CLINICAL PATH FOR: RADICAL PERINEAL PROSTATECTOMY

TODAY'S DATE: _____   POST OP DAY 1

Addressograph

MULTIDISCIPLINARY DOCUMENTATION TOOL

	INITIAL/TIME TO INDICATE GOAL/ PROCESS MET, NOT MET *If not applicable, write NA beside goal		
	Met	Met	Not Met
1. PATIENT PROBLEM: Pain Management EXPECTED OUTCOMES: Patient verbalizes pain management is satisfactory.			
• Patient transitions to IM or PO pain meds and verbalizes adequate pain control.			
• Patient verbalizes pain management options and techniques			
•			
2. PATIENT PROBLEM: Potential alteration in respiratory status. EXPECTED OUTCOMES: Patient maintains effective airway clearance and absence of respiratory infection.			
• Patient performs TCDB/incentive spirometry every 2 hours independently			
• Lung sounds remain clear.			
3. PATIENT PROBLEM: Perineal healing EXPECTED OUTCOMES: Patient's perineum will show progressive healing and bowel function.			
• Patient tolerates sitz baths BID			
• Perineal incision is intact, without redness, oozing, or foul odor.			
•			
4. PATIENT PROBLEM: Patient/family education EXPECTED OUTCOMES: Patient will be able to verbalize medication use, activity, diet, treatments, incision care, wound healing and leg bag use.	SEE PATIENT EDUCATION RECORD		

Progress Notes:

Initial	Signature/Title	ID #	Initial	Signature/Title	ID #

POST OP DAY 3

DUKE UNIVERSITY MEDICAL CENTER CLINICAL PATH FOR:
RADICAL PERINEAL PROSTATECTOMY   POST OP DAY 3
TODAY'S DATE: _____

Addressograph

MULTIDISCIPLINARY DOCUMENTATION TOOL

	INITIAL/TIME TO INDICATE GOAL/ PROCESS MET, NOT MET *If not applicable, write NA beside goal		
	Met	Met	Not Met

1. PATIENT PROBLEM: Perineal Healing
EXPECTED OUTCOMES: Patient's perineum will show progress toward healing and normal bowel function.
- Patient taking sitz baths BID independently
- Perineum remains intact without redness, oozing, or foul odor.
-

2. PATIENT PROBLEM: Patient/family education
EXPECTED OUTCOMES: Patient verbalizes understanding of home care instructions and signs and symptoms to report to MD. Performs home care activities independently. Able to demonstrate leg bag change independently.

SEE PATIENT EDUCATION RECORD

3. PATIENT PROBLEM:
EXPECTED OUTCOMES:

	INITIAL/TIME TO INDICATE GOAL/ PROCESS MET, NOT MET *If not applicable, write NA beside goal		
	Met	Met	Not Met

-
-
-

4. PATIENT PROBLEM:
EXPECTED OUTCOMES:

	INITIAL/TIME TO INDICATE GOAL/ PROCESS MET, NOT MET *If not applicable, write NA beside goal		
	Met	Met	Not Met

-

Progress Notes:

Initial	Signature/Title	ID #

Initial	Signature/Title	ID #

POST OP DAY 2

DUKE UNIVERSITY MEDICAL CENTER CLINICAL PATH FOR:
RADICAL PERINEAL PROSTATECTOMY   POST OP DAY 2
TODAY'S DATE: _____

Addressograph

MULTIDISCIPLINARY DOCUMENTATION TOOL

	INITIAL/TIME TO INDICATE GOAL/ PROCESS MET, NOT MET *If not applicable, write NA beside goal		
	Met	Met	Not Met

1. PATIENT PROBLEM: Potential alteration in respiratory function
EXPECTED OUTCOMES: Patient will maintain effective airway clearance.
- Patient demonstrates turn, cough, deep breathing exercises and incentive spirometry q2 h independently
- Lungs remain clear

2. PATIENT PROBLEM: Perineal healing
EXPECTED OUTCOMES: Patient will show progress with perineal healing and normal bowel function.

	INITIAL/TIME TO INDICATE GOAL/ PROCESS MET, NOT MET *If not applicable, write NA beside goal		
	Met	Met	Not Met

- Patient tolerates sitz baths BID
- Perineal incision healing without redness, oozing, or foul odor.
-
-

3. PATIENT PROBLEM:
EXPECTED OUTCOMES:

	INITIAL/TIME TO INDICATE GOAL/ PROCESS MET, NOT MET *If not applicable, write NA beside goal		
	Met	Met	Not Met

-
-
-

4. PATIENT PROBLEM: Patient/family education
EXPECTED OUTCOMES: Patient will be able to verbalize medication use, activity, diet, treatments, incision care, wound healing, sitz bath use, and leg bag use.

SEE PATIENT EDUCATION RECORD

Progress Notes:

Initial	Signature/Title	ID #

Initial	Signature/Title	ID #

Exhibit 18–A–3 Patient Version: Clinical Path

Welcome to Duke Medical Center. You are scheduled to have your prostate removed on _____.

The "Clinical Path" has been developed for you and your family. The "Clinical Path" is a guide of what to expect before, during, and after your surgery. It identifies people you will meet, tests, treatments, medications, diet, activity, and teaching. Changes may be made in the plan based on your personal health care needs.

We hope to make your stay at Duke a positive experience. We encourage you to ask your doctor, nurse, or other team member any questions about your care, procedure, or hospital stay. Please use the space below to list questions you may have.

Questions

1. _____

2. _____

3. _____

4. _____

Radical Perineal Prostatectomy: Patient's Clinical Path

Path	Before Surgery	Day of Surgery	Day 1 After Surgery	Day 2 After Surgery	Day 3 After Surgery	Explanation
People You Will Meet	• Physician Team • Nurses • Business Office personnel	• Physician Team • Nurses • Respiratory Therapists if needed	• Physician Team • Nurses	• Physician Team • Nurses • If needed: Social worker or Discharge planner	• Physician Team • Nurses • If needed: Social worker or Discharge planner	• Nurses and physicians will explain the procedure and daily treatment plan. • Respiratory therapists will assist with oxygen and breathing therapy if needed. • Discharge planners and social workers will be available to help you plan your recovery and assist with equipment, transportation, or home health needs. • Business Office personnel will review insurance information with you and your family and ensure that all necessary forms are signed.
Tests	• Blood tests • Urine sample • Chest X-Ray • Electrocardiogram (EKG) if needed • Other diagnostic tests as needed	• Blood tests	• Blood tests	• Blood tests if needed	• Blood tests if needed	• Blood tests will be checked regularly.
Treatments	• Vital signs • Height and weight • Shower with antibacterial soap • Mechanical bowel prep • Enema if needed	• Use incentive spirometer every 2 hours • Vital signs will be monitored frequently • Intravenous (IV) • Dressing reinforcement • Bladder tube • White leg hose and leg pumps • Do breathing exercises every 2 hours • Nothing per rectum • Perineal drain	• Use incentive spirometer every 2 hours • Vital signs • Intravenous (IV) • Remove perineal dressing • Bladder tube • White leg hose and leg pumps • Do breathing exercises every 2 hours • Sitz baths once drain removed • Apply gauze fluffs with spandex pants	• Vital signs • White leg hose • Bladder tube • Dressing change • Do breathing exercises every 2 hours • Nothing per rectum • Sitz baths after first bowel movement and drain is removed	• Vital signs • White hose • Dressing change • Do breathing exercises every 4 hours • Bladder tube • Sitz baths • Glans stitch removed • Nothing per rectum	• You will be given medication and/or an enema to clear out your bowels before surgery. • Your rectal dressing may need to be changed frequently the first day or 2 after surgery due to oozing from perineal drain. Drainage will be bloody. • Vital signs (blood pressure, heart rate, breathing rate and temperature) are checked routinely. • Medication and fluids will be given through IVs. Do not insert anything into your rectum for 6 weeks after surgery. Notify your nurse as soon as possible after your first bowel movement. • Coughing & deep breathing exercises will be taught to prevent pneumonia. • A bladder tube will be used to drain urine.

continues

Exhibit 18–A–3 continued

Path	Before Surgery	Day of Surgery	Day 1 After Surgery	Day 2 After Surgery	Day 3 After Surgery	Explanation
						• A drain will be inserted into the surgical area, which helps drain blood and fluid away from the surgical site. This is removed once you have a bowel movement and may vary from person to person.
Medications	• Continue current prescribed medications unless specified by your doctor • You will be told which medication to take the morning before surgery	• Oxygen if needed • Antibiotic • Pain medicine • IV for fluids and medication • Medication for nausea if needed • Medication for bladder spasms if needed	• Antibiotic • Pain medicine • Medications for nausea if needed • Medication for bladder spasms if needed • Stool softener if needed	• Pain medication; IV capped off (Saline Lock); • Medication for bladder spasms if needed • Medication at night to help sleep if needed • Stool softener if needed • Oral antibiotic	• Pain medication • Medication for bladder spasms if needed • Medication at night to help sleep if needed • Stool softener if needed • Oral antibiotic • Saline lock removed	• You will be given medication and/or an enema to clean out your bowels before surgery. • Antibiotics will be given to help prevent infection. • Taking pain medication will enable increased activity and deep breathing. • Addiction to pain medication when used for short periods of time is unlikely. • Taking pain medication 30–45 minutes before physical activity will allow you to participate more easily. • Medications to help ease nausea and vomiting are called antiemetics. • A laxative will be given if you are unable to have a bowel movement after 2–3 days. • Medication is available to help you sleep at night if needed. • Medication for bladder spasms (antispasmotic) is available.
Diet	• Clear liquids day before surgery • Nothing to eat or drink after midnight (NPO) or time specified by anesthesiologist	• Nothing to eat or drink except medications and ice chips	• Resume regular diet if not nauseated. Begin by taking clear liquids and advancing to full liquids and then a regular diet.	• Regular diet	• Regular diet	• The stomach should be empty prior to the operation. • Food/liquids will be withheld several hours prior to surgery. • After surgery the diet will be advanced as tolerated starting with sips of liquids. • We will be checking to see if you are passing gas rectally and if your bowel sounds have returned prior to allowing a regular diet.
Activity	• No restrictions on activity	• Bedrest. Turn from side to side every 2 hours. • You may dangle at bedside this evening if you desire	• You will begin getting out of bed today. Sit in chair on horseshoe cushion • You may walk in hall with assistance	• Walk in hall 4 times a day	• Walk in hall 4 times a day	• Activity is important after surgery and instructions will be provided. Active participation in physical activity will assist you in being able to be discharged to your home faster. • We ask you to always request assistance the first few times out of bed and when walking in the hall at first. • You should always sit on your horseshoe cushion when out of bed sitting. Take this home with you.
Teaching	• Pre-operative teaching	• Incentive spirometer use • Pain management • Activity • Diet advancement • Breathing exercises	• Incentive spirometer use • Pain management • Activity advancement • Diet advancement • Incision care/ wound healing • Sitz baths • First bowel movement	• Pain management • Leg bag teaching • Care of foley catheter • Review discharge teaching sheet • Sitz baths • Wound/incision care; Home care needs • Home equipment needs	• Pain management • Leg bag teaching • Care of foley catheter • Review discharge teaching sheet • Sitz baths • Discharge medication • Signs and symptoms to report to your doctor • Follow-up appointment arranged	• Discharge teaching will include activity progression, diet, incision care, medications, pain control, and follow-up medical care. • Please feel free to ask questions of the medical team at any time. Teaching is an ongoing process. • Written prescriptions and instructions will be provided. Discharge by 11 AM if possible.

Note: How you progress through this pathway will depend on your doctor's preference and your medical condition.

Source: Copyright © Duke University Medical Center.

Exhibit 18–A–4 Generic Variance Tool

MULTIDISCIPLINARY CLINICAL PATH QUALITY IMPROVEMENT MONITOR

Clinical Path: _____

Date Path Initiated: _____

Date Path Cancelled: _____

Date Admitted: _____

Date Discharged: _____

ADDRESSOGRAPH

PATH DAY/PHASE	DESCRIBE VARIANCE	OCCURRENCE CODE	REASON CODE	ACTION TAKEN	DESCRIPTION OF INTERVENTION/ACTION(S) (IF NECESSARY)	INITIALS/ TITLE

OCCURRENCE CODE:
1. Did not occur
2. Additional
3. Early
4. Late

REASON CODE:
See reverse side

ACTION(S) TAKEN/INTERVENTIONS
1. Done as ordered
2. Physician contacted
3. Rescheduled
4. Conference with oncoming caregiver
5. Contacted appropriate dept.
6. None
7. No longer needed
8. Other (describe)

Instructions for Exception/Variance Tracking for Care Path QI Monitor to be filled out by *all* health care team members.
1) A copy of this form is attached to the care path for each patient.
2) Fill out information in header and addressograph.
3) For any exception/variance to the map, record the information in appropriate column. Use code as indicated. Describe if necessary.
4) Write legibly.
5) Attach additional forms if necessary.
6) When patients are discharged, remove this form from the care path and send to:
   Care Management Office, P.O. Box 3634 DUMC, Attn: Care Map Coordinator. If you have questions, please call 416-5220.

ACTIONS TAKEN/INTERVENTIONS
1) Done as ordered
2) Physician contacted
3) Rescheduled
4) Conferred with oncoming caregiver
5) Contacted appropriate dept.
6) None
7) No Longer Needed
8) Other (describe)

REASONS CODE LIST FOR QI MONITOR—CARE PATH

REASONS CODE:

Patient
1. Patient condition
2. Patient developed complication
3. Patient demonstrates too much pain/discomfort/fatigue to comply
4. Patient decision
5. Patient availability
6. Patient mental status limited
7. Patient pre-existing condition alters progress (different from admitting diagnosis or reason for hospitalization)
8. Patient progressing ahead of map
9. Patient does not require expected intervention

Family
10. Family decision
11. Family availability

Provider
12. Medication not administered
13. Treatment/assessment/etc. not done—specify
14. Response time delay

15. Additional tests ordered
16. Modified care map related to admitting diagnosis
17. Modified care map per MD preference/order
18. Physician decision
19. Not ordered
20. No data
21. Not needed
22. Not available/delivered (specify)
   A) Equipment
   B) Extended Care Facility
   C) Home Health Agency
   D) Medication
   E) Reports from DUMC
   F) Reports from source outside DUMC
   G) Transfer Bed
   H) Other (specify)
23. Other (please specify)

Exceptions and variations are generally expected as paths are customized to meet individual patient needs. Revisions to the path should be made as needed.

SUGGESTIONS/COMMENTS:

Note: QI, quality improvement.

Source: Copyright © Duke University Medical Center.

Exhibit 18–A–5 Current Variance Tool for Prostatectomy Path

MULTIDISCIPLINARY CLINICAL PATH QUALITY IMPROVEMENT MONITOR

Path #: **52**   Radical Perineal Prostatectomy_____

Date Admitted: _____

Date Discharged: _____

Date Patient Removed from path (if appropriate): _____   ADDRESSOGRAPH

CARE PATH DAY/PHASE	DESIRED OUTCOMES	MET PLACE ✓ If not met, note reason	REASON CODE	ACTION TAKEN	COMMENTS FOR REASONS/ACTION(S) IF OUTCOME NOT MET	INITIALS
Day of Surgery	Pt admitted for same day surgery	❑ Yes ❑ No				
Day of Surgery	Pt verbalizes adequate pain level **Also, check all that apply	❑ Yes ❑ No ❑ Epidural ❑ IV narcotics ❑ IV toradol ❑ PO meds				
POD #1	Pt verbalizes adequate pain level **Also, check all that apply	❑ Yes ❑ No ❑ Epidural ❑ IV narcotics ❑ IV toradol ❑ PO meds				
POD #1	Pt tolerating PO Fluids	❑ Yes ❑ No				
POD #1	Pt ambulating in hall	❑ Yes ❑ No				
POD #2	Pt. OOB to hall QID	❑ Yes ❑ No				
POD #2	Pt verbalizes adequate pain level **Also, check all that apply	❑ Yes ❑ No ❑ IV narcotics ❑ PO meds				
POD #2	Pt can change leg bag	❑ Yes ❑ No				
POD #2	Pt able to use sitz bath	❑ Yes ❑ No				
POD #2	Diet progress to Regular	❑ Yes ❑ No				
POD #3	Patient discharged on POD #3	❑ Yes ❑ No With home health ❑ Without home health ❑				
POD #3	Pt afebrile at discharge	❑ Yes ❑ No				
POD ____	If off map, indicate reason/action					

Events: **Please Note if these occur**	❑ Excessive bleeding post-op ❑ Fever ❑ Poor nutrition ❑ Elevated WBC/ESR System issues: _____	❑ Failure to have a BM ❑ Hematoma ❑ PE	❑ CHF ❑ Superficial infection ❑ Obesity ❑ Wound drainage	❑ CVA ❑ Ileus ❑ Pneumonia ❑ MRSA+	❑ DVT ❑ Mental status changes ❑ PUD ❑ Other_____	❑ Electrolyte abnormalities ❑ MI ❑ UTI

Instructions for Exception/Variance Tracking for Care Path QI Monitor to be filled out by *all* health care team members.
1) A copy of this form is attached to the care path for each patient.
2) Fill out information in header and addressograph.
3) For any exception/variance to the map, record the information in appropriate column. Use code as indicated. Describe if necessary.
4) Write legibly.
5) Attach additional forms if necessary.
6) When patients are discharged, remove this form from the care path and send to:
   Care Management Office, P.O. Box 3634 DUMC, Attn: Care Map Coordinator. If you have questions, please call 416-5220.

ACTIONS TAKEN/INTERVENTIONS
1) Done as ordered   2) Physician contacted   3) Rescheduled   4) Conferred with oncoming caregiver
5) Contacted appropriate dept.   6) None   7) No Longer Needed   8) Other (describe)

REASONS CODE LIST FOR QI MONITOR—CARE PATH

REASONS CODE:

Patient
1. Patient condition
2. Patient developed complication
3. Patient demonstrates too much pain/discomfort/fatigue to comply
4. Patient decision
5. Patient availability
6. Patient mental status limited
7. Patient pre-existing condition alters progress (different from admitting diagnosis or reason for hospitalization)
8. Patient progressing ahead of map
9. Patient does not require expected intervention

Family
10. Family decision
11. Family availability

Provider
12. Medication not administered
13. Treatment/assessment/etc. not done—specify
14. Response time delay

15. Additional tests ordered
16. Modified care map related to admitting diagnosis
17. Modified care map per MD preference/order
18. Physician decision
19. Not ordered
20. No data
21. Not needed
22. Not available/delivered (specify)
    A) Equipment
    B) Extended Care Facility
    C) Home Health Agency
    D) Medication
    E) Reports from DUMC
    F) Reports from source outside DUMC
    G) Transfer Bed
    H) Other (specify)
23. Other (please specify)

Exceptions and variations are generally expected as paths are customized to meet individual patient needs. Revisions to the path should be made as needed.

SUGGESTIONS/COMMENTS:

Note: BM, bowel movement; CHF, congestive heart failure; CVA, cardiovascular accident; DVT, deep-vein thrombosis; ESR, erythrocyte sedimentation rate; IV, intravenous; MI, myocardial infarction; MRSA, methicillin-resistant *Staphylococcus aureus*; OOB, out of bed; PE, pulmonary embolus; PO, oral; POD, postoperative day; PUD, peptic ulcer disease; QI, quality improvement; QID, four times a day; UTI, urinary tract infection; WBC, white blood cell count.

Source: Copyright © Duke University Medical Center.

■ 19 ■

Achieving Outcomes through Interdisciplinary Path-Based Care for Burn Patients

Janice Steele Allwood, Kim Curry, and Virginia C. Campbell

Approximately 1.25 million people are burned annually in the United States. Of these, 50,000 people are admitted to the hospital, and 600,000 are treated in hospital emergency rooms.[1-4] Burn injuries account for more than 1 billion health care dollars allocated to direct hospital care costs. The indirect costs due to lost work days and rehabilitation have reached $3 billion annually. Additional costs can be attributed to psychological trauma of burn victims, expertise of highly skilled medical teams, and insufficient reimbursement, which grossly affect the financial bottom line.

As managed care penetrates the health care industry, hospitals have been forced to find ways to reduce costs. To survive in these changing economic times, burn centers must implement creative strategies to ensure effective management of costs, while preserving optimal patient outcomes and high-quality, specialized burn care.[3]

In 1993, a case management task force was developed at Tampa General Healthcare (TGH). Populations of patients in major diagnostic categories (MDCs) were identified based on excessive costs. The burn diagnosis-related groups (DRGs) in MDC 22 were some of the high-cost groups. Numerous measures were implemented to lower costs while preserving high-quality patient outcomes at the TGH Tampa Bay Regional Burn Center.

First, routinely used supplies, such as burn dressings, were evaluated and changed to less costly items where appropriate. Redesigning and streamlining the use of resources and provision of services through a case management process was the second strategy used and is the focus of this chapter.

The Committee on the Organization and Delivery of Burn Care of the American Burn Association (ABA) began to prepare for the impact of changes in health care economics on burn centers. This committee, with a great deal of input from members from each discipline, identified acceptable outcome criteria expected for each phase of burn injury: burn shock, wound surgery, and rehabilitation. Once these outcomes were identified, our burn team met to develop critical pathways.[5]

PATHWAY

One of the first activities undertaken was to survey other burn centers to find out what critical pathways were currently developed and/or used. The surveys proved useful, and we avoided reinventing the wheel. Few responses were received; this indicated that other centers were only in the initial stage of critical pathway development. One burn center, at Vanderbilt University, did have a completed pathway, and this information proved helpful to us.

The next step was to review our preprinted standard orders. The burn team, including the attending plastic surgeon, nursing, physical therapy, occupational therapy, social work, recreational therapy, the pain management physician, the dietitian, the chaplain, and psychiatry, worked together to streamline the preprinted orders. This process was very effective. For example, we decreased laboratory blood draws from every six hours to one a day after agreement that changes in electrolytes in the immediate postburn phase (first 24 hours) are expected and that these do not typically affect medical management. We added guidelines in the orders for the physician residents to select patients who would receive chest X-ray and electrocardiogram, rather than ordering these tests on all patients admitted. Many pharmacological agents were replaced by less costly but equally effective alternatives. Some agents were determined unnecessary and removed from the standard orders. Guidelines were also developed for the use of specialized air beds. These beds would be used only on burns greater than 40 percent of the total body surface area (TBSA).

Before actually writing critical pathways, we compared our length-of-stay (LOS) data with all other Florida hospitals having burn centers and with Medicare's standard days and reimbursement under the prospective payment system. We identified our target average LOS, and the template for our first critical pathway was developed for Burn DRG 460, nonextensive burn without operating room (OR) procedures. This DRG was the least complicated diagnosis to standardize. We identified an acceptable average LOS of four days. This was one-half of the state's average and 2.3 days less than TGH's average in 1993. We were confident that using the pathway and working more closely and communicating more effectively would enable us to work more efficiently and provide optimal safe care. This would allow patients to return home at more appropriate times.

The ABA's outcome indicators were divided into categories based on either body system or burn management. These categories were used and formatted to fit the institution's critical pathway template. This would facilitate ease of use for all disciplines.

We then met twice a week, and each team member participated in completing discipline-specific processes on the critical path within an agreed-upon time frame.

After receiving input from all disciplines, we concluded that the pathway could easily be adapted for use for a patient with nonextensive burns with OR procedures, such as debridement under anesthesia (DUA; DRG 459). The DUA would have to be performed on day 1 or 2 to meet defined outcomes by day 4. This required acceptance from the surgeons as well as hospital administration for appropriate resource allocation. We also developed an admission debridement flowsheet. The combined pathway for DRG 459/460 was in effect as of March 1994 on a pilot basis (see Exhibit 19–A–1 in Appendix 19–A).

Streamlined admission orders, corresponding to the pathway, were then implemented to make it easier for the admitting physician or nurse practitioner to write orders consistent with the critical pathway and individualized for the type of burn.

The next pathway developed was for the nonextensive burn with skin grafts, or DRG 458 (see Exhibit 19–A–2 in Appendix 19–A). Because patients in this category vary somewhat in the number of days they spend in each phase, we found it more appropriate to use the preoperative, postoperative, and rehab phases as the time frames. This allowed for some individualization and slight variation while remaining on the path. The same steps were followed by the burn team, designing their process

steps on the timeline to meet the outcomes within the established time frame of 10 days. This pathway was implemented in March 1995.

Pathway development for the most complicated burn, DRG 472, extensive burn with skin grafts, was undertaken in early 1996 (see Exhibit 19–A–3 in Appendix 19–A). The same steps were used for developing this pathway. Because this DRG covers everything from 20 percent TBSA full-thickness burns, to 100 percent TBSA full-thickness burns, there is a wide range of expected outcomes for patients in this category. However, we developed our path on the average patient, reflecting a basic principle of critical path development. Depending on the size of burn, a patient may need to go to the OR for excision and grafting several times, and the critical pathway allows for this by simply repeating the pre- and postoperative phases as needed. This was completed in March 1996.

CASE MANAGEMENT IMPLEMENTATION

The burn center advanced registered nurse practitioner (ARNP) provides leadership in implementing a case management approach and critical pathways in the center. The ARNP/case manager role has evolved from a clinical nurse specialist role during the course of implementing the hospitalwide case management program. The combined role works well in a small, contained unit such as the burn center, where the ARNP is involved with care planning and implementation for every patient treated.

All staff are accountable for ensuring that patient care is consistent with the developed pathways. Each discipline is responsible for providing and documenting the care agreed upon by the interdisciplinary team and for making ongoing suggestions for pathway improvement. Although the critical pathway form was designed for interdisciplinary documentation, in reality it has been used primarily for nursing documentation. Efforts to integrate the documentation of all disciplines are ongoing. The ARNP monitors use of the pathway, provides staff education concerning path implementation, and coordinates variance tracking as described below.

Documentation consists of several integrated pieces. As mentioned previously, medical order sheets are preprinted and correspond to the pathway, thus saving time in writing initial orders and avoiding orders for unnecessary services. The critical pathway (see Exhibits 19–A–1 through 19–A–3 in Appendix 19–A) contains blank areas next to each action or outcome so that staff can initial the completed items. This allows easy tracking of items to be completed. Flowsheets and narrative notes are used to supplement the critical path for any items not covered by the path and for vital signs, fluid monitoring, and other observations. Finally, a patient/family version of the path (see Exhibit 19–A–4 in Appendix 19–A) is used for patient education and mutual goal setting. This tool is used to improve communication and to recreate realistic patient and family expectations of the care provided.

OUTCOMES

Measurement of Outcomes

Outcomes measured monthly include patient clinical outcomes, LOS, costs, and charges. Clinical outcomes are summarized through the use of the National TRACS/ABA Burn Registry, which provides for detailed tracking of several patient parameters. LOS, costs, and charges are measured using standard reports from the hospital financial information system. The system provides monthly reports of the above parameters at the DRG level, as well as lists of patients coded into each DRG. This allows a comparative review by the case manager for hospital coding accuracy.

While the system has been helpful in providing this basic information, limitations such as lack of ability to query the system and lack of timeliness of report availability have kept it from providing the flexibility needed for a sophisticated and timely system of case management. A new decision support system is planned for installation in the near future to allow for more timely and specific feedback.

As shown in Tables 19–1 and 19–2, average LOS and charges have come down substantially each year since 1992. One exception to this trend occurred in 1995. It is also important to note that the average case mix has steadily risen from 1992–1996.

This situation in 1995 was analyzed using National TRACS/ABA Burn Registry,[6] and it was determined that 11 percent of the patients in 1995 had burns greater than 50 percent TBSA and/or greater than 20 percent full-thickness burns. Comparatively, in 1994, only 5 percent of burn patients had burns of this extent with the same extreme risk of mortality. Numerous complications

Table 19–1 Burns, MDC 22

Calendar Year	# Burn Cases	Average LOS*	Average Charge	Average Case Mix
92	208	13.6	36,638	2.85
93	222	12.3	33,181	2.89
94	189	9.8	26,680	2.77
95	189	11.6	39,741	3.11
96	173	8.2	27,289	3.33

Note: LOS, length of stay; MDC, major diagnostic category.
Source: Tampa General Healthcare.

such as septicemia, respiratory failure, and paralytic ileus accounted for many physiologic variances. Although these patients had an extreme risk of mortality, 75 percent survived.

Variance Analysis

Following the initiation of a pathway for a given patient, the Burn Center ARNP/case manager and burn center nursing staff track pathway use and variances from the pathway. A standard system of variance analysis is employed throughout the institution. Variances are categorized into four major categories for ease of classification and analysis: *patient variances* (including physiological, educational, and family), *clinician variances* (including decision making, response time, and skill performance), *institutional variances* (including bed, service, or supply availability), and *community variances* (including placement options, home care, and financial arrangements).

The ARNP/case manager assists staff in recording variances, then collects and analyzes variances monthly. When trends are identified, the interdisciplinary team addresses these issues. The results of the variance analysis are then reported to the hospitalwide outcomes program manager for consolidation with other case manager reports. Overall trends are identified, and common problems within the institution can then be addressed.

Table 19–2 DRGs Currently Managed on Critical Paths

DRG	1992 LOS	1992 CHG	1993 LOS	1993 CHG	1994 LOS	1994 CHG	1995 LOS	1995 CHG	1996 LOS	1996 CHG
458	19.5	46,537	15.2	38,742	14.2	36,749	13.3	36,463	11.2	29,599
459	9.0	21,080	7.8	12,991	4.5	8,521	5.6	12,801	4.8	12,053
460	5.9	12,860	6.3	10,235	3.5	6,688	3.8	8,097	3.3	6,800
472	27.0	119,274	27.1	116,431	22.5	98,955	39.8	190,271	21.3	106,271

Note: CHG, charge; DRG, diagnosis-related groups; LOS, length of stay.
Source: Tampa General Healthcare.

Variance analysis is also used in the hospital's peer review process. Any unexpected outcomes are handled via a medical record review to identify the nature of the variance. If a professional practice issue is identified, the information is entered into the hospitalwide quality improvement system, and review questions are generated for peer investigation by the professional discipline involved.

Patient/Family Pathway

A version of the burn critical path has been developed for use in patient and family education (see Exhibit 19–A–4 in Appendix 19–A). Ideally, this type of pathway is implemented in the preadmission phase to serve as a foundation for patient education, elicit questions from the patient and family, and create realistic expectations of the treatment experience. Obviously, anticipatory implementation is not possible in the case of burn trauma. However, the pathway has proven to be beneficial in similar ways. It assists patients and family members in developing an understanding of the treatment plan and is useful for questions and answers related to specific aspects of burn treatment.

Customer Satisfaction

It is imperative to evaluate the responses of customers who receive services guided by the pathway. Practicing according to critical pathways is often a new process and concept for the practitioner as well as the patient and family. Therefore, it is important to educate customers to the pathway model and establish early and consistent parameters to assess customer perceptions regarding the appropriateness of the pathway. The feedback gained is helpful in guiding future revisions and adaptations to the model.

The groups that should be assessed include the patient and significant other, physicians, and staff. Patients and significant others should be educated about the pathway upon admission and their needs assessed throughout the implementation process. The tool used to measure the level of satisfaction with the pathway should contain fundamental language and statements that can be accurately understood by respondents. The data should be grouped and summarized to solicit their perceptions. These data can be identified by diagnostic group and compared to those for other diagnostic groups to ascertain similarities and/or differences.

Assessing and evaluating customer satisfaction affects the refinement and use of the critical pathway and guides operational practices that influence the organization's strategic plans and marketing endeavors.

It is essential to involve physicians throughout the development and evaluation process. Physician orders guide the pathway model through concurrent written and verbal feedback. As more physicians participate in the pathway mode, there is growing acceptance and support from physician colleagues practicing in a variety of specialties. Acceptance and compliance from physicians are essential to the integration of critical pathways and the attainment of optimal outcomes. Ongoing input from physicians and other burn team members is continuously sought through their leadership and involvement in the interdisciplinary pathway development teams.

Managed Care

Managed-care organizations establish and maintain stringent standards and expectations regarding patient care, which health care organizations must embrace to ensure participation in the managed-care plan. Reliable communications and compliance with expected utilization of resources ensures the provision of appropriate care within the desired time frame and in the appropriate health care setting.

The managed-care organizations focus on cost-effectiveness, utilization of resources, services provided, service area, and provider's ability to respond to the needs of the health care organization and enrolled members. The managed-care agencies want to know that the organization has a reliable program that ensures the provision and measurement of quality care.

Critical pathways guide health care providers (hospital-based case managers) in providing substantiated statements of outcomes when discussing and/or negotiating with managed-care case managers. The hospital case manager is able to specify the achievement of outcomes and identify what desired goals need to be achieved. Tracking patients and patterns of variance enable the organization to correlate findings and analyze the effectiveness of services provided as well as the responsiveness of patients on pathways. This gives credibility to outcomes according to specific patient conditions.

The integration of financial data with the pathway is useful to the organization when negotiating managed-care contracts. In addition, the organization can use outcome data from critical pathways in designing and marketing materials for targeted patient care markets. For example, the data obtained from tracking burn patient outcomes, based on the pathway format, play a major role in determining the expansion of the burn program and attainment of managed-care contracts.

Critical paths developed in the institution have been provided to the managed-care department for use in demonstrating that a well-thought-out and efficient plan for patient care is in effect. This approach may not be necessary for highly specialized programs such as the burn center, since only a few similar dedicated burn units exist in the state. Most severe burn injuries are referred to specialty centers by necessity. However, documentation of a structured care approach may prove useful in the future for less serious injuries in which the payer has greater latitude in choice of providers. Critical paths can be used to document a standard of quality as well as efficiency, which can be used to highlight the desirability of a given institution.

CONCLUSION AND FUTURE CHALLENGES

The critical pathway model and format are successful when there is effective interdisciplinary communication. Participation from each discipline ensures the creation and implementation of processes that guide the provision of appropriate services. Equally important is the need to ensure that consistent and concurrent methods are in place to assess compliance with the pathway from each discipline. Administrative support is essential to the successful use of critical pathways. Administrative staff must instill and reinforce the expectation that staff must provide care as guided by the pathway, monitor the patient's compliance with the pathway, and report variances as needed. Compliance from all members of the team ensures effective patient outcomes and integrity of the pathway.

Collaborative practice models using critical pathways support utilization review and quality improvement programs within the organization. Multidisciplinary groups should assess patterns of variance to determine barriers that influence pathway compliance. In turn, they can collaborate in the redesign and evolution of the pathway on an ongoing basis. Through this process, the organization will be able to meet changes in health care practices and be responsive to reimbursement issues.

A valued expectation of pathways is the reduction in the patient's LOS at each hospitalization. Decreasing inpatient LOS serves the organization well when negotiating with managed-care organizations. The key is to ensure that quality care is provided through concurrent measurement of established outcome criteria.

Imperative to the measurement and effectiveness of critical pathways is the integration of financial data, tracking systems, and computerization. The ability to computerize pathways and outcome findings will enhance integration of pathways as an essential guide to ensuring the provision of quality and efficient health care services. Reliable and integrated data allow the organization and providers to make sound decisions regarding the utilization of resources, patient outcomes, and strategic plans.

REFERENCES

1. Burn Foundation. *Fact Sheet*. Philadelphia, PA: Burn Foundation; 1997.

2. Kongstvedt PR. *The Managed Health Care Handbook*. 2nd ed. Gaithersburg, MD: Aspen Publishers; 1993.

3. Mathews JJ, Supple K, Calistro A, Gamelli RL. A burn center cost-reduction program. *J Burn Care Rehab*. 1995;18:358–363.

4. Allwood J. The primary care management of burns. *Nurs Pract*. 1995;20:74–87.

5. Committee on the Organization and Delivery of Burn Care. *Burn Care Outcomes and Indicators*. Chicago, IL: American Burn Association; 1995.

6. American College of Surgeons. *National TRACS/ABA Burn Registry Version 1.0, User and Training Manual*. Chicago, IL: American College of Surgeons; 1995.

■ Appendix 19–A ■
Clinical Pathways

The following is a key to abbreviations used in the appendix exhibits.

ABG, arterial blood gas
ADLs, activities of daily living
BICU, burn intensive care unit
BID, twice daily
BLE, burned lower extremity
C, carbohydrate
CBC, complete blood count
CCMS, clear catch midstream
CM, case manager
CMS, circulation, motion, sensation
CXR, chest X-ray
D/C, discontinue, discharge
D5½NS, 5½% dextrose in normal saline
D/T, due to
DUA, debridement under anesthesia
EKG, electrocardiogram
FT, feeding tube
FU, follow-up
GI, gastrointestinal
HL, heparin lock
Hx, history
I/O, intake and output
IS, incentive spirometer
IV, intravenous
LOC, laxative of choice
MR, may repeat
MVI, multivitamin
NgT, nasogastric tube
Nsg, nursing
OOB, out of bed
OR, operating room
OT, occupational therapy

Ox, oximetry
P, protein
PO, oral
prn, as needed
PT, physical therapy
PT, prothrombin time
PTT, partial thromboplastin time
QOD, every other day
ROM, range of motion
RL, Ringer's lactate
RT, respiratory therapy
Rx, therapy
SaO$_2$, oxygen saturation
SEs, side effects
SH, shift
SMA, serum
SO, significant other
SW, social worker
T, temperature
TBSA, total body surface area
TCDB, turn, cough, deep-breathe
TFTs, thyroid function tests
TID, three times daily
TOL, tolerates
Tx, treatment
UA, urinalysis
UDS, urine drug screen
UE, upper extremity
UUN, urine, urea, nitrogen
VS, vital signs
W/A, while awake
WCU, wound care unit

Exhibit 19–A–1 Critical Path for Nonextensive Burn without Operating Room Procedures (DRG 459/460)

DATE INITIATED: DRG #459/460: NONEXTENSIVE BURN WITHOUT OR <50% TBSA: not requiring skin grafts		EXPECTED LOS: 4 DAYS		
MEASUREMENT PHASE	**DAY 1**	**DAY 2**	**DAY 3**	**DAY 4**	**OUTCOME CRITERIA**

MEASUREMENT PHASE	DAY 1	DAY 2	DAY 3	DAY 4	OUTCOME CRITERIA
CONSULTS	___ RT consult & follow ___ OT Eval & Tx ___ PT Eval & Tx ___ Consult Anesth for DUA if needed ___ Consult Social Services for D/C planning ___ Consult Dietitian to follow ___ Consult Psychiatry to follow ___ If <18 yo: Pediatric consult & CMS consult	___ RT follow up ___ OT follow up ___ PT follow up ___ Anesth follow up ___ Social Services follow up ___ Dietitian follow up ___ Psychiatry follow up ___ Pediatric follow up	___ RT follow up ___ OT follow up ___ PT follow up ___ Anesth follow up ___ Social Services follow up ___ Dietitian follow up ___ Psychiatry follow up ___ Pediatric follow up	___ RT follow up ___ OT follow up ___ PT follow up ___ Anesth follow up ___ Social Services follow up ___ Dietitian follow up ___ Psychiatry follow up ___ Pediatric follow up	___ All consults done
TESTS	___ SMA.6, CBC ___ CCMS UA ___ PT, PTT if there is any hx of bleeding ___ CXR & EKG if > 50 yo	___ Repeat any abnormal lab			___ All labs within acceptable limits
INTERVENTIONS					
Cardiovascular	___ VS q4h x2 then q8h ___ H.L.IV, flush Q sh ___ Monitor I/O Q4h ___ Elevate burned extremities and head above heart ___ UE Protocol ___ Check peripheral pulses Q2h to burned extremities ___ Weigh pt ___ D/C foley by end of 1st 24 hrs	___ VS q8h ___ Flush HL q SH ___ I&O q8h ___ Elevate burned ext's and head above heart level at all times ___ UE Protocol ___ Check peripheral pulses 98° burned ext's ___ Administer antipyretic for T>101.1 and notify MD ___ Weigh pt if pediatric	___ VS q8h ___ D/C HL ___ I/O q8h ___ Continue elevation of burns ___ Check peripheral pulses q8h ___ Weigh pt ___ Obtain urine, blood, sputum and wound cultures if T > 102 ___ Administer antipyretic for T > 101.5 and notify MD	___ VS q8h ___ D/C HL ___ I/O q8h ___ Continue elevation of burns ___ Check peripheral pulses q8h ___ Weigh pt ___ Obtain urine, blood, sputum and wound cultures if T > 102 ___ Administer antipyretic for T > 101.5 and notify MD	___ Hemodynamic stability ___ Peripheral pulses palpable x4 extremities ___ Absence of T > 102 sustained for 24 hrs
Respiratory	___ TCDB q2h w/a	___ TCDB q2h w/a	___ TCDB q2h w/a	___ TCDB q2h w/a	___ Respiratory rate appropriate for age/pre-burn status
Wound Care	___ Obtain consent for DUA if indicated ___ Perform admission debridement ___ Institute wound care protocols per orders ___ Document wound care on wound doc sheet ___ Check during and after initial debridement	___ Assess/document wound on wound doc sheet BID	___ Assess/document wound on wound doc sheet BID	___ Assess/document wound on wound doc sheet BID	___ Burned/grafted areas healing/healed without complications

Init.	Signature	Init.	Signature	Addressograph
_____	_____	_____	_____	
_____	_____	_____	_____	
_____	_____	_____	_____	
_____	_____	_____	_____	

continues

Exhibit 19–A–1 continued

MEASUREMENT PHASE	DAY 1	DAY 2	DAY 3	DAY 4	OUTCOME CRITERIA
Comfort	___ Medicate with IV and/or PO analgesics prn ___ Medicate with IV/PO anxiolytics prn	___ Medicate with IV/PO analgesics prn ___ Medicate with IV/PO anxiolytics prn	___ D/C IV analgesics use PO only prn ___ D/C IV anxiolytics start antihistamines PO prn	___ Instruct pt. on home Rx	___ Pt attains a self-identified level of acceptable comfort
MEDICATIONS	___ Administer antipyretic for T > 102 ___ See comfort section ___ Follow meds per orders	___ Administer antipyretic for T > 102 ___ See comfort section ___ Follow meds per orders	___ Administer antipyretic for T > 102 ___ See comfort section ___ Follow meds per orders	___ Administer antipyretic for T > 102 ___ See comfort section ___ Follow meds per orders	___ Pt verbalizes knowledge of med, dose, frequency, SE's, food or drug interactions, of any medications prescribed
DIET/ NUTRITION	___ Pt to feed self with assist ___ Start age approp. diet (↑C↑P)	___ Pt to feed self without assist ___ Start age approp. diet (↑C↑P)	___ Pt to feed self without assist ___ Start age approp. diet (↑C↑P)	___ Feeds self ___ Start age approp. diet (↑C↑P)	___ TOL > 80% daily caloric needs
DISCHARGE PLANNING	___ Identify possible D/C needs	___ Eval need for home OT/PT ___ Eval need for home health	___ Determine need for home OT/PT & notify CM SW to arrange home health ___ Anticipate supplies as needed at home & notify home health agency and/or pt.	___ Home health arranged	___ Discharge not to be delayed D/T system variance
ACTIVITY	___ Institute positioning protocols OT/PT ___ Determine any physical limitations to ADLs and notify case manager ___ Place on fall prevention protocol if indicated ___ If no identified problems, up ad lib with ace wraps to burned BLE	___ Follow positioning protocols with OT/PT ___ Pt. performs ADLs without assist. ___ Place on fall prevention protocol if indicated ___ Pt performs active ROM exercises	___ Follow positioning protocols with OT/PT ___ Pt performs ADLs without assist. ___ Place on fall prevention protocol if indicated ___ Pt performs active ROM exercises	___ Follow positioning protocols with OT/PT ___ Pt. performs ADLs without assist. ___ Place on fall prevention protocol if indicated ___ Pt performs active ROM exercises	___ Has optimal level of functional status within the limits of physical capabilities ___ Demonstrates home exercise program ___ Pt or caregiver assumes ADLs
TEACHING	___ Teach importance of nutrition ___ Encourage family and friends to bring in pts' home-cooked favorites ___ Orient pt & family to unit and unit routines—give visitors guide ___ Give Burn Discharge Teaching Booklet to pt/family	___ Teach importance of nutrition ___ Encourage family and friends to bring in pts' home-cooked favorites ___ Review some of the D/C teaching booklet with pt/family	___ Teach importance of nutrition ___ Encourage family and friends to bring in pts' home-cooked favorites ___ Cont. D/C teaching & review of burn teaching booklet ___ Demonstrate wound care to pt/family/SO and observe return demonstration of wound care	___ Teach importance of nutrition ___ Encourage family and friends to bring in pts' home-cooked favorites ___ Cont. D/C teaching & review of burn teaching booklet ___ Demonstrate wound care to pt/family/SO and observe return demonstration of wound care	___ Pt/caregiver communicates understanding of Tx and projected goals ___ Pt/caregiver demonstrates dressing change appropriate for home care

Init.	Signature		Init.	Signature	Addressograph
_____	_____		_____	_____	
_____	_____		_____	_____	
_____	_____		_____	_____	
_____	_____		_____	_____	

continues

Exhibit 19–A–1 continued

MEASUREMENT PHASE	DAY 1	DAY 2	DAY 3	DAY 4	OUTCOME CRITERIA
PSYCHOSOCIAL	___ Encourage pt/family to verbalize needs and concerns and notify case manager ___ Discuss daily plan of care with pt/family	___ Encourage pt/family to verbalize needs and concerns and notify case mgr ___ Recreational therapy set up age approp. activities ___ Discuss daily plan of care with family/pt ___ Eval. need for home health	___ Encourage pt/family to verbalize needs and concerns and notify case mgr ___ Recreational therapy set up age approp. activities ___ Discuss daily plan of care with family/pt ___ Eval. need for home health	___ Encourage pt/family to verbalize needs and concerns and notify case mgr ___ Recreational therapy set up age approp. activities ___ Discuss daily plan of care with family/pt ___ Eval. need for home health	___ Coping appropriate for phase of injury ___ Communicates needs/ concerns

Init.   Signature   Init.   Signature   Addressograph

_____   _____   _____   _____

_____   _____   _____   _____

_____   _____   _____   _____

_____   _____   _____   _____

Source: Copyright © Tampa General Healthcare.

Exhibit 19–A–2 Critical Path for Nonextensive Burn with Skin Grafts (DRG 458)

DATE INITIATED:		EXPECTED LOS: 10 DAYS		
DRG #458:	NONEXTENSIVE BURN WITH SKIN GRAFTS <50% TBSA & <20% full thickness			

MEASUREMENT PHASE	PRE-OP PHASE 1–2 DAYS	POST-OP PHASE 3–4 DAYS	REHAB PHASE 5–8 DAYS	OUTCOME CRITERIA
CONSULTS	___ OT Eval & Tx ___ PT Eval & Tx ___ RT Eval & Tx ___ Social Work for discharge planning ___ Dietitian Eval & Tx ___ Psychiatry Eval & Tx ___ Pain Mgmt Eval & Tx ___ Anesthesia Consult PRE-OP	___ OT follow up ___ PT follow up ___ RT follow up ___ Social Work follow up ___ Dietary follow up ___ Psychiatry F/U ___ Pain Mgmt F/U ___ Anesthesia follow up	___ OT follow up ___ PT follow up ___ RT follow up ___ Social Work follow up ___ Dietary follow up ___ Psychiatry F/U ___ Pain Mgmt F/U ___ Anesthesia follow up	___ All consults done
TESTS	___ SMAC, CBC ___ CCMS UA ___ PT, PTT if Hx of bleeding ___ CXR & EKG if > 50 yo ___ Wound biopsy to full thickness burns	___ SMA 7 ___ CBC	___ SMA 7 ___ Pan Culture for T > 102	___ All Labs within acceptable limits
INTERVENTIONS				
Cardiovascular	___ IV RL, 1st 24 hrs ___ VS q4h x2, then q8h ___ I's & O's q4h x2, then q8h ___ Elevate burned ext's & head ___ Follow UE protocol if indicated ___ Weigh pt QOD ___ D/C Foley by end of 1st 24 hr if present	___ Change IV site Day 3 ___ D5½ NS—add K+ if indicated until PO well ___ I&Os q4h x2, then q8h ___ Elevate grafted areas & donor sites ___ Weigh pt QOD ___ D/C Foley if present	___ D/C IV by Day 6 ___ I&Os q8h ___ Weigh pt QOD	___ Hemodynamic stability ___ Peripheral pulses palpable x4 ext's ___ Absence of T > 102 sustained for 24 hrs
Respiratory	___ TCDB q 2° W/A ___ Instruct pt on use of IS & instruct to use 2° W/A	___ TCDB q 2° W/A ___ Reinforce importance of IS use 2° W/A	___ TCDB q 2°	___ Respiratory rate appropriate for age/pre-burn status
Wound Care	___ Admission debridement per protocol (Obtain consent if indicated) ___ Wound care protocol ___ Document wound care on wound doc. sheet	___ Continue wound care for nongrafted areas per protocol ___ Keep grafted areas positioned per protocol ___ Remove outer covering to donor sites 24° postop & follow protocol for donors ___ Document wound care on wound doc. sheet	___ Wound care per protocol	___ Burned/grafted areas healing/healed without complications
Comfort	___ Medicate with IV & OR PO analgesic/anixolytics prn ___ Start Pain Mgmt per protocol/Pain MD	___ Medicate with IV for wound care or PO background pain analgesics prn ___ Cont. Pain Mgmt per Pain MD ___ Teach relaxation techniques/provide distraction therapy	___ PO analgesics prn ___ Relaxation techniques/distraction therapy	___ Pt attains a self-identified level of acceptable comfort

Init.	Signature	Init.	Signature	Addressograph
_____	_____	_____	_____	
_____	_____	_____	_____	
_____	_____	_____	_____	
_____	_____	_____	_____	

continues

Exhibit 19–A–2 continued

MEASUREMENT PHASE	PRE-OP PHASE 1–2 DAYS	POST-OP PHASE 3–4 DAYS	REHAB PHASE 5–8 DAYS	OUTCOME CRITERIA
MEDICATIONS	___ Administer antipyretic for T° > 102 ___ Meds per MD orders	___ MVI, MYLANTA, COLACE ___ L.O.C. ___ PERCS/VICODIN/T#3	___ MVI ___ MYLANTA PRN ___ COLACE ___ LOC PRN ___ PERCOCETS/VICODIN/T#3 ___ BENADRYL PRN	___ Pt verbalizes knowledge of med, dose, frequency, SEs, food or drug interactions, of any medications prescribed
DIET/ NUTRITION	___ Pt to feed self with assist ___ Start age approp. diet (↑C↑P) ___ Assess need for FT	___ Feed self with min assist ___ Adv. to ↑C↑P as TOL ___ Burn shakes TID ___ FT if indicated	___ Feeds self ___ ↑C↑P diet ___ Burn shakes PO TID ___ D/C Ft by postop Day 3	___ TOL > 80% daily caloric needs
DISCHARGE PLANNING	___ Identify possible D/C needs ___ Obtain data for Tracs National Burn Registry	___ Reinforce info in pt teaching ___ Update Tracs	___ Home care arranged ___ D/C Tracs	___ Discharge not to be delayed D/T system variance
ACTIVITY	___ Eval for splints before surgery ___ Institute positioning protocols per OT/PT sign above bed ___ Determine any physical limitations to ADLs & notify CM ___ Fall prevention protocol if indicated ___ Up ad lib with ace wraps per PT condition ___ Instruct on active ROMs	___ Follow OT/PT instructions on signs above bed for positioning ___ Notify CM of limitations to ADLs ___ Cont. fall prevention protocol if indicated ___ Hold ROM to grafted areas only ___ OOB per protocols	___ Cont OT/PT instructions on sign above bed for positioning ___ Notify CM of limitations ___ ROM/ambulation per PT/OT bid ___ OOB per protocols	___ Has optimal level of functional status within the limits of physical capabilities ___ Demonstrates home exercise program ___ Pt or caregiver assumes ADLs
TEACHING	___ Teach importance of nutrition ___ Encourage visitors to bring in home-cooked favorites ___ Orient pt/fam to unit ___ Give visitors guide ___ Give burn teaching booklet to pt & family ___ OR consent ___ HIV consent (MD to get on admit) ___ Med photo consent	___ Teach importance of nutrition ___ Encourage visitors to bring in home-cooked favorites ___ Read teaching booklet over with pt/fam ___ Teach use of IS	___ Teach wound care to pt/fam ___ Teach home exercise program ___ Complete pt teaching booklet ___ Make F/U appt at burn clinic	___ Pt/caregiver communicates understanding of Tx and projected goals ___ Pt/caregiver demonstrates dressing change appropriate for home care
PSYCHOSOCIAL	___ Encourage pt/family to verbalize needs & concerns ___ Discuss daily plan of care with pt/family ___ Give "Family Critical Path" when available	___ Encourage pt/family to verbalize needs & concerns ___ Discuss daily plan of care with pt/family	___ Encourage pt/family to verbalize needs & concerns ___ Discuss daily plan of care with pt/family	___ Coping appropriate for phase of injury ___ Communicates needs/concerns

Init.   Signature   Init.   Signature   Addressograph

_____ _____   _____ _____

_____ _____   _____ _____

_____ _____   _____ _____

_____ _____   _____ _____

Source: Copyright © Tampa General Healthcare.

Exhibit 19–A–3 Critical Path for Extensive Burn with Skin Grafts (DRG 472)

DATE INITIATED: DRG #472: EXTENSIVE BURN WITH SKIN GRAFTS >50% TBSA OR >20% full thickness		EXPECTED LOS: 10 DAYS		
MEASUREMENT PHASE	***PRE-OP PHASE** **1–3 DAYS**	***POST-OP PHASE** **4–8 DAYS**	**REHAB PHASE** **9–10 DAYS**	**OUTCOME CRITERIA**
CONSULTS	___ OT Eval & Tx ___ PT Eval & Tx ___ RT Eval & Tx ___ Social Work for discharge planning ___ Dietitian Eval & Tx ___ Psychiatry Eval & Tx ___ Pain Mgmt Eval & Tx ___ Anesthesia Consult PRE-OP	___ OT follow up ___ PT follow up ___ RT follow up ___ Social Work follow up ___ Dietary follow up ___ Psychiatry F/U ___ Pain Mgmt F/U ___ Anesthesia follow up	___ OT follow up ___ PT follow up ___ RT follow up ___ Social Work follow up ___ Dietary follow up ___ Psychiatry F/U ___ Pain Mgmt F/U ___ Anesthesia follow up	___ All consults done
TESTS	___ SMA7 ___ CBC ___ CCMS UA ___ PT, PTT ___ CXR & EKG if > 50 yo ___ Wound biopsy to full thickness burns ___ Start port CXR ___ UDS with history if not done in ER ___ TFTs if > 50 yo ___ Candida Ag titre q mon ___ HIV test per protocol ___ 12 lead EKG if > 40 yo ___ Pregnancy test for females 12–50 yo ___ Sickle cell prep on black pts	___ SMAc, Mg, transferrin 24 hr UUN q mon ___ SMA7 @ 0200 stat qd ___ CBC @ 0200 stat qd ___ ABG q am while intubated ___ Portable CXR qd while intubated ___ Pulse O x continuous	___ Pan Culture for T > 102 ___ D/C pulse O x when off O$_2$	___ All labs within acceptable limits
INTERVENTIONS	___ Institute BICU nursing protocol	___ Cont BICU nursing protocol	___ To WCU	
Cardiovascular	___ IV RL, 1st 24 hrs rate per MD orders ___ VS q 15 min until stable ___ I's & O's q h, then q 1 h ___ Elevate burned ext's & head ___ Follow UE protocol if indicated ___ Weigh pt q Mon AM and document ___ Foley to gravity drainage with urimeter	___ Change IV site Day 3 ___ D5½ NS—add K+ if indicated until PO well ___ I&Os q4h x2, then q8h ___ Elevate grafted areas & donor sites ___ Weigh pt Q mon & document ___ D/C Foley if present	___ D/C IV by Day 6 ___ I&Os q8h ___ Weigh pt Q mon & document	___ Hemodynamic stability ___ Peripheral pulses palpable x4 ext's ___ Absence of T > 102 sustained for 24 hrs
Respiratory	___ O$_2$ therapy as ordered ___ Suction if intubated q 2° ___ TCDB q 2° W/A if not intubated ___ Instruct pt on use of IS and instruct to use 2° ___ Resp therapy as ordered	___ O$_2$ therapy per orders ___ TCDB q 2° when extubated ___ Reinforce importance of IS use q 2° W/A if not intubated	___ Extubate ASAP ___ TCDB q 2° when extubated ___ Wean 0° SaO$_2$ > 92%	___ Respiratory rate appropriate for age/pre-burn status

*May repeat Pre and Post OP phase as needed.

Init.	Signature	Init.	Signature	Addressograph
_____	_____	_____	_____	
_____	_____	_____	_____	
_____	_____	_____	_____	
_____	_____	_____	_____	

continues

Exhibit 19–A–3 continued

MEASUREMENT PHASE	*PRE-OP PHASE 1–3 DAYS	*POST-OP PHASE 4–8 DAYS	REHAB PHASE 9–10 DAYS	OUTCOME CRITERIA
Wound Care	___ Admission debridement per protocol (Obtain consent if indicated) document on Admission Debridement flow sheet ___ Wound care protocol ___ Document wound care on wound doc. sheet	___ Continue wound care for nongrafted areas per protocol ___ Keep grafted areas positioned per protocol ___ Remove outer covering to donor sites 24° postop & follow protocol for donors ___ Document wound care on wound doc. sheet	___ Wound care per protocol	___ Burned/grafted areas healing/healed without complications
Comfort	___ Medicate with IV & OR PO analgesic/anixolytics prn ___ Start Pain Mgmt per protocol OR Pain MD ___ Low pressure air bed	___ Medicate with IV for wound care or PO background pain analgesics prn ___ Cont. Pain Mgmt per Pain MD ___ Teach relaxation techniques/provide distraction therapy	___ PO analgesics prn ___ Relaxation techniques/distraction therapy ___ D/C air bed and change to comforter	___ Pt attains a self-identified level of acceptable comfort
MEDICATIONS	___ Administer pyretic for T > 102° ___ Meds per MD orders ___ Tetanus Tox 0.5 cc/ml	___ MVI, MYLANTA, COLACE ___ L.O.C. ___ PERCS/VICODIN/T#3	___ MVI ___ MYLANTA PRN ___ COLACE ___ LOC PRN ___ PERCOCETS/VICODIN/T#3 ___ BENADRYL PRN	___ Pt verbalizes knowledge of med, dose, frequency, SEs, food or drug interactions, of any medications prescribed
GI/DIET/NUTRITION	___ NqT for > 30% TBSA ___ After tube for feeding verified in stomach or lower in GI tract, start TFs ___ Pt to feed self with assist ___ Start age approp. diet (↑C↑P) ___ FT if > 30% burn within 2° of admission. Nsg to insert.	___ Feed self with min assist ___ Adv. to ↑C↑P as TOL ___ Burn shakes TID ___ FT continue tube feeds	___ Calorie count prn ___ Feeds self ___ ↑C↑P diet ___ Burn shakes PO TID ___ D/C Ft by postop Day 3	___ TOL > 80% daily caloric needs
DISCHARGE PLANNING	___ Identify possible D/C needs ___ Obtain data for Tracs National Burn Registry	___ Reinforce info in Pt teaching ___ Update Tracs	___ Home care arranged ___ D/C Tracs	___ Discharge not to be delayed D/T system variance
ACTIVITY	___ Eval for splints before surgery ___ Institute positioning protocols per OT/PT sign above bed ___ Determine any physical limitations to ADLs & notify CM ___ Fall prevention protocol if indicated ___ OOB daily when stabilized ___ ACE wraps to BLEs when OOB ___ Instruct on active ROMs	___ Follow OT/PT instructions on signs above bed for positioning ___ Notify CM of limitations to ADLs ___ Cont. fall prevention protocol if indicated ___ Hold ROM to grafted areas only until further MD orders ___ OOB per protocols	___ Cont OT/PT instructions on sign above bed for positioning ___ Notify CM of limitations ___ ROM/ambulation per PT/OT BID ___ OOB per protocols	___ Has optimal level of functional status within the limits of physical capabilities ___ Demonstrates home exercise program ___ Pt or caregiver assumes ADLs

*May repeat Pre and Post OP phase as needed.

Init.   Signature   Init.   Signature   Addressograph

___   _____   ___   _____

___   _____   ___   _____

___   _____   ___   _____

___   _____   ___   _____

continues

Exhibit 19–A–3 continued

MEASUREMENT PHASE	*PRE-OP PHASE 1–3 DAYS	*POST-OP PHASE 4–8 DAYS	REHAB PHASE 9–10 DAYS	OUTCOME CRITERIA
TEACHING	___ Teach importance of nutrition ___ Encourage visitors to bring in home-cooked favorites ___ Orient pt/fam to unit ___ Give visitors guide ___ Give burn teaching booklet to pt & family ___ OR consent ___ HIV consent (MD to get on admit) ___ Med photo consent	___ Teach importance of nutrition ___ Encourage visitors to bring in home-cooked favorites ___ Read teaching booklet over with pt/fam ___ Teach use of IS	___ Teach wound care to pt/fam ___ Teach home exercise program ___ Complete pt teaching booklet ___ Make F/U appt at burn clinic	___ Pt/caregiver communicates understanding of Tx and projected goals ___ Pt/caregiver demonstrates dressing change appropriate for home care
PSYCHOSOCIAL	___ Encourage pt/family to verbalize needs & concerns ___ Discuss daily plan of care with pt/family ___ Give "Family Critical Path" when available	___ Encourage pt/family to verbalize needs & concerns ___ Discuss daily plan of care with pt/family	___ Encourage pt/family to verbalize needs & concerns ___ Discuss daily plan of care with pt/family	___ Coping appropriate for phase of injury ___ Communicates needs/concerns

*May repeat Pre and Post OP phase as needed.

Init.   Signature   Init.   Signature   Addressograph

_____ _____   _____ _____

_____ _____   _____ _____

_____ _____   _____ _____

_____ _____   _____ _____

Source: Copyright © Tampa General Healthcare.

Exhibit 19–A–4 Patient/Family Version of Critical Path for Burn Care

DATE INITIATED:		EXPECTED DAYS IN THE HOSPITAL: 16 days for average patient		
DRG #472:	1ST DEGREE BURNS (RED SKIN), 2ND DEGREE BURNS (BLISTERED SKIN), and 3RD DEGREE BURNS (FULL SKIN THICKNESS) REQUIRED SKIN GRAFTS			

Hospital Days	1–3 Days Pre Surgery	4–8 Days Post Surgery	9–16 Days Rehabilitation	Outcomes
CONSULTS are the members of the Burn Team that might be asked to evaluate, treat, and follow up (more than one visit)	Respiratory Therapy (RT) Physical Therapy (PT) Occupational Therapy (OT) Recreational Therapy (RecT) Pain Management Dietary (Nutrition) Psychiatry to follow Anesthesia consult for presurgery Social Services (SS) for discharge	Physicians and/or services that may continue for follow up are: RT PT OT RecT Dietary Pain Management Psychiatry Anesthesia SS	Physicians and/or services that may continue for follow up are: RT PT OT RecT Dietary Pain Management Psychiatry Anesthesia SS	All of the physicians and/or services that were asked did their evaluation, treatment and follow up.
TESTS routine admitting test(s) that may be done	Blood samples Urine samples Chest X-rays Other X-rays EKGs on persons with heart problems or are older than 40 Wound biopsy to the full thickness burns Fungus test every Mon. Pregnancy test for females 12–50 years old Sickle cell prep on all Blacks	The physician may order more tests or may repeat tests to: – recheck a test(s) after receiving medication to correct the problem – make sure that tests are remaining normal Certain blood tests will be done every day at 2 AM while on the respirator: – special blood test (ABGs) abd chest X-rays are done every day – pulse oximeter (measure the oxygen saturation) is worn on a finger, toe or possibly an earlobe continuously.	The physician may continue to reorder tests for ongoing monitoring of blood, urine, X-rays or EKGs. Cultures (these are tests to determine if or what type of bacteria is present) will be done if temperature is greater than 102.0°F. When no longer receiving any type of oxygen therapy: – the pulse oximeter will be removed – daily ABGs and chest x-rays will be stopped – daily 2 AM blood tests may or may not be stopped	Blood, urine, X-rays or EKG tests are all within normal limits before they go home.
RESPIRATORY SYSTEM (Lungs and air exchange in your body) Deep breathing any coughing oxygenates (air exchange) the body and exercises the lungs.	The nurses will assess the lungs by listening, watching how patients breathe and the color of their skin. Oxygen therapy will be ordered and be monitored by RT (ex. respiratory, breathing mask or nasal cannula—little prongs that are just inside the nose that deliver the oxygen). If not on a respirator, patient must turn, take several deep breaths and cough every few hours. Will be taught the use of an inspirometer (used to expand and exercise the lungs) every 2 hours while awake if not on the respirator.	The nurses will continue assessing the lungs by frequently listening to lung sounds. Respiratory rate (number of breaths per minute), rhythm (breathing patterns) and their skin color. Oxygen therapy will continue if needed. Must continue to turn, deep breathe and cough every few hours while awake if not on the respirator. Continue use of the inspirometer every 2 hours while awake if not on the respirator.	Will wean off any oxygen therapy and keep their oxygen saturation above 92% (pulse oximeter). Nurse will continue assessing respiratory rate, rhythm and skin color. Continue with the oxygen therapy and to get off the respirator as soon as possible. Continue to turn, deep breathe and cough every few hours while awake if not on the respirator. Continue to teach and use inspirometer every 2 hours while awake if not on the respirator.	The respiratory system (lungs) will be normal or back to their normal level before they were burned.
CIRCULATORY SYSTEM/VITAL SIGNS (Heart and blood vessels) The nurses will assess heart and blood vessels to make sure they are functioning properly	BP (Blood Pressure) checked frequently the first few hours, then once a shift. IV fluids (fluids being given in the bloodstream). All fluids (IV or by mouth) will be measured. Weigh every Mon and Thurs to keep track of fluids, nutrition. A foley catheter (a tube in the bladder to drain urine) the first 24 hours to measure urine. The burn areas (arms and legs) and the head will be elevated to slow down or prevent swelling. The burn area will be watched closely for changes during daily dressing changes.	BP and vital signs (pulse, respirations, and temperature) taken every shift (may be taken every 2 hours). Tylenol will be given for temperatures over 101.1°F (it is normal for the temperature to rise slightly). Continue to measure all fluids (IV or by mouth). The foley catheter (if present) to be removed and the nurses will continue to measure the urine. The graft areas and the donor sites (where the grafts come from) will be elevated as possible. Continue to elevate the head and burn areas.	BP and vital signs (pulse, respirations, and temperature) taken every shift. Tylenol will be given for temperatures over 101.1°F (it is normal for the temperature to rise slightly). Continue to weigh MR. IV fluids will probably be stopped by day 6. Continue to measure all fluids (IV or by mouth). Continue to measure the urine. Continue to elevate the graft and donor sites. Continue to elevate the burn area and head.	The following will be stable and normal for at least 8 hrs before going home: – Temperature will be less than 102°F – BP and vital signs (pulse rate and respirations) – Able to go to the bathroom without problems (bowel and bladder functions return to normal)

continues

Exhibit 19–A–4 continued

Hospital Days	1–3 Days Pre Surgery	4–8 Days Post Surgery	9–16 Days Rehabilitation	Outcomes
WOUND CARE is the treatment of the burn area. Wound care is ordered by the physician to prevent infections and promote healing with the least amount of scarring by using medicines and special burn dressings.	Debridement (removal of the burnt skin) upon admission, consent or permission will be needed before any surgical procedure can be performed. Daily wound care will be started.	The nurses will continue to assess the burns and give appropriate wound care as ordered by the physician. The grafted areas will be kept positioned as posted above their bed. Outer dressings on the donor sites will be removed after 24 hours.	Nurse will continue to assess the burns. Continue to give appropriate wound care. Continue to check all burn areas.	The burns, grafts, and donor sites are healing or are healed without any problems or infections
COMFORT is very important to the Burn Center	Medications will be given for comfort (pain, anxiety, restlessness, sleeplessness and fears) either by mouth or IV as needed or requested. Pain management will be started for comfort. May be placed on a special bed.	Will be taught different ways and techniques to remain comfort-able. Will be medicated for comfort before any dressing changes are done.	All comfort medications will be given by mouth. Nurses will continue teaching different ways and techniques to remain comfortable, may include Biofeedback.	Will have acceptable level of comfort for them
MEDICATIONS **PLEASE** tell the nurses the names, dosage (amounts) and frequency (how many times a day taken) of **ALL** the medications taken (recently) before being admitted to the hospital. And **ALLERGIES**	The physician may prescribe several kinds of medicines including medications for pain, itching, fever and current prehospital medications. They may prescribe medicines that prevent or stop infections. A tetanus toxoid shot to prevent "lockjaw" (See Comfort section)	The physician may prescribe multivitamins, mylanta, and a laxative of choice (LOC) for their comfort (many pain medicines are constipating). Continue current pre-hospitalization medication. Continue medications for comfort, itching, fever, and infections (see Comfort section).	Continue to receive multivitamins, mylanta, and LOC for comfort. Continue current pre-hospitaliza-tion medication. Continue medications for comfort, itching, fever and infections (see Comfort section).	Able to state the following reasons for all their medication: – for what – when to take – how much to take – side effects and/or reactions to watch for – what drugs and/or foods not to be taken with the medications
NUTRITION (diet) **PLEASE** tell the nurses of any special diet. Will be asked about the food **ALLERGIES,** likes, dislikes, when and how much, including snacks. A diet high in protein and calories is needed to aid in wound healing.	Burns that are more than 30% of the body, a special feeding tube (NgT) may be passed from the nose to the stomach. Special liquid diet will be given by the NgT. A special diet (Burn diet) of extra carbohydrates and protein to help heal the burns. Patients are encouraged to feed self as much as possible to encourage independence.	Milk shakes given three times a day (called Burn shakes) provide extra calories needed to heal the burns. Feeding tubes (NgT) may be removed after the 3rd day of surgery. Encouraged to eat with little help. Continue Burn shakes three times a day. Continue Burn diet.	Should be feeding themselves without any help. Continue Burn shakes three times a day. Continue Burn diet.	Should be able to feed themselves and eat at least 80% of their daily requirement
DISCHARGE PLANNING (plans being made to go home) Social Work (SW) will work very closely to identify needs—special needs for the home and the needs of the persons involved in the care. The SW will contact you by the post-surgery phase.	Special equipment may or may not be needed for the home. Please think of what needs or what concerns you might have to ask the SW during discharge planning. Will be placed on the *Tracs National Burn Registry.*	PT/OT will evaluate the need of PT/OT in the home. Home Health may be asked to evaluate the need for help with self care.	If PT, OT and Home Health has determined a need for their services, SW will arrange for the services and order the supplies that will be needed for their care.	The discharge planning is completed. Be able to go home without delays from the hospital or from the services not arranged. The services are set up to include supplies and their schedules of who and when they will visit.

continues

Exhibit 19–A–4 continued

Hospital Days	1–3 Days Pre Surgery	4–8 Days Post Surgery	9–16 Days Rehabilitation	Outcomes
ACTIVITY	May be fitted for splints before surgery. Will have PT/OT instruction posted above their bed on how to position themselves. PT/OT will instruct and help with daily exercise programs in or out of bed. Will be able to get out of bed with ace wraps to the burn area(s) if there are no problems.	The grafted areas will not be exercised during this time. Should be able to perform their own daily personal needs with help. Continue to follow posted instructions above bed. PT/OT continue to help with daily exercise programs in or out of bed. Continue to get out of the bed with ace wraps to the burn area(s).	Continue to not exercise the grafted areas. Continue to perform their own daily personal needs with little help. Continue to follow posted instruction above bed. PT/OT continue to help with daily exercise programs in or out of the bed. Continue to get out of the bed with ace wraps to the burn area(s).	Should be able to take care of themselves within the limits of their physical abilities. Should be able to perform all of their activities. Should be able to perform all of their activities of daily living and should be able to demonstrate their home exercise programs.
TEACHING	Pt and family are taught the importance of nutrition. Encouraged to bring favorite home-cooked foods. Told about the Burn Unit and their routines. Given a guide/tour. Taught about surgery. Consent/permission to photograph burns. Asked to give consent/permission to do HIV testing.	Taught how and why to use the inspirometer. Continue to be taught the importance of nutrition. Continue to be encouraged to bring in favorite home-cooked foods.	Family member may be demonstrated wound care by you and asked for return demonstra-tion (questions and concerns are encouraged). An appointment will be made at the Burn Clinic. Continue to teach and review the inspirometer. Review the *Burn Discharge Teaching Booklet.* Continue teaching the importance of nutrition. Continue to bring in favorite food.	Be able to understand their treatments, how to take care of them, and what to expect when they go home. Be able to demonstrate dressing changes for home care.
MENTAL ATTITUDE	The nurses will encourage you to state your needs and concerns. The nurses may direct your needs/concerns to other members of the Burn Team (e.g., Social Worker, Chaplain, etc.). The nurses will discuss daily plan of care.	Recreational Therapy (RecT) to set up activities according to age and needs may be evaluated for home health. Continue to encourage to state your needs and concerns. Continue to discuss daily plan of care.	Continue to have RecT to set up activities according to age and needs. Continue to evaluate for home health needs. Continue to encourage to state your needs and concerns. Continue to discuss daily plan of care.	Be able to communicate your needs and concerns. To be able to cope appropriately. The behaviors, fears, and anxiety are appropriate for the injury.

The following area is for you to write down any questions or concerns you may have that you would like to address with any member of the Burn Team.

■ Part V ■

Orthopaedics

■ 20 ■

A Clinical Pathway for Calcaneal Fracture: Process and Outcomes

Lynne Nemeth, Jack S. Olsen, Nanette B. Fricke, Barbara E. Parlotz, and Johnese Spisso

Harborview Medical Center (HMC) is a Level I trauma center that serves as a referral institution for a four-state region, including Washington, Alaska, Montana, and Idaho. Wyoming was recently added as a component of this regional affiliation with the University of Washington (UW). Because of the regional trauma mission that is supported at HMC, the nature of our orthopaedic service differs from many community and academic medical centers. We see many more emergent and urgent cases than those hospitals that serve patients for elective surgery. Typically, clinical pathways are more effective in populations that have a predictable nature.

One of the initial products of our clinical pathway program at HMC was a calcaneal fracture pathway. Calcaneal fracture was selected for pathway development at the onset of our program in 1994 because of the increasing number of patients referred to HMC for surgical intervention. The predictable nature of the treatment plan for calcaneal fractures and the strong participation of the interdisciplinary team made this problem an excellent starting point for developing clinical pathways. At HMC, the calcaneal fracture pathway gave us an opportunity to identify the best practices of the interdisciplinary team and to achieve positive patient outcomes, improve hospital length of stay (LOS), and attain financial goals. This

chapter discusses the development, implementation, and evaluation of the clinical pathway for the calcaneal fracture patient population at HMC.

BACKGROUND

Calcaneal fractures are usually the result of axial loading, or falls from heights.[1] Fractures of the calcaneus have traditionally been treated by nonoperative methods.[2] At HMC in the past several years, two of our orthopaedic surgeons have developed a strong referral base for an operative repair that has resulted in decreased disability for this patient population. This became the impetus for developing a coordinated effort in managing care for this group.

A clinical pathway–authoring team was formed in December 1994 to develop the process. Members of the authoring team included representation from all participants across the continuum of care for a fractured calcaneus. The pathway was implemented in May 1995 and remains in place. The implementation has facilitated the process for more pathway development within the orthopaedic service.

PATHWAY DEVELOPMENT

Before the pathway initiative, the department of orthopaedics had developed a protocol with the occupational therapy (OT) and physical therapy (PT) departments for the management of patients with calcaneal fracture. This protocol addressed the need for splinting, pre- and postoperative mobilization, and early postoperative active range of motion (AROM). Additionally, the rehabilitation phase for follow-up to one year after surgery had been defined in terms of the therapy goals and activities.

We acknowledge the hard work and dedication of the following team members who developed the clinical pathway: Stephen Benirschke, MD, Orthopaedics; Ann Boatman, RN, Orthopaedic Clinic; Susan Bytner-Peyou, RN, Emergency Department (ED); Susan Carton, Utilization Review (UR); Carmella Franz, RN, Ambulatory Surgery and Post Anesthesia Care Unit (PACU); Nanette Fricke, PT; Tiffany Giesler, OTR/L; Cathy Graham, RN, Operating Room (OR); Carol Metcalf, RN, Acute Pain Service; Lynne Nemeth, RN, Clinical Pathways; Jack Olsen, RN, Orthopaedics; Ann Parker, RD; Barbara Parlotz, MSW; Margie Peyovich, RN, Orthopaedics; Jeff Purcell, Pharm D; Bruce Sangeorzan, MD, Orthopaedics. We also acknowledge the expert assistance of Joan Knecht and Tarek Salaway, MA, MPH, MHA, in the preparation of this manuscript.

The protocol was employed by two of the surgeons in the department who specialize in the treatment of calcaneal fractures and the staff of the OT and PT departments at HMC. The two surgeons have slightly different approaches to the rehabilitation for this injury; therefore, it was necessary to combine some of these approaches in developing the clinical pathway.

At the first meeting of the authoring team, the project manager presented an overview of clinical pathway development processes, outlined goals for the project, and introduced the format that had been developed by the interdisciplinary design team. This was one of the first pathways in development at HMC. The core process, through the initial work of this authoring team, was a template for further pathway development at our medical center. One of the surgeons outlined the key processes of care involved in the surgical management for calcaneal fracture. The team had the opportunity to listen to the overview by this orthopaedic surgeon and ask questions regarding patient care management goals and optimal processes of care. We discussed potential variances that we could expect and identified approaches that would minimize potential problems. Team members engaged in a very productive dialogue that presented the perspective of each interdisciplinary team member who played an important role in the delivery of care.

Once the team received the initial overview at this first meeting, a structure emerged that considered the patient's progression through the episode of care. Team members needed to learn their new responsibilities in defining their disciplines' approach to developing and implementing a pathway. HMC's clinical pathway design team had determined that the pathway would replace the usual narrative, redundant documentation process, which often contained numerous lengthy assessments by each discipline. The pathway evolved as the interdisciplinary plan of care. Assessments were combined into this format, and key interventions and outcomes to be achieved along the timeline of the patient's inpatient and outpatient care were mapped. Variances would be documented within the narrative section of the pathway as well as tracked separately. In this manner, the processes of care could be analyzed by the team and process improvements identified and implemented.

The psychosocial and discharge planning process was reviewed quite closely by the social work representatives on the design team and authoring team. HMC had consultation with the Center for Case Management's principal, Kathleen A. Bower, RN, DNSc, in the development of the clinical pathway process. Work was done with the social work staff to develop this discipline's approach toward integration of their documentation within a more abbreviated clinical pathway model that was emerging.

A social work supervisor in the medical surgical area was assigned to participate with most of the pathway-authoring teams at that time so that a consistent format could be developed with the input and approval of the department. The social work department had numerous protocols that defined the assessment content required by the social worker for specific situations, concerns, or problems. The Joint Commission on Accreditation of Healthcare Organizations' guidelines were used to ensure that the assessments met the standards, including such information as patient and family participation in their plan of care, education, and discharge planning. Testing the pathway as the documentation format enabled an interactive process with the social work staff to develop comfort with the ongoing development and refinement process used within the pathway program.

The following assumptions were made by social work early in the developmental process for the calcaneal fracture pathway:

- The pathway would provide better continuity of care from all services and disciplines on the team.
- Because these patients would be hospitalized for approximately four postoperative days, they would have limited social work needs.
- Discharge planning processes would provide for limited home support and transportation problems.
- Since calcaneal fracture is frequently a work-related injury, it would be covered by labor and industry compensation. With such a comprehensive funding source, medical needs would be adequately covered, and financial issues would be less of a concern, requiring fewer social work interventions, especially with those having supportive families.

THE CALCANEAL FRACTURE PATHWAY: KEY PROCESSES

The emergent phase of injury was reviewed closely. Because the surgical treatment is not performed until the initial swelling related to the traumatic injury has subsided,[1] elements of care during this preoperative phase were defined to minimize the need for an inpatient stay. Elevation of the foot and bed rest are critical interventions to definitively reduce the swelling. The lack of a safe environment can preclude the effective achievement of that goal. Patients are initially seen in our emergency department or orthopaedic clinic. If the providers and/or patient have concerns about safety regarding ambulation with crutches, an assessment by a physical therapist (PT) is performed at this time. A posterior foot splint is applied by an occupational therapist (OT) at the time of this initial assessment. Patient education is provided by nursing, focusing on reducing the expected swelling, as well as observing and protecting skin from excess pressure. Radiographs and computed tomography imaging studies are obtained and reviewed. The patient's pain is assessed, and referrals are made to our pain relief service if pain is at an unacceptable level for the patient.

The patient is admitted the day of surgery into our ambulatory surgery unit and is prepared for the procedure. Pain management is a major postoperative issue and can delay the patient's progress toward early AROM. Therefore, a sciatic nerve block is performed, either during the operative case or at the conclusion of the case in the postanesthesia care unit (PACU), to minimize the perceptions of pain during this initial postoperative period. Patient-controlled analgesia (PCA) is used from the immediate postoperative period. On the second postoperative day, the patient is usually switched over to oral analgesia. During the first year of the pathway's implementation, an anesthesiologist from HMC studied postoperative analgesia after repair of calcaneus fractures, comparing the effectiveness of morphine alone and morphine with perioperative sciatic nerve blockade. The results of the study indicated that patients who had the sciatic nerve blockade and morphine PCA achieved more effective pain relief and were generally more satisfied.[3]

An assessment by PT occurs on the first postoperative day. This includes a brief screening of strength and range of motion (ROM) of uninvolved joints. In addition, architectural barriers and needed assistance at home are discussed with the patient. Bed exercises begin on postoperative day 1, and the patient is assisted out of bed to a chair, with the involved leg elevated if this is tolerated. On the second postoperative day, non–weight-bearing ambulation begins with PT. Passive range of motion (PROM) to the toes also begins on day 2, since the block has usually worn off by this point and the patient should now be able to feel his or her toes. Tibiotalar AROM may begin on day 3, depending on the surgeon's preference. Much of the recent literature supports early postoperative motion in the management of calcaneal fracture.[1,4–8] Since subtalar motion may stress the surgical wound, this does not begin until the surgical wound is deemed sealed by the attending physician.[4] Patients on the pathway are often instructed by the PT in AROM on the uninvolved foot. Specific directions are given to the patient by the surgeon regarding when to start AROM on the uninvolved foot after discharge. Before the implementation of the pathway, many of the patients stayed in the hospital until their wounds were dry enough to begin AROM. This may account for the dramatic decrease we have seen in our LOS in pathway and nonpathway patients since this pathway was implemented.

The protocol for outpatient PT, whether guided by the PT in the orthopaedic clinic at HMC or in a formal outpatient PT setting, begins at the time of discharge from the hospital and continues for up to 6 to 12 months postoperatively. This protocol is outlined in the outpatient section of the pathway, which follows the inpatient stay.

CARE/CASE MANAGEMENT

The orthopaedics unit at HMC provides an integrated, interdisciplinary approach to managing inpatient care. At HMC, interdisciplinary team members are usually assigned to work with a specific medical service's patient population to meet the distinct needs of each patient population and provide continuity of care. The physicians, nurses, therapists, social workers, nutritionists, utilization review, and patient care coordinators develop an individualized plan of care for each patient. Pathways are used by all disciplines in their daily interdisciplinary rounds, as well as in the nursing report between shifts. The pathway is used as a benchmark for clinical outcomes. The responsibility for coordination of the discharge plan is assigned to the social worker. However, it is a process shared by all of the involved disciplines.

Nursing care is provided using a model of total patient care. We employ a predominantly registered nurse (RN) staff, with few assistive personnel. Our clinical pathways are implemented at the entry point into our health care system, which, in the case of calcaneal fractures, is the emergency department or the orthopaedic clinic.

When the patient arrives in the ambulatory surgery unit on the day of surgery, key interventions are completed and signed off during the perioperative process by operating room (OR) and PACU nursing staff. Outcomes are evaluated during this phase of care. The RN on the inpatient unit is primarily accountable for making sure that the patient is on a pathway and checks to see that the resident physician places the appropriate patients on the pathway. A preprinted physician order sheet facilitates this process. The admitting RN completes the admission assessment, which is a component of the pathway, and completes the documentation indicated in the appropriate pathway time frame. At each point on the timeline, nurses document on the pathway, including the emergency department, clinic, ambulatory surgery, OR, PACU, and inpatient unit staff. PTs, OTs, and social workers also document using the pathway at the appropriate time, as indicated by the progression of the patient through the phases of care. Physicians ideally document key milestones in an addendum to the pathway. However, we have had difficulty in ensuring that this is consistently done by the physicians. Our team continues to work with this issue. The clinical pathway is kept in the active medical record.

The staff nurse role in care management is to evaluate the patient's progress on the pathway and to keep the patient on track with the plan by facilitating the involvement of other disciplines. The nurse requests the appropriate orders from the physicians needed to individualize care and contributes to the discharge plan. It was originally intended that the pathway be shared with the patient and family to enable them to be active participants in the plan of care. As a result of the need to consider and stage the change processes at our facility, we have not yet implemented a patient version of the pathway. Patients do not receive a copy of their pathway. Plans are currently underway to develop a version that is oriented to the patient's perspective. This patient pathway incorporates graphic elements as well as text and can be developed in different languages to meet the needs of our diverse ethnic population.

Clinical nurse specialists (CNSs) serve as resources for care management. They provide consultation to the nursing staff and patients as needed. The CNS educates the team regarding pathways and participates in the development of pathways.

Pathways are reviewed and revised on the basis of user feedback, trends seen in the analysis of pathway variances, and advances in the knowledge based upon the scientific literature. We have revised the pathways once or twice a year since the initial implementation in 1995 and are currently making major changes in our current design on the basis of staff evaluations. The clinical pathway–authoring team members and other clinicians using the pathway have contributed to these changes. The clinical pathway design team is currently addressing format changes, which will streamline the current, more lengthy, complete documentation system.

CLINICAL OUTCOMES

Intermediate and discharge outcomes are measured by the team members participating in the plan of care. There are specific outcomes on which designated team members sign off, but the majority of outcomes are evaluated by the staff nurses providing daily patient care. The outcomes are listed on the pathways in bold, italic format (see Exhibit 20–A–1 in Appendix 20–A). Some examples of outcomes evaluated by therapists are

- adequate fit of the posterior foot splint for appropriate positioning, with no pressure on bony prominences
- ability to don and doff the posterior foot splint independently or with the help of an identified caregiver

- ability to perform an independent exercise program for the uninvolved extremities
- ability to ambulate with an assistive device safely and independently and to maintain the weight-bearing restrictions per the pathway
- ability to perform PROM to the toes independently or with the help of an identified caregiver
- ability to perform AROM of the ankle and subtalar joints (if indicated at the time of discharge) independently

In the rehabilitation phase of the pathway, AROM, PROM, strength, and functional mobility are measured for up to one year postoperatively.

On an inpatient basis, the outcomes listed above are measured by the OT or PT as appropriate to their clinical focus. Each discipline maintains standards of practice that are integrated into the evaluation of patient outcomes. In the outpatient environment, the surgeon and the PT follow the patient in our clinic, and the patient may receive ongoing PT at an outside facility if necessary. Either of these therapists may follow and document ROM of the ankle and subtalar joints and functional mobility as indicated. The pathway allows for a progression of weight bearing and slightly more aggressive exercises at about three or four months after surgery, depending on the type of surgery performed and how well the fracture is healing. The progression of therapy is determined by the surgeon. At six months after surgery, the patient may begin PROM and aggressive strengthening exercises. At this postoperative stage, the patient will benefit from instruction in the exercises and precautions in the next phase of rehabilitation by a PT in the orthopaedic clinic. Some patients will require outpatient PT for assistance with their rehabilitation progression. If ongoing outpatient PT is initiated, the therapist takes the primary role in measuring ROM, strength, and functional mobility.

HOSPITAL OUTCOMES

Our LOS outcomes have improved since this pathway began in 1995. Before pathway implementation, in the beginning of 1995, the median LOS was six days for this patient population. With the pathway, the LOS has been reduced to a median of four days (a 33 percent reduction in median LOS). Our mean LOS has shown consistent reduction, with our prepathway mean LOS at 9.2 days. Currently, our mean LOS for pathway and nonpathway patients with calcaneal fracture (ICD 9-CM principal diagnosis codes of 825.0 and 825.1 and principal procedure code of 79.37) combined is 4.8 days. The standard deviation for LOS has been reduced from 4.4 days to 3.0 days in 1996 alone.

In reviewing our financial outcomes, we have tracked total hospital charges, since we do not currently have a costing methodology in place. We are currently working on developing our costs as we shift our financial databases over to an HBO&C (Atlanta, Georgia) system. Our total hospital charges were a median of $20,341 in 1995 and were quickly brought down to a median of $16,830 after the first six months of using this pathway. We have sustained this 18 percent drop in median total charges with this program, as shown in Figure 20–1.

SATISFACTION

Patient satisfaction outcomes have not been specifically measured for this pathway population. In general, HMC uses Picker surveys to assess patient satisfaction. However, we have not yet stratified these survey data for the different case types on pathways versus nonpathway groups.

Initial staff satisfaction regarding the pathway implementation process was mixed. System change in general

		Count (N)	Mean	Median	SD	Maximum	Minimum
Nonpathway	Length of Stay (days)	22	7.18	5.00	4.91	20.00	3.00
	Total Charges ($)	22	23,710.58	18,792.69	11,105.81	55,800.99	10,773.26
Pathway	Length of Stay (days)	50	5.00	4.00	2.98	17.00	2.00
	Total Charges ($)	50	16,954.69	16,701.63	4,581.12	37,939.81	10,208.07

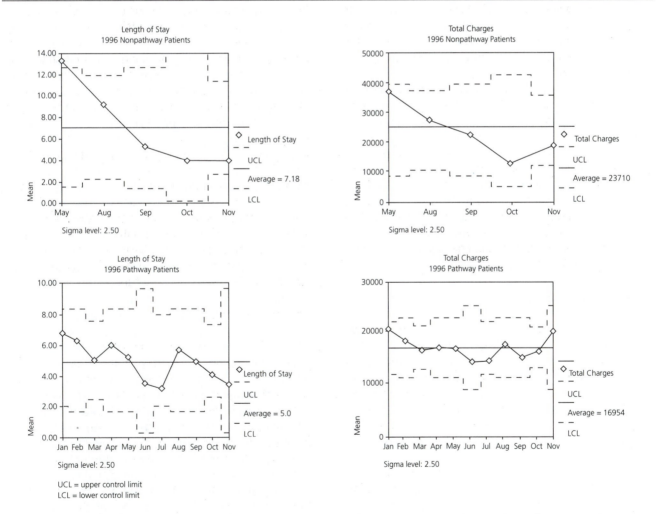

UCL = upper control limit
LCL = lower control limit

Figure 20–1 Length-of-Stay and Financial Outcomes Control Charts for Pathway and Nonpathway Calcaneal Fracture Patients, 1996 (Principal ICD-9 Diagnoses 825.0, 825.1, Principal Procedure = 79.37). *Source:* Copyright © Harborview Medical Center.

is always a difficult process for staff. As we moved to a comprehensive interdisciplinary format for our documentation, we experienced some confusion by various team members regarding who was to document where in the pathway. Since we did not convert all documentation over to this format and since we maintained the usual documentation processes for all other nonpathway groups, we had some difficulty maintaining the new process once it was learned by the staff.

In this new interdisciplinary format, since we had the goal to be concise in our documentation patterns, a short narrative section was built into the pathway format. This served as a place to document patient-related variances or modifications in the plan of care that were needed to individualize care for the patient. Social workers' and therapists' notes in this area were not consistently referred to by the physicians. This became a source of dissatisfaction for these interdisciplinary team members. Social work staff in general tended to be more descriptive of their interventions and patient social history and felt that the checklist nature of the pathway did not capture for them the substantial detail needed for discharge planning purposes. With this feedback, modifications have been made to address these issues, and staff satisfaction is improving. We are currently changing this by deleting the progress note section of the pathway and reverting to the standard progress note used institutionally.

We surveyed the staff in late 1996 to identify specific issues of concern. Overall, staff are clear on the purpose and benefit of the pathway. Staff are unclear on how to complete the documentation that would be indicated if LOS was shorter or longer than the pathway. This causes ambiguity in the process; consequently, staff are somewhat more dissatisfied with patient "detours" from the pathway. These detours are primarily patient-related variances, which require individualization to their plan of care, as shown in Figure 20–2.

Physician satisfaction was surveyed at the same time as staff satisfaction, in late 1996. Since two primary surgeons specialize in this surgical procedure, the clinical pathway program has maintained good communication. They are satisfied with the reduced use of resources realized as we have implemented this pathway. No adverse patient consequences have been noted as a result of the streamlined approach we have taken.

ADVANTAGES AND DISADVANTAGES OF THE PATHWAY

With the clinical pathway program, all patients receive the same standard of care. This core interdisciplinary staff with expertise in trauma orthopaedic practice is very familiar with the recovery and rehabilitation process for re-

pair of a calcaneal fracture. If, for reasons of excess capacity, a patient with a calcaneal fracture is routed to another clinical unit, the staff in that area now have an explicit guide to the processes and outcomes of care that are expected. This provides for increased quality and consistency of care.

Authoring team members learned a great deal about what each discipline contributes to the overall plan of care. This better understanding has enhanced team function, which in turn has effected better, coordinated service for the patient.

Disadvantages to using the pathway were primarily noted in the use of the progress note as a communication source. The free-text, narrative process learned by most disciplines as an effective and professional method of describing subjective and objective data—that is, problem-oriented charting with narrative assessments and evaluation—tends to repeat what we have attempted to describe concisely in the pathway. It has been part of the medical legal record for decades. This documentation style is tightly connected to each discipline's self-image as professional members of the health care team. With the pathway process representing only a fraction of our professional practice at HMC, it has been very hard to make the transition to an abbreviated documentation style that attempts to describe variances to the pathway in the narrative portion of the record.

VARIANCE AS A SOURCE OF OUTCOME DATA

Variance is defined as any clinical pathway intervention or outcome that does not happen or happens later or sooner than defined by the pathway. Variances are reported by all clinical staff who participate with patient care on the pathway. Since there is no one designated staff person to collect the variances, this ensures a broad interdisciplinary perspective.[9] However, because of the voluntary nature of this reporting process, we do not always have complete data. Interrater reliability may be compromised through this data collection method. There is still a learning curve present for staff. We plan to continue to develop staff expertise in the area of variance reporting and to move toward a key indicator tracking form, rather than collecting the entire universe of potential variances. The data we have reported over the two years of using this pathway have provided an assessment of the usual problem areas and have allowed us to focus on improvement of key processes in the future.

Our system for entering our variance data into an automated database has greatly facilitated reporting data to the clinicians involved in the process of care. Utilization review (UR) coordinators code the narrative variances and enter the data into the UR system used by the medi-

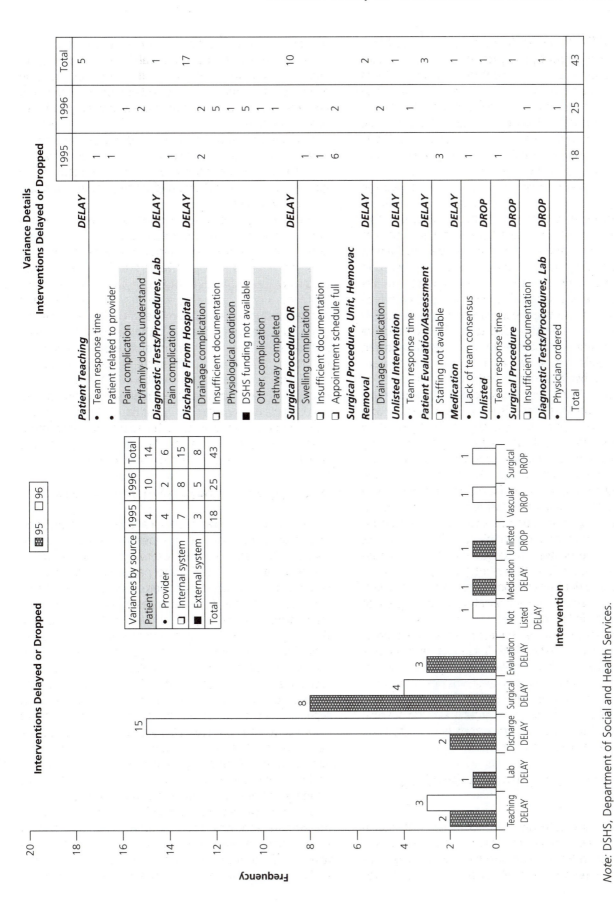

Note: DSHS, Department of Social and Health Services.

Figure 20–2 Calcaneal Fracture Variance Analysis, May 1995 to December 1996. *Source:* Copyright © Harborview Medical Center.

cal center. This UR database integrates with our master medical information networked database for clinical data within the two academic medical centers that constitute the UW system—HMC and the UW Medical Center.

The clinical pathway project manager and clinical reporting specialist extract data from multiple information sources within our medical centers' information systems to provide clinicians and administrative staff with feedback related to our pathway and nonpathway patients on a biannual basis. We use trends seen within six-month periods, since the numbers of these patient groups are not large. The pathway reports include variances from the pathway, which are categorized by source, and a temporal element: delayed, early, dropped, discontinued, or continued beyond the pathway timeframe. Financial data pertaining to total hospital charges and LOS are presented using statistical process control methods to facilitate interpretation. The authoring team members, managers, and faculty physicians review the data, interpret results, and develop priorities for improvement based upon the trends noted in the report. We have been able to provide pathway users with reports since the first six-month period that the pathway was put into place, due to major emphasis on this component of our pathway program.

Our pathway system is currently not automated, but plans are underway to integrate the clinical pathway and variance-tracking methodology into the computerized clinical information system that our medical centers are developing. The ability to automate our pathways was weighted heavily in the selection process of a vendor for these two medical centers who would assist us in achieving our information system requirements.

PERSPECTIVES ON OUTCOMES

Clinical outcomes for patients have historically been measured in terms of primary outcome indicators, such as hospital LOS, readmission or recidivism rates, presence or absence of expected complications, mortality, and infection rates. In more recent years, the focus has changed to include outcome indicators that assess the quality of care and appropriate utilization of resources to provide the level of care needed. More important, efforts have been made to return the patient to previous and/or highest level of functional status, shortening the inpatient LOS or converting the encounter to an outpatient episode of care and reducing the cost of care while maintaining desired outcomes. When variances or outcomes are not acceptable, cases are reviewed at department mortality and morbidity conferences by medical and leadership staff quality improvement committees and from feedback provided

to the appropriate disciplines. Authoring teams for clinical pathways are expected to incorporate desired changes when feedback and recommendations have been provided by the disciplines involved.

At HMC, patient outcomes are measured in a variety of ways, including the clinical pathway analysis, Trauma Registry (American College of Surgeons) outcome indicators, and other decision support tools. Outcomes assessment and management are beneficial when used effectively to identify opportunities for improvement, guide practice patterns, and promote cost-effective models of care. Findings of the outcome analysis are shared with the health care team to allow for a process to evaluate effectiveness and implement necessary changes in practice.

CHALLENGES WITH MANAGED-CARE PAYERS

We had the opportunity to review this pathway with a payer who was using Milliman & Robertson (M&R) criteria[10] for LOS, which did not correlate with our pathway. The M&R criteria for foot fracture, calcaneus or talus—open reduction, internal fixation (S-490), recommends a goal of a one-day postoperative LOS. Due to the complex psychosocial factors present in our patient population, as well as the regional trauma caseload that we serve in this five-state region, early discharge with home care is not feasible for the majority of our patients. Our findings in the pain management study of calcaneal fracture surgical patients led us to provide a higher level of pain management interventions, including the use of sciatic nerve block and PCA. The team believes that the recovery and rehabilitation process that we have outlined in the pathway provides an opportunity to return our patients to a higher functional ability than that experienced by patients who are treated conservatively in some other settings.

CONCLUSION

Orthopaedics has been an effective clinical service to test our clinical pathway process at HMC. We have been successful at demonstrating the model at our institution through the efforts of this interdisciplinary team. This early success has paved the way for future pathway development within the orthopaedic service and other specialty areas. In this very interdisciplinary care process, which spans the continuum of care within this episode of injury, we have developed a method whereby the patients' functional abilities are systematically assessed and interventions are aimed toward improving strength and mobility needed for the successful return to productive and independent premorbid level of function.

It is expected that within a few months a professional graphic design format will be developed for our patient pathways. Once these are implemented, patient satisfaction measures will be added as part of the evaluation of our outcomes. It has been a rewarding process to increase our efforts to provide cost-effective care that is designed to promote the patient's input into and satisfaction with the processes of care.

The ability to sustain a reduced level of resource utilization has been significant. The reductions have not compromised the clinical quality of care. We are now able to review patient variation from the ideal plan of care and evaluate areas where we can improve our systems or approaches to patient care. Our future electronic medical record will further promote a more consistent and standardized approach to pathways and variance tracking. There will be increased possibilities for retrospective clinical outcomes analysis with a large clinical database to draw directly from.

Since the calcaneal fracture pathway was developed, another pathway in orthopaedics has been developed and implemented. The foot reconstruction pathway has been in place since September 1996. Plans are currently underway to develop pathways for spine fracture without neurologic deficits and for acetabular fracture. With more pathways within the orthopaedic service, the process and practice of team-based, outcome-oriented clinical practice become a greater reality.

REFERENCES

1. Sangeorzan BJ. Foot and ankle joint. In: Hansen ST Jr, Swiontkowski MF, eds. *Orthopaedic Trauma Protocols.* New York, NY: Raven Press; 1993:354–359.

2. Kundel K, Funk E, Brutscher M, Bickel R. Calcaneal fractures: operative versus nonoperative treatment. *J Trauma.* 1996;41:839–845.

3. Cooper JO, Benirschke SK, Sangeorzan BJ, Bernards CM, Edwards WT. Postoperative analgesia after operative repair of calcaneus fractures: an outcome study comparing morphine alone with morphine and one-shot perioperative sciatic nerve blockade. Presented at the Orthopaedic Trauma Association Annual Meeting; September 27, 1996; Boston, MA.

4. Stephenson JR. Surgical treatment of displaced intraarticular fractures of the calcaneus: a combined lateral and medial approach. *Clin Orthop.* 1993;290:68–75.

5. Benirschke SK, Sangeorzan BJ. Extensive intraarticular fractures of the foot: surgical management of calcaneal fractures. *Clin Orthop.* 1993;292:128–134.

6. Carr JB. Surgical treatment of the intra-articular calcaneus fracture. *Orthop Clin North Am.* 1994;25:665–675.

7. Tornetta P III. Open reduction and internal fixation of the calcaneus using minifragment plates. *J Orthop Trauma.* 1996;10:63–67.

8. Buckley RE, Meek RN. Comparison of open versus closed reduction of calcaneal fractures: a matched cohort in workmen. *J Orthop Trauma.* 1992;6:216–222.

9. Brown S, Nemeth L. Developing a variance reporting system to facilitate quality improvement. *Outcomes Manage Nurs Pract.* 1998;2:10–15.

10. Doyle R. *Healthcare Management Guidelines. Volume 1: Inpatient and Surgical Care.* Seattle, WA: Milliman & Robertson; 1996.

■ Appendix 20–A ■
Clinical Pathways

The following is a key to abbreviations used in the exhibit in this appendix:

ADLs, activities of daily living
AP, anterior posterior
AROM, active range of motion
ASLR, active straight leg raise
BM, bowel movement
C&DB, cough and deep-breathe
CMS, circulation, motion, sensation
CT, computed tomography
CTA, clear to auscultation
DC, discharge
d/c, discontinued, discharge
DF, dorsiflexion
DVT, deep-vein thrombosis
EBL, estimated blood loss
ER, emergency room
EV, eversion
FWB, full weight bearing
HCT, hematocrit
HO, house officer
hx, history
INV, inversion
IS, incentive spirometer
IV, intravenous
LE, lower extremity
LB, lower body
MR, may repeat
NC, nasal cannula
NPO, nothing by mouth
NSAID, nonsteroidal anti-inflammatory drug
NWB, non–weight-bearing
OOB, out of bed

OR, operating room
OT, occupational therapy/therapist
PACU, postanesthesia care unit
PCA, patient-controlled analgesia
PCP, primary care physician
PF, plantarflexion
PFS, posterior foot splint
PO, oral
POD, postoperative day
prn, as needed
PROM, passive range of motion
PRON, pronation
PRS, pain relief service
PT, physical therapy/therapist
QS, quantity sufficient
RA, room air
ROM, range of motion
RTC, round the clock
SCD, sequential compression device
S/SX, signs and symptoms
SUP, supination
SW, social work(er)
TEDs, antiembolic stockings brand name
TKE, terminal knee extension
UB, upper body
UE, upper extremity
UO, urinary output
VS, vital signs
WB, weight bearing
WNL, within normal limits

Exhibit 20–A–1 Calcaneal Fracture Clinical Pathway

<table>
<tr><td colspan="2" align="center">CALCANEAL FRACTURE
EMERGENCY DEPARTMENT
OR
ORTHOPAEDIC CLINC</td></tr>
<tr><td colspan="2">DATE</td></tr>
<tr>
<td valign="top">

Assessment and Monitoring
• Initiate clinical pathway assessment

• Arrange for hx/physical completion

Medications
[Analgesic prescribed and given]

Pain Management

• Pain assessed using 0–10 scale (or alternative as defined by procedures) initially and at appropriate intervals, and after treatment.

Pain scale introduced (0–10) current level _____
[Pt identifies level of pain within acceptable range]

Consults/Diagnostic Studies
• Obtained and reviewed X-rays and CT
• If pain at unacceptable level, PRS consultation initiated.
• Hematocrit

</td>
<td valign="top">

Nutrition
• Instruct patient of NPO status prior to OR

Activity/Safety
• Non–weight-bearing affected extremity
• Crutches provided by ER/clinic staff
• Consult PT if problems exist affecting safety with gait/activity

• If mod. to severe edema, discharge pt w/appt. to return

• Bedrest w/foot elevated

Occupational Therapy
• Posterior foot splint fit/or fabricated on patient.
• Dermal pad added when needed to ensure no pressure at heel.

• Patient/family instructed in splint fit/wearing schedule/purpose.

Patient Education
• Patient instructed:
  1) maintain elevation of affected limb
  2) importance of bedrest
  3) skin integrity
  4) crutches non–weight-bearing
• Teach patient NPO before surgery.
• Smoking cessation teaching if applicable
• No smoking after midnight prior to OR

Discharge Planning
• Assess ability to manage at home. Care referral if needed.

• Pt discharged from ER or Clinic with plans to return for surgery when swelling decreased

</td>
</tr>
<tr>
<td valign="top">

ADDRESSOGRAPH

</td>
<td valign="top">

UNIVERSITY OF WASHINGTON MEDICAL CENTERS
Harborview Medical Center—UW Medical Center
Seattle, Washington
CLINICAL PATHWAY—CALCANEAL FRACTURE
pds 4002   1

continues

</td>
</tr>
</table>

Exhibit 20–A–1 continued

DAY OF SURGERY—AMBULATORY SURGERY AND INTRAOPERATIVE

DATE	DATE
Assessment and Monitoring • Intraop: Temp. every 30 min. • Intraop discharge temp _____ • Bair hugger & thermal hat applied	*Consults/Diagnostic Studies* • Pre-op protocol reviewed and all required studies available
Cardiovascular • TED/SCDs on unaffected extremity if at risk: • Obesity • Previous hx DVT • Pregnancy • Oral contraceptives • Other_____ **[Intraop blood loss in OR is monitored] EBL:** _____	*Fluid/Volume/Nutrition* • NPO confirmed (except oral medications) • IV fluids started
Neurological • CMS evaluation of affected extremity • Block in effect _____ level	*Elimination* Gastrointestinal • Last BM
	Genitourinary **INTRAOP:** • Foley inserted prn **[Urine output WNL]**
Integumentary • Screen for allergies to latex, betadine, tape • Intraop: All pressure points created during lateral positioning are padded. • Postop: skin pressure points assessed **[Skin integrity intact]**	*Activity/Safety* Musculoskeletal • Affected extremity elevated above heart **Intraop:** **[Pt remains free of injury secondary to lateral positioning in OR]** Fall Risk/Safety **[Pt safety is maintained in OR]**
Respiratory **[Lungs clear to auscultation]** **[Pt demonstrates ability to cough/deep breathe]**	**Occupational Therapy** • Posterior foot splint reassessed for proper fit **[No areas of skin redness/breakdown]** **[Dermal pad in place]** • Reinforce purpose and wearing schedule
Medications • Anesthesia: IV antibiotics administered 20 min. before incision **[Analgesic regimen discussed and initiated]** **[Side effects effectively treated]**	**Patient Education** • Teach frequency and use of routine post-VS monitoring • OR procedure explained and consent obtained • Instruct C&DB and/or use of I.S. **[Patient/family understands clinical plan of care]** **[Patient demonstrates understanding of Med-Surg unit environment and protocols]**
Pain Management	

ADDRESSOGRAPH Pathways reproduced with special permission. All rights reserved.	**UNIVERSITY OF WASHINGTON MEDICAL CENTERS** Harborview Medical Center—UW Medical Center Seattle, Washington **CLINICAL PATHWAY—CALCANEAL FRACTURE** pds 4002 2

continues

Exhibit 20–A–1 continued

Post-op/Day of Surgery PACU/Floor	
DATE	DATE
Assessment and Monitoring • Intraop: end temp • PACU VS q 15" • Temp on admit and d/c to PACU ***[PACU d/c temp within 1° of preop]*** <u>FLOOR</u> • Routine post-op VS monitoring. Notify HO if T > 38.5 *Cardiovascular* • TED/SCDs if at risk ***[Capillary refill of toes on affected extremity < 3 sec; pulses palpable]*** *Neurological* • CMS checks q 1 x 4 then q shift and prn *Integumentary* • Hemovac to bulb suction ***[Post-op splint is dry and intact]*** *Respiratory* ***[Lungs clear to auscultation]*** • C&DB &/or IS q 1° • Oxygen 2L NC x 48 ° *Pain Management* ***[Block in effect]*** ***[Patient comfortable at rest]*** • PCA instructions reinforced • As block resolves, loading doses are utilized to maximum prescribed or until level of pain is acceptable to patient ***[Pt identifies level of pain within acceptable range]*** **Pain Scale**	*Medications* • DVT prophylaxis if at risk • IV antibiotics q 8 hours till drain dc'd • Post-op analgesic regimen initiated ***[Side effects effectively treated]*** • Adjunct (NSAID) administered RTC x 48 hrs. unless contraindicated *Fluid/Volume* • IV fluids continued until taking adequate p.o. ***[IV patent without redness or tenderness]*** *Nutrition* **<u>FLOOR</u>** • May advance diet as tolerated *Elimination* ***[Absence of active vomiting]*** Gastrointestinal • Assess return of bowel function ***[Bowel sounds present all four quadrants]*** Genitourinary ***[Bladder is without palpable distention]*** ***[Urine output WNL]*** *Activity/Safety* Musculoskeletal • Foot of bed gatched w/affected lower extremity above the level of the heart • Bedrest

N		
D		
E		

UNIVERSITY OF WASHINGTON MEDICAL CENTERS
Harborview Medical Center—UW Medical Center
Seattle, Washington
CLINICAL PATHWAY—CALCANEAL FRACTURE
pds 4002   3

continues

Exhibit 20–A–1 continued

POD #1	POD #2	POD #3	POD #4
Assessment and Monitoring *Initial VS* **[VSS afebrile]** • VS q shift • TED/ SCDs as indicated	**[VSS afebrile]** • VS q shift • TED/SCDs **[No calf tenderness]**	**[VSS afebrile]** • VS q shift • TED/SCDs **[No calf tenderness]** **[Duplex scan WNL]**	**[VSS afebrile]** • VS q shift **[No calf tenderness]**
Cardiovascular *Neurological* **[CMS WNL]** • CMS ✓ Q shift	 **[CMS WNL]** • CMS ✓ Q shift	 **[CMS WNL]** • CMS ✓ Q shift	 **[CMS WNL]**
Integumentary **[Post-op splint dry and intact]** • Hemovac to bulb suction	**[Splint dry, skin integrity intact]** • Hemovac d/c'ed	• Dressing changed **[Skin integrity remains intact]** **[Splint properly positioned]**	**[Skin integrity remains intact]** **[Splint properly positioned]**
Respiratory **[Lungs CTA]** • Oxygen continued • C&DB and/or use I.S. q 1 hr while awake	• D/C oxygen if RA sats > 92% **[Lungs CTA]** • C&DB and/or use I.S. q 1 hr while awake	**[Lungs CTA]** • C&DB and/or use I.S. q 1 hr while awake	**[Lungs CTA]**
Medications • IV antibiotics until drain dc'd • Bowel Medications • PCA continued • Pain scale q 4° • Pain with PCA regimen	• IV antibiotics until drain dc'd • Bowel Medications • Pain scale q 4° • Patient medicated prior to increased activity • PCA dc'd after oral analgesics are effective **[Pain controlled with oral analgesic regimen]**	• Bowel Medications • Pain scale q 4° • Patient medicated prior to increased activity • DC analgesic prescriptions discussed **[Pain controlled with oral analgesic regimen]**	• Bowel Medications • Pain scale q 4° • Patient medicated prior to increased activity • Patient discharged with sufficient supply of analgesics **[Pain controlled with oral analgesic regimen]**
PAIN SCALE	**PAIN SCALE**	**PAIN SCALE**	**PAIN SCALE**
N	N	N	N
D	D	D	D
E	E	E	E
Consults/Diagnostic Studies • HCT **[HCT WNL]**		Duplex scan completed if at risk: • Obesity • Previous hx DVT • Pregnancy • Oral contraceptive usage	

ADDRESSOGRAPH

Pathways reproduced with special permission. All rights reserved.

UNIVERSITY OF WASHINGTON MEDICAL CENTERS
Harborview Medical Center—UW Medical Center
Seattle, Washington
CLINICAL PATHWAY—CALCANEAL FRACTURE
pds 4002   4

continues

Exhibit 20–A–1 continued

POD #1	POD #2	POD #3	POD #4
Fluid/Volume • IV fluids continued until taking adequate PO ***[Patient is adequately hydrated]*** ***[IV site without redness, tenderness or drainage]*** *Nutrition* ***[Patient tolerating regular diet]*** *Elimination* ***[No nausea/vomiting]*** ***[Passing flatus. Bowel sounds present in 4 quadrants]*** ***[U.O. q.s. clear/yellow]*** • D/C Foley after 24° post-op • Straight cath if no void • MR x 2 then notify HO *Activity/Safety* • Mobilize per PT	• Peripheral IV site capped if PO intake q.s. ***[Patient is adequately hydrated]*** ***[IV site without redness, tenderness or drainage]*** ***[Eating 50% of meals]*** • If not, contact Dietitian ***[No nausea/vomiting]*** ***[Passing flatus. Bowel sounds present in 4 quadrants]*** • Mobilize per PT	***[Patient is adequately hydrated]*** ***[Eating 50% of meals]*** • If not, contact Dietitian • Weight _____ If > 10% wt. loss from stated weight, contact dietitian ***[No nausea/vomiting]*** ***[BM with normal pattern consistency]*** ***[Patient voiding QS]*** • Mobilize per PT	***[Patient is adequately hydrated]*** ***[Eating 50% of meals]*** ***[No nausea/vomiting]*** ***[Eliminates as per baseline status]*** • Mobilize per PT
Psychosocial/Emotional ***[Appearance and affect are appropriate to situation]*** • Assess anxiety level of patient and family • SW high risk screen completed • Needs additional assessment for _____ • No current Psycho/social issues. Contact SW if pt needs change **Additional notes:**	**SOCIAL WORK** (If additional assessment needed, indicate location of note.) **Psychosocial Assessment** Living Situation: _____ _____ Family/Support System/Service Providers ❑ Adequate ❑ Marginal ❑ Stressed ❑ None ❑ Unknown Name Relationship Address/Phone _____ _____ _____ Legal Consent Authority: _____ Guardianship ❑ Y ❑ N ❑ Needed Power of Attorney: _____ Other Legal Issues: _____ Psychiatric/Involuntary Treatment Act (ITA) ❑ Y ❑ N Orientation/Cognitive Issues: _____ Coping Skills ❑ Pt WNL ❑ Unable to assess ❑ Needs Assistance ❑ Family WNL ❑ Needs Assistance ❑ N/A Issues: _____ Pre-morbid functioning: ❑ Independent ❑ Req. Assistance with: _____ _____ _____ Etoh/Substance Abuse: ❑ Ass'mt completed ❑ Cannot assess ❑ N/A ❑ Medical barrier to Rx ❑ Does not agree with Rx Add'l Assess. _____ Income: _____ Health Care Coverage: ❑ Insurance: _____ Case Manager:_____ Phone: _____ PCP: _____ ❑ Has Medicare A&B ❑ A only ❑ Has Medicare Supplement: _____ ❑ Has Medicaid Active #_____ PIC _____ ❑ Medicaid Pending Plan: _____ _____	***[Pt and family adjusting to impact of illness]*** ***[Pt and family are aware of support systems available]*** **Additional notes:**	***[Pt and family adjusting to impact of illness]*** ***[Pt and family are aware of support systems available]*** **Additional notes:**

ADDRESSOGRAPH 	**UNIVERSITY OF WASHINGTON MEDICAL CENTERS** Harborview Medical Center—UW Medical Center Seattle, Washington **CLINICAL PATHWAY—CALCANEAL FRACTURE** pds 4002 5

continues

Exhibit 20–A–1 continued

POD #1	POD #2	POD #3	POD #4

Occupational Therapy
 	
OT signature _____
	• Posterior foot splint refit on patient
	[Proper fit]
	[No areas of skin redness]
	[Dermal pad in place]
		Posterior foot splint reassessed
		[Proper fit]
		[No areas of skin redness]
		[Dermal pad in place]
		• After crutch ambulation initiated by PT evaluation by OT as needed:
		• Toilet transfer
		    with _____ assist
		    with _____ device
		    with _____ device
		• Shower/tub
		    with _____ assist
		    with _____ device
		    with _____ device
		• Dressing
		    with _____ assist
		    with _____ assist
		• Endurance:
		    _____ minutes
			• Posterior foot splint reassessed
			[Proper fit]
			[No areas of skin redness]
			[Dermal pad in place]
			If needed:
			• Review toilet transfers with device
			    _____ assist
			• Review bathing precautions
			    _____ plastic wrap at splint _____ NWB
			• Equipment recommended: reacher, bath sponge, raised toilet seat, grab bars, tub set, other: _____
			  _____
			[Pt/family verbalizes understanding use of home equipment/adaptations]
			• Adaptive equipment ordered to be delivered
			by: _____
			to: _____
			on: _____
			• Equipment issued/in pt's possession.
			OT/DC Goals:
			[Independent toilet transfer and tub transfer NWB involved extremity with assistive device(s)]
			[Independent UB/LB dressing, including don/doff of post. foot splint]
			• Family training given if patient not independent as above.
			See recommended equipment issued/ordered

Patient/Family Education	• Oral pain management regimen reviewed • Activity plan • Elevation of extremity maintained to decrease swelling	• Oral pain management regimen reviewed • Activity plan • Elevation of extremity maintained to decrease swelling	• Oral pain management regimen reviewed • ADLs & mobility limitations reviewed per Pt and OT

ADDRESSOGRAPH

UNIVERSITY OF WASHINGTON MEDICAL CENTERS
Harborview Medical Center—UW Medical Center
Seattle, Washington
CLINICAL PATHWAY—CALCANEAL FRACTURE
pds 4002   6

continues

Exhibit 20–A–1 continued

POD #1	POD #2	POD #3	POD #4
Physical Therapy Assessment • Initiate PT assessment • OOB to chair w/leg elevated if tolerated. NWB on involved extremity. No ambulation. (If block has not worn off, will need more assistance) Transfer status: _____ assist **Physical Therapy Assessment** • Chart reviewed for history of present illness as well as premorbid history which could affect PT program. Comments: Precautions: • Screen UE/LE ROM and strength, except involved splinted joints ❑ • Instruct in AROM and strengthening of upper extremities and uninvolved lower extremity ❑ • Instruct in involved extremity quad sets, glut sets, AROM of hip and knee and toes as tolerated. Progress to ASLR and TKE ❑ Abnormal findings: Baseline mobility status: Independent with all mobility ❑ Required assist _____ ❑ Equipment pt owns/rents: _____ Equipment ordered/issued: _____ Architectural barriers at home: Stairs ❑ yes ❑ no Railing ❑ yes ❑ no Other: Assistance available at home for home program: ❑ yes ❑ no Describe _____ _____ _____ _____	• Begin ambulation with NWB involved extremity with assistive device (limited time w/leg dependent) • Begin PROM to involved toes when block has worn off • Continue to monitor and advance bed exercise program Transfer status: _____ assist Ambulation status: Level surface: _____ assist _____ device _____ distance Stairs: _____ _____	• Consult OT for ADL assessment PRN ❑ Yes ❑ No • If cleared by attending MD, begin tibiotalar AROM (see physician documentation sheet) • Continue PROM to involved toes *[Independent exercise program for UE/LE strengthening]* • Continue ambulation per POD #2 Ambulation status: Level surface: _____ assist _____ device _____ distance Stairs: _____ _____	• When wound sealed and cleaned by attending MD (see physician documentation sheet), begin tibiotalar and subtalar AROM (DF, PT, INV, EV, PRON, SUP, ankle circles) • If wound not sealed, by discharge, see attending note for options • No motion until return to clinic ❑ • Teach AROM on uninvolved foot; pt instructed *when* to start AROM by attending MD ❑ • Other: _____ _____ Discharge Goals: *[Pt verbalizes precautions and demonstrates compliance with precautions]* *[Patient independently ambulates NWB involved extremity on level surface and stairs with assistive device]* *[Patient independent with donning/doffing PFS, PROM toes, AROM ankle/subtalar joints]* • Family training given if pt not independent in exercise programs as above (assistance often required with PFS & PROM toes) _____ _____ _____ • Arrange for home or outpatient PT, if needed for safety check or assistance with exercise programs per the discretion of primary therapist. _____ _____ _____
Discharge Planning		• SW confirm d/c plan: Date of d/c: _____ To: _____ Time: _____ By: _____ Facility contact: _____ _____ D/C summary requested of: _____ _____ Family notified ❑ Pt/Family approve ❑	• Inpatient DC planning—Clinic RN • Provide Clinic D/C instructions A. Time for Visit B. Plan for Visit C. Parking D. Physical Therapy? E. Medications? • Given written clinic D/C handout Coordinate D/C planning with Med. Staff

ADDRESSOGRAPH **Pathways reproduced with special permission. All rights reserved.**	**UNIVERSITY OF WASHINGTON MEDICAL CENTERS** Harborview Medical Center—UW Medical Center Seattle, Washington **CLINICAL PATHWAY—CALCANEAL FRACTURE** pds 4002 7

continues

Exhibit 20–A–1 continued

Clinical Pathways provide guidance in the management process for a specified case type. Using Clinical Pathways in actual practice requires consideration of individual patient needs.	INITIAL	SIGNATURE	HOURS OF CARE	DATE	TITLE
Legend: • = task intervention [...] = outcome ✓ = normal * = abnormal NA = not applicable Ø = not done					
RN Registered Nurse					
DT Dietitian/Nutrition Asst.					
OT Occupational Therapy					
CNS Clinical Nurse Specialist					
RCP Respiratory Care Practitioner					
PT Physical Therapist					
PH Pharmacist					
USC Unit Secretary					
MSW Social Worker					
MD Physician					

ADDRESSOGRAPH

UNIVERSITY OF WASHINGTON MEDICAL CENTERS
Harborview Medical Center—UW Medical Center
Seattle, Washington
CLINICAL PATHWAY—CALCANEAL FRACTURE
pds 4001   8

continues

Exhibit 20–A–1 continued

	CLINIC FOLLOW-UP			
	1st Visit Date: _____	6 Weeks Date: _____	3 Months Date: _____	6 Months/9 Months/1 Year
Wound Assessment	• Educate re: wound care **[Incision clean and dry.]** **[Incision without redness, tenderness or drainage]** **[Sutures removed and steri-strips applied]**	• Wound assessed **[No s/sx infection]** **[Incision healing WNL]** • X-rays of affected limb (AP lateral and axial)	• Wound assessed **[Incision healed]** • X-rays of affected limb	**6 Months Date: _____** X-rays of affected limb Monitor pain control **Physical Therapy** Orthotic assessment **[Gait shows progressive improvement]** **[Balance strength, & ROM improves]** Ankle, subtalar isometric/isotonic strengthening with tubing/Theraband (No free weights) Soft tissue mobilization
Pain Control	• Pain assessed using 0–10 scale or alternative **[Pt identifies pain level within acceptable range]** • Supply of analgesic prescriptions sufficient • Educated about use of pain meds. • PRS consult if pain control unsatisfactory	• Pain assessed using 0–10 scale or alternative **[Pt identifies pain level within acceptable range]** • Supply of analgesic prescriptions sufficient • Educated about use of pain meds. • PRS consult if pain control unsatisfactory • Monitor pain control	• Monitor pain control	**9 Months Date: _____** • X-rays of affected limb • Monitor pain control **Physical Therapy** • Ankle, subtalar PROM • Joint mobilization • Ankle, subtalar isokinetic assessment, strengthening, endurance training • Advanced balance, gait training as indicated. • Monitor orthotic status
Nutrition	**[Nutrition adequate for optimal wound healing]** • Assess for adequacy of healing and nutrient intake. If not healing, contact nutrition services.			**1 Year Date: _____** Monitor pain control **Physical Therapy** Advanced strengthening ROM, endurance training Closely monitor gait & orthotic status
Elimination	**[BM within normal pattern and consistency]**			**S/P Hardware Removal** **Physical Therapy** • Progress gait training • Home strengthening ROM program prn
Physical Therapy	**Physical Therapy** • Initiate and/or reinforce early AROM ankle, subtalar toes. • Progress involved extremity knee strengthening, progress body conditioning • Desensitization techniques prn	**Physical Therapy** • Continue progress early AROM ankle, subtalar, toes • Progress home strengthening program	**Physical Therapy** • Begin gradual increase WB at 20 lbs, increase over 1 month to FWB. • Gradual wean from assistive device as patient tolerates. • Pool therapy if available • Gait training, re-education • Desensitization techniques as needed • Initiate balance, proprioceptive training. • Ankle, subtalar, AAROM/Isometric strengthening • Low impact endurance training • Soft tissue mobilization • Orthotic initiation prn	Comments:

ADDRESSOGRAPH

UNIVERSITY OF WASHINGTON MEDICAL CENTERS
Harborview Medical Center—UW Medical Center
Seattle, Washington
CLINICAL PATHWAY—CALCANEAL FRACTURE
pds 4002   Clinic Visit

■ 21 ■

Improving the Outcomes of Cervical Spinal Cord Injury: Developing a Clinical Pathway Model

Karen March, Lynne Nemeth, Norma J. Cole, Karol Wilson, and Tarek A. Salaway

As a Level I trauma center and a component of the Northwest Regional Spinal Cord Injury Model Center, Harborview Medical Center (HMC) admits an average of 60 tetraplegic spinal cord–injured patients annually. The scope of services provided at HMC includes prehospital air transport and emergency medical care through rehabilitation. The University of Washington (composed of HMC and the University of Washington Medical Center) joined the model spinal cord injury (SCI) project in 1990. Funding for the participation in the model center project is provided by the National Institute on Disability and Rehabilitation Research, a division of the U.S. Department of Education. This project was developed to provide the highest quality of emergency-through-rehabilitation care to patients with SCI, effective systems of care delivery, and improved outcomes for patients. A major component of the program focuses on the provision of data to the National SCI Statistical Center at the University of Alabama at Birmingham and the conduct of research on SCI.[1]

We acknowledge the hard work and dedication of the following authoring team members who developed the clinical pathway: Missy Armstrong, Therapeutic Recreation; Asad Awan, PharmD; Vickeri Barton, RD, Clinical Nutrition; JoAnne Bastion, PT (Physical Therapy), Acute Care; Charles Bombardier, PhD, Rehabilitation Psychology; Wendy Cohen, MD, Neuroradiology; Norma Cole, MSW, Northwest Regional Spinal Cord Injury System; M. Sean Grady, MD, Neurosurgery; Kelly Hazel, Speech Pathology; Bob Hochter, OT (Occupational Therapy) Rehabilitation; Carol Kasper, Respiratory Care; Lynn Krog, RN, Trauma Rehabilitation Coordinator; Debbie Levin, PT, Rehabilitation; Karen March, RN, Neuroscience Clinical Nurse Specialist; Anthony Margherita, MD, Physical Medicine and Rehabilitation; Kathy Michael, RN, Nurse Manager, Rehabilitation; Paula Minton-Foltz, RN, Nurse Manager, Neurology/Neurosurgery; Lynne Nemeth, RN, Clinical Pathways Project Manager; Colleen Redeker, RN, Neurosurgery ICU; Lawrence Robinson, MD, Physical Medicine and Rehabilitation; Mary Royce, RN, Emergency Department (ED); Roseann Sealy, OT, Acute Care; Karol Wilson, PT, Acute Care. We also acknowledge the expert assistance of Joan Knecht in the preparation of this manuscript and Susan Pilcher, RN, Trauma Registry Coordinator, for the injury-specific data used for this chapter.

Because of the high-cost and high-risk nature of the cervical spine–injured patient population, the concept of using clinical pathways to coordinate patient care delivery first emerged in 1992 at HMC. The neuroscience nursing quality assurance (QA) subcommittee discussed high-risk indicators for measurement that would be relevant to our patients throughout the neuroscience units, which included a neurosurgery intensive care unit (ICU), a neuroscience unit, a rehabilitation unit, and an epilepsy unit. There was great concern regarding the care of these patients. They were extremely labor intensive, and they required a great deal of planning and coordination between disciplines and the family. Unless concepts of early rehabilitation were promoted,[2] serious complications could occur for the patient. We identified variations in care delivered to the cervical spine–injured patients who were not cared for on a neuroscience unit.

INTEGRATING PATHWAYS WITH QUALITY IMPROVEMENT INITIATIVES

Because of the volume of trauma patients admitted to HMC, we are not always able to place patients with a specific diagnosis in the optimal specialty unit. Patients are placed in another ICU, if a bed is not available, within the appropriate specialty unit at the time of their admission, according to a matrix for patient admission. The staff working in the other specialty ICU may not be as familiar with the approach of early rehabilitation for SCI that we have used in our neurosurgery ICU. This unfamiliarity can increase patient complications, length of stay (LOS), and cost of care. Accordingly, this has been a major focus for our pathway implementation, since patients may be admitted to a variety of clinical units within HMC. Development of a consistent assignment process to the pathway throughout the different clinical units has been a focus of our quality initiative. The implementation of clinical pathways for cervical spine injuries has been a key piece of our overall quality improvement efforts, whereby key indicators such as LOS, rates of complications and readmission, treatment variances, and comprehensive compliance with measures limiting nosocomial infections are evaluated. These indicators are examined and monitored on a timely basis to enable us to improve our care processes described in the development of our pathway below. In due course, these measures have provided the means for HMC to provide the most appropriate care most efficiently and effectively and thus to respond to the demand for our services.

The neuroscience nursing QA subcommittee proposed to nursing administration that a pilot clinical pathway for the cervical SCI patient be developed. This initial effort eventually paved the way for the clinical pathway program to begin at HMC in 1994. An interdisciplinary team met regularly, first to develop and design the format to be used and later to develop the content. The format chosen for this project was a document divided into phases, including the emergency department, prestabilization, poststabilization, and rehabilitation phases. To transfer the patient to the various phases of care, designated outcome goals were established, specifically regarding when the patient would transfer to the rehabilitation unit. The criteria for direct transfer from the ICU to the rehabilitation unit were also outlined. The pathway was designed as a guide for care to be delivered by the nurses, social workers, respiratory therapists, nutritionists, and the rehabilitation team. The physicians supported the process but were not actively involved. In the spring of 1994, the pilot pathway was disseminated for use as a guide. It would not be used as a documentation tool in patient care at this time.

PATHWAY DEVELOPMENT

In 1994, HMC appointed a project manager to develop the clinical pathways program. Consultation with the Center for Case Management (S. Natick, MA) helped get the program started, with the full administrative commitment of the organization. A design team was appointed to develop the philosophy and structure for the clinical pathway initiative at HMC. The design team met on a biweekly basis during the first half-year to develop the format and policies for our pathway documents, develop an educational process for staff, and design an approach for variance data collection and aggregation. The design team examined multiple formats for pathways and identified core values for the project. It was decided that the pathway would be an interdisciplinary documentation

tool with a matching order set. All documentation except for vital signs, the neurologic assessment, intake and output, and medication record would be on the pathway. The design team felt it was important that the pathway be not only a guide but also a documentation tool for treatment interventions and outcomes throughout the continuum.

The design team believed there were common questions asked by all disciplines in the admission assessment process. We shared a goal of decreasing redundant patient inquiries and documentation of the same data. An assessment form was developed by the team that was later implemented housewide as a nursing admission assessment form. The admitting nurse is responsible for completing this assessment, and the team members use it as a database.

Although developing a pathway for the cervical spine–injured patient was a pilot project, development of the final pathway did not occur for another two years. The final clinical pathway–authoring team was appointed in March 1996. The team consisted of members from the original group, in addition to many new members from the acute care and rehabilitation teams.

The neurosurgeon team member reviewed the key elements of the medical care required for these patients and the expected LOS in the acute care phase. The neuroradiologist outlined the radiographic studies required and the optimal timing of those studies. The rehabilitation physician discussed key outcomes of care and particularly focused on the patient's requirements to participate in a structured rehabilitation plan. The pilot pathway was used as the framework from which the current pathway was developed. After all team members had a clear understanding of the process required to develop a seamless plan of care, the team split into acute care and rehabilitation teams. Much of the work for the acute care phase had been developed previously and needed only to be refined. The emergent, prestabilization, and poststabilization phases of this pathway were implemented in August 1996. The rehabilitation phase was more complicated and required more work. At this time, the rehabilitation component has been written but not implemented.

Developing Specific Components of the Pathway

The social work department at HMC had a well-defined system of protocols and procedures for psychosocial assessments and discharge planning. Due to these preestablished guidelines, the process of adapting them to the clinical pathway format was much less time consuming for the social work department than for the disciplines without well-integrated protocols and procedures.

With the cervical spine patients, it was necessary to broaden the standardized social work assessment to include issues related to the patients pending rehabilitation and long-term care needs, as well as financial needs. It was apparent that as each new pathway was developed, the social services aspect of the pathways became more fine tuned and could be carried over and used in all pathways regardless of diagnosis. It also became clear early in this process that the pathway was best developed by having the team members take responsibility for writing their unique interventions and outcomes within the pathway. Having the entire authoring group work on each section became tedious and overly time consuming. Setting deadlines for each discipline to complete a rough draft of their section, which could then be evaluated by the entire team, was a much more efficient use of everyone's time. The need to respect discipline-specific boundaries and trust in the professionalism of each team member was foremost in this process.

Phases of Care

The cervical spine injury clinical pathway is divided into phases of care: emergency department (ED), prestabilization, poststabilization, and rehabilitation. The patient is started on the pathway in the ED, and the admission assessment is initiated. The ED phase includes the initial assessment, diagnostic workup, and stabilization of the spine. In consultation with the spine team, the spine is reduced by placement of the patient in cervical traction. Methylprednisilone, 30 milligrams per kilogram of body weight, is administered as a bolus, followed by 5.4 milligrams per kilogram per hour over the next 23 hours.[3] Additional investigational drug protocols may be used as well, depending on the current research activities being undertaken at HMC.

The prestabilization phase begins when the patient is discharged from the ED to the ICU or intermediate care unit. During this phase, stabilization of the patient's medical status continues, and evaluation by the rehabilitation team members is initiated. The spine team, composed of neurosurgery and orthopaedic surgeons, determines whether the patient requires surgical intervention with external orthosis to stabilize the spine or external orthosis alone. Occupational therapy (OT) evaluates and sets up a call light system for the patient. Speech pathology assesses the intubated patient's ability to communicate and initiates an augmented communication system as needed. For the nonintubated patient, speech pathology assesses communication and swallowing, advancing the diet as appropriate to a modified oral diet. Physical therapy (PT) performs an initial evaluation of joint mobility and strength and measures the patient for a custom wheelchair that will be used to mobilize the patient after

stabilization. The Northwest Regional Spinal Cord Injury System social worker makes an initial evaluation of the patient's and family's psychosocial and financial needs during this phase. Nursing care focuses on maintaining cardiovascular, neurologic, and pulmonary function, skin integrity, function of the bowels and bladder, and patient comfort. This phase lasts one to three days, with the philosophy of early mobilization.

The poststabilization phase begins following either a combination of internal and external orthosis or just external orthosis. Once the spine is stabilized, the rehabilitation process begins. Interventions focus on maintaining function of all systems and facilitating adaptation to disabilities. PT and nursing begin to mobilize the patient once the patient is permitted to get out of bed. This permission is given by the resident physician on the patient's primary service (usually neurosurgery) after interpretation of upright cervical spine radiography.[4] The PT customizes the equipment needed by the patient to ensure maximal function and safety. OT assesses the patient's needs for splinting and positioning to prevent contractures or deformities and evaluates the patient's ability to participate in self-care tasks. Speech pathology continues to assess and intervene with communication and swallowing. The social worker continues providing psychosocial support and financial consultation to the patient and family as well as preparing them for discharge to a rehabilitation facility. The clinical pathway outlines the sequence of key interventions and goals that move the patient toward rehabilitation within three to five days after spine stabilization.

CARE/CASE MANAGEMENT

An interdisciplinary team approach is required when caring for complex neurosurgical patients, who often develop long-term disabilities. Care of the cervical spine–injured patient is an excellent example of complex care that requires a coordinated approach to ensure that interventions are sequenced and resources are available when needed to optimize the patient's outcome. Currently, interdisciplinary team rounds occur once a week on the acute care floor but should probably occur more often. Efforts at prioritizing multidisciplinary team meetings are underway. This is an area where we could improve our systems of care coordination. The needs of the patient are discussed by the physician, PT, OT, speech pathology, the social worker, utilization review, the clinical nurse specialist (CNS), the trauma rehabilitation coordinator, and bedside nurses. An individualized plan of care is developed by customizing the pathway to move the patient forward to the rehabilitation phase. The clinical pathway is used to determine if the patient is progressing toward discharge in a timely manner or if he or she requires additional interventions before rehabilitation. Patient care conferences occur as issues arise requiring more intense coordination between disciplines and the patient or family.

Nursing delivers care using a total patient care delivery system. In the ICU, the nurse-to-patient ratio is 1:1 or 1:2 and is 1:2 in the intermediate care unit. As the acuity of the patient decreases, the patient may be cared for on the neurosurgical acute care unit by primary nurses.

The CNS consults on complex patients to ensure that quality care is provided. For patients in the nonspecialty units, the CNS, as well as the trauma rehabilitation coordinator, monitors the progress of the patients and facilitates the care coordination. When necessary, the CNS will arrange for an interdisciplinary team conference to discuss the progress of the patient and to develop an individualized plan of care to meet the defined outcomes.

The clinical pathway (Exhibit 21–A–1 in Appendix 21–A) serves as the tool that outlines day-to-day care, key interventions, and outcomes to be met throughout the hospitalization. The health care team, except for the physicians, document the patient's progress on the pathway. The nurse is accountable for ensuring that the patient is placed on the appropriate clinical pathway and that the order set is signed. If the pathway was not started in the ED, the nurse admitting the patient to the unit starts it, and a variance is recorded. Orders and the pathway are individualized to progress the patient toward meeting defined outcomes. The pathway is kept in the front of the patient's medical record.

Pathways are reviewed and revised by the authoring teams on the basis of feedback from the users, analysis of the pathway variances, and changes in clinical practice based on scientific advances. No revisions have been made to the content at this time. The clinical pathway design team is currently changing the layout of the pathway to streamline it further.

CLINICAL OUTCOMES

In each phase of care, several critical outcomes are measured by the staff providing direct care to the patient on the pathway. The following are some outcomes evaluated in the ED phase (Exhibit 21–A–1 in Appendix 21–A): the patient is off the spinal immobilization backboard within two to three hours after placement on the board; skin is intact; and the patient is placed on a kinetic treatment table (Rotorest bed) before admission to the unit. In the prestabilization phase, outcomes evaluated include that

- the patient maintains normothermia
- the patient shows no bradycardia during suctioning or turning
- the patient participates in spinal assessment
- neurologic status is stabilized
- lungs are clear
- the patient is able to communicate basic needs

- the patient has no deformities or contractures
- skin remains intact

Poststabilization includes the above outcomes and begins to focus on rehabilitative outcomes, such as

- range of motion of all joints within normal limits or no deformity
- orthosis in place and properly fitting
- sitting tolerance of 30 to 60 minutes
- skin intact after sitting for _____ minutes
- patient positioned well in wheelchair and bed

These outcomes are measured on a daily basis by the responsible team member. If the patient meets the outcomes, he or she can progress to the next phase of care. If an outcome is not met at the specified time, patient care is individualized to assist the patient in meeting the outcome at the earliest possible time. A variance is recorded only when the inability to meet the outcome impedes the progression to the next phase of care.

The discharge planning and psychosocial outcomes, developed primarily by the social work participant on the authoring team, have often appeared a bit vague. How do you adequately measure a patient's or family's coping skills or level of support? With the development of clinical pathways, social workers have been challenged to address this issue and have constructed intended outcome statements. Some of these are:

- Patient and/or family has adequate housing.
- Patient has adequate social supports.
- Patient is satisfied with his/her level of chemical use.
- Patient has medical coverage.
- Patient is receiving mental health services as needed.
- Patient has an adequate discharge plan for physical support or rehabilitation needs.

These are very broad outcome statements, which can be answered "yes" or "no." However, when dealing with individuals, family systems, and diverse coping skills, it is impossible to make these statements more precise. Since the path to these "yes" or "no" answers involves complex and extremely diverse discharge plans for each patient, it seemed most appropriate to keep them broad and achievable. The goal is to have each outcome statement answered with a "yes"; however, the route to that "yes" answer remains as diverse and individual as our patient population.

FINANCIAL OUTCOMES

Review of the 10 months after implementation of our pathway in August 1996 showed that our average LOS had decreased to 13.8 days. Even in this short time frame, the difference between pathway and nonpathway patients is striking. Our average LOS for pathway patients is 11.6 days as compared to 15.1 days for patients in the nonpathway group. Consistent with that difference, the mean total hospital charge for pathway patients during this period was $55,400, compared to $65,100 for our nonpathway patients (Figure 21–1). Clearly, implementation of the clinical pathway for cervical SCI has yielded positive financial results.

A limitation of this comparison is that we have not been able to move all eligible patients to the pathway process due to inconsistency among our staff involved in

	All Patients		
	Count (N)	Mean	Median
LOS (days)	44	13.78	10.50
Total charges ($)	44	62430.07	41550.54

	Nonpathway			Pathway		
	Count (N)	Mean	Median	Count (N)	Mean	Median
LOS (days)	33	15.09	11.00	11	11.64	9.00
Total charges ($)	33	65,107.55	49,482.99	11	55,425.63	34,934.10

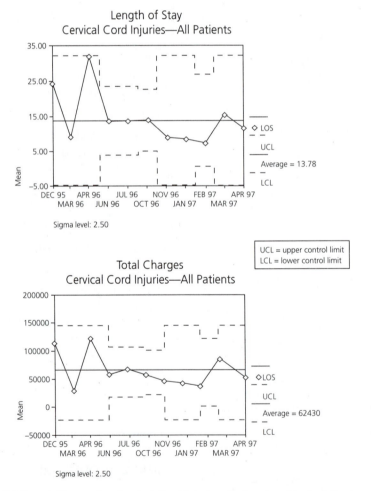

Figure 21–1 Length of Stay (LOS) and Charges for Cervical Cord Injuries, Harborview Medical Center, January 1996 to June 1997 (Principal ICD-9 Diagnoses, 806.00–806.19 or 952.00–952.08). *Source:* Copyright © Harborview Medical Center.

the admitting process. Originally, our intent was to adopt the pathway for use with all cervical spinal cord–injured patients without multiple injuries. Patients may be admitted to a variety of clinical units within HMC, and though we have not accomplished a consistent assignment of the pathway to all eligible patients, our aim in the development and evaluation process is to standardize the assignment process throughout the institution.

SATISFACTION

Patient satisfaction outcomes have not been specifically measured for this pathway population. In general, HMC uses the Picker Survey to assess patient satisfaction. However, we have not yet used these survey data to analyze differences between our patients on pathway versus nonpathway groups. We have not used the pathway as a patient educational tool at this time, but plans to do that are underway.

Initial staff satisfaction regarding the pathway implementation process was mixed. Change is always difficult, and this was no exception. The change to a comprehensive interdisciplinary documentation tool and a less detail-oriented form of documentation has been very stressful for some staff members. Staff have been accustomed to writing free-text narrative documentation and often have used their own flowsheets to distinguish their work from others. Locating the pathway within the medical record has been difficult at times due to the increased numbers of staff using the same document to record their discipline-specific interventions and outcomes. The pathway has not always been found in the patient's chart because it has been removed from the chart while in use by another team member. With automation of the pathway, these issues will improve considerably. Other issues include the lengthiness of the pathway, staff documenting in the wrong place, and the lack of physician documentation on the pathway.

In late 1996, staff were surveyed to identify specific concerns. Staff appeared to understand the purpose and benefits of the pathway but were unclear on how to document when the patient deviated from the pathway. This became a source of dissatisfaction and poor utilization of the pathway. Such deviations are usually related to patient variances requiring an individualized plan of care.

With spinal cord–injured patients, due to variation in the level of the injury sustained and the potential for comorbidities and complications, it is difficult to predict accurately the entire plan of care that is needed. There is a strong need to individualize care that is needed and also to stay focused on meeting overall goals to reach the rehabilitation phase of care. Many staff who are learning to work with a pathway seem to feel that there is a black and white standard and to feel uncomfortable with the individualization that is often needed. This challenges the staff to coordinate closely and carefully the many dimensions of care required for the cervical spine–injured patient. Rather than saying, "The patient fell off the pathway; let's abandon the pathway," staff must recognize that this patient needs the pathway even more to get on track with the intended outcomes he or she is seeking to achieve.

Advantages and Disadvantages of the Pathway

Utilization of the clinical pathway ensures that all patients receive a consistent approach to their care, regardless of whether they are cared for on a specialty unit or by a staff member familiar with this type of patient. The pathway ensures that the staff caring for the patient know which intervention needs to be performed at which time. This improves patient progression toward the desired goals. Moreover, it increases the quality and consistency of care provided and reduces redundancy or omission.

The authoring teams have learned a great deal about other disciplines' roles in the overall care of the patient. Team members also have had an opportunity to discuss the rationale for practice issues and to question processes that perhaps could be changed. This has facilitated understanding of each other's practices and demonstrated that communication leads to a more coordinated approach to care. The advantages to patient care delivery have been that less time is spent writing reports and thus more time can be spent with patients or families. It appears that patients get a more comprehensive treatment plan from all disciplines. A clear plan of care with deferred expectations enables staff to provide the appropriate care in a more timely manner.

Disadvantages of the pathway center on documentation and the process of change. For ICU nurses, a big obstacle was reverting to a paper-based documentation system after making the transition to an electronic medical record. For others, it was the use of an abbreviated system rather than the lengthy narrative system they were familiar with. Narrative documentation with the pathway was designed to be limited to variances. This was a tremendous shift in philosophy for staff, since we have maintained the use of traditional methods of narrative documentation for our nonpathway patients. For social workers, it was particularly difficult at times to document in a "yes/no" fashion because of the complexity of the psychosocial situation.

Initially, until all disciplines were familiar with the pathways, it was difficult to get them to look beyond the original chart. Staff expectations regarding the legal aspects of the medical record made it difficult to get be-yond the ingrained "If it wasn't charted, it wasn't done" mind-set. The pathway does not convey the same message about the patient as an individual that a narrative report does. This is especially true for long-term patients who will need weeks, sometimes months, of rehabilitation. For instance, a check box that states that a patient was homeless on admit does not go into the personal ramifications or tell the complete story as a narrative that explains that the patient's farm was repossessed when floods destroyed his crops.

VARIANCE AS A SOURCE OF OUTCOME DATA

A detailed description of the variance-tracking methodology and aggregation process that is used at HMC can be found in Chapter 20 of this book, on calcaneal fracture, in the section "Variance as a Source of Outcome Data." We have just begun to review the first year's experiences with this pathway and are currently analyzing the variances that have been collected.

CONSIDERATIONS ON OUTCOMES AS A REGIONAL PROVIDER OF TRAUMA CARE

With trauma center designation, HMC has served as the only regional Level I trauma center for Washington, Alaska, Montana, Idaho, and now Wyoming. Since our designation process within the state of Washington, effective in 1995, we have seen a significant increase in our overall volume of trauma patients, including those with SCI. With the 75 percent increase in the cervical spine–injured population at HMC (Figure 21–2), this growth in volume has brought us a more complicated patient population. We have reduced specific complications overall from 1990 to 1996, consistent with the time frame of the early pilot work of the initial pathway program. However, consistent with the growth in our volume, the proportion of this group who experience complications has similarly increased (Figure 21–3) reflecting our role in the extended trauma community we are serving.

The etiologies of the injuries seen at HMC are primarily motor vehicle accidents (35 percent) and falls (33 percent); this correlates closely with national data.[1] Over 90 percent of these patients are clinically defined as having major trauma, with an injury severity score (ISS) greater than 15,[5,6] and over 55 percent of this population have an ISS over 24. Despite a trend in recent years of these injuries increasing in the elderly (older than 60 years of age), a population with greater comorbidities, from 4.7 percent in the 1970s to 9 percent in 1990,[1] our overall mortality rate has remained steady at approximately 19 percent for this severely injured patient population. Because we are part of the Model Spinal Cord Injury Systems, the outcomes for our patients can be compared to the National SCI Database.

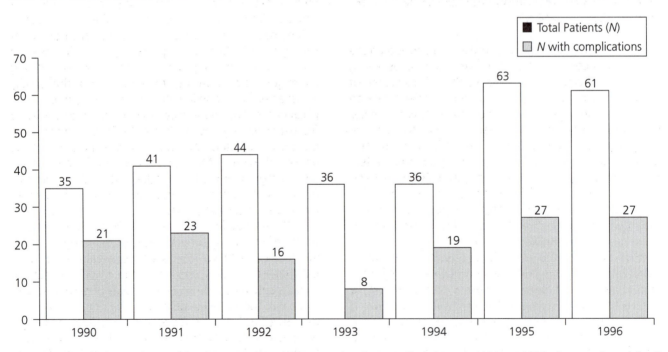

Figure 21–2 Numbers of Cervical Spine–Injured Patients at Harborview Medical Center, 1990 to 1996. *Source:* Copyright © Harborview Medical Center.

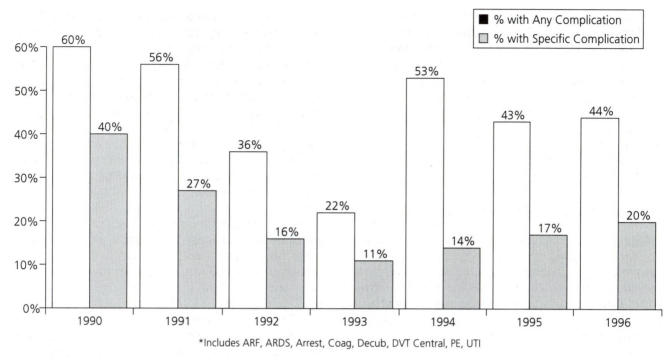

*Includes ARF, ARDS, Arrest, Coag, Decub, DVT Central, PE, UTI

Note: ARG = Acute Renal Failure; ARDS = Adult Respiratory Distress Syndrome; Arrest = Cardiac Arrest; Coag = Coagulation Problems; Decub = Decubitus Ulcers; DVT = Deep-Vein Thrombosis; PE = Pulmonary Embolus; UTI = Urinary Tract Infection.

Figure 21–3 Complication Rate for Cervical Cord Injuries, Harborview Medical Center, 1990 to 1996. *Source:* Copyright © Harborview Medical Center.

COMPARING HARBORVIEW OUTCOMES WITH THE MODEL CENTER DATA

In the development of this pathway, we considered the reduction of complications to be paramount in assessing the quality of care provided. Before the official launch of our pathway program, we had started to collect data on the rate of complications within our patients with cervical spine injury. We recognized early on that for most of these patients, pneumonia was the most common complication, and efforts at reducing such pulmonary infections, pulmonary emboli, and decubitus ulcers have been the focus of our overall care improvement process. With careful evaluation of our program, we have been able to track our success in achieving reductions in other key indicators for our spinal cord population.

The National Institute on Disability and Rehabilitation Research has demonstrated the importance of the Model Spinal Cord Injury Systems. These systems embody a comprehensive, interdisciplinary service delivery in which medicine, rehabilitation, engineering, and the social sciences work with the spinal cord–injured person to achieve maximum potential through individual empowerment, independence, and community integration. The

medical management and rehabilitation of SCI requires complex, interdisciplinary collective action, since acutely injured patients routinely experience multiple trauma resulting from high-impact injuries. Accordingly, they are often prone to develop medical complications involving all major body systems, thus hindering the timely achievement of rehabilitation goals. Among these complications, pulmonary and cardiovascular conditions are a frequent source of morbidity for SCI. These are often the most common cause of death in the acute and chronic phases of injury. HMC has integrated a highly coordinated program of medical surgical and allied health professionals to address the medical and surgical management of such complex cases.

HMC has developed a tracking system to study outcomes of our most complex cases of neurologic impairment. Since 1990, we have seen a gradual and substantial increase in our volume of patients with tetraplegic impairment. When we compared rates of cardiovascular conditions for our SCI patients over the period from 1990 to 1996 (*N* = 271) with those of the Model Systems, we were able to identify areas where we had achieved substantial success in limiting these complications. Specifically, we report an incidence of 5.5 percent for developing deep-venous thrombosis among tetraplegic patients over the seven-year period from 1990 to 1996, compared to 11.9 percent for those with tetraplegia reported by the Model Systems.[7] Similarly, we have substantially reduced our reported rates of pulmonary embolism complications to approximately 1.8 percent compared to 3.9 percent reported by the Model Systems for patients with tetraplegia. As for other cardiovascular conditions, we have reported an incidence of 1.5 percent for myocardial infarction (MI) over the same period, which is comparable to the 1.1 percent complication rate reported by the Model Systems for patients with the same category of neurologic impairment. Cardiopulmonary arrest was a slightly more common complication than MI among patients with cervical SCI at HMC; we report a 5.2 percent incidence compared to 4.5 percent among tetraplegic patients from the Model Systems.

Pulmonary complications remain our most significant problem area for SCI patients, a finding consistent with the findings of the Model Systems, in which higher incidences of pneumonia are reported among patients with the most severe neurologic impairment. The reported incidence of pneumonia over the 1990 to 1996 period is 34.7 percent at HMC, which is lower than the rate reported by the Model Systems for patients with complete tetraplegia (37.8 percent). However, the rates may not be comparable, since we have not split out this complication rate among complete and incomplete tetraplegia. Pneumonia has remained the single most common complication reported each year among SCI patients at HMC, and careful efforts to limit these complications have resulted in a reduction in the prevalence of pneumonia among SCI patients in recent years. Though we have succeeded in reducing our complication rate for pneumonia in recent years, these rates are also associated with a reported increased incidence of SCI among the elderly, who experience a higher incidence of pneumonia and other comorbidities.

CONCLUSION

Cervical SCI is a major life-changing event with catastrophic consequences. These consequences may include the personal impact of limitations in function and role,

the financial impact of disability, and the long-term potential for medical complications associated with the injury. Prompt and immediate provision of coordinated care can have a major impact on long-term outcomes related to this devastating injury.

Functional independence measure total scores can estimate the impact of disability and predict the safety issues and level of dependence on others, as well as technological devices.[8] Developing systems of care that factor in the optimal timing of interventions and striving for outcome achievement within specified time frames can help care providers refine the plan of care. With variance-tracking systems and the use of trauma registry data to help identify system problems in care delivery, health care systems can prioritize the improvement of inefficient processes to achieve the most desirable outcomes.

REFERENCES

1. University of Alabama at Birmingham Model Regional Spinal Cord Injury Care System/National SCI Database Spinal Cord Injury Facts and Figures at a Glance, August 1997. Downloaded August 21, 1997, from the Internet at http://www.spinalcord.rehabm.uab.edu/doc/facts96.htm

2. Nemeth L, Kiljanczyk H. Intensive care of the spinal cord injured: a focus on early rehabilitation. *Crit. Care Nurs Q.* 1988;11:79–84.

3. Bracken MB, Shepard MJ, Holford TR, et al. Administration of Methylprednisolone for 24 or 48 hours or Tirilizad Mesylate for 48 hours in the treatment of acute spinal cord injury: results of the third national acute spinal cord injury randomized controlled trial. *JAMA.* 1997;277:1597–1604.

4. Pierce B, March K. Spine clearance: a standardized approach to assist caregivers. *J Trauma Nurs.* 1996;3:47–50.

5. Champion HR, Copes WS, Sacco WJ, et al. The major trauma outcome study: establishing norms for trauma care. *J Trauma.* 1990;30:1356–1365.

6. Gerber BG, Hebert PC, Wells G, Yelle JD. Validation of trauma and injury severity score in blunt trauma patients by using a Canadian trauma registry. *J Trauma.* 1996;40:733–737.

7. Stover SL, DeLisa JA, Whiteneck GG. *Spinal Cord Injury Clinical Outcome from the Model Systems.* Gaithersburg, MD: Aspen Publishers, Inc; 1995.

8. Maynard FM Jr, Bracken MB, Creasey G, et al. International standards for neurological and functional classification of spinal cord injury. *Spinal Cord.* 1997;35:266–274.

■ Appendix 21–A ■
Clinical Pathway

The following is a key to the medical symbols, abbreviations, and codes used in the exhibit of this appendix:

' = minute
° = hour
✓ = normal
* = abnormal
Ø = not done
• = task intervention
[...] = outcome
AHA = American Heart Asssociation
AROM = active range of motion
AAROM = active assist range of motion
ABG = arterial blood gases
ADL = activities of daily living
APS = adult protective services
ASA = aspirin (acetyl salicylic acid)
Aug./Alt = augmented/alternative
BP = blood pressure
BM = bowel movement
BUE = bilateral upper extremities
C&DB = cough and deep-breathe
cc = cubic centimeter
CBC = complete blood count
CHF = congestive heart failure
CK—MB = Creatine Kinase—Myocardial Band (% of CK)
CMS = circulation, motion, sensation
CNS = clinical nurse specialist
CPS = child protective services
CT = computerized axial tomography
CTA = clear to auscultation
CXR = chest X-ray
d/c = discontinued; discharge
D5½NS = 5½% dextrose in normal saline
DC = discharge
DSHS = Department of Social and Health Services

DVT = deep-vein thrombosis
dx = diagnosis
EBL = estimated blood loss
ECG = electrocardiogram
EOB = edge of bed
ER = emergency room
ETOH = ethyl alcohol
ETT = endotracheal tube
FPIP = facility placement inquiry process
FT = feeding tube
GCS = Glasgow Coma Score
GED = general equivalency diploma
H & P = history and physical
Hct = hematocrit
Hg = mercury
HO = house officer
HR = heart rate
HS = high school
hx = history
ICU = intensive care unit
I/O = input/output
IPPB = intermittent positive pressure breathing
IS = incentive spirometer
ITA = Involuntary Treatment Act
IV = intravenous
KCl = potassium chloride
LB = lower body
LE = lower extremity
M7 = chemistry blood test
MI = myocardial infarction
MR = may repeat
MSO_4 = morphine sulfate
MVI = multivitamins

Na = sodium
NA = not applicable
NC = nasal cannula
NHP = nursing home placement
NPO = nothing by mouth
NSAID = nonsteroidal anti-inflammatory drug
NSR = normal sinus rhythm
NT = nasotracheal
NTG = nitroglycerine
NWB = non–weight-bearing
OBRA = Omnibus Budget Reconciliation Act of 1987
OOB = out of bed
OR = operating room
OT = occupational therapist
PACU = postanesthesia care unit
PCA = patient-controlled analgesia
PCP = primary care provider
PEG = percutaneous endoscopic gastrostomy
PERL = pupils equal and reactive to light
PIC = Medicaid case number
PO = by mouth
prn = as required, when necessary
PROM = passive range of motion
PRS = Pain Relief Service
Pt = patient
PT = prothrombin time
PTT = partial thromboplastin time
Pulm = pulmonary
q = every or each
q.s. = quantity sufficient
Quad = quadriplegic

RA = room air
RD = registered dietitian
RN = registered nurse
RT = respiratory therapist
RTC = round the clock
SaO_2 = oxygen saturation
SCD = sequential compression device
SCI = spinal cord injury
SL = sublingual
SBP = systolic blood pressure
SP = speech pathologist
ST = sinus tachycardia
SW = social work
Sx = symptoms
T = temperature
TBC = total body cultures
TEDs = antiembolic stockings brand name
TF = tube feeding
tx = treatment
UB = upper body
UE = upper extremity
u.o. = urine output
VNS = visiting nurse service
VS = vital signs
VSS = vital signs stable
w/ = with
w/u = workup
WBC = white blood count
WC = wheelchair
WNL = within normal limits

Exhibit 21–A–1 Cervical Spinal Cord Injury Pathway

CERVICAL SPINAL CORD INJURY PATHWAY ICU DAY 1 PRE-STABILIZATION PHASE	
DATE	

Assessment and Monitoring
• VS per ICU standard
• A line
• Cardiac monitoring
[Maintains normothermia]
 	
Cardiovascular
[No bradycardia during suctioning or turning]
• SCDs/TEDs
 	
Neurological
• Neuro ✔ q 1° (GCS)
• SCI assessment q 4°
• Strict spine precautions
[Pt can participate in spinal assessment]
[Pt's GCS = 13–15]
[Neurological status stabilized]
[No changes in sensori-motor assessments]
 	
Integumentary
• Skin assessed q 4° (open rotobed hatch)
• Rotobed in rotation for 20°/24°
• Assess occiput for areas of breakdown
• Foam position wedges placed as needed
• Pin care
[Skin intact]
 	
Respiratory/Airway
• Baseline resp. assessment q 1° and prn
[O₂ Sat ≥92%]
[Lungs clear]
[Spontaneously breathing]
• Quad cough q 4° prn
• IS q 2°
[VC ≥1000cc]
• RT assess q am/pm
  • Tidal volume
  • Vital Capacity
  • Inspiratory effort
  • Rate
• Provide ventilatory support as needed
• IPPB as per RT parameters
• NT suction PRN with hyperoxygenation
Medications
• Stress ulcer prophylaxis   • Bowel prog
• MSO₄   • SCI study drugs
• Ativan   • Electrolytes prn
• Atropine prn   • Reglan
 	
Pain Management
• Pain scale q 4°
[Pt obtains a self-identified acceptable level of pain relief]
Fluid/Volume
• Maintain volume nomotensive
• D5½NS@125cc/hr
Nutrition
• RN: Current nutrition support
  Stop rotobed to side for meals
• RD: Nutrition assessment w/in 24–72° of admit.
 	
[Pt receiving nutrition support within 24–72° of admit]
	Consults/Diagnostic Studies
	• Pulm. Med • Social Work
	• Spine team • Nutrition
	• SCI Study RN • Speech Pathology
	• Rehab Medicine • ABG if intubated
	• Occupational Therapy • M7, Hct, WBC
	• Physical Therapy
	• Therapeutic Recreation (for pediatric pts)
	
	Elimination/Reproductive
	[Bowel elimination as per baseline]
	• Bowel tones q 4°
	• Bowel program
	  • Colace
	  • Metamucil
	  • Ducolax
	    (followed by digstim q 15 min until results)
	
	[Abd. soft]
	[Bowel tones ⊕]
	
	[Bladder elimination as per baseline]
	• Foley catheter
	
	Activity/Safety
	• Spine precautions
	• Restraints to protect tubes
	• Footplates on/off q 2°
	• OT to assess positioning on rotobed
	• OT initiate PROM BUE's (except shoulders)
	
	Communication
	<u>Vent-Dep. Pts.</u>
	• SP: Assess communication w/in 48° if intubated
	• SP: Provide Aug./Alt commun. system as indicated
	
	[Pt is able to communicate for assistance when needed]
	[Pt is able to communicate basic needs]
	
	Psychosocial/Emotional
	• Provide emotional support to patient/family
	
	Patient/Family Education
	• Trauma Rehab Coordinator notified

ADDRESSOGRAPH **Pathways reproduced with special permission. All rights reserved.**	**UNIVERSITY OF WASHINGTON MEDICAL CENTERS** Harborview Medical Center—UW Medical Center Seattle, Washington **CLINICAL PATHWAY—CERVICAL SPINAL CORD INJURY** 1

continues

Exhibit 21–A–1 continued

CERVICAL SPINAL CORD INJURY PATHWAY ICU DAY 2 PRE-STABILIZATION PHASE	
DATE	

Assessment and Monitoring
• VS per ICU standard
• A line
• Cardiac monitoring
[Maintains normothermia]
 	
Cardiovascular
[No bradycardia during suctioning or turning]
• SCDs/TEDs
 	
Neurological
• Neuro ✔ q 1° (GCS)
• SCI assessment q 4°
• Strict spine precautions
[Pt can participate in spinal assessment]
[Pt's GCS = 13–15]
 	
Integumentary
• Skin assessed q 4° (open rotobed hatch)
• Rotobed in rotation for 20°/24°
• Assess occiput for areas of breakdown and need for special cushion or support to be acquired from KCI
• Foam position wedges placed as needed
• Pin care
[Skin intact]
 	
Respiratory/Airway
• Baseline resp. assessment q 1° and prn
[O₂ Sat ≥92%]
[Lungs clear]
[Spontaneously breathing]
• Quad cough q 4° prn
• IS q 2°
[VC ≥ 1000cc]
• RT assess q am/pm
  • Tidal volume
  • Vital Capacity
  • Inspiratory effort
  • Rate
• Provide ventilatory support as needed
• IPPB as per RT parameters
• NT suction PRN with hyperoxygenation
 	
Medications
• Methylprednisolone per standard   • Bowel prog
• Stress ulcer prophylaxis   • SCI study drugs
• MSO₄   • Electrolytes prn
• Ativan   • Reglan
• Atropine prn
 	
Pain Management
• Pain scale q 4°
[Pt obtains a self-identified acceptable level of pain relief]
 	
Fluid/Volume
• Maintain volume nomotensive
• D5½NS@125cc/hr
 	
Nutrition
• RN: Current nutrition support
  Stop rotobed to side for meals
[Pt receiving nutrition support within 24–72° of admit]
	Consults/Diagnostic Studies
	• Coag Studies
	• RN to notify any disciplines, not previously up for consult
	
	Elimination/Reproductive
	[Bowel elimination as per baseline]
	• Bowel tones q 4°
	• Bowel program
	  • Colace
	  • Metamucil
	  • Ducolax
	  (followed by digstim q 15 min until results)
	[Abd. soft]
	[Bowel tones ⊕]
	
	[Bladder elimination as per baseline]
	• Foley catheter
	
	Activity/Safety
	• Spine precautions
	• Footplates on/off q 2°
	• Restraints to protect tubes
	
	Communication
	• OT: assess call light and instruct and recommend needs to nursing
	
	  Equipment required _____
	
	• SP: Assess communication w/in 48° if intubated
	• SP: Reassess Augmented/Altered communication system as indicated
	
	[Pt is able to communicate for assistance when needed]
	[Pt is able to communicate basic needs]
	
	Psychosocial/Emotional
	• Provide emotional support to pt/family
	[Coping effectively]
	
	Patient/Family Education
	• Spinal cord injury packet given to pt/family
	• Plan for stabilization of spine discussed with pt/family by MD

ADDRESSOGRAPH Pathways reproduced with special permission. All rights reserved.	**UNIVERSITY OF WASHINGTON MEDICAL CENTERS** Harborview Medical Center—UW Medical Center Seattle, Washington **CLINICAL PATHWAY—CERVICAL SPINAL CORD INJURY** 2

continues

Exhibit 21–A–1 continued

OR/POST-OP *OR* SPINE STABILIZATION	
DATE	

Assessment and Monitoring
- VS per unit standard
- A line
- Cardiac monitoring

[Maintains normothermia] or *[Fever w/u, source identified, tx started]*
- Daily wt

Cardiovascular
[No bradycardia during suctioning or turning]
[No sign of dysreflexia]

- Monitor for orthostasis
[Normotensive when OOB]

Neurological
- Neuro ✔ q 1°
- SCI assessment q 4°

[Alert and oriented]
[Baseline neuro status maintained or improved]

Integumentary
- Regular hospital bed
- Skin assessed, repositioned q 2°
- Reverse trendelenberg
- Braces & specific care, assessing tolerance q 4°
- External orthosis

[Skin integrity intact]
[Dressing dry and intact]

- If any sign of skin breakdown, consult Burn/Plastics CNS

Respiratory/Airway
- Portable vent ordered

[SaO$_2$ ≥92%]
[Lungs CTA]

Medications
- Stress ulcer prophylaxis
- MSO$_4$
- Ativan
- Atropine prn
- Bowel prog
- SCI study drugs
- Electrolytes prn
- Reglan
- Heparin

Pain Management
- Pain scale q 4°

[Pt obtains a self-identified acceptable level of pain relief]

Consults/Diagnostic Studies
- TBC Temp > 38.5
- M7, Hct, WBC prn

Fluid/Volume
- IV fluids to maintain normotensive
- D5½NS with 20 mEq KCl/liter@125cc/hr

Nutrition
- NPO
- If anterior approach fusion, insert feeding tube in OR
- If pt to receive trach., evaluate if PEG appropriate

Elimination/Reproductive
- Bowel tones q 4°
- Bowel program
  - Colace
  - Metamucil
  - Ducolax
    (followed by digstim q 15 min until results)

[Abd. soft]
[Bowel tones ⊕]

[Bladder elimination as per baseline]
- Foley catheter

Activity/Safety
- OR position protocol
- Fitted for appropriate orthosis

Psychosocial/Emotional
- Provide updates to family
- MD to speak with family post-op

ADDRESSOGRAPH	UNIVERSITY OF WASHINGTON MEDICAL CENTERS
Pathways reproduced with special permission. All rights reserved.	Harborview Medical Center—UW Medical Center Seattle, Washington **CLINICAL PATHWAY—CERVICAL SPINAL CORD INJURY** 3

continues

Exhibit 21–A–1 continued

STABILIZATION PHASE (POD #1)	
DATE _____ POD #1	

Assessment and Monitoring
- VS per unit standard

[Maintains normothermia] or *[Fever w/u, source identified, tx started]*
- Daily wt

Cardiovascular
[No bradycardia during suctioning or turning]
[No sign of dysreflexia]

- Monitor for orthostasis
[Normotensive when OOB]

Neurological
- Neuro ✔ q 1°
- SCI assessment q 4°

[Spines cleared by primary service]
[Alert and oriented]
[Baseline neuro status maintained or improved]

Integumentary
- Regular hospital bed
- Skin assessed, repositioned q 2°
- Reverse trendelenberg
- Braces & specific care, assessing tolerance q 4°
- Build skin tolerance; OOB time of 15'; ↑ 15' per day
- Skin check when returned to bed
- Pressure release q 15' when up
- External orthosis
- If any sign of skin breakdown, consult Burn/Plastics CNS

[Skin intact after sitting _____ minutes]
[All other areas of skin intact]
[Dressing dry and intact]

Respiratory/Airway
- Resp. assessment q 1° and prn
- Quad cough, IS q 1–2°
- IPPB per RT parameters
[SaO$_2$ ≥92%]
[Lungs CTA]

Medications
- Stress ulcer prophylaxis
- MSO$_4$
- Ativan
- Atropine prn
- Bowel prog
- SCI study drugs
- Electrolytes prn
- Reglan
- MVI with minerals if PO diet
- Heparin

Pain Management
- Pain scale q 4°
[Pt obtains a self-identified acceptable level of pain relief]

Consults/Diagnostic Studies
- Upright C-spine before OOB
- M7, Hct, WBC prn
- TBC Temp > 38.5

Fluid/Volume
- IV fluids d/c'ed
- Saline lock if adequate fluid intake PO/TF
- If tube feeding, assess need for supplemental fluids via FT
[Pt adequately hydrated]

Nutrition
- SP: Swallow eval (after extubation/trach), esp. if s/p anterior fusion

[Pt on a modified p.o. diet]

- RN: Current nutrition support _____
  _____
  - Insert FT if ordered
  - Encourage OOB for meals
- RD: Assess as needed for adequacy of intake.
  - Order PO diet _____
  - Order MVI w/minerals
  If vent. dependent:
  - Order tube feeding:
    Formula _____
    Goal Rate _____
- Start at 40cc, progress in 20cc increments to goal rate.

Elimination/Reproductive
[Bowel elimination as per baseline]
- Bowel tones q 4°
- Bowel program
  - Colace
  - Metamucil
  - Ducolax
    (followed by digstim q 15 min until results)

[Abd. soft]
[⊕ bowel tones]

[Bladder elimination as per baseline]
- Foley catheter

Communication
- OT reassess effectiveness of call light system

Equipment changes: _____

- SP: Reassess Aug./Alt commun. system as indicated

[Pt is able to communicate for assistance when needed]
[Pt is able to communicate basic needs]

Psychosocial/Emotional
- Provide update regarding progress to pt/family
[Coping effectively]

Patient/Family Education
- Spinal cord injury packet reviewed as needed with pt/family
- Plan for stabilization of spine discussed with pt/family by MD

ADDRESSOGRAPH	UNIVERSITY OF WASHINGTON MEDICAL CENTERS
	Harborview Medical Center—UW Medical Center Seattle, Washington **CLINICAL PATHWAY—CERVICAL SPINAL CORD INJURY** 4

continues

Exhibit 21–A–1 continued

STABILIZATION PHASE (POD #1)

DATE

SW Critical Care Screen
- Rehab/Discharge Planning

Needs Funding:	Y ❑	N ❑	Unknown ❑
Needs Rehab/Discharge Coordination	Y ❑	N ❑	Unknown ❑

- Legal Problems

Unidentified Pt	Y ❑	
Locate next of kin	Y ❑	N ❑

- Pt w/regulatory requirements   Y ❑   N ❑   APS ❑   CPS ❑   Jail Hold ❑
- Pt/family in crisis   Y ❑   N ❑

SW Psychosocial Assessment
SW Assessment
- Pre-injury living situation
  - ❑ Lives alone
  - ❑ Lives w others _____
  - ❑ Health Care Facility _____
  - ❑ Other _____
- Family Relationships
  - ❑ Adequate   ❑ Dysfunctional   ❑ Stressed   ❑ None
- Social Support
  - ❑ Adequate   ❑ Marginal   ❑ None   ❑ Unknown
- Pre-injury functioning
  - ❑ Independent
  - ❑ Required Assistance _____
  - Psychiatric Hx   Y ❑   N ❑ _____
- Coping
  - ❑ Pt WNL   ❑ Unable to Assess   ❑ Needs Assistance
  - ❑ Family WNL   ❑ Needs Assistance
- ETOH/Substance Abuse

Assessment Completed	Y ❑	N ❑	Unable to Assess ❑
Significant Hx	Y ❑	N ❑	
Treatment recommended	Y ❑	N ❑	

- Legal

Guardianship	Y ❑	N ❑	Needed ❑
Power of Atty	Y ❑	N ❑	Needed ❑

- Highest Formal Educational Level
  - ❑ < High school diploma   ❑ HS diploma or GED   ❑ College Degree
  - ❑ Other _____   ❑ Unknown
- Vocational Status
  - ❑ Employed   ❑ Unemployed   ❑ Student   ❑ Retired
  - ❑ Other _____   ❑ Unknown

Discharge Planning Assessment
- Financial
  - Needs Funding   Y ❑   N ❑
  - ❑ Private Insurance _____   ❑ Labor and Industries
  - ❑ Medicare   ❑ Medicaid Pending   Y ❑   N ❑   Other _____
  - Case Mgr. Y ❑   N ❑ _____
- Social Security Disability/Supplemental Security Income application initiated   Y ❑   N ❑
- Needs Rehab/D/C Coordination _____   Y ❑   N ❑
- Does Pt have funding for:Rehab _____   Y ❑   N ❑

Sub-Acute	Y ❑	N ❑	NA ❑	Nursing Home	Y ❑	N ❑	NA ❑

- Discuss Rehab/Discharge options with pt/family:
  - If NHP or Sub-Acute required complete OBRA & FPIP   Not needed ❑
  - If Medicaid pt, contact DSHS screener regarding anticipated d/c date
  - DSHS screen completed
- Pt/family selected Rehab/alternate placement @ _____

- **SW Signature** _____

ADDRESSOGRAPH	**UNIVERSITY OF WASHINGTON MEDICAL CENTERS**
	Harborview Medical Center—UW Medical Center
Pathways reproduced with special permission. All rights reserved.	Seattle, Washington
	CLINICAL PATHWAY—CERVICAL SPINAL CORD INJURY 5

continues

Exhibit 21–A–1 continued

STABILIZATION PHASE (POD #2)	
DATE	

Assessment and Monitoring
- VS per unit standard

[Maintains normothermia] or *[Fever w/u, source identified, tx started]*
- Daily wt

Cardiovascular
[No bradycardia during suctioning or turning]
[No sign of dysreflexia]
- Monitor for orthostasis
[Normotensive when OOB]

Neurological
- Neuro ✔ q 1°
- SCI assessment q 4°
[Alert and oriented]
[Baseline neuro status maintained or improved]

Integumentary
- Regular hospital bed
- Skin assessed, repositioned q 2°
- Reverse trendelenberg
- Braces & specific care, assessing tolerance q 4°
- Build skin tolerance; OOB time of 15'; ↑ 15' per day
- Skin check when returned to bed
- Pressure release q 15' when up
- External orthosis
- If any sign of skin breakdown, consult Burn/Plastics CNS
[Skin intact after sitting _____ minutes]
[All other areas of skin intact]
[Dressing dry and intact]

Respiratory/Airway
- Resp. assessment q 1° and prn
- Quad cough, IS q 1–2°
- IPPB per RT parameters
[SaO$_2$ ≥92%]
[Lungs CTA]

Medications
- Stress ulcer prophylaxis
- MSO$_4$
- Ativan
- Atropine prn
- Bowel prog
- SCI study drugs
- Electrolytes prn
- Reglan
- MVI with minerals if PO diet
- Heparin

Pain Management
- Pain scale q 4°
[Pt obtains a self-identified acceptable level of pain relief]

Consults/Diagnostic Studies
- M7, Hct, WBC prn
- TBC Temp > 38.5

Fluid/Volume
[Pt adequately hydrated]

Nutrition
- SP: Swallow eval (after extubation/trach), esp. if s/p anterior fusion

- SP: Continue swallow mgmt.

[Pt on a modified p.o. diet]
- RN: Current nutrition support _____
  - Insert FT if ordered
  - Encourage OOB for meals
- RD: Assess as needed for adequacy of intake.
- OT evaluate for adaptive equipment needs for self feeding

Elimination/Reproductive
[Bowel elimination as per baseline]
- Bowel tones q 4°
- Bowel program
  - Colace
  - Metamucil
  - Ducolax
    (followed by digstim q 15 min until results)

[Abd. soft]
[⊕ bowel tones]

[Bladder elimination as per baseline]
- Foley catheter

Communication
- OT to reassess call light system
- SP: Reassess Aug./Alt commun. system as indicated
[Pt is able to communicate for assistance when needed]
[Pt is able to communicate basic needs]

Activity/Safety
Precautions:

_____
_____

- OT reassess bed positioning and splint needs
- OT evaluate oral hygiene, grooming. Communicate with nursing re patient adaptive equipment needs
- OT evaluate self feeding (when tolerating 15 min in WC). Communicate with nursing re patient adaptive equipment needs
- OT to perform UE strengthening activities
- OT to provide progressive ADL training with adaptations as needed

- PT—assess patient and provide ongoing treatment for:
  - P/AA/ROM
  - Bed mobility
  - Sitting, balance training
- PT mobilize to wheelchair
- PT assist nsg with patient to cardiac chair with cushion, etc

Equipment
- PT measure and order solid seat wheelchair
- PT issues sliding board for patient transfers

Psychosocial/Emotional
[Coping effectively]
- If not rehab psych consult

Patient/Family Education
- Reinforce educational packet materials
- Answer questions as needed

ADDRESSOGRAPH	UNIVERSITY OF WASHINGTON MEDICAL CENTERS
	Harborview Medical Center—UW Medical Center Seattle, Washington **CLINICAL PATHWAY—CERVICAL SPINAL CORD INJURY** 6

continues

Exhibit 21–A–1 continued

STABILIZATION PHASE (POD #3)		
DATE		

Assessment and Monitoring
- VS per unit standard

[Maintains normothermia]

or

[Fever w/u, source identified, tx started]
- Daily wt

Cardiovascular

[No bradycardia during suctioning or turning]
[No sign of dysreflexia]
- Monitor for orthostasis

[Normotensive when OOB]

Neurological
- Neuro ✔ q 1°
- SCI assessment q 4°

[Alert and oriented]
[Baseline neuro status maintained or improved]

Integumentary
- Regular hospital bed
- Skin assessed, repositioned q 2°
- Reverse trendelenberg
- Braces & specific care, assessing tolerance q 4°
- Build skin tolerance; OOB time of 15'; ↑ 15' per day
- Skin check when returned to bed
- Pressure release q 15' when up
- External orthosis
- If any sign of skin breakdown, consult Burn/Plastics CNS

[Skin intact after sitting _____ minutes]
[All other areas of skin intact]
[Dressing dry and intact]

Respiratory/Airway
[Respiratory status stable]
- Evaluate intubated pt's need for trach
- Resp. assessment q 1° and prn
- Quad cough, IS q 1–2°
- IPPB per RT parameters
[SaO$_2$ ≥92%]
[Lungs CTA]

Medications
- Stress ulcer prophylaxis
- MSO$_4$
- Ativan
- Atropine prn
- Bowel prog
- SCI study drugs
- Electrolytes prn
- Reglan
- MVI with minerals if PO diet
- Heparin

Pain Management
- Pain scale q 4°

[Pt obtains a self-identified acceptable level of pain relief]

Consults/Diagnostic Studies
- M7, Hct, WBC prn
- TBC Temp > 38.5

Fluid/Volume
[Pt adequately hydrated]

Nutrition
- SP: Swallow eval (after extubation/trach), esp. if s/p anterior fusion

- SP: Cont. swallow mgmt.
[Pt on a modified p.o. diet]
- RN: Current nutrition support _____

- Insert FT if ordered
- Encourage OOB for meals
- RD: Assess as needed for adequacy of intake.
[Pt receiving 100% of estimated nutritional needs]

Elimination/Reproductive
[Bowel elimination as per baseline]
- Bowel tones q 4°
- Bowel program
  - Colace
  - Metamucil
  - Ducolax
    (followed by digstim q 15 min until results)

[Abd. soft]
[⊕ bowel tones]

[Bladder elimination as per baseline]
- Foley catheter

Communication
- OT to reassess effectiveness of call light system

Equipment required: _____

- SP: Reassess Aug./Alt commun. system as indicated
[Pt is able to communicate for assistance when needed]
[Pt is able to communicate basic needs]

Activity/Safety
- OT recheck bed positioning and splint needs
- OT to perform UE strengthening activities
- OT to provide progressive ADL training with adaptations as needed

[PROM of all joints WNL and no deformity]

[Orthosis in place & properly fitting]

[Pt positioned well in bed/wc]
[Sitting tolerance of 30–60 min]

Mobility
- Pt requires _____ @ for bed ⇔ w/c with sliding board
- ICU bed
- RN/PT bed mobility
- Mobility
  - Rolling _____ @
  - Sup ⇔ sit _____ @
  - Bed ⇔ w/c _____ @
  - Static sit bal _____ @

Psychosocial/Emotional
[Coping effectively]

Patient/Family Education
- Reinforce educational packet materials
- Answer questions as needed

UNIVERSITY OF WASHINGTON MEDICAL CENTERS
Harborview Medical Center—UW Medical Center
Seattle, Washington
CLINICAL PATHWAY—CERVICAL SPINAL CORD INJURY 7

continues

Exhibit 21–A–1 continued

STABILIZATION PHASE (POD #4)		
DATE _____		

Assessment and Monitoring
- VS per unit standard
[Maintains normothermia]
or
[Fever w/u, source identified, tx started]
- Daily wt

Cardiovascular
[No bradycardia during suctioning or turning]
[No sign of dysreflexia]

- Monitor for orthostasis
[Normotensive when OOB]

Neurological
- Neuro ✔ q 1° (GCS)
- SCI assessment q 4°
[Alert and oriented]
[Baseline neuro status maintained or improved]

Integumentary
- Regular hospital bed
- Skin assessed, repositioned q 2°
- Reverse trendelenberg
- Braces & specific care, assessing tolerance q 4°
- Build skin tolerance; OOB time of 15'; ↑ 15' per day
- Skin check when returned to bed
- Pressure release q 15' when up
- External orthosis
- If any sign of skin breakdown, consult Burn/ Plastics CNS
[Skin intact after sitting _____ minutes]
[All other areas of skin intact]
[Dressing dry and intact]

Respiratory/Airway
- Resp. assessment q 1° and prn
- Quad cough, IS q 1–2°
- IPPB per RT parameters
[SaO$_2$ ≥ 92%]
[Lungs CTA]

Medications
- Stress ulcer prophylaxis
- MSO$_4$
- Ativan
- Atropine prn
- Bowel prog
- SCI study drugs
- Electrolytes prn
- Reglan
- MVI with minerals if PO diet
- Heparin

Pain Management
- Pain scale q 4°
[Pt obtains a self-identified acceptable level of pain relief]

Consults/Diagnostic Studies
- TBC Temp > 38.5
- M7, Hct, WBC prn

Fluid/Volume
[Pt adequately hydrated]

Nutrition
- SP: Swallow eval (after extubation/trach), esp. if s/p anterior fusion

- SP: Cont. swallow mgmt.

[Pt on a modified p.o. diet]
- RN: Current nutrition support _____
  _____

  - Insert FT if ordered
  - Encourage OOB for meals
  RD: Assess as needed for adequacy of intake.
[Pt receiving 100% of estimated nutritional needs]

Elimination/Reproductive
[Bowel elimination as per baseline]
- Bowel tones q 4°
- Bowel program
  - Colace
  - Metamucil
  - Ducolax
    (followed by digstim q 15 min until results)

[Abd. soft]
[⊕ bowel tones]

[Bladder elimination as per baseline]
- Foley catheter

Communication
  SP: Reassess Aug./Alt commun. system as indicated
[Pt is able to communicate for assistance when needed]
[Pt is able to communicate basic needs]

Activity/Safety
- PT start family training re: PROM and tilt backs
- PT—assess patient and provide ongoing treatment for:
  - P/AA/ROM
  - Bed mobility
  - Sitting, balance training
- PT mobilize to wheelchair
[PROM of all joints WNL and no deformity]

[Orthosis in place & properly fitting]

[Pt positioned well in bed/wc]
[Sitting tolerance of 30–60 min]

- OT reassess bed and/or out of bed positioning and splint needs
- OT recheck self feeding and communicate with nursing re appropriate assist plan

Psychosocial/Emotional
- Social Work continues to monitor pt/family coping as needed
- Provide update regarding progress to pt/family
[Coping effectively]

Patient/Family Education
- Reinforce educational packet materials
- Answer questions as needed

ADDRESSOGRAPH

UNIVERSITY OF WASHINGTON MEDICAL CENTERS
Harborview Medical Center—UW Medical Center
Seattle, Washington
CLINICAL PATHWAY—CERVICAL SPINAL CORD INJURY   8

continues

Exhibit 21–A–1 continued

STABILIZATION PHASE (POD #5)		
DATE _____		

Assessment and Monitoring
- VS per unit standard

[Maintains normothermia]

or

[Fever w/u, source identified, tx started]
- Daily wt

Cardiovascular

[No bradycardia during suctioning or turning]

[No sign of dysreflexia]

- Monitor for orthostasis

[Normotensive when OOB]

Neurological
- Neuro ✔ q 1°
- SCI assessment q 4°

[Alert and oriented]

[Baseline neuro status maintained or improved]

Integumentary
- Regular hospital bed
- Skin assessed, repositioned q 2°
- Reverse trendelenberg
- Braces & specific care, assessing tolerance q 4°
- Build skin tolerance; OOB time of 15'; ↑ 15' per day
- Skin check when returned to bed
- Pressure release q 15' when up
- External orthosis
- If any sign of skin breakdown, consult Burn/Plastics CNS

[Skin intact after sitting _____ minutes]

[All other areas of skin intact]

[Dressing dry and intact]

Respiratory/Airway
- Resp. assessment q 1° and prn
- Quad cough, IS q 1–2°
- IPPB per RT parameters
- SP: Begin GPB training (nonventilator dependent pts)

Vent-Dep. Pts w Trachs
- RT: Cuff deflation trials as tolerated w SP
- RT: ↑ cuff deflation trn. for commun

[Tolerates cuff deflation 15–30 min] (If no trach, N/A)

[SaO_2 ≥92%]

[Lungs CTA]

Medications
- Stress ulcer prophylaxis
- MSO_4
- Ativan
- Atropine prn
- Bowel prog
- SCI study drugs
- Electrolytes prn
- Reglan
- MVI with minerals if PO diet
- Heparin

Pain Management
- Pain scale q 4°

[Pt obtains a self-identified acceptable level of pain relief]

Consults/Diagnostic Studies
- Upright C-spine
- TBC Temp > 38.5
- M7, Hct, WBC prn

Fluid/Volume

[Pt adequately hydrated]

Nutrition
- SP: Swallow eval (after extubation/trach), esp. if s/p anterior fusion
- SP: Cont. swallow mgmt.

[Pt on a modified p.o. diet]
- RN: Current nutrition support _____

_____
- Insert FT if ordered
- Encourage OOB for meals

RD: Assess as needed for adequacy of intake.

[Pt receiving 100% of estimated nutritional needs]

Elimination/Reproductive

[Bowel elimination as per baseline]
- Bowel tones q 4°
- Bowel program
  - Colace
  - Metamucil
  - Ducolax
    (followed by digstim q 15 min until results)

[Abd. soft]

[⊕ bowel tones]

[Bladder elimination as per baseline]
- Foley catheter

Communication

Vent-Dep Pts w ETT Tubes

SP: Cuff deflation trials as tolerated w RT

[Pt tolerates cuff deflation 15–30 min for communication]

SP: Reassess Aug./Alt commun. system as indicated

[Pt is able to communicate for assistance when needed]

[Pt is able to communicate basic needs]

Activity/Safety
- OT reassess bed and/or out of bed positioning and splint needs
- OT recheck self feeding and communicate with nursing re appropriate assist plan

[PROM of all joints WNL and no deformity]

[Orthosis in place & properly fitting]

[Pt positioned well in bed/wc]

[Sitting tolerance of 30–60 min]

- PT—assess patient and provide ongoing treatment for:
  - P/AA/ROM
  - Bed mobility
  - Sitting, balance training
- PT mobilize to wheelchair

Psychosocial/Emotional
- Social Work continues to monitor pt/family coping as needed
- Provide update regarding progress to pt/family

[Coping effectively]

Patient/Family Education
- Reinforce educational packet materials
- Answer questions as needed

ADDRESSOGRAPH	UNIVERSITY OF WASHINGTON MEDICAL CENTERS
	Harborview Medical Center—UW Medical Center
Pathways reproduced with special permission. All rights reserved.	Seattle, Washington
	CLINICAL PATHWAY—CERVICAL SPINAL CORD INJURY 9

continues

Exhibit 21–A–1 continued

STABILIZATION PHASE (POD #6)		
DATE _____		

Assessment and Monitoring
- VS per unit standard

[Maintains normothermia]

or

[Fever w/u, source identified, tx started]
- Daily wt

Cardiovascular

[No bradycardia during suctioning or turning]

[No sign of dysreflexia]

- Monitor for orthostasis

[Normotensive when OOB]

Neurological
- Neuro ✔ q 1°
- SCI assessment q 4°

[Alert and oriented]

[Baseline neuro status maintained or improved]

Integumentary
- Regular hospital bed
- Skin assessed, repositioned q 2°
- Reverse trendelenberg
- Braces & specific care, assessing tolerance q 4°
- Build skin tolerance; OOB time of 15'; ↑ 15' per day
- Skin check when returned to bed
- Pressure release q 15' when up
- External orthosis
- If any sign of skin breakdown, consult Burn/Plastics CNS

[Skin intact after sitting _____ minutes]

[All other areas of skin intact]

[Dressing dry and intact]

Respiratory/Airway
- Resp. assessment q 1° and prn
- Quad cough, IS q 1–2°
- IPPB per RT parameters

Vent-Dep. Pts w Trachs
- RT: Cuff deflation trials as tolerated w SP
- RT: ↑ cuff deflation trn. for commun

[Tolerates cuff deflation 15–30 min] (If no trach, N/A)

[SaO2 ≥92%]

[Lungs CTA]

Medications
- Stress ulcer prophylaxis
- MSO4
- Ativan
- Atropine prn
- Bowel prog
- SCI study drugs
- Electrolytes prn
- Reglan
- MVI with minerals if PO diet
- Heparin

Pain Management
- Pain scale q 4°

[Pt obtains a self-identified acceptable level of pain relief]

Consults/Diagnostic Studies
- TBC Temp > 38.5
- M7, Hct, WBC prn

Fluid/Volume

[Pt adequately hydrated]

Nutrition
- SP: Swallow eval (after extubation/trach), esp. if s/p anterior fusion
- SP: Cont. swallow mgmt.

[Pt on a modified p.o. diet]
- RN: Current nutrition support _____

_____
- Insert FT if ordered
- Encourage OOB for meals
RD: Assess as needed for adequacy of intake.

[Pt receiving 100% of estimated nutritional needs]

Elimination/Reproductive

[Bowel elimination as per baseline]
- Bowel tones q 4°
- Bowel program
  - Colace
  - Metamucil
  - Ducolax
    (followed by digstim q 15 min until results)

[Abd. soft]

[⊕ bowel tones]

[Bladder elimination as per baseline]
- Foley catheter

Communication
SP: Reassess Aug./Alt commun. system as indicated

[Pt is able to communicate for assistance when needed]

[Pt is able to communicate basic needs]

Vent-Dep Pts w ETT Tubes
SP: Cuff deflation trials as tolerated w RT

[Pt tolerates cuff deflation 15–30 min for communication]

Activity/Safety
Precautions:

_____

_____
- OT to reassess/provide with splinting and positioning as indicated
- OT to assess self feeding when tolerating 15 min in w/c
  ❑ Dependent
  ❑ With setup
  ❑ Use of adaptive devices and

_____ assist
- OT to perform UE strengthening
- OT to provide progressive ADL training with adaptations as needed

- PT continues ongoing treatment
  - P/AA/AROM
  - Bed mobility
  - Sitting, balance training
  - Mobilize to wc
  - Family training as appropriate
    - PROM
    - Tilt backs

[PROM of all joints WNL and no deformity]

[Orthosis in place & properly fitting]

[Pt positioned well in bed/wc]

[Sitting tolerance of 30–60 min]

Mobility
- Pt requires _____ @ for bed ⇔ w/c with sliding board
- ICU bed
- RN/PT bed mobility
- Mobility
  - Rolling _____ @
  - Sup ⇔ sit _____ @
  - Bed ⇔ w/c _____ @
  - Static sit bal _____ @

Psychosocial/Emotional
- Social Work continues to monitor pt/family coping as needed
- Provide update regarding progress to pt/family

[Coping effectively]

Patient/Family Education
- Prepare for transfer to rehab

ADDRESSOGRAPH

UNIVERSITY OF WASHINGTON MEDICAL CENTERS
Harborview Medical Center—UW Medical Center
Seattle, Washington
CLINICAL PATHWAY—CERVICAL SPINAL CORD INJURY   10

continues

Exhibit 21–A–1 continued

STABILIZATION PHASE (POD #7)		
DATE _____		

Assessment and Monitoring
[Stable VS]
[Afebrile]

Cardiovascular
[Cardiovascular system w/in expected norms]
[No active DVT]

[Normotensive when OOB]

Neurological
• Neuro ✔ q 1°
• SCI assessment q 4°
[Alert and oriented]
[Baseline neuro status maintained or improved]

Integumentary
[Skin integrity is intact]
[Skin intact after sitting _____ minutes]
[All other areas of skin intact]

Respiratory/Airway
<u>Vent-Dep. Pts w Trachs</u>
• RT: Cuff deflation trials as tolerated w SP
• RT: ↑ cuff deflation trn. for commun
[Tolerates cuff deflation 15–30 min] (If no trach, N/A)
[Pulmonary status stable]

Medications
• Stress ulcer prophylaxis
• MSO$_4$
• Ativan
• Atropine prn
• Bowel prog
• SCI study drugs
• Electrolytes prn
• Reglan
• MVI with minerals if PO diet
• Heparin

Pain Management
• Pain scale q 4°
[Pt obtains a self-identified acceptable level of pain relief]

Consults/Diagnostic Studies
• TBC Temp > 38.5
• M7, Hct, WBC prn

Fluid/Volume
[Pt adequately hydrated]

Nutrition
• SP: Swallow eval (after extubation/trach), esp. if s/p anterior fusion
• SP: Cont. swallow mgmt.
[Pt on a modified p.o. diet]
• RN: Current nutrition support _____

  • Insert FT if ordered
  • Encourage OOB for meals
• RD: Assess as needed for adequacy of intake. Recommendations in progress notes: _____

[Pt receiving 100% of estimated nutritional needs]

Elimination/Reproductive
[Bowel elimination as per baseline]
• Bowel tones q 4°
• Bowel program
  • Colace
  • Metamucil
  • Ducolax
    (followed by digstim q 15 min until results)

[Abd. soft]
[⊕ bowel tones]

[Bladder elimination as per baseline]
• Foley catheter

Communication
<u>Vent-Dep Pts w ETT Tubes</u>
  SP: Cuff deflation trials as tolerated w RT
[Pt tolerates cuff deflation 15–30 min for communication]

  SP: Reassess Aug./Alt commun. system as indicated
[Pt is able to communicate for assistance when needed]
[Pt is able to communicate basic needs]

Activity/Safety
Precautions:
_____
_____
_____
• OT to reassess/provide with splinting and positioning as indicated
• Positioning Devices:
• OT to assess UE function:

[PROM of all joints WNL and no deformity]
[Strength and ROM maximized at all innervated levels]

• OT to assess self feeding when tolerating 15 min in w/c
  ❑ Dependent
  ❑ With setup
  ❑ Use of adaptive devices and _____ assist
• OT to perform UE strengthening
• OT to provide progressive ADL training with adaptations as needed

• PT continues ongoing treatment
  • P/AA/AROM
  • Bed mobility
  • Sitting, balance training
  • Mobilize to wc
  • Family training as appropriate
    • PROM
    • Tilt backs

[PROM of all joints WNL and no deformity]
[Orthosis in place & properly fitting]
[Pt positioned well in bed/wc]
[Sitting tolerance of 30–60 min]

Mobility
• Pt requires _____ @ for bed ⇔ w/c with sliding board
• ICU bed
• RN/PT bed mobility
• Mobility
  • Rolling _____ @
  • Sup ⇔ sit _____ @
  • Bed ⇔ w/c _____ @
  • Static sit bal _____ @

Psychosocial/Emotional
• Social Work continues to monitor pt/family coping as needed
• Provide update regarding progress to pt/family
[Coping effectively]

Patient/Family Education
• Continue with reinforcing of rehabilitation objectives

ADDRESSOGRAPH	UNIVERSITY OF WASHINGTON MEDICAL CENTERS
	Harborview Medical Center—UW Medical Center
Pathways reproduced with special permission. All rights reserved.	Seattle, Washington
	CLINICAL PATHWAY—CERVICAL SPINAL CORD INJURY 11

continues

Exhibit 21–A–1 continued

Clinical Pathways provide guidance in the management process for a specified case type. Using Clinical Pathways in actual practice requires consideration of individual patient needs.	INITIAL	SIGNATURE	HOURS OF CARE	DATE	TITLE
RN Registered Nurse					
DT Dietitian/Nutrition Asst.					
OT Occupational Therapist					
CNS Clinical Nurse Specialist					
RCP Respiratory Care Practitioner					
PT Physical Therapist					
SP Speech Pathologist					
USC Unit Secretary					
MSW Social Worker					
MD Physician					
TR Therapeutic Recreation					

ADDRESSOGRAPH

UNIVERSITY OF WASHINGTON MEDICAL CENTERS
Harborview Medical Center—UW Medical Center
Seattle, Washington
CLINICAL PATHWAY—CERVICAL SPINAL CORD INJURY   12

Source: Copyright © Harborview Medical Center.

■ 22 ■

Case Managing To Maximize Quality and Cost Outcomes: A Total Hip Replacement Pathway

Mary G. Nash, Cynthia Barginere-Urquhart, Lisa Karen Brown, and John M. Cuckler

With the tremendous growth in managed care and capitated reimbursement, many organizations are making dramatic changes in care delivery to control and monitor cost per case. However, many of the work redesign and restructuring changes have been highly criticized in relation to perceived negative effects. To ensure that organizational changes do not produce negative effects, the impact of changes associated with patient care delivery processes must be carefully evaluated before any large-scale changes are made. Unfortunately, most organizations either do not take this into consideration or lack the systems and processes necessary for evaluating the impact of work redesign and/or restructuring on quality of care and other outcomes.

One successful strategy that many acute care hospitals are implementing is case management. According to Zander,[1] the most successful case management programs are multidisciplinary and begin with an understanding of the organization's mission and philosophy. In addition, a well-organized case management program provides a comprehensive approach for supporting the financial and clinical goals associated with patient care delivery by preventing fragmented care delivery and increasing efficiency in planning care.[2]

Since case management focuses on collaboration, continuous quality improvement, and seamless care delivery, it appears an ideal strategy to achieve significant cost reduction and quality outcomes management. Many of these outcomes have been realized with the advent of a case management program at the University of Alabama at Birmingham (UAB) University Hospital, a 908-bed, tertiary medical center. Case management at UAB Hospital is designed as a multidisciplinary team approach to provide cost-effective, quality care across the continuum.

There is a long history of support for case management at UAB Hospital, which is sustained as part of the mission for providing a continuum of health services of the highest quality supported through the most advanced scientific and technological clinical practice. The initial development of case management started at the hospital during the 1980s and evolved over the years throughout the various clinical departments. It was not, however, until 1994 that the hospital made the decision to coordinate case management efforts throughout acute, home, and long-term care. In 1994, a case management coordinator was hired who reported to the chief nursing officer. To begin focusing on a coordinated case management approach, the chief nurse, in collaboration with the chief of staff, initiated the case management steering committee (CMSC), consisting of multidisciplinary departments and physicians.

The following chapter describes the pathway development and outcome measures for the total hip replacement patient. This pathway development was one of the initial efforts of the CMSC and served, later on, as a template for other disciplines. Part of the success of this approach and the pathway implementation stemmed from the strong desire, on the part of the orthopaedic divisional medical director, to change the way things were being done and to demonstrate more cost-effective outcomes. In addition, the orthopaedic division has a dedicated and knowledgeable clinical care coordinator (CCC) who identified the processes and disciplines necessary to effect meaningful change.

CASE MANAGEMENT PROGRAM

At UAB Hospital, case management is defined as a team approach to ensure seamless delivery of quality care in the most effective manner for our patients and their families. Patients and families are actively included in the plan of care from the time of admission and are assured that they will be knowledgeable and prepared and have appropriate support available at discharge.

The goals of the case management program are to decrease the length of stay (LOS), decrease cost per case, improve or maintain the quality of care, ensure appropriate use of resources, and increase customer satisfaction (patients, physicians, employees, and payers). A major aspect of the strategy for the accomplishment of these goals is the use of patient care pathways (PCPs).

The PCP is an interdisciplinary tool designed to facilitate case management by providing a detailed description of the activities and outcomes that must be accomplished for a specific patient group. This ability to plan the care, based upon time frames for critical activities and outcomes, provides for more predictable outcomes and higher quality care. The PCP is intended to outline the care delivery of patients who experience few or no complications related to their illness. At UAB Hospital, the general case management principles support the belief that 60 to 80 percent of the patient population will be effectively managed using the PCP strategy.

PATHWAY DEVELOPMENT

Development of a patient care pathway will not succeed unless everyone who is directly involved in the delivery of the care participates in the development process. Physician involvement to ensure adherence to the pathway process is essential to success. At UAB Hospital, much of the emphasis for interdisciplinary orthopaedic case management began with the arrival of the new division director in April 1994.

In determining which case types were appropriate for PCP development, the division director of the orthopaedic surgery division considered patient homogeneity, cost, and volume. One example that resulted from this collaborative approach with the CCC and other members of the orthopaedic care delivery team was the development of a PCP for patients undergoing total hip replacement.

In the development process for the total hip replacement pathway, the division director, along with the team, reviewed current practice and identified opportunities for improvement. The outcome was the development of a pathway with a four-day LOS, a decrease in the amount of routine laboratory work ordered, the elimination of the use of antiembolic stockings postoperatively, and the initiation of coumadin for deep-vein thrombosis (DVT) prophylaxis. The use of antiembolic stockings was replaced by routine active range of motion, and incentive spirometry was replaced by coughing and deep-breathing exercises. For the total hip replacement PCP, revisions were also made regarding the frequency of laboratory studies, such as the packed cell volume and prothrombin times.

In making such seemingly radical changes in clinical practice, the team believed that the patients' cost per case for these direct cost items would show a substantial decrease. The team also recognized that these changes in practice, although having a positive impact on cost, could potentially have a negative impact on quality of care if pathway implementation was not successful. Therefore, clinical outcomes associated with incidents of pneumonia and DVT were carefully monitored.

The total hip replacement PCP was designed to address the patient's needs across the continuum of care, including the clinic, hospitalization, and rehabilitation. Therefore, it was very important to develop a comprehensive pathway that addressed both the clinical and the teaching needs of patients.

Once the initial draft of the pathway was developed, it was tested for completeness and accuracy by the division director and the CCC. After several months of use and appropriate editing and refinement, additional physicians providing similar care began using the pathway (Exhibit 22–A–1 in Appendix 22–A).

CARE MANAGEMENT

The hip pathway is used by all personnel involved in the care of the orthopaedic patient. It is the ultimate responsibility of the CCC as the "quarterback" of the health care team to ensure that the pathway is properly implemented. The system is designed in conjunction with computerized physician order sets that correlate with the key aspects of the pathway. These order sets are generated by medical residents when the patient arrives in the postanesthesia care unit (PACU) following surgery. Once the patient is placed on the total hip replacement pathway, it is the responsibility of the PACU staff to initiate these orders. Key aspects of the order sets include laboratory studies, physical therapy, medications, and discharge planning orders.

All orders become a permanent part of the medical record and are indeed a "skeleton" of the complete orthopaedic clinical hip pathway. The order sets also guide the staff nurses, medical residents/faculty, physical therapists, pharmacists, social workers, and the rehabilitation care team in the care of the patient. Upon the patient's arrival in the orthopaedic unit, all pathway orders are verified by the unit secretaries. The unit secretaries are well versed regarding the pathway and order sets and often serve as monitors for strict adherence to the protocols. In general, if there has been an omission or oversight in pathway orders, the secretaries will notify the CCC or medical resident for corrections or additions to orders.

The total hip replacement patient's daily adherence to and progress along the pathway are closely followed by the CCC. Any changes or advancements in the patient's pathway progress are communicated to all involved in the care of the patient. This approach to pathway management has been found to be very beneficial in that it enhances continuity of care. Although currently the PCP is not a permanent part of the medical record, the pathway order sets are. Efforts are underway to include the pathway as a permanent part of the medical record so that all disciplines will document their patient interventions and outcomes directly onto the pathway.

SATISFACTION

The staff, including nursing, physical therapists, social services, pharmacists, and rehabilitative medicine, have all been receptive to and participatory in the total hip pathway processes. Their receptiveness was enhanced by their being able to provide input during the design of the pathway. The staff also stated that they appreciated a more standardized approach to patient care, since it decreased fragmentation and took the "guesswork" out of providing care.

Initially, there was physician apprehension regarding the thought of standardizing the hip patient's care in a pathway format, primarily due to fear over loss of autonomy and individualization of patient care. However, after six months, all of the joint replacement surgeons were eager to have their hip patients placed on the pathway. The medical residents found the total hip PCP to be a very useful tool in the management of their patients. Since the pathway begins in the preoperative phase and continues during the postoperative phase, the residents can obtain a concise overview of how to manage the patient from the clinic setting through hospitalization to discharge. As noted earlier, the order sets, initiated by the residents, expedite and complement the pathway process. Due to the overwhelming success and outcomes of the hip pathway process, several other orthopaedic pathways have been developed by other orthopaedic surgeons and their respective clinical care coordinators.

Implementing patient pathways that could potentially increase patient satisfaction, through increased coordination and education, was also very important to the team. It was felt that fear of the unknown and lack of knowledge were two of the major deficits facing preoperative hip patients. During their preoperative clinic visit, the pathway process is explained verbally, and they are also given written instructions. Although the pathway itself is not formally given to the patients, the patients appear to appreciate a verbal synopsis of what is to occur in the hospital, along with a demonstration of some of the equipment used and a viewing of a total hip arthroplasty video.

After discharge, all patients receive a comprehensive patient satisfaction questionnaire. Figure 22–1 illustrates the levels of satisfaction with nursing care on the orthopaedic surgery unit over a four-year period.[3] During this time period, the unit experienced many changes, including the appointment of a new nurse manager. Although it is difficult to isolate all the variables that affect patient satisfaction, the individual positive responses from patients are very encouraging to the staff. The reinforcement of the pathway process and preoperative information occurs on a daily basis by all involved members of the health care team. There is a strong belief among team members that the consistent use of the pathway by all disciplines has led to decreased patient anxiety levels. The care team also believes that the knowledge level of patients relative to their procedure and discharge planning has increased significantly.

VARIANCES

Deviations or variations from the total hip pathway are addressed on an individual basis by the CCC. Often clinician variances, such as the omission of an order, are simply an oversight and may be attributed to human error. These are generally noted by the nursing staff and are corrected by the medical residents and/or clinical care coordinator by initiating the correct order from the hip pathway order sets. System variances, such as the failure of the computer system to forward an order set to rehabilitative medicine services, can be corrected technically by the appropriate support personnel. Patient variances, such as a decrease in packed cell volume, are to be expected and are managed on an individual basis according to the patient's response and pathway protocols. For example, if a hip patient presents on postoperative day 1 with a packed cell volume of 22 percent but is otherwise asymptomatic, a decision is made to withhold autologous transfusion. However, if the patient is symptomatic and/or elderly, the option for transfusion is initiated.

PATHWAY UTILIZATION

Since the total hip pathway is not a mandatory component of the UAB Hospital documentation process, it is currently not used as the formal report mechanism on the orthopaedic unit. However, components of the pathway are conveyed during report, such as laboratory values and discharge planning. Postoperative and discharge teaching for the hip patient are also documented on the UAB patient instruction record, which is a permanent form of documentation. During team rounds and reports, which occur with the CCC, house staff, and the

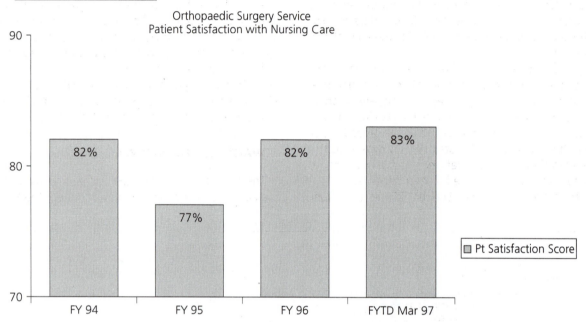

Orthopaedic Surgery Service
Patient Satisfaction with Nursing Care
FY 94–97

	PT Satisfaction Score
FY 94	82%
FY 95	77%
FY 96	82%
FYTD Mar	83%

Figure 22–1 Patient satisfaction with nursing care has remained relatively constant since pathway implementation. *Source:* Copyright © University of Alabama, Birmingham, University Hospital.

attendings, the components of the pathway are addressed. Other professions, such as the physical therapists, readily convey the patient's progress on the pathway through phone calls to the CCC and/or the patient's progress notes in the permanent chart.

The CCC is responsible for the orientation of any new nursing staff and medical residents to the total hip PCP. Introduction to the pathway is provided through an informal lecture regarding pathways and case management, along with copies of the actual pathway. All medical residents are given copies of the pathway along with resident orientation guidelines that provide an overview of its usage and importance. It is stressed to all staff that strict adherence to the pathway is mandatory to achieve the quality outcomes desired.

OUTCOMES

Within the orthopaedic service, the divisional director is responsible for outcome measurement. Both clinical and financial information is reviewed by the division di-

rector, orthopaedic surgeons, and CCCs. This process is very important to the physician staff to ensure that quality care is maintained.

The PCPs are designed to identify both the clinical activities necessary and the clinical outcomes expected for the patient group. These outcomes are expressed in specific, measurable clinical terms and placed on the pathway at appropriate intervals. They may be incremental or discharge outcomes. For example, the clinical process or activity of ambulating the patient on the first postoperative day would be accompanied by a corresponding clinical outcome stating the desired criteria for the completion of the activity (e.g., "The patient is able to ambulate 15 feet"). This not only guides the clinicians in what needs to be done for the patient but also states what outcomes are expected as a result of that activity. The discharge outcomes outline similar criteria necessary in preparing and evaluating patients for discharge to the rehabilitation facility or home.

Both clinical activities and outcomes are important in the implementation and evaluation of the pathway. The clinical outcomes are used to determine whether the goals of quality care are maintained. In addition, monitoring clinical activities assists in the evaluation of patient care delivery processes and the impact of those processes on the case management goals related to cost, LOS, and quality care.

The coordinator for case management provides a monthly report, by surgeon, that outlines number of cases, patient days, and total and direct cost by payer and diagnosis-related group (DRG). These reports allow the CCCs to have consistent feedback regarding the cost of care for a patient group. In addition, patient satisfaction data are provided for the nursing unit responsible for the care of this patient group.

The evaluation process for the total hip patient care pathway has been based on both clinical and financial criteria. Incidents of pneumonia and pulmonary embolus are monitored to ensure clinical quality. As illustrated in Figure 22–2, the orthopaedic service experienced a decrease in these complication rates.[4] This indicates to the clinical team that they are meeting their goal to improve or maintain the quality of care for their patients.

Indicators such as average LOS (ALOS) and cost per case assist in analyzing the financial impact of the pathway. Because the PCP was initiated late in fiscal year (FY) 94, a comparison of these indicators was made for FY 94 with FY 95. As noted in Figures 22–3 and 22–4, after implementation of the PCP, the ALOS decreased by 19 percent, and the cost per case decreased by 12 percent.[4] During FY 96 and 97, the organization changed accounting methods and implemented a new cost accounting system changing from analysis of cost-to-charge ratios to analysis of direct and indirect costs. As a result of these methodology changes, it is not possible to compare cost-per-case data accurately for this time period. However, comparisons indicate that ALOS has remained relatively stable over the past two years, with a 4 percent increase in FY 96 and a 1 percent increase for FY to date at March 97. It is estimated that these increases in ALOS were due to the increased number of higher acuity patients.

ADVANTAGES OF PATHWAY USE

The advantages of using the total hip pathway are numerous. The standardization and the use of the PCP align

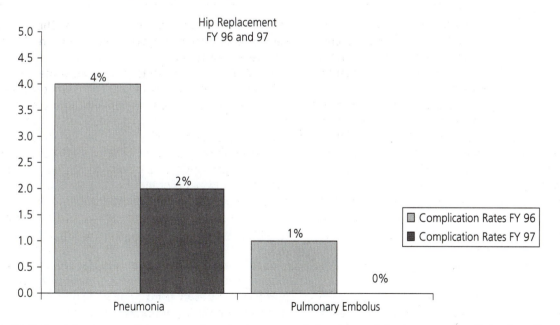

Hip Replacement
FY 96 and 97

	Complication Rates	
	FY 96	FY 97
Pneumonia	4%	2%
Pulmonary Embolus	1%	0%

Figure 22–2 Complication rates for pneumonia and pulmonary embolus show a slight decrease after elimination of incentive spirometry and the modification of antiembolic therapy. *Source:* Copyright © University of Alabama, Birmingham, University Hospital.

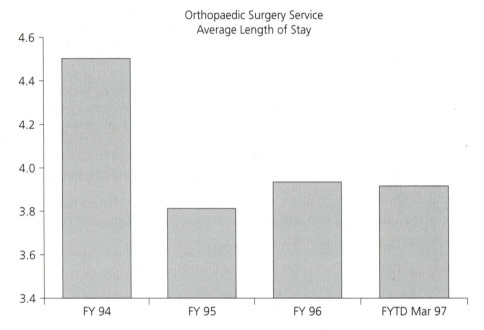

Orthopaedic Surgery Service
Average Length of Stay
FY 94–97

FY 94	4.56
FY 95	3.83
FY 96	3.98
FYTD Mar 97	3.94

Figure 22–3 Average length of stay decreased by 19 percent with the implementation of the pathway from FY 94 to FY 95. *Source:* Copyright © University of Alabama, Birmingham, University Hospital.

patient care processes for managing care along the continuum. Since the pathway is initiated in the clinic, discharge planning begins before the patient ever arrives in the hospital. This tends to enhance and expedite the patient's hospitalization. As a result, there has been a dramatic decrease in LOS, along with a reduction in total costs.

Another advantage of the use of pathways is the ability to define better the costs associated with the provision of care for a specific patient group. Defining costs more precisely assists in preparing the organization for managed-care contract negotiations. To remain competitive, it is imperative that organizations be aware of the costs and quality outcomes associated with their patient management processes.

DISADVANTAGES OF PATHWAY USE

There is a major disadvantage when the PCP is not a permanent part of the patient's medical record. Although the

clinical staff use the pathway as a reference in the care of their patients, they are not currently required to document on the pathway, since this would result in unnecessary duplication. In addition, the lack of documentation on the pathway by all clinical and ancillary staff limits the ability to pull data retrospectively on the patient group in the aggregate. Plans are underway to make the pathway a permanent part of the record. To include the pathway in the documentation process, it has been necessary to identify current documentation that can be replaced or deleted. There is a strong belief that to achieve consistent compliance, there must be minimal duplication.

CONCLUSION

Case management uses a tightly focused, collaborative team approach to patient care, with planned oversight of every aspect of care from preadmission to home care and an emphasis on continuity and efficiency. The outcomes

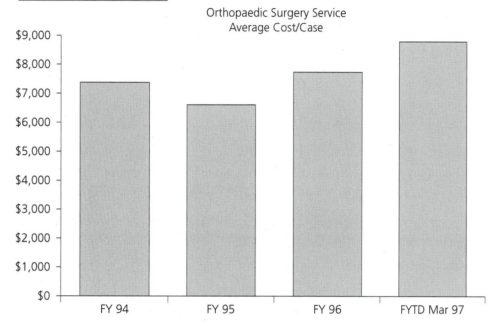

Orthopaedic Surgery Service
Cost per Case Summary
FY 94–97

Fiscal Period	Cost/Case
FY 94	$ 7,377.15
FY 95	$ 6,610.75
FY 96	$ 7,739.00
FYTD Mar 97	$ 8,800.00

Figure 22–4 Average cost per case decreased by 12 percent with the implementation of the pathway from FY 94 to 95. The implementation of a new costing methodology in FY 96 limits the ability to make accurate cost comparisons for FY 96 and 97. *Source:* Copyright © University of Alabama, Birmingham, University Hospital.

of a successful case management program can assist with managed-care contract negotiation, cost containment, quality improvement, support of the patient care delivery model, and improved patient satisfaction.

One of the most successful aspects of the case management approach at UAB Hospital has been the use of the PCP developed by clinical divisions, using the most common DRGs. Successful implementation of pathways has been enhanced by including all staff in the design of the pathways. Pathways are effectively used in the management of total hip replacement patients and provide a detailed road map with constant checkpoints throughout the hospitalization. The pathways also assist in providing a mechanism for identifying and measuring expected patient outcomes.

One of the key ingredients for successful patient care management is a well-designed approach that emphasizes multidisciplinary collaboration. This approach pro-

vides the foundation for effectively managing complex patients. For example, patients who undergo total hip replacement are generally a complex population primarily because of age and other confounding medical problems. Our experience at UAB Hospital reveals that total hip patients who have undergone such complex procedures are better managed and have better outcomes due to a structured case management program and the use of a PCP. For example, use of coumadin for DVT prophylaxis is now routine, along with the elimination of unnecessary products such as antiembolic stockings and incentive spirometry. Additionally, there has been a marked decrease in the LOS and use of routine lab work.

The early success of the UAB Hospital case management program has provided a framework for many other clinical departments in developing and implementing patient care pathways. The spirit of collaboration has led to improvements in patient outcomes and operational indicators.

Yearly review of the results of the pathways with regard to patient satisfaction, LOS, and efficiency of the care processes will involve all users of the pathway or administrators of the pathway in an effort to maintain or improve compliance. Future goals include the refinement of outcome measurement related to clinical interventions such as coumadin therapy for DVTs. Finally, it is imperative that all disciplines be able to document interventions and outcomes effectively and efficiently. Plans are in progress to include pathways as a permanent part of the medical record to reduce duplication in charting and improve documentation.

REFERENCES

1. Zander K. Nursing case management strategic management of cost and quality outcomes. *J Nurs Adm.* 1988;18:22–30.

2. Zander K. Case management in acute care: making the connections. *Case Manager.* 1991;2:39–43.

3. *University of Alabama at Birmingham Patient Report Card Data Base (1994–1997).* Birmingham, AL: Office of Director of Health Systems; 1997.

4. *University of Alabama at Birmingham Data Base (1994–1997).* Birmingham, AL: Data Resources Department; 1997.

■ Appendix 22–A ■
Orthopaedic Patient Care Pathway

Exhibit 22–A–1 Orthopaedic Patient Care Pathway

Patient Care Path is not to be used as medical orders, order set, or protocol risk assessment score

THE UNIVERSITY OF ALABAMA AT BIRMINGHAM HOSPITAL
ORTHOPAEDIC PATIENT CARE PATHWAY
TOTAL HIP ARTHROPLASTY/REVISION
RIGHT/LEFT

Keyplate

Encounter # ____________

	KIRKLIN CLINIC VISIT Date: / /	DAY OF SURGERY Date: / /	POST OP DAY 1 Date: / /	POST OP DAY 2 Date: / /	POST OP DAY 3 Date: / /	POST OP DAY 4 Date: / /	OUTCOMES Date: / /	CONCURRENT DISEASES SURGERIES Date: / /	COMMENTS
	MET UNMET	MET UNMET	MET UNMET	MET UNMET	MET UNMET	MET UNMET	MET UNMET	MET UNMET	MET UNMET
ASSESSMENTS/CONSULTS	❑ DR's H&P ❑ ❑ Clinical care coordinator (CCC) assess/teaching	❑ NSG./DR. review of systems (ROS) assessment ❑ ❑ Physical therapy (PT) consult ❑ ❑ Social services (SS) consult ❑ ❑ Spain rehab. consult ❑ ❑ UAB orthotics ❑	❑ NSG/DR. ROS assess. ❑ ❑ Pt visit x 2 or ______ ❑ ❑ SS visit ❑ ❑ SP. rehab. visit ❑ ❑ UAB orthotics ❑	❑ NSG/DR. ROS assess. ❑ ❑ Pt visit x 2 ❑ ❑ SS visit ❑ ❑ SP. rehab. visit ❑	❑ NSG/DR. ROS assess. ❑ ❑ Pt visit x 2 ❑ ❑ SS visit ❑ ❑ SP. rehab. visit ❑	❑ NSG/DR. ROS assess. ❑ ❑ SS visit ❑ ❑ ______ Home health care visit/arrangements ❑ ❑ DME visit ❑ ❑ D/C to sp. rehab. ❑ ❑ D/C to ______ ❑	❑ All assessments/ consults will be completed by d/c ❑	❑ ETOH abuse ______ ❑ ❑ Rec. drugs ______ ❑ ❑ Other: ______ ❑	
NEURO/CV *Potential alter. tissue perfusion *Potential confusion/alter. in thought processes	❑ Baseline VS: ❑ ❑ Temp. ______ ❑ ❑ Pulse ______ ❑ ❑ Resp. ______ ❑ ❑ B/P ______ ❑ ❑ Baseline NVC: ______ ❑ ❑ Baseline neuro: ______ A&O X ______ ❑	❑ VS/NVC Q1 hr. x 4, ❑ ❑ Q4 hr. x 4 ❑ Then Q shift TMAX = ______ ❑ Notify if: temp. > 102, ❑ Pulse <50 >110, ❑ Resp. <15 >30 ❑ B/P <90/60, >200/100 ❑ ❑ ✔ LOC/orient Q shift ❑	❑ VS/NVC Q shift ❑ TMAX = ❑ ✔ orient Q. shift ❑	❑ VS/NVC Q shift ❑ TMAX = ______ ❑ ✔ orient Q. shift ❑	❑ VS Q shift ❑ TMAX = ______ ❑ D/C NVC ❑ ❑ ✔ orient Q. shift ❑	❑ VS Q shift ❑ TMAX = ______ ❑ ✔ orient Q. shift ❑	❑ VS accord to pt. baseline ❑ ❑ Periph. pulses + ___ ❑ ❑ Color/temp of extrem. accord to baseline ❑ ❑ ___ sec. cap. refill ❑ ❑ +D/P flex. feet ❑ ❑ A&O x 3 or accord. to baseline ❑	❑ Angina ______ ❑ ❑ MI ______ ❑ ❑ Card. arrhy. ______ ❑ ❑ HTN ______ ❑ ❑ DVT/PTE ______ ❑ ❑ PVD ______ ❑ ❑ Dementia ______ ❑ ❑ Other: ______ ❑	
PULMONARY *Potential ineffective airway clearance *Potential impaired gas exchange	❑ Baseline pulm. ______ ❑ Assess: ______	❑ Cough/deep breathe x 4/hr. w/a ❑ ❑ O2 at 2 lit. n/c & ❑ ❑ Pulse ox. (x 12 hrs. on pt's >60 yo) ❑	❑ Cough/deep breathe x 4 w/a ❑ ❑ D/C O2 & pulse ox. ❑	❑ Cont. pulm. exercises ❑	❑ Cont. pulm. exercises ❑	❑ Cont. pulm. exercises ❑	❑ Lungs clear or accord to baseline ❑ ❑ O_2 sat. > 90% or baseline of ___% ❑	❑ Asthma ______ ❑ ❑ COPD ______ ❑ ❑ Smoker ______ ❑ ❑ Other: ______ ❑	
GI/DIET & NUTRITION *Potential N/V *Potential situational constipation	❑ Baseline GI assess: ___ ❑ ❑ Baseline nutri. assess: ______ ❑ ❑ NPO after mn. ______ ❑	❑ Diet as tolerated (DAT) ______ ❑ ❑ Encourage juices ❑ ❑ I&O Q 8 hrs ❑ ❑ If n/v: give compazine ❑ ❑ Provide with cool compresses for n/v	❑ DAT: ______ ❑ ❑ I&O Q 8 hrs ❑ ❑ ✔ for BM ❑	❑ DAT: ______ ❑ ❑ D/C I&O after 48 hrs. if adeq. PO intake ❑ ❑ ✔ BM: if none give 1/2 prune 1/2 apple juice ❑	❑ DAT: ______ ❑ ❑ ✔ BM: if none give MOM with cascara ❑	❑ DAT: ______ ❑ ❑ ✔ BM: if none give dulcolax suppository ❑	❑ + Bowel sounds ❑ ❑ Adeq. intake ❑ ❑ + BM ❑ ❑ No nausea/vomiting ❑	❑ Peptic ulcer dz. ______ ❑ ❑ Hiatal hernia ______ ❑ ❑ Hx. constipation ______ ❑ ❑ Laxative abuse ______ ❑ ❑ Other: ______	
GU *Potential urinary retention	❑ Baseline GU assess: ___ ❑	❑ Foley/insert if no void 8 hrs. postop; leave in up to 48 hrs. ❑	❑ Foley remains ❑ ❑ D/C Foley ❑	❑ D/C Foley ❑ ❑ Monitor for retention/UTI ❑	❑ Monitor for retention/UTI ❑	❑ Monitor for retention/UTI ❑	❑ Adeq. output ❑ ❑ No retention/UTI ❑	❑ BPH ______ ❑ ❑ Urinary retention ______ ❑ ❑ Urinary incontinence ______ ❑ ❑ UTI hx. ______ ❑ ❑ Kidney dz. ______ ❑ ❑ Other: ______ ❑	
MUSCU/ACTIVITY EXER/MOBILITY (ADL) *Immobility *Potential injury	❑ Baseline musc. assess: ______ ❑ ❑ Harris Hip Score: ___ ❑	❑ Overhead frame/ trapeze bar ❑ ❑ Knee immob./ abd pillow; THA ❑ ❑ ABD brace: THR; 30°–60° flex./15° abd. ❑ ❑ Turn PRN both sides ❑ ❑ HOB ↑ no > 60° ❑ ❑ Towel roll under operative ankle ❑ ❑ +D/P flex. feet bilat. x 10/hr. w/a ❑ ❑ Ice pack to operative site ❑	❑ Cont. previous treatments ❑ ❑ Asst. total bath ❑ ❑ ABD brace/knee immobilizer pillow intact ❑ ❑ ↑ BID 2 hrs. in chair ❑ ❑ ↑ to BSC ❑ ❑ Pt visit x ___: initial travel via stretcher, return via w/c ❑ ❑ Pt instruc. ❑ ❑ ___able to verbalize THA precautions ❑	❑ Cont. previous treatments ❑ ❑ Mod. asst with bath ❑ ❑ D/C knee immob. pillow ❑ Soft pillow between legs (always) ❑ ↑ in chair 2 hrs. BID ❑ ❑ ↑ to BSC ❑ ❑ Pt x 2: ❑ ❑ ___able to verbalize demo THA/THR precautions & prescribed wt bearing status ❑	❑ Cont. prev. treatments ❑ ❑ Minimal asst. bath ❑ ❑ Soft pillow between legs ❑ ❑ ABD brace intact ❑ ❑ Pt visit x 2 ❑ ❑ ___able to get in/out of bed with min. asst./indep. ❑ ❑ ___able to perform 10 reps of prev. taught exercises ❑ ❑ ___initiate stair climbing ❑	❑ Cont. prev. treatments ❑ ❑ Cont. to work on pt: will d/c to SRC or to home/output pt ❑ ❑ Pt x ___: demo; indep. with all prev. pt activities ❑ ❑ Bath independently with asst feet/back ❑	❑ Amb. 100–200' or using ______ asst device with ______ assist. ❑ ❑ Amb. ↑ & ↓ stairs ❑ ❑ Demonstrate proper THA precautions ❑ ❑ Remain injury free ❑	❑ RA ______ ❑ ❑ DJD/OA. ______ ❑ ❑ Lupus ______ ❑ ❑ Gout ______ ❑ ❑ Other: ______ ❑	

	KIRKLIN CLINIC VISIT Date: / /	DAY OF SURGERY Date: / /	POST OP DAY 1 Date: / /	POST OP DAY 2 Date: / /	POST OP DAY 3 Date: / /	POST OP DAY 4 Date: / /	OUTCOMES Date: / /	CONCURRENT DISEASES SURGERIES Date: / /	COMMENTS
MUSCU/ACTIVITY EXER/MOBILITY (ADL) (continued)		□ able to sit on side of mat & transfer with ___ asst., ___ amb. 15' or ___' or ___' with TDWB/50%/NWBAT with ___ asst. device		□ Able to perform the following: □ ___ quad sets □ ___ terminal knee extensions □ ___ hip flexion in standing □ ___ ABD standing □ ___ gluteal sets □ Able to get in/out of bed with ___ asst □ AMB/25'–50' or ___ with presc. wt. bear status/asst. device	□ ___ amb. 100'–200' □ or ___ □ ___ asst. device for home ordered on chart				
INTEG/DRSSG./DRAINS *Skin integrity impairment *Potential infection	□ Baseline skin assess: □ Depilatory cream rx. given	□ Postop. drssg.: 6" xero-form. 4x8's, abd, ace. □ May reinforce with abd's □ Hemovac: notify if output >500cc/8 hrs. HVC =	□ Post op surg. drssg. remains HVC =	□ 1st drssg. change by DR. □ 4x8's/tape □ Paint incisions/stables with betadine □ PRN drssg. changes by NSG. □ D/C HVC <80 CC/ 8 hrs. HVC =	□ Cont. betadine PRN drssg. changes □ Staples intact □ Neosporin ointment PRN for tape irritation □ Dressing D/C by DR	□ Drssg. D/C by DR □ Cont. betadine to incision if draining □ Staples d/c: Benzoin & steristrips applied □ Staples remain: to be d/c later	□ Incision approximated □ No. sx incisional infection □ No skin impairment	□ Skin impairment ___ □ Diabetes ___ □ Other: ___	
DIAGNOSTICS/LABS & PROCEDURES *Potential hypovolemia/ hypervolemia	□ Autologous/direct donor □ T&C 2 units-primary □ T&C 3 units-revision □ CBC, PT, PTT, SMA-12 □ CXR, EKG if > 40 □ Other:	□ HCT in PACU = Notify if <23%	□ HCT = □ SMA-6 (if abnml. preop; repeat on POD #3) □ Other:	□ HCT = □ PT = □ INR = □ Other =	□ HCT = □ PT = □ INR = □ Other =	□ PT = □ INR = □ Other:	□ Lab values within normal limits or accord. to pt. baseline: □ No SX of hypo/ hypervolemia	□ Sickle cell ___ □ Liver dz. ___ □ Hemophiliac ___ □ Thyroid dz. ___ □ Cancer ___ □ HIV ___ □ Other: ___	
PAIN CONTROL *Pain management	□ Baseline pain mgmnt. Assess:	□ MSO4 PSA ___ mg Q ___ min., not to exceed ___ mg Q ___ hrs. □ Demerol PCA ___ mg Q ___ min., not to exceed ___ mg Q ___ hrs. □ Other:	□ Cont. PCA pump □ Supplement PCA pump with tylenol i-ii tabs Q4 hrs. PRN	□ D/C PCA pump □ Tylenol #3 i-ii tabs PO Q 4 hrs PRN pain □ Other:	□ Cont. tylenol #3 OR ___ Q4 hrs. PRN pain □ May supplement PO PRN with tylenol 325 mg tabs i-ii PO Q4 hrs. PRN	□ Cont. tylenol #3 OR ___ Q 4 hrs. PRN □ Provide with D/C prescription of above	□ Pain controlled comfortably with D/C meds of	□ Pain mgmnt. deficit ___ □ Other ___ □ Allergies: ___	
MEDICATIONS/IVF'S & TRANSFUSIONS *Potential hypersensitive skin rxn	□ Baseline home meds: 1. 2. 3. 4. 5. 6.	□ For THA: vancomycin 1 gm IV Q12 hrs. x 3 doses (1st dose in OR) □ For THR: vancomycin 1 gm IV Q12 hrs x 4 doses & gentamicin 80–100 mg (*adjust dosage according to body weight) Q 12 hrs x 4 doses (1st loading dose 100–120 mg in OR) *If allograft give above 2 meds x 5 days □ D5/.45NS @ ___ cc/hr OR □ ___ @ ___ cc/hr □ Colace 100 mg. po BID □ ASA 325 mg. TABS po OR □ Coumadin ___ mg po hs tonight □ Halcion, 125 mg po hs PRN sleep □ Tylenol 650 mg po 4 hrs. PRN	□ Cont. antibiotics □ Cont. IVF □ Cont. sched./home meds.	□ Cont. ASA OR □ Coumadin ___ mg. □ PO HS tonight □ Cont. sched./home meds □ Convert IV to heplock when antibiotics □ D/C antibiotics	□ Cont. heplock □ Cont. ASA OR □ Coumadin ___ mg PO HS tonight □ Cont. sched./home meds	□ D/C heplock □ ASA i Q D OR □ Coumadin ___ mg at home □ Cont. sched./home meds □ Other D/C meds:	□ Will maintain compliance with med. therapy upon D/C with prescribed medications of ___ □ No SX. of hypersensitive skin RXN	□ Hx. transfusion rxn. ___ □ Other ___	

continues

Exhibit 22-A-1 continued

	KIRKLIN CLINIC VISIT Date: / /	DAY OF SURGERY Date: / /	POST OP DAY 1 Date: / /	POST OP DAY 2 Date: / /	POST OP DAY 3 Date: / /	POST OP DAY 4 Date: / /	OUTCOMES Date: / /	CONCURRENT DISEASES SURGERIES Date: / /	COMMENTS
	MET / UNMET	MET / UNMET	MET / UNMET	MET / UNMET	MET / UNMET	MET / UNMET	MET / UNMET	MET / UNMET	MET / UNMET
		□ Compazine 10 mg Q 6 hrs. IM/IV PRN nausea □ Maalox 30 cc po Q 6 hrs. PRN indigestion □ MOM with cascara 30 cc po PRN constipation □ Dulcolax supp. PRN constipation □ Zantac 150 mg po BID PRN GI upset □ Begin home med: ___							
PATIENT EDUCATION *Knowledge deficit *Potential noncompliance	□ Review pathway/preop & postop routine □ Review patient/family goals □ Review THA video □ Provide with written teaching materials	□ Review interventions to prevent ROS complications □ Review postop routines □ Emphasize pain control □ Review patient/family goals and expectations	□ Reinforce pathway process □ Review THA/THR protocols & impt. of indep. with ADLS □ Reinforce pain control □ Reinforce pt instructions & pt/family goals	□ Reinforce pathway process □ Reinforce THA/THR protocols □ Review PO pain control □ Reinforce pt instructions	□ Provide with sp. rehab. info □ Review wound care □ Reinforce all prev. teachings	□ Reinforce all previous teachings □ Review wound care □ Provide with D/C instruction sheet and HIP card □ Review home goals	□ Pt/family verbalizes/demonstrates the following: □ THA precautions □ Use of asst. device(s) □ Wound care □ Home exercises □ Pot. injury factors in the home (ie, throw rugs, low chairs/bed) □ Compliance in all of the above	□ Mental/neuro impair. □ Other: ___	
D/C PLANNING *Potential alter. in family dynamics	□ Baseline info re: □ Home environment ___ □ Equipment ___ □ Mode of transport ___ □ Home care provider ___ □ Financial concerns ___ □ Other ___	□ Review D/C needs again with family □ Collaborate with SS re: these needs	□ Continue collaboration with SS/SP rehab & pt/family re: D/C needs	□ Cont. to discuss D/C plan with interdisc. team members & patient/family	□ Arrange D/C needs with SS/SP rehab.	□ Finalize D/C plans □ Equipment with pt./sent to home: □ Home via ambulance	□ Pt/family will have a definitive plan of home care □ Necessary THA/THR equipment	□ Family dynamics ___ □ Name/# caregiver: ___	

Source: Copyright © University of Alabama, Birmingham, University Hospital.

■ 23 ■

Outcomes-Based Practice: A Lumbar Laminectomy Model

Terri Hawkins Pigg

The exciting and contemporary topic of outcomes-based practice is one with which modern nursing leaders must become more familiar. Since the creation of new Medicare regulations and diagnosis-related groups (DRGs) in the 1980s, caregivers have been forced to be more cognizant of resource utilization, lengths of stay, and, most of all, quality patient care. In the late 1980s and early 1990s, changes in patient care payment systems caused a closer look at the delivery of patient care. Managed care swept parts of the nation, bringing with it a "per-episode" payment system that was a real contrast to the traditional "per diem" system. It soon became obvious that doing things the "same old way" would no longer suffice. The Clinton administration made health care reform a key component of its agenda. As patients became more knowledgeable about reform, their expectations of the health care system increased. Health care leaders began to search for new ways to care for patients in a more cost-effective manner. Although many nurses were educated in best patient care practices, few were taught to be mindful of resource utilization and cost containment.

As managed care has increased, so has the nursing and allied health literature increased. Karen Zander, one of the foremost experts on the subject, published one of the first case management articles describing a case management model at New England Medical Center in Boston.[1] This model was successful in the face of many obstacles such as a high patient census, lack of funding for the project, lack of computerized documentation and tracking, and lack of nursing leaders such as clinical nurse specialists (CNSs). Other authors described differing approaches to case management processes. It became obvious that each institution needed to individualize the model to best suit its needs. But one common theme that did not vary was the need for better utilization of re-

sources while improving quality of care.[2] Outcomes became a prevalent topic, with a focus on patient satisfaction and patient functional ability.

At Saint Thomas Hospital in Nashville, Tennessee, the outcomes-based practice journey began in the late 1980s. A multidisciplinary leadership group consisting of physicians, nurses, and finance experts explored the concept to determine what it meant to Saint Thomas. Critical pathways, a result of these early discussions, were created to assist in organization of care delivery and outcomes analysis. These tools were updated, and the name was changed to CareMap® at a later date. Saint Thomas leaders believed that the multidisciplinary approach to critical pathway development should be a standard. Physician involvement throughout every phase of the process was critical to success. Physicians were thought to have been significantly involved throughout development. In retrospect, however, their involvement, buy-in, and support might have been better nurtured along the way.

One of the first critical pathways developed at Saint Thomas Hospital was for lumbar and cervical laminectomy patients. Development of a subsequent separate lumbar laminectomy CareMap® and the outcomes-based practice process is the focus of this chapter.

LAMINECTOMY PATHWAY DEVELOPMENT

In 1989, a steering group consisting of various leaders sought to identify diagnoses that could be standardized with best-practice interventions, best financial outcomes, and best patient outcomes as the goal. The neuroscience head nurse at that time was instrumental in developing a first draft of a combined cervical and lumbar laminectomy critical pathway. At a later date, the critical pathway was divided into separate documents for cervical and lumbar laminectomy cases.

The laminectomy patients were chosen for pathway development for a couple of reasons. First, laminectomies were a frequently admitted DRG, so outcomes for a large number of patients could be affected. Because DRGs 214 and 215 are broad and include many procedures, the decision was made to limit the inclusion criteria for the pathway to what we loosely termed "simple" laminectomies. This excluded procedures such as fusion or decompression, which generally are more complicated and require longer lengths of stay. It was felt that the relatively uncomplicated pre- and postoperative care for simple laminectomy patients made this procedure a good choice for the neuroscience team's first pathway. Second, the physicians admitting these patients were a cohesive group who were approachable and agreeable to reviewing practice patterns. The goal became to create a tool and a system that would help to improve the quality of care delivered to laminectomy patients. It was obvious that there was much to be learned along the way. Any tool created as a guideline for care was subject to change as knowledge and expertise levels increased. As within any system, a change in one area of the hospital affected staff and usually demanded changes in other areas. Any change, positive or negative, is stressful as learning takes place and people internalize the new process. It was also agreed that this undertaking would require a massive educational endeavor.

In September 1993, a neuroscience CNS was hired, and the laminectomy project was continued. A multidisciplinary group of caregivers reviewed the previously drafted outline of care for cervical and lumbar laminectomy patients. This group gathered order sets from physicians to review laboratory tests, treatments, medications, activity orders, and so forth. An updated literature review focused on practice throughout the country. This core group, consisting of representatives from social services, pastoral care, nursing, pharmacy, dietary, and rehabilitation, prepared subsequent drafts that were reviewed for relevancy and content. The neurosurgical division chief and other medical staff reviewed the pathway and gave input related to content and clarification of concepts.

Caregivers were continuously educated and reeducated about managed care and case management. Information was shared at staff meetings, articles were posted, and impromptu opportunities were captured to spark interest. The physicians, while accustomed to examining their practice and care habits for trends and quality improvement opportunities, were initially cautious and guarded about case management. The ever-increasing involvement of private insurance companies,

as well as state and federally funded government insurances, seemed overbearing and invasive. Overall, the physicians supported the lumbar critical pathway's development. This might have been attributed to the fact that once the pathway was written and approved, it actually affected the physician's daily practice very little and probably was viewed as a nursing document, much like a nursing care plan.

PATHWAY DESIGN

The original lumbar laminectomy pathway was divided into sections created in landscape design. Rows were entitled "Tests," "Diet," "Medications and IV Fluid," "Vital Signs/Monitoring," "Pulmonary Care/Monitoring," "Treatments," "Activity," "Discharge Planning/Teaching," and "Outcomes." Columns represented time frames such as preoperative and postoperative day 1. The pathway served as a guide for care and included important interventions such as preoperative and postoperative teaching, pain management techniques, wound/dressing assessment, and body mechanics. Expected goals were entitled "Outcomes" and appeared as the final row. Each time frame had corresponding outcomes. At that time, the pathway was not a permanent part of the medical record.

In mid-1994, there were no standardized guidelines at Saint Thomas for the creation, ratification, or utilization of critical pathways. As new pathways were developed, each looked slightly different. Caregivers were educated about pathway utilization in various ways, depending on the group creating them. This lack of standardization soon created confusion for the caregivers expected to manage patients' care. Staff raised questions such as "What exactly is a critical pathway?" "How do I use this?" and "What does this mean to me?"

In 1994, an orthopaedics-neuroscience case manager (CM) was hired to assume part of the responsibility for staff nurse and ancillary department education regarding the case management process. Lack of a computerized system forced the CM to extract variance data manually from critical pathways. This information, albeit crude, was shared at nursing and rehabilitation staff meetings and with the neurosurgical division chief. The orthopaedics-neuroscience CM and neuroscience CNS worked closely together to continue education with staff while correlating variance data with patient care. Caregiver education regarding critical pathway usage was vital at this point because variance data became available only if the staff completed the pathways. Monitoring of CareMap® utilization continues to be done.

During this early critical pathway time period, a patient pathway was created to mimic the caregiver pathway for laminectomy. The patient pathway listed, in simple terminology, procedures and expected outcomes to be reached at different times. This patient pathway was presented during preadmission testing or upon admission. Although no formal feedback was obtained, informal feedback was positive as patients and families expressed appreciation for the tool.

EQUIPMENT STANDARDIZATION

At about the same time, a group of operating room (OR) staff, materials management staff, and physicians began to consider standardization and streamlining of OR instrumentation and equipment for lumbar laminectomy cases. The group reviewed and categorized surgeon preferences for supplies. The team reached an agreement on equipment, including a standardized procedure and device for positioning patients. The resulting OR equipment changes produced a one-time $25,000 cost reduction.

SAINT THOMAS HOSPITAL CASE MANAGEMENT SYSTEM

In mid-1995, a change in nursing leadership brought a new focus to critical pathways and to case management in general. A closer inspection of the current system revealed several deficits and many opportunities for improvement. Up to this point, only patients with selected diagnoses had been placed on critical pathways. Nursing care plans had been utilized for the remainder of patients. A documentation committee was formed to oversee creation of a standardized format for pathways and to look at how pathways should interface with other documentation.

The vice-president for nursing services at that time was very interested in case management. Through her efforts, a group from the Center for Case Management in Boston analyzed our current system. Suggestions were made to formulate a plan for improvement. Saint Thomas Hospital became customers of the Center for Case Management, purchasing the right to call our tools "CareMaps®." With the new standard format, members of the oversight documentation committee assisted all areas of the hospital with revision of existing pathways. It was agreed that all patients in the hospital would be placed on a CareMap® and that CareMaps® would become a permanent part of the medical record. Because only a limited number of diagnosis-specific maps were ready at that time, a number of "generic" maps were created for use with various patients. This was done to attempt to decrease schizophrenia of documentation by having one documentation system. The "generic" maps are very nonspecific and may have

created another set of problems. Generic CareMaps® do not lend themselves to focused care needs, thus creating the need for major individualization. The documentation committee made suggestions and changes with other areas of documentation to create a system of manual charting that would decrease duplication. Compliance continues to be monitored.

CASE MANAGER ROLE

A CM is assigned to patients according to attending physician. The system has two goals: staff nurses "manage" the care of all patients, and CMs consult for patients who have multiple problems, need help maneuvering through the system, or have many variances. The CareMaps® are initiated during the preadmission testing process for elected admissions or upon admission for all others. The clerical associate (unit clerk) consults with the nurse to decide the most appropriate map for the patient. The staff nurses manage patient care and notify the CM of problems. CMs, social workers, staff nurses, and physicians work with other members of the multidisciplinary team to care for patients through the continuum. One goal is to improve the frequency of appropriate CM consultations for assistance with difficult patient issues. The orthopaedics-neuroscience CM position was vacated in early 1996, and the neuroscience CNS assumed responsibilities for neuroscience case management.

CAREMAP® FORMAT AND PROCESSES

Current CareMaps® (Exhibit 23–A–1 in Appendix 23–A) are formatted with time frames guiding each phase or step and may be several pages long. Outcomes at the bottom of the page must be addressed and guide the caregiver toward goals of that particular step. Interventions in categories entitled "Nutrition," "Activity," "IVs/Medication," "Assessment," "Treatments," "Diagnostics," "Teaching," "Discharge Planning," and "Psychosocial" guide the caregiver. At this time, there are no standard orders that correlate with the Lumbar Laminectomy CareMap®, so physicians' orders must drive patient care. Consults for caregivers such as physical therapists or social workers are suggested on the CareMap®, and a space is provided to communicate with others when the consultant has seen the patient. Separate sections for individual patient needs and individual patient outcomes allow patient-specific information to be communicated. Outcomes are marked as having been met or not met by the end of the specified time frame, usually midnight. The CareMap® is retained in a special section of the medical record.

CMs and CNSs have played a vital role in the creation of CareMaps® and in the education of caregivers and physicians. Nurses in both these roles continue to be liaisons to physicians to provide and analyze variance information and assist with quality improvement initiatives. There has been some ambiguity in the roles of the CNS, CM, and social worker throughout this process. Many CNSs were asked in mid-1996 to assume a dual role and become CNSs/CMs for their particular areas. Clarification continues to be sought as all parties struggle to ensure this system's success while also reducing duplication of work and effort.

Although all pertinent disciplines were involved with the creation of the CareMaps®, the document is still seen predominantly as a nursing tool and its completion as a nursing responsibility. A revision of CareMaps® was completed and implemented in February 1997, and education was repeated to involve other disciplines. At present, no formal guideline exists for frequency of CareMap® revisions; rather, changes are made as practices change and variance analyses warrant. All CareMaps® are approved by the appropriate medical divisions and by the medical executive committee.

OUTCOMES

Studies have been done to analyze the outcomes of low back pain patients and those with spinal disorders.[3,4] In 1994, to determine patient satisfaction, members of the orthopaedics-neuroscience team began to make patient phone calls post discharge. A simple questionnaire was designed to determine patients' postoperative status. Patients were selected and telephoned at random and included anyone discharged from the neurosurgical or orthopaedic unit, not just laminectomy patients. By mid-1994, the orthopaedics-neuroscience continuous improvement council, a multidisciplinary quality team, searched for scientifically valid and reliable tools to use for patient interviews. The goal was to determine patient functional and satisfaction data as well as to link these data to critical pathway variance information. After searching the literature from the back pain patient outcomes research team study, the decision was made to use tools from the Health Outcome Institute's Outcomes Management System. It was determined that no other measures had been identified that had been widely used for these purposes.[5] This system was developed in the late 1980s by working with leading researchers, clinicians, professional organizations, and government agencies. The instruments were designed to be administered to the patient at two points in time: upon admission to the hospital before surgery and at six months post discharge. This battery also included a physician baseline and diagnosis questionnaire that abstracted information from the patient's chart. In addition to these question-

naires, selected single-item questions were used from the Medical Outcomes Study (MOS-6) functional health status questionnaire.[6]

Since beginning the patient questionnaire study in 1995, 327 patients have been surveyed. Major findings from data collected six months post discharge are presented. As can be seen in Table 23–1, the sample consisted primarily of male patients (55 percent) with an average age of 48 years and an average of 1.39 comorbidities. The results displayed in Figure 23–1 show that most patients experienced few symptoms at six months post discharge. For the overall sample, between 63 and 83 percent reported no or occasional symptoms of back pain, leg pain, numbness, tingling, or weakness when asked, "During the past week, how often have you had each of the following symptoms?" The response categories from which they could choose were *never, occasionally, every day, many times a day,* and *all of the time.* No significant differences by DRG were observed.

As shown in Figure 23–2, when patients were asked at six months post discharge, "During the past week how bothersome were the following symptoms?" more variation existed between symptoms. The response categories to which patients could respond were *not at all bothersome, slightly bothersome, moderately bothersome,* and *extremely bothersome.* The percentage of patients reporting not being bothered or occasionally being bothered ranged from 49 percent for back pain to 77 percent for weakness. Chi-square testing revealed significant differences by DRG ($\chi = 5.52$, $p < .02$) for back pain. Patients in DRG 214 reported not being bothered only 40 percent of the time, whereas patients in DRG 215 reported not being bothered 54 percent of the time. Thus,

Table 23–1 Demographic Characteristics of the Patients Who Responded to the Six-Month Postdischarge Questionnaire (N = 327)

	Total Sample	DRG 214	DRG 215
Gender			
Female	45%	48%	42%
Male	55%	52%	58%
Age			
<35	17%	12%	19%
36–45	29%	20%	35%
46–55	24%	27%	22%
56+	28%	37%	23%
Average age	48 years	51 years	46 years
Average comorbidities	1.39	2.14	1.03
DRG			
214	37%		
215	63%		

Source: Copyright © St. Thomas Health Services.

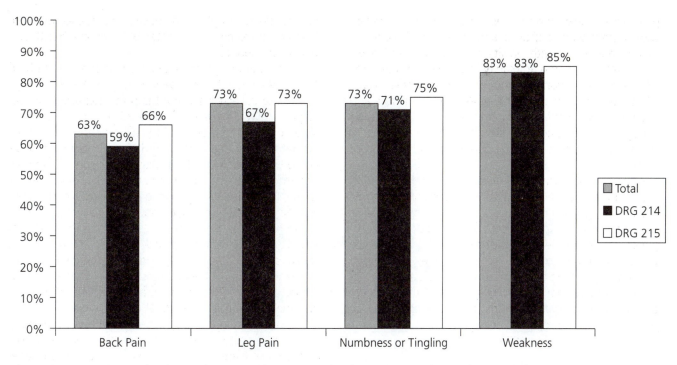

Figure 23–1 Percentage of Patients Who Reported No or Occasional Symptoms at Six Months post Discharge. *Source:* Copyright © St. Thomas Health Services.

a greater percentage of patients in DRG 214 experienced discomfort with those symptoms than did patients in DRG 215. This might be expected because DRG 214 includes patients with back or neck procedures having comorbidities. No other significant differences by DRG were noted.

Patients were asked to rate their satisfaction with the information that staff gave them regarding their condition, surgery, and recovery. As can be seen in Figure 23–3, the majority of patients (81 percent) were satisfied with the results of their surgery. When asked, "If you had to spend the rest of your life with your back condition as it is now, how would you feel about it?" 73 percent of the patients responded that they were delighted, pleased, or mostly satisfied. Only 27 percent of the patients responded that they felt equally satisfied and dissatisfied, mostly dissatisfied, unhappy, or terrible. As noted in Figure 23–4, there were no significant differences by DRG, but slightly more patients in DRG 215 (75 percent) were satisfied than were patients in DRG 214 (69 percent). This might be explained by the fact that patients in DRG 214 had other medical conditions (comorbidities) that possibly affected their overall recovery and satisfaction from surgery.

The majority of patients (84 percent) who were employed full or part time preoperatively reported having returned to work when surveyed six months postoperatively. As can be seen in Figure 23–5, there were differences by DRG. Chi-square testing revealed that significantly more employed patients in DRG 215 (92 percent) were able to return to work than patients in DRG 214 (67 percent; $\chi = 13.7$, $p < .001$). Again, there is a correlation between the number of comorbid conditions a patient has and the likelihood that patient will be able to return to work.

Currently, questionnaire results are correlated with CareMap® variance data. Outcomes on the CareMap® have been chosen as important milestones to be achieved. Through experience, it was learned that the large *number* of outcomes on the CareMap® led to a scattered approach to variance analysis. In the future, we will strive to focus on key outcomes that will help to improve care, rather than to track many different outcomes. It has been noted that lumbar laminectomy patients consistently meet the outcomes chosen, with very few variances documented. Variance data worth monitoring include the ability and timing of patients' postoperative ambulation as well as the effectiveness of pain control in correlation to length of stay. Tracking of these outcomes continues.

Data for patients surveyed with the six-month follow-up questionnaire showed a change in length of stay and

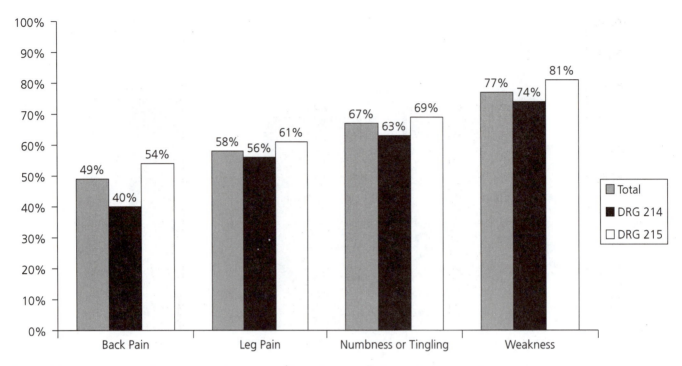

Figure 23–2 Percentage of Patients Who Reported Not Being Bothered or Being Only Slightly Bothered by Symptoms at Six Months post Discharge. *Source:* Copyright © St. Thomas Health Services.

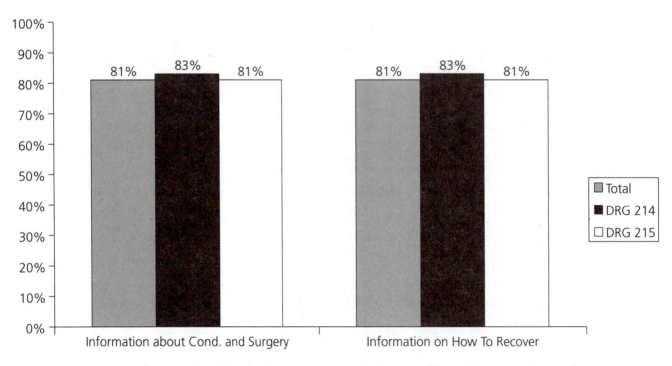

Figure 23–3 Percentage of Patients Satisfied with Information Given Regarding Condition and Surgery and How To Recover. *Source:* Copyright © St. Thomas Health Services.

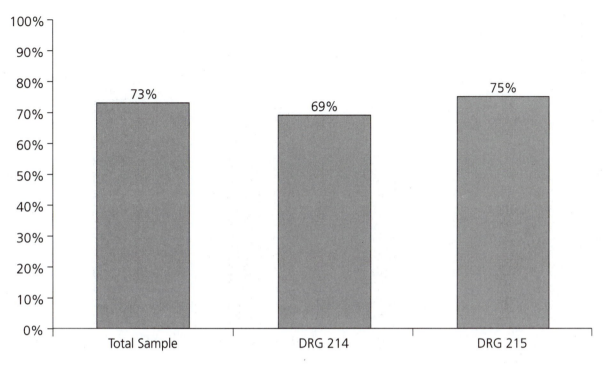

Figure 23–4 Percentage of Patients Satisfied with Back Condition at Six Months post Discharge. *Source:* Copyright © St. Thomas Health Services.

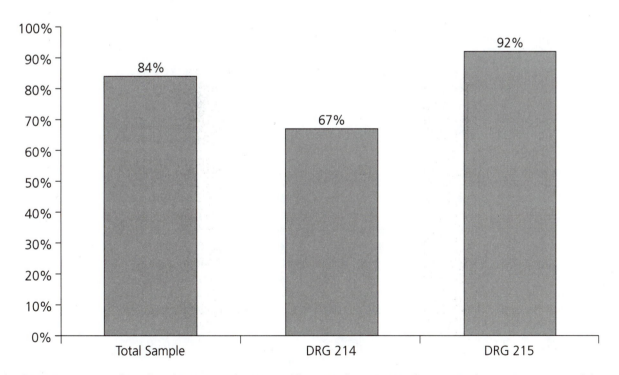

Figure 23–5 Percentage of Employed Patients Who Were Able to Work at Six Months post Discharge. *Source:* Copyright © St. Thomas Health Services.

financial outcomes. Length of stay decreased from 1.9 days in quarter 4 of fiscal year 1995 to 1.4 days in quarter 4 of fiscal year 1996. Patient charges decreased by 8.7 percent from quarter 4 of fiscal year 1995 to quarter 4 of fiscal year 1996. These data continue to be tracked, and trends are shared with caregivers and physicians.

VARIANCE TRACKING

Caregivers are responsible for CareMap® and variance documentation. Over time, staff have arrived at an increased comfort level with these tools. Compliance with documentation and utilization of the CareMaps® has improved since implementation. Nurses use the CareMaps® to give shift reports, with a focus on interventions and outcomes. The CareMaps® may also be used when consulting with physicians about patient concerns, although this is not done on a consistent basis. Ongoing formal and informal reeducation is done regarding outcomes-based practice, case management, and CareMaps®. Periodic audits of CareMap® documentation, CareMap® variance data, and patient questionnaire information provide the basis for dialogue with staff and physicians. It is believed that staff involvement and their understanding of the case management process and their role are vital to success.

Variance documentation occurs directly on the CareMap® as each outcome is met or not met at the end of each designated time frame. Outcomes not achieved require documentation on the "Outcome Tracking Tool" located at the front of each CareMap®. Within this section, each outcome is listed with choices of common reasons that patients might not achieve the outcome expected. Any outcome not met and requiring intervention necessitates documentation in the medical record. An "other" blank allows the caregiver to document a brief reason for lack of outcome achievement that might not be listed as a variance choice. Variances marked "Other" are coded as such, and specific causes are not tabulated. The presence of a number of "other" variances might alert the CM to investigate the possible need for additional variance choices.

Variance data are entered into a computer system by an individual in medical records. Variance reports are provided for CMs by a data specialist. The challenge to obtain meaningful data since the 1995 creation of CareMaps® has been hindered in part by the large number of outcomes and variance choices on all the CareMaps®. Variance data are analyzed for positive and negative trends. Opportunities for improvement are shared with staff and physicians. Data are reviewed quarterly or as they become available. Follow-up patient questionnaires continue to be analyzed despite changes in personnel and difficulty in questionnaire collections. At this point, only outcomes and variances are tracked by computer. In the future, hopefully within the next five years, CareMaps® and other documentation will be computerized.

PATIENT CAREMAP®

Although a patient pathway was used in the original critical pathway system, at present a correlating patient CareMap® for lumbar laminectomy is not available. As existing pathways were reformatted, a patient pathway standardized format was not created. This will probably be done in the near future and will probably mimic the standardized format for caregiver CareMaps®. At present, various educational materials and verbal instructions are given preoperatively, and patients and families are encouraged to be involved in all aspects of their care. Postoperative instructions are given by physicians and nurses. Patients are instructed to notify their physician for any problems once they are discharged.

OUTCOMES PROCESSES

All outcomes are currently tracked by the CNS/CM, who is the liaison to all the other disciplines. These outcomes, which were agreed upon by the multidisciplinary team, focus on key elements of patient care, such as understanding of the plan of care, ability to ambulate and void, and satisfaction with pain control. Over time, different outcomes may be chosen as variances are tracked and as focus of care changes. Ever-shorter lengths of stay continue to challenge caregivers to educate and care for patients effectively while simultaneously monitoring for variances.

ADVANTAGES AND DISADVANTAGES

Case management and outcomes monitoring of lumbar laminectomy patients have been well received at Saint Thomas. The feedback that physicians receive regarding lengths of stay, charges, patient satisfaction surveys, and CareMap® outcomes has created the basis for healthy dialogue between hospital staff and physicians. This ongoing monitoring of care helps to ensure that best-practice habits are maintained by all caregivers. While physician orders drive patient care, physician-approved CareMaps® are utilized on all laminectomy patients and alert staff to common care practices.

Although it is becoming more common in parts of the country, at present, managed-care companies do not review the lumbar laminectomy CareMap®. The Nashville area is approximately 26 percent managed-care infiltrated. Saint Thomas continues to negotiate managed-care contracts with companies for various DRGs.

Though not necessarily a disadvantage, the massive change from the use of nursing care plans to outcomes management and CareMaps® has required a great deal of commitment, time, and energy by many people. Education must be viewed as a continual process. Staff have different degrees of knowledge about outcomes processes, and new caregivers are continuously being hired. A change such as the one that has been undertaken at Saint Thomas Health Services must be driven by a well-educated, committed, multidisciplinary team. This group must be supported by administration and must be given the authority and resources to oversee such a change.

CONCLUSION

Outcomes assessment and management is a vital part of patient care. In this time of vigorous resource scrutiny, health care providers must be able to prove that the services they provide are relevant, appropriate, and satisfactory. Outcomes management provides a means to view quality of care through the eyes of those receiving care. Outcomes management also allows for the tracking of objective data of predetermined patient populations. Through the process of creating a case management system at Saint Thomas Health Services, successful outcomes have been demonstrated for the lumbar laminectomy population.

Lumbar laminectomy patients who are surveyed six months postoperatively express overall satisfaction with symptoms and functional ability. Most patients, if working preoperatively, report being able to work six months after surgery. Patients report being satisfied with information given by caregivers, even in the face of shorter lengths of stay.

Several lessons have been learned as we have moved through the process of outcomes-based practice. This new system has been implemented in the face of a high patient census and increased patient acuity on the neurosurgical floor. The lack of a computerized documentation system necessitated a careful, lengthy creation and implementation of the new system. The need for increased monitoring and reeducation during simultaneous housewide implementation challenged the committee and required much time and energy.

Saint Thomas Health Services and the multidisciplinary care team are committed to providing quality patient care. The pledge to monitor resource utilization and patient satisfaction is no longer just an option. Health care providers have the obligation to provide the best care possible. Outcomes management is a reflection of customers, including patients, physicians, third-party payers, and caregivers. Effective, ongoing communication and teamwork are the keys to positive outcomes.

REFERENCES

1. Zander K. Nursing case management: resolving the DRG paradox. *Nurs Clin North Am.* 1988;23:503–520.
2. Petryshen PR. The case management model: an innovative approach to the delivery of patient care. *J Adv Nurs.* 1992;17:1188–1194.
3. Long DM, Zeidman SM. Outcomes of low back pain therapy. *Perspect Neurol Surg.* 1994;5:41–52.
4. Keller RB, Atlas SJ, Singer DE, et al. The Maine lumbar spine study, part I. *Spine.* 1996;21:1769–1794.
5. *Publications of the Patient Outcomes Research Teams (PORTS).* Washington, DC: Agency for Health Care Policy and Research; 1994.
6. Stewart AL, Ware JE. *Measuring Functioning and Well-Being: The Medical Outcomes Study Approach.* Durham, NC: Duke University Press; 1992.

■ Appendix 23–A ■

CareMaps®

Exhibit 23–A–1 CareMap® for Lumbar Laminectomy

CareMap®: Lumbar Laminectomy

Inclusion: Any patient admitted for a "simple" lumbar laminectomy/discectomy
Exclusion: Any patient admitted for a decompressive lumbar laminectomy or lumbar fusion

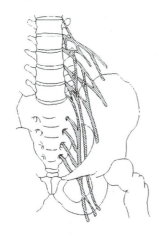

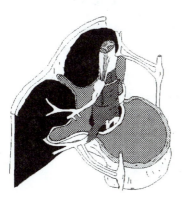
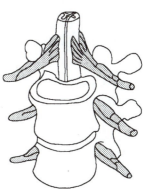

Printed Name	Initial	Printed Name	Initial	Printed Name	Initial

All signatures are those individuals recording information in the CareMap®

continues

Exhibit 23–A–1 continued

Outcome Tracking Tool CareMap®: Lumbar Laminectomy

If outcome unmet, check reason(s); if reason is not listed, check "Other" and document in Progress Notes. Actions taken as a result of unmet outcomes also are charted in the Progress Notes.

#1 Patient verbalizes understanding of plan of care
- ❑ Patient too sick for discussion
- ❑ Communication barrier
- ❑ Information not provided at this time
- ❑ Refusal to participate
- ❑ Other _____

#2 Patient verbalizes understanding of plan of care
- ❑ Patient too sick for discussion
- ❑ Communication barrier
- ❑ Information not provided at this time
- ❑ Refusal to participate
- ❑ Other _____

#3 Dressing dry/intact
- ❑ Dressing saturated: reinforced
- ❑ Dressing saturated: changed
- ❑ Other _____

#4 Patient verbalizes satisfaction with pain control
- ❑ Patient states intervention ineffective
- ❑ Medication held
- ❑ Patient refuses medication
- ❑ Knowledge deficit
- ❑ Communication barrier
- ❑ By patient report pain level is acceptable
- ❑ Med not ordered
- ❑ Pain score not obtainable due to patient condition
- ❑ Other _____

#5 Physiological stability
(lungs clear, HR 60–110; BP 90/50–160/100; Temp < 100.5°; Respirations 10–20; UOP > 240 cc/8 hr)
- ❑ Abnormal breath sounds
- ❑ Heart rate below baseline or > 110
- ❑ Temperature > 100.5°
- ❑ Blood pressure: SBP < 90 or > 160 DBP < 50 or > 100
- ❑ Other _____

#6 Neuro status stable
(Alert & Oriented, moves all extremities, speech clear, swallows without difficulty)
- ❑ Patient disoriented or confused
- ❑ Weakness, numbness, or tingling in extremity not present prior to surgery
- ❑ Difficulty clearing oral secretion or swallowing
- ❑ Communication barrier
- ❑ Prior condition
- ❑ Other _____

#7 Patient ambulates independently
- ❑ Unable due to physical condition
- ❑ Refuses to participate
- ❑ Pre-existing condition
- ❑ Unable due to patient's verbalization of pain
- ❑ Other _____

#8 Physiological stability
(lungs clear, HR 60–110; BP 90/50–160/100; Temp < 100.5°; Respirations 10–20; UOP > 240 cc/8 hr)
- ❑ Abnormal breath sounds
- ❑ Heart rate below baseline or > 110
- ❑ Temperature > 100.5°
- ❑ Blood pressure: SBP < 90 or > 160 DBP < 50 or > 100
- ❑ Other _____

#9 Incision clean, dry, well-approximated
- ❑ Incisional drainage
- ❑ Irregularity in wound approximation
- ❑ Incision red and tender
- ❑ Documented infection
- ❑ Other _____

#10 Patient verbalizes satisfaction with pain control
- ❑ Patient states intervention ineffective
- ❑ Medication held
- ❑ Patient refuses medication
- ❑ Knowledge deficit
- ❑ Communication barrier
- ❑ By patient report pain level is acceptable
- ❑ Med not ordered
- ❑ Pain score not obtainable due to patient condition
- ❑ Other _____

#11 Patient verbalizes understanding of D/C instructions
- ❑ Patient too sick for discussion
- ❑ Communication barrier
- ❑ Information not provided at this time
- ❑ Other _____

#12 Discharge disposition
- ❑ Home with self/family care
- ❑ Home with Home Health
- ❑ Home with Hospice
- ❑ Rehab facility
- ❑ Skilled nursing facility
- ❑ Patient expired
- ❑ Other _____

continues

Exhibit 23–A–1 continued

CareMap®: Lumbar Laminectomy	
Time Interval	**Pre-Admission (1–4 weeks)/Pre-Op** **Date**_____
Nutrition	Clear Liquid for CT and myelogram
Activity	As ordered
IVs/ Medication	As ordered Evaluate for food/drug interactions
Assessment	Vital signs and Neuro Checks as ordered Pain Assessment every 8 hours and PRN Skin Assessment Routine Post-myelogram assessment (puncture site)
Treatments	Obtain permit for myelogram
Diagnostics	CT, myelogram, MRI and/or EMG, as ordered Lab as ordered
Teaching	Evaluate learning needs; document on Education Records CT STEPS Myelogram STEPS MRI STEPS Pre/Post-Op STEPS
Discharge Planning	Evaluate Discharge Planning needs Social Service Consult Indicated: **YES** ☐ **NO** ☐ Requested: **Date**_____ **Initial**_____ Consult Initiated: **Date**_____ **Consultant's Initial**_____
Psychosocial	Pastoral Care Consult Indicated: **YES** ☐ **NO** ☐ Requested: **Date**_____ **Initial**_____ Consult Initiated: **Date**_____ **Consultant's Initial**_____
Individual Patient Needs	

Individual Patient Outcomes		Date	Time	Initial

Outcomes		Date	Time	Initial
	1. Patient verbalizes understanding of plan of care **YES** ☐ **NO** ☐ 1			

Treatments documented on the CareMap® do not need to be duplicated on the Flowsheet.

continues

Exhibit 23–A–1 continued

CareMap®: Lumbar Laminectomy				
Time Interval	**PAT/EMA Day 1 (before surgery)** Date_____			
Nutrition	NPO			
Activity	As ordered			
IVs/ Medication	As ordered Evaluate for food/drug interactions			
Assessment	Vital signs and Neuro Checks as ordered Pain Assessment every 8 hours and PRN Skin Assessment daily and PRN Pre-op anesthesia evaluation DVT Risk Assessment			
Treatments	Obtain operative permit			
Diagnostics	Lab according to PAT orders			
Teaching	Evaluate learning needs; document on Education Records Lumbar Laminectomy STEPS Discuss treatment plan, expected outcomes, post-op progression			
Discharge Planning	Evaluate Discharge Planning needs Social Service Consult Indicated: **YES** ❏ **NO** ❏ Requested: **Date**_____ **Initial** _____ Consult Initiated: **Date**_____ **Consultant's Initial**_____			
Psychosocial	Pastoral Care Consult Indicated: **YES** ❏ **NO** ❏ Requested: **Date**_____ **Initial** _____ Consult Initiated: **Date**_____ **Consultant's Initial**_____			
Individual Patient Needs				
Individual Patient Outcomes		Date	Time	Initial
Outcomes		Date	Time	Initial
	2. Patient verbalizes understanding of plan of care YES ❏ NO ❏ 2			

Treatments documented on the CareMap® do not need to be duplicated on the Flowsheet.

continues

Exhibit 23–A–1 continued

CareMap®: Lumbar Laminectomy	
Time Interval	**Day 1 continued PACU-2359** **Date**_____
Nutrition	Diet as ordered after surgery Advance as tolerated
Activity	Out of bed with assistance as tolerated Ambulate in room—progress to hall as tolerated PT Consult Indicated: **YES** ❑ **NO** ❑ Requested: **Date**_____ **Initial** _____ Consult Initiated: **Date**_____ **Consultant's Initial**_____
IVs/ Medication	IV and Medications as ordered Evaluate for food/drug interactions
Assessment	Vital signs and Neuro Checks every 4 hours x first 24 hours Monitor intake and output x 24 hours Pain Assessment every 8 hours and PRN Skin Assessment daily and PRN Assess wound drainage/dressing
Treatments	Cough/deep breathing 10 breaths every two hours while awake for first 24 hours post-op Dressing, check for drainage every 2–4 hours, reinforce/change dressing as needed TNS unit, as ordered
Diagnostics	Lab and tests as ordered
Teaching	Evaluate learning needs; document on Education Records Teach positioning and transfer techniques Teach proper body mechanics Initiate discharge instructions regarding medications, wound care, and activity Medrol STEPS, if applicable Teach use of TNS unit, if applicable
Discharge Planning	Evaluate Discharge Planning needs Social Service Consult Indicated: **YES** ❑ **NO** ❑ Requested: **Date**_____ **Initial** _____ Consult Initiated: **Date**_____ **Consultant's Initial**_____ CNS/CM Consult Indicated: **YES** ❑ **NO** ❑ Requested: **Date**_____ **Initial** _____ Consult Initiated: **Date**_____ **Consultant's Initial**_____
Psychosocial	Pastoral Care Consult Indicated: **YES** ❑ **NO** ❑ Requested: **Date**_____ **Initial** _____ Consult Initiated: **Date**_____ **Consultant's Initial**_____
Individual Patient Needs	

Individual Patient Outcomes		Date	Time	Initial

Outcomes		Date	Time	Initial
3. Dressing dry/intact YES ❑ NO ❑	3			
4. Patient verbalizes satisfaction with pain control YES ❑ NO ❑	4			
5. Physiological stability (lungs clear; HR 60–110; BP 90/50–160/100; **Temp < 100.5°; Respirations 10–20; UOP > 240 cc/8 hr)** YES ❑ NO ❑	5			
6. Neuro Status Stable (Alert & oriented, moves all extremities, **speech clear, swallows without difficulty)** YES ❑ NO ❑	6			

Treatments documented on the CareMap® do not need to be duplicated on the Flowsheet.

continues

Exhibit 23–A–1 continued

CareMap®: Lumbar Laminectomy				
Time Interval	**Day 2 2400–2359** Date_____			
Nutrition	Regular diet as ordered			
Activity	Ambulate in hall PT Consult Indicated: **YES** ❑ **NO** ❑ Requested: **Date**_____ **Initial** _____ Consult Initiated: **Date**_____ **Consultant's Initial**_____			
IVs/ Medication	D/C IV fluids if PO intake adequate and antibiotics completed Evaluate for food/drug interactions			
Assessment	Vital signs and Neuro Checks every 12 hours or PRN Pain Assessment every 8 hours and PRN Skin Assessment daily and PRN Assess incision, change dressing as needed			
Treatments	Cough/deep breathing TNS unit, if ordered			
Diagnostics	Lab and tests as ordered			
Teaching	Evaluate learning needs; document on Education Records Reinforce proper positioning and transfer techniques Reinforce discharge instructions regarding medications, wound care, and activity Give patient 4 x 4s and 3 occlusive dressings for wound care post discharge			
Discharge Planning	Evaluate Discharge Planning needs			
Psychosocial	Patient/Family support			
Individual Patient Needs				
Individual Patient Outcomes		Date	Time	Initial
Outcomes		Date	Time	Initial
	7. Patient ambulates independently YES ❑ NO ❑ 7			
	8. Physiological stability (lungs clear; HR 60–110; BP 90/50–160/100; Temp < 100.5°; Respirations 10–20; UOP > 240 cc/8hr) YES ❑ NO ❑ 8			
	9. Incision clean, dry, well-approximated YES ❑ NO ❑ 9			
	10. Patient verbalizes satisfaction with pain control YES ❑ NO ❑ 10			
	11. Patient verbalizes understanding of D/C instructions YES ❑ NO ❑ 11			
	12. Discharge disposition (refer to Outcome Tracking Tool) YES ❑ NO ❑ 12			

Treatments documented on the CareMap® do not need to be duplicated on the Flowsheet.

Note: BP, blood pressure; CM, case manager; CNS, clinical nurse specialist; CT, computed tomography; DBP, diastolic blood pressure; D/C, discharge, discontinue; DVT, deep-vein thrombosis; EMG, electromyelogram; HR, heart rate; IV, intravenous; MRI, magnetic resonance imaging; PACU, post anesthesia care unit; PAT/EMA, pre-admission testing/early morning admission; PO, by mouth; PRN, as needed; PT, physical therapy; SBP, systolic blood pressure; STEPS, Saint Thomas Education for Patients; TNS, transcutaneous nerve stimulation; UOP, urinary output.

CareMap® reproduced with special permission. All rights reserved. Certain questions are shaded to indicate key outcomes that should be met before the patient is discharged.

■ 24 ■

Total Joint Replacement Outcome Improvement

Shelly C. Anderson, Patricia A. Soper, Shannon Ericson, and Danny Gurba

At Saint Luke's Hospital of Kansas City, a 642-bed urban tertiary teaching hospital in Kansas City, Missouri, quality of care is the focus of our mission, and continual improvement of outcomes is a major organizational strategy. In 1992, development of a collaborative care program with associated tools such as clinical paths was instituted as a means to ensure appropriate clinical processes and to improve outcomes. This chapter describes the development of the collaborative care program for joint replacement patients, highlights the associated tools of a knee replacement clinical path and patient path, and describes the outcomes from this effort.

In 1992, managed care was increasing its community penetration, and it was apparent that for Saint Luke's Hospital to retain its quality reputation and remain competitive in the marketplace, changes were required. Clinical outcomes needed to be maintained or enhanced and financial results improved. This required careful scrutiny and appropriate revisions of clinical care processes. In response to this challenge, development of a collaborative care program began in April 1992 and involved broad stakeholder representation. This program was developed to adapt the principles of internal case management through implementation of a collaborative interdisciplinary care delivery model designed for specific patient populations.

Collaborative practice teams were formed to focus on the care of certain patient groups, defined generally by diagnosis-related groups (DRGs). Two of the first patient groups selected were total joint replacement DRGs 209 (unilateral hip or knee) and 471 (bilateral hip or knee), chosen because they met our initial criteria of high volume, high risk, high cost/loss, and clinician readiness.

- The joint replacement patient population represents a high volume within the hospital. In 1992 at Saint Luke's Hospital, DRG 209 ranked fourth in volume by DRG. Hip and knee replacements/revisions accounted for 65 percent of the orthopaedic unit volume and for 21 percent of the operating room cases.
- These patients were at high risk for developing debilitating pain, decreased progressive mobility, embolus and thrombus, and a negative perception of their hospital experience. Referrals were high for joint revisions, as well as for initial joint replacements, and the clinical and social situations were often complex. Many patients and their families lived a distance away from the hospital and were sometimes unfamiliar with the area. Since these are elective procedures with admission delayed until the morning of surgery, an opportunity existed to provide assessment, education, and intervention before hospitalization.
- Patient, physician, and staff satisfaction scores were lower than targeted.
- Costs were higher than those of benchmarked hospitals, and there was a significant monetary loss associated with caring for this patient population.
- The average length of stay (ALOS) for these patients was 9.0 days, below the Health Care Financing Administration's mean length of stay of 10.2 days but higher than that of benchmarked hospitals.
- The orthopaedic program at Saint Luke's Hospital continues to be respected in the region as a center of excellence, drawing referrals from a wide geographic radius. At the onset of the collaborative care program, the orthopaedic clinical staff and physicians demonstrated a readiness to chart a course with a new program designed to improve outcomes and processes. They had an established team approach on a dedicated orthopaedic unit, and they

showed an interest and willingness to commit the time and energy necessary to design an improved care delivery system.

PATHWAYS AND PROTOCOLS

The newly formed orthopaedic collaborative practice team consisted of clinicians dedicated to the orthopaedic department: surgeons, staff nurses, physical and occupational therapists, a nutritionist, a social worker, a chaplain, and a nurse manager. Others participated with the team: the medical/surgical clinical nurse specialist, the collaborative care coordinator, a utilization review specialist, a medical records specialist, and a financial specialist. Consultants to the team included a pharmacist, a pathologist, and laboratory and radiology personnel. Members of the team were selected by the orthopaedic surgeons, the orthopaedic nurse manager, and the orthopaedic nursing professional practice committee. Team members agreed to a two-year commitment, with the goal of initiating the collaborative care program in orthopaedics and ensuring its continued success. While the initial focus was inpatient care, the team soon realized the value of the outpatient component and included the prehospital and home health practitioners in the collaborative care process.

Goals of the collaborative care program were to

- maintain/improve patient outcomes
- increase continuity of care while allowing patient individualization
- increase patient and family understanding of, participation in, and satisfaction with care
- identify opportunities and provide mechanisms for quality improvement
- promote appropriate, cost-effective resource utilization
- operationalize the value-added concept

Team roles were carefully defined and periodically revised. The collaborative care director provided extensive education and rationales for the team regarding the current and projected local and national health care financial environment, case management concepts, data collection and analysis tools, clinical path models, and effective team skills. The nurse manager provided necessary resources (i.e., time available for staff to participate, support of the program and the team). The surgeons and the medical/surgical clinical nurse specialist contributed clinical expertise, scientific literature, support, and assistance with team process, including consensus building and conflict resolution. The clinical staff provided their knowledge of this patient population and the clinical care processes involved. The medical records, utilization review,

and finance members provided data necessary to define accurately the current state related to patient volume, case mix, demographics, utilization of resources, ALOS, clinical outcomes, cost and reimbursement figures, and marketplace demands.

An all-day meeting was held off-site to facilitate this process and to define goals and timelines. At this meeting, the team reviewed the case mix of the joint replacement patients at Saint Luke's Hospital. They determined that knee patients possessed enough differences in treatment modalities from hip patients that a separate path for each would be beneficial. Bilateral joint replacement care, however, was felt to be similar enough to unilateral replacement that they could be combined. The team then drafted a knee replacement clinical path that outlined their current practice, using chart audits and empirical data. At subsequent meetings, the involved disciplines reviewed the literature for examples of "best practice," examined clinical paths obtained from other hospitals, and collaboratively revised the path to reflect improved outcomes, starting with discharge goals and then adding daily patient goals and clinical interventions to achieve those goals within the desired timeline. The expectation was that the path would predict the needs of 80 percent of the knee replacement patient population. The process of review, revision, and validation continued, and each draft was sent to the orthopaedic surgeons and the staff for comment. When the team reached consensus, the final draft was prepared as the interdisciplinary plan of care for knee replacement patients. Final approval from various groups was required before implementation of the path in November 1992.

The initial format changed over time as the staff suggested improvements. The front page of the clinical path (Exhibit 24–A–1 in Appendix 24–A) defines the general plan of care. This begins with specific patient information, such as surgeon and date of surgery. The targeted length of stay was defined by the team and written on the first path as eight days (one day less than actual ALOS). On later revisions, the targeted length of stay was omitted from the path because physicians believed this hindered earlier discharge in some cases.

Anticipated patient problems and nursing diagnoses for knee replacement patients are listed next. These include knee pain/acute pain related to operative procedure and range-of-motion limitation or impaired physical mobility related to operative procedure. Initially, the path included only nursing diagnoses, but it was revised to include a problem statement, since we found that non-nursing colleagues related better to a problem list than to nursing diagnoses. For each problem, the outcomes expected for these patients and the interdisciplinary staff interventions required to meet those outcomes are defined.

Pages 2 through 4 of the clinical path outline the general plan of care in more detail. The vertical columns refer to designated time frames. In the case of the knee replacement path, the measured time frame is days, as opposed to some paths for other diagnoses, in which the time elements might be minutes, stages, or weeks. The last column on page 4 defines patient outcomes required for discharge. The horizontal rows refer to care categories (i.e., assessment, consults, etc.). This format is standardized throughout the hospital, with minor variations to accommodate specific patient populations. Within this grid, daily patient outcomes are defined to ensure that the discharge outcomes are achieved within the defined time frame. Also included are the physician and staff interventions required to assist the patient in achieving each goal. For example, to resolve the mobility problem and to achieve the discharge outcome of independent mobility, the patient must begin physical therapy on postoperative day 2.

In developing the clinical path, we found it most effective first to define the general care plan for the knee replacement patient, then to determine expected discharge goals, and finally to fill in the grid each day, working toward each discharge goal. Over time, the goals and interventions in the grid changed as the clinicians and patients determined more effective and efficient processes to achieve improved outcomes.

The clinical path is used as a guide rather than as a standard of care and is expected to be individualized for each patient. As such, interventions on the path that require a physician order are not instituted until the order is written. To standardize and streamline that process, the surgeons developed standing orders reflecting the path interventions.

Using the path for interdisciplinary documentation was an initial goal of the team, but it was not achieved until the path had been in use for a number of months. Because this was a change in charting procedures for all disciplines, the new process required negotiation and consensus from the practice committees of each discipline.

To standardize procedures and to facilitate documentation, various protocols were developed by the team. For example, in the medications row for each hospital day, "bowel protocol" is listed; in the teaching row for operative day, "instruct post-op interventions per knee protocol" is indicated. All protocols are available on the unit for staff reference, but the content is not repeated in the path.

Originally, the team shared the clinical path with patients and families, but they soon found that it was too detailed for patients to comprehend. With assistance from patients and families, the team developed a patient path (Exhibit 24–A–2 in Appendix 24–A) as an educa-

tional tool for patients. It outlines the plan of care and indicates what the patient can expect from the staff and the facility, as well as what responsibilities are expected of the patient. This tool has been revised periodically on the basis of changes in processes and as a result of patients' feedback regarding their learning needs and their desire for involvement, accountability, and satisfaction with the care experience.

In mid-1996, the home health service of Saint Luke's nurses and physical and occupational therapists became members of the orthopaedic team. They developed a clinical path for knee replacement patients in home care. This path continues the hospital path but is not used as a documentation tool.

CARE/CASE MANAGEMENT

Clinical nurses at the bedside assume responsibility for the care management of all knee replacement patients. All of these patients are managed through the use of the clinical path. Progress on the path is evaluated each shift, additional interventions are initiated if necessary to reduce variation, and path progress is discussed in the shift report. The patient path is provided to and reviewed with the patient and family before hospitalization, often in the surgeon's office or in the Preoperative Assessment Center. It is again reviewed with the patient on at least a daily basis during hospitalization.

For ease of use by the interdisciplinary care team, the clinical path is located in the patient's chart at the bedside. Physicians may refer to the path during daily rounds. Nurses, the social worker, the nutritionist, and the physical and occupational therapists use the path as a care plan, shift report, and documentation tool. The utilization review specialist analyzes the path daily as part of concurrent chart review and puts pertinent information into a database via laptop computer. The clinical nurse specialist reviews the path when intervening for a specific patient. Nurses who "float" to the orthopaedic unit rely on the path to define the individualized plan of care, and nurses orienting to the orthopaedic unit use it as a learning tool.

In the past, a multidisciplinary clinical team on the orthopaedic unit held weekly interdisciplinary discharge planning rounds to discuss each patient's progress toward discharge. In 1994, as length of stay declined and acuity increased, this process was replaced by a newly formed orthopaedic care coordination team (care team). This care team consists of the medical/surgical clinical nurse specialist as leader, the orthopaedic social worker, and the orthopaedic utilization review specialist. They communicate daily with each other and with the orthopaedic physicians and clinical providers to review the progress and needs of each orthopaedic patient and to assist the direct care providers to meet specific patient needs. The care team formally makes rounds on the orthopaedic unit at least three times per week and is particularly valuable in cases in which patients are not progressing according to path criteria. The home health nurse coordinator works closely with the staff and the care team. Knee replacement patients planning home assistance are visited and assessed by the nurse coordinator and assist in developing their plan of care.

OUTCOMES

Outcome measurement is both a concurrent and a retrospective process that commences with the outpatient Preoperative Assessment Center (PAC) visit and continues through hospitalization and home care. Four categories of outcomes are measured throughout the continuum of care for the total knee replacement patient: discharge outcomes, clinical outcomes, patient satisfaction outcomes, and fiscal outcomes. In addition to patient outcomes, satisfaction of the patient care team with the collaborative care process is periodically assessed, and the clinical path/patient path is reviewed quarterly and revised as needed. The team reviews all data and reports its findings to the department of orthopaedics for practice recommendations.

Discharge Outcomes

Discharge outcome criteria are established to ensure patient safety and provide for continuing care based on the patient's assessed needs at the time of discharge from a setting. These criteria focus on activity; nutrition; living environment; mental, emotional, and spiritual components; and physical limitations and changes.

The discharge outcome criteria for the acute care setting (Exhibit 24–1) address pain control and the patient's understanding of pain management in the home setting. Patient safety is measured by assessing the patient's status with bed-to-chair mobility and ambulation. The discharge outcome indicators dictate that independent bed-to-chair mobility and ambulation with minimal assistance be achieved. The patient must be able to verbalize and demonstrate understanding of safety measures and physical limitations. To rule out future potential complications, before discharge a patient must be afebrile for at least 24 hours, the surgical site must be free of signs and symptoms of infection, and the patient must verbalize key changes to report to the physician that might indicate the development of a complication. The discharge outcome criteria are located on the clinical path, and the patient is monitored on a daily basis for meeting criteria.

The clinical nurse documents the achievement of these outcomes during the patient's hospital stay. A quarterly aggregated summary is prepared to examine the correlation between acute lengths of stay and the length of time required to meet the discharge outcomes. The team analyzes the data to determine the necessity of practice changes and clinical path revisions.

Discharge criteria from the home health setting are determined by the home health therapist in collaboration with the patient and physician. The desired outcomes for discharge from the home health component of the total knee joint replacement program are listed in Exhibit 24–2. In cases in which the patient's status changes to one of no longer being homebound, referral for outpatient therapy in the clinic setting may be indicated in order to meet the discharge criteria.

Clinical Outcomes

Fourteen clinical indicator outcomes are assessed for total knee replacement patients (Exhibit 24–3). These indicators are defined by the department of orthopaedic surgery and are based on professional practice standards. Clinical review specialists conduct 100 percent concurrent chart review of the total knee replacement patients to capture occurrences related to these clinical indicators. These data are reviewed quarterly by the team, the orthopaedic department quality committee, and the hospital performance improvement committee to detect trends and need for change.

Exhibit 24–1 Acute Stay Discharge Criteria

Temperature less than 100 degrees for 24 hours
Skin integrity intact
Pain controlled by oral medication
Verbalizes understanding of medications/schedules
Wound edges well approximated without redness or drainage
In and out of bed/chair independently
Ambulates with minimal assistance
Verbalizes/demonstrates ability to perform care at home and/or with support or community resources
Verbalizes understanding of discharge diet/nutrition goals
Elimination patterns reestablished
Verbalizes/demonstrates understanding of safety measures and physical limitations
Identifies support systems
Verbalizes signs and symptoms of infection and demonstrates preventative measures

Exhibit 24–2 Home Health Program Desired Discharge Outcomes

Car transfer: standby assist
Tub transfer: contact guard assist to independent
Stairs: standby assist to independent
Active knee flexion: 110 to 115 degrees
Active knee flexion: 0–5 degrees
Pain is effectively managed by patient
Patient performs home exercise program independently

Source: Copyright © Saint Luke's Hospital of Kansas City. Reproduction of the pathway is by permission of Saint Luke's Hospital of Kansas City. All Rights Reserved.

Patient Satisfaction Outcomes

The patient-focused care approach at Saint Luke's Hospital means more than just being nice to patients. It requires that staff meet the patient's needs as the patient perceives them, not just as the professionals do. Continuous monitoring of patient satisfaction is accomplished through the use of several mechanisms. A concurrent review of patient satisfaction is conducted by the hospital's patient advocates. They are responsible for visiting each patient throughout the hospital stay to assess satisfaction and to intervene if issues arise. The advocate visits on the first full day of hospitalization and on the fourth day of hospitalization; further visits are made any time by request of the patient, family, physician, or employee. Unit

Exhibit 24–3 1996 Clinical Indicator Outcomes

Readmit rate within 72 hours: 0/100 admissions
Unplanned removal/injection of organ/structure: 0.31 (per 100 discharges)
Unplanned return to operating room/special procedures room: 0.15 (per 100 discharges)
Excessive blood loss: 0.53 (per 100 discharges)
Cardiac or respiratory arrest: 0.08 (per 100 discharges)
Unplanned transfers to intensive care unit: 0.61 (per 100 discharges)
Hospital-incurred patient incident: 0.15 (per 100 discharges)
Death: 0.69 (per 100 discharges)
Neurological deficit not present on admission: 0.38 (per 100 discharges)
Pulmonary problems: 1.45 (per 100 discharges)
Deep-vein thrombosis: 0.61 (per 100 discharges)
Cardiac problems: 0.23 (per 100 discharges)
Urinary tract infection: 0.23 (per 100 discharges)
Gastrointestinal complication requiring blood/procedure: 1.15 (per 100 discharges)

Source: Copyright © Saint Luke's Hospital of Kansas City. Reproduction of the pathway is by permission of Saint Luke's Hospital of Kansas City. All Rights Reserved.

staff and managers receive immediate feedback from the advocate.

Patient satisfaction is assessed on a quarterly basis using a written survey process conducted by the system marketing and research department. The acute care experience is separated from home health care and/or outpatient experiences. Patients are selected at random for the confidential survey process. The survey return rate is approximately 30 percent.

The inpatient survey instrument consists of a 5-point Likert scale that assesses the patient's perception of hospital departments, admissions process, physicians, nursing staff, home health staff, facility operations, home care services operations, and the discharge process. Questions are phrased to measure the concepts associated with satisfaction: overall experience, reliability, responsiveness, competence, access, and empathy. Questions related to hospital physicians inquire about a physician's skill and competence, kind and caring mannerisms, availability, and ability to provide clear and complete explanations regarding the patient's condition or experience. Questions related to patient care staff focus on the sensitivity of staff members, provision of skilled treatment, responsiveness, availability, and ability to provide good explanations of care and daily routine.

The Home Health of Saint Luke's Hospital Survey, structured to measure the same satisfaction concepts, also uses the 5-point Likert scale. Survey questions assess the home care nurse, therapist, aide, and equipment delivery person. Year-end 1996 data reveal a 96.8 percent overall satisfaction rate with home health services. Friendliness and courtesy, together with prompt service provision, were the two indicators identified by surveyed patients as the most important. Overall satisfaction with these indicators was 95.2 percent in 1996.

Patient satisfaction data from the orthopaedic service line are reviewed quarterly by all stakeholders and presented to the team, orthopaedic staff and surgeons, and hospital administration. The orthopaedic unit has appointed a patient satisfaction representative who is charged with facilitating and evaluating action plans to continually improve upon patient satisfaction. An upward trend in overall patient satisfaction by the orthopaedic patient population has been realized in the past three years (Figure 24–1).

Patient focus study sessions are scheduled annually for small random groups of orthopaedic patients to elicit perceptions of their experiences with their care. The patients and significant others are invited to a meeting site away from the hospital. A meal is usually served, and discussion is facilitated by a professional moderator. The meeting is recorded in audio and/or video format for later analysis. Using an open-discussion format, participants

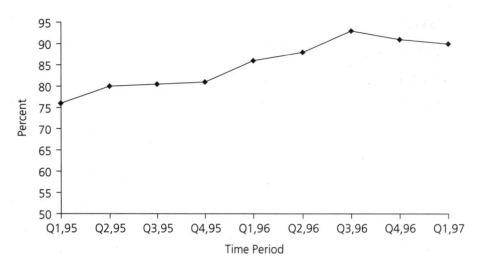

Figure 24–1 Overall Patient Satisfaction, 1995–1997. *Source:* Copyright © Saint Luke's Hospital of Kansas City. Reproduction of the pathway is by permission of Saint Luke's Hospital of Kansas City. All Rights Reserved.

are encouraged to share openly their experiences: how they were satisfied and how they were disappointed with service. Discussion is directed to specific areas/staff functions with which they probably came into contact as orthopaedic patients at Saint Luke's Hospital, including physicians, patient care staff, food service, transportation, admitting, billing, discharge, and home health services. Participants are asked to identify specific changes that could have been made to improve their satisfaction. Results from these meetings are summarized and presented to the patient care staff on the orthopaedic unit, the orthopaedic surgeons, and hospital administration. The videotape is shown to the patient care staff when possible to emphasize that the feedback is from patients cared for on our unit, not merely generic, unfamiliar patients. The unit's patient satisfaction representative uses this feedback in improvement action planning.

The unit nursing staff makes postoperative phone calls to total knee replacement patients discharged to home. These calls are targeted for the patient's second week in the home setting. Open-ended questions are posed, and the patient is asked to describe experiences with patient care team members, effectiveness of the preoperative education program, and experiences with discharge planning. Information from these calls is summarized and presented to the team. One result of such feedback was production of a video now used in the preoperative education program to assist the patient to visualize physical therapy exercises.

Patient perception is also collected independently by the National Research Corporation® (NRC).[1] In NRC's *1996 Healthcare Market Guide V,* the number one consumer preference for orthopaedics care among Kansas City hospitals was Saint Luke's Hospital of Kansas City (Figure 24–2). The second-choice hospital, 5 percentage points below, was Shawnee Mission Medical Center, our other metro hospital partner in the Saint Luke's-Shawnee Mission Health System. Saint Luke's Hospital is strongly committed to ensuring patient satisfaction while rendering high-quality care, as evidenced by its achievement of the Missouri Quality Award in 1995 and the National Quality Health Care Award in 1997.

Fiscal Outcomes

The clinical path for the total knee replacement patient was implemented in November 1992. Since then, inpatient cost has sustained a steady decline (Figure 24–3). These declining costs are attributable to several interdisciplinary efforts, including increased awareness of resource consumption, standardization and contracting with vendors for joint components, decreasing length of acute hospital stay, development of a skilled nursing unit at the hospital, and development of home health protocols to enhance rehabilitation outcomes following hospital discharge. Economies of scale were realized as patient volume increased from 207 patients in 1992 to 251 patients in 1996. Of note is a 17 percent savings of over $940,000 (inflation adjusted) realized in 1996 as the result of contracting efforts in purchase of joint prosthesis components.

Compared to national and regional data, data for Saint Luke's Hospital show a shorter length of stay at a lower cost. In 1992, the ALOS for a total knee replacement patient at Saint Luke's Hospital was 9.0 days, with an average cost of $13,200. By 1996, ALOS had decreased to

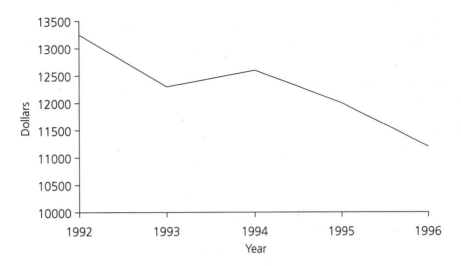

Figure 24–2 Mean Cost for Total Knee Replacement Patients, 1992–1996. *Source:* Copyright © Saint Luke's Hospital of Kansas City. Reproduction of the pathway is by permission of Saint Luke's Hospital of Kansas City. All Rights Reserved.

6.1 days, and average cost to approximately $11,000. The Prospective Payment Assessment Commission (ProPAC) Report to Congress data for 1995 revealed a total knee replacement ALOS of 7.59 days at an approximate cost of $18,500.[2] The 1995 data for the West North Central Region (including Iowa, Kansas, Minnesota, Missouri, Nebraska, North Dakota, and South Dakota) reported an ALOS of 6.85 days and an average cost of approximately $14,500.

As a member of the care coordination team, the utilization review specialist monitors the patient's ongoing progress during the hospital stay. Direct communication with third-party payers is established to relay information

regarding the patient's mobility and safety throughout the acute care stay. Progress toward the discharge criteria is reviewed with third-party case managers to negotiate for continuing benefits on behalf of the patient. The utilization review specialist also monitors for potentially avoidable hospital days and works with the social worker and the clinical nurse specialist to minimize these occurrences. Careful planning and early anticipation of discharge needs have led to a decrease in the potentially avoidable days since 1995.

Home Health of Saint Luke's Hospital has been a longtime participant in the care of the total knee replacement patient. In 1996, 58 percent of total knee replacement pa-

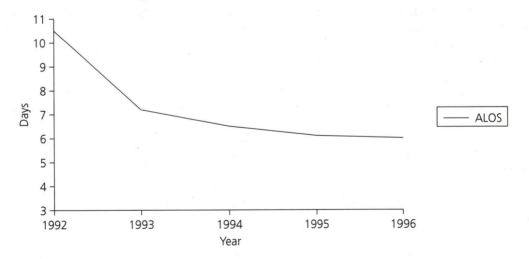

Figure 24–3 Average Length of Stay (ALOS) for Total Knee Replacement Patients, 1992–1996. *Source:* Copyright © Saint Luke's Hospital of Kansas City. Reproduction of the pathway is by permission of Saint Luke's Hospital of Kansas City. All Rights Reserved.

tients were referred to the home health program, the average length of episode per beneficiary was 35.1 days, and the average number of visits per beneficiary was 17.5. In comparison, national data on hospital-based home health programs contained in the 1996 ProPAC reveal a 35.9 percent referral rate, an average length of episode per beneficiary of 44.9 days, and 24.4 average visits per beneficiary. While the ProPAC Report does not enumerate average visits for all services for DRG 209 (including the total knee joint replacement patient population), it does state that 29 percent (7.08 visits) of the 24.4 visits were made by registered nurses and that 13.2 percent (3.22 visits) were made by home health aides. In the Home Health of Saint Luke's Hospital program, 12.9 percent (2.27 visits) of the 17.5 visits were by registered nurses, and 9.8 percent (1.72 visits) were by home health aides. Physical therapy visits averaged 77.3 percent (13.51 visits) of our home health visits compared to 57.8 percent (14.1 visits) nationally.

VARIANCE ANALYSIS

Variance analysis of total knee replacement patients occurs on both a concurrent and a retrospective basis. The care team conducts rounds to evaluate patient progress and path variation that might interfere with the rehabilitation process. Concurrent variance analysis allows the patient plan of care to be individualized and updated appropriately to ensure the most optimal patient outcomes in the most efficient manner. Members of the care team collaborate with the surgeon, therapists, and nurses to ensure that patient needs are continuously emphasized.

Cumulative variances from the clinical pathway for total knee replacement patients are analyzed on a quarterly basis. Data accumulation is a manual process, with the clinical nurse specialist retaining accountability for tabulation and summation of the data. The team then reviews the data for trends and occurrences. The variance analysis process is a dynamic one, and the variations measured have changed as practices have changed. Initially, all variances found were recorded and analyzed, but this system created a data overload and hindered identification of variances that actually made a difference in patient outcomes or clinical processes. Six major variance categories are now reviewed across the continuum of care: pain control, mobility, skin integrity, infection potential, bowel and bladder elimination, and blood transfusions.

A variance is defined as any intervention or expected outcome that is not completed or realized due to the behavior of the patient, provider, or system. A variance can occur for several reasons, including if the patient exceeds the expectations or does not require the intervention on the clinical path, if the physician does not order a test on the path or the nurse does not perform a given treat-

ment, or if an equipment failure occurs that interferes with a patient's progress on the clinical path. Approximately 58 percent of the variations experienced with the total knee replacement patient population are patient variances, while system variances constitute approximately 10 percent. In 1993, system variances accounted for nearly 18 percent of total variances. This reduction is attributed to the earlier placement of interdisciplinary consults and improved preoperative patient assessment, planning, and education.

Variance collection is the responsibility of everyone who uses the clinical path. A variance report sheet is placed with the clinical path during the patient's hospital stay. It is printed on colored paper to readily identify it as a quality monitoring tool. The care provider first noting a variance records it on the form, using a series of codes and abbreviations to keep documentation time to a minimum. Upon discharge from the hospital, the variance report is removed from the medical record and used by the clinical nurse specialist and the unit quality improvement representative to categorize quarterly variances for team analysis. Trends may be detected in the quarterly reviews that require further study and follow-up.

Variance analysis is an ongoing opportunity for the team. In 1993, variance data showed a problem with the timing of blood transfusions that resulted in increased patient discomfort and prolonged length of stay. A formal interdisciplinary research study was completed, using a revised clinical path as the data collection tool. Since initiation of variance analysis in 1993, changes in the timing of blood transfusions, decreased preoperative lab work, restructured preoperative education efforts, standardization of postoperative rehabilitation protocols, enhanced use of pain management interventions, preoperative discharge planning, and standardization of a postoperative bowel protocol have all been implemented.

Implementation of the orthopaedic home health program is currently underway. Interviews were conducted with physicians, key patient care and social service staff, home health agency staff, payers, and consumers. While no one was dissatisfied with the current practice of care to orthopaedic patients, most parties felt that a specialized orthopaedic home care program had value and would improve quality and continuity of care. Home Health of Saint Luke's Hospital believes that educating home care staff in all areas of orthopaedic care, coupled with the use of clinical paths, will decrease the cost of care by providing timely and efficient rehabilitative care in the home. With comprehensive discharge planning, adequate preparation, and home monitoring, orthopaedic care services can be provided in the home to improve the patient's clinical outcomes and functional status while providing a cost savings for the payers, patients, and health system.

PATIENT EDUCATION AND PATIENT PATHWAYS

An opportunity to enhance patient outcomes was realized in 1993. Knee replacement patients were no longer being admitted the evening before surgery, so valuable preoperative teaching time was lost. Preoperative education needs were at an all-time high, since patients were expected to be willing participants in their postoperative care but were unaware of what the postoperative period entailed. Saint Luke's Hospital opened a PAC in 1993 to facilitate the collection of lab specimens and complete other preoperative diagnostics before hospital admission. Visits to the PAC were scheduled approximately one week before surgery.

Recognizing a window of opportunity, the team participated in the preparation of educational materials to be reviewed with patients during PAC visits. The clinical path was revised to include a column to aid the PAC staff with documentation and promote continuity of care. Staff nurses in the PAC reviewed the packet of information with each patient and used the clinical path to explain the postoperative hospital stay. However, early evaluations of this teaching opportunity disclosed that patients were confused by the clinical path. Consequently, a patient path was developed to assist patients in understanding the routines and expectations surrounding postoperative recovery (Exhibit 24–A–2 in Appendix 24–A). This path is reviewed with patients at the PAC, including explanations of daily outcomes and activities. A brief description of the health team members is included to acquaint patients with the different roles. Postoperative discharge instructions are reviewed to encourage patients' active participation in the discharge planning process before the actual surgery. After surgery, the patient path is reviewed daily with the patient and family to facilitate understanding of the recovery process.

In 1995, a social service screening tool was implemented at the PAC to assist the care coordination team in identifying those patients who may present with discharge planning or care need challenges. Patients are referred to the orthopaedic social worker or the clinical nurse specialist for a preoperative phone call to identify special needs. This screening tool has now been amended to include criteria for a preoperative home visit by home health for thorough assessment and planning before hospital admission.

STAFF AND PHYSICIAN SATISFACTION

The collaborative care process has facilitated a high level of satisfaction with the care delivery process on the orthopaedic unit. Before the revision of the clinical path for documentation, the path was merely another piece of paper in the medical record. The transformation of the path to a major interdisciplinary documentation tool has increased staff understanding of the collaborative process, decreased the amount of time used to document patient care interventions and outcomes, and freed the caregiver to have more time at the bedside with the patient. Nursing staff were appreciative that the clinical path also helped to eliminate the closed-chart nursing process quality audits that had been required. The clinical path has streamlined the care process for knee joint replacement patients so that new staff members quickly learn the routine care patterns. Use of the patient path and teaching protocols strengthens patient education efforts and further reduces documentation requirements. Physicians have recognized that the patient care staff know the care routine and are prompting for timely receipt of orders. Caregivers now have more time to spend with the patient. All staff members have taken a proactive role in discharge planning. Satisfaction levels of staff and physicians have increased since the implementation of the paths in 1992.

CONCLUSION

Addressing the future of the collaborative care process at Saint Luke's Hospital is as challenging as the development of the initial program. This program is very dynamic and capable of changing in response to time-sensitive requests of all stakeholders. Demands by third-party payers and regulators will continue to play an active role in the success of the program. It is anticipated that efficiencies and new standards of practice will continue to be introduced and will necessitate changes. Partnering with primary third-party payers to create a win-win opportunity and enhance patient outcomes is desired. A patient-focused information system is being developed at Saint Luke's Hospital to automate the clinical path, documentation, and variance analysis. Standardization of the collaborative care process among the many providers of orthopaedic care in the Saint Luke's-Shawnee Mission Health System could result in increased efficiency and cost savings.

Regardless of what the future brings for health care as an industry, it is guaranteed that the total knee replacement patients at Saint Luke's Hospital of Kansas City will continue to receive the level of high-quality care to which the community has become accustomed.

REFERENCES

1. National Research Corporation®. *1996 Healthcare Market Guide V.* Lincoln, NE: NRC; 1997.
2. *The Prospective Payment Assessment Commission Report to Congress.* Washington, DC: PPAC/US House of Representatives; 1997.

■ Appendix 24–A ■
Clinical Pathway and Patient Pathway

The following is a key to the acronyms used in the clinical pathway and patient pathway that follow:

ADL—activities of daily living
BID—twice daily
BRP—bathroom privileges
CBC—complete blood count
CMS—circulation, movement, and sensation
CPM—continuous passive motion
DC—discontinue
D/C—discharge
EKG—electrocardiogram
hgb—hemoglobin
H&H—hemoglobin and hematocrit
HO—house officer
I&O—intake and output
I.S.—incentive spirometry
IV—intravenous
NPO p MN—nothing by mouth after midnight

Nursing dx—nursing diagnosis
OT—occupational therapy
PAC—preoperative assessment center
PACU—postanesthesia care unit
PCA/IM—patient-controlled analgesia/intramuscular
PO—oral
PRN—if needed
PT—physical therapy
PT/OT—physical therapy/occupational therapy
SNU—skilled nursing unit
TID—three times daily
Verb—verbalizes
VS—vital signs
wa—while awakeW/C—wheelchair
W1—0700–1900 shift
W2—1900–0700 shift

Exhibit 24–A–1 Clinical Pathway for Total Knee Replacement

SAINT LUKE'S HOSPITAL OF KANSAS CITY
Clinical Path—Total Knee Replacement Patient

ADMISSION DATE _____ DISCHARGE DATE _____

Medical diagnosis: _____

Surgeon: _____

Primary Replacement   Revision   Bilateral

Problem: **KNEE PAIN**

Nursing Diagnosis: Pain, Acute Related To Operative Procedure   Modified: _____

Initiated Date _____   Resolved: _____

Time: _____ Initials: _____

Expected Outcome: 1. Patient will communicate relief of pain to a tolerable level (scale 0–10).
                      2. Patient will be able to perform expected activities.

Assessment/   1. Assess comfort level every 2 hours x 24 then every 4 hours.
Interventions   2. Administer analgesics or treatments as indicated.
(Nursing Practice   3. Instruct and/or provide alternative comfort measures as appropriate.
  Standards)   4. Reassure on expected pain and relief measures.
                      5. Instruct to rate pain on a scale of 0–10.
                      6. Collaborate with pt to develop a plan for pain control.

Problem: **RANGE OF MOTION LIMITATION**

Nursing Diagnosis: Impaired Physical Mobility Related To Operative Procedure   Modified: _____

Initiated Date _____   Resolved: _____

Time: _____ Initials: _____

Expected Outcome: 1. Mobility is maintained or increased.
                      2. Patient will demonstrate an increase in strength/balance.

Assessment/   1. Support and instruct to perform activity as condition allows daily.
Interventions   2. Instruct and/or assist patient/significant other with optimal level of mobility.
(Nursing Practice   3. Assess functional capacity daily.
  Standards)   4. Assess for complications of immobility daily and prn.

Problem: _____   Modified: _____

Initiated Date _____   Resolved: _____

Time: _____ Initials: _____

Expected Outcome: 1. _____

Assessment/   1. _____
Interventions   2. _____
(Nursing Practice   3. _____
  Standards)   4. _____

❑ Patient path reviewed with patient/significant other.

Date: _____ RN Signature: _____

PATIENT IMPRINT   **SAINT LUKE'S HOSPITAL OF KANSAS CITY**
                                       COLLABORATIVE CARE PROGRAM

The suggested plan represents the initial desired course of treatment and goals of recovery. These are representative or average guidelines only and should be reviewed periodically by the attending physician and other involved disciplines. Deviations are generally expected and revisions to the plan should be made as warranted.

continues

Exhibit 24–A–1 continued

	DATE: _____ DAY: _____ PRE-HOSP/PRE-OP W1 W2	PACU	DATE: _____ DAY: _____ OPERATIVE DAY/ADMIT W1 W2	DATE: _____ DAY: _____ POST-OP DAY 1 W1 W2
ASSESSMENT	___ Verbalizes concerns	❏ ❏ ❏	❏ VS PACU routine ❏ ❏ Call temp > 101 ❏ ❏ Lung sounds q 12° ❏ ❏ ✔ CMS q1° x 12, then q2° x 12, then q4° ___ ___ Skin pressure points clear bid (1) ___ ___ Verb pain at tolerable level q2° (2) ___ ___ Verbalizes concerns	❏ ❏ Call temp > 101 ❏ ❏ Lung sounds q12° ❏ ❏ ✔ CMS q4° ___ ___ Skin pressure points clear bid (1) ___ ___ Verb pain at tolerable level q2–4° (2) ___ ___ Verbalizes concerns
CONSULT	❏ Medical clearance prn		❏ PT consult ❏ Social service	❏ SNU evaluation
TESTS/LABS	❏ Pre-op lab	❏ ❏ ❏	H&H in PACU X-ray in PACU ❏ Call HO hgb < 8.5	❏ Protime (5) _____ ❏ Hgb _____ ❏ CBC, lytes ❏ Call HO hgb < 8.5
MEDICATIONS		❏ ❏ ❏ 	❏ ❏ ? Transfuse blood (4) ❏ ❏ Antibiotic q8° x 6 ❏ ❏ Pain med (PCA/IM) (2) ❏ ❏ Antiemetic—PRN ❏ Coumadin (5) ❏ ❏ Bowel protocol (3)	❏ ❏ ? Transfuse blood (4) ❏ ❏ Pain med (PCA/IM) (2) ❏ ❏ Antiemetic—PRN ❏ ❏ Bowel protocol (3)

TREATMENTS	❏ Ted hose measurement (1)
	   Length _____
	   Calf _____
	❏ Instruct on shower at home the morning of surgery
		❏
		❏
		❏
		❏
		❏
		
			❏ ❏ Trapeze
			❏ ❏ Clamp drain > 75cc
			❏ ❏ I.S. 10x/hr w.a.
			❏ ❏ Teds/cuffs (off am/pm) (1)
			❏ ❏ I&O
			❏ ❏ IV _____
			❏ ❏ CPM ___° flexion per protocol (2)
			❏ ❏ Ice pack to knee (2)
			___ ___ Dressing dry & intact (6)
				❏ ❏ Trapeze
				❏ ❏ Drain
				❏ ❏ I.S. 10x/hr w.a.
				❏ ❏ Teds/cuffs (off am/pm) (1)
				❏ ❏ I&O
				❏ ❏ IV _____
				❏ ❏ CPM ___° flexion (2)
				❏ ❏ Ice pack to knee (2)
				___ ___ Dressing dry & intact (6)

MOBILITY/ADLs		❏ 	❏ ❏ Bedrest ___ ___ Turns q2° with assist (1)	___ ___ Chair 15 min bid (2)
NUTRITION	❏ NPO p MN ❏ Nutrition screen		❏ ❏ Clear liquids p OR	❏ ❏ Diet _____
ELIMINATION	❏ Instruct on laxative at home prn	❏	❏ ❏ Foley cath ❏ ❏ Straight cath prn q8° ❏ ❏ Bedpan/Urinal	❏ ❏ Foley cath ❏ ❏ Straight cath prn q8° ❏ ❏ Bedpan/Urinal
TEACHING AND DISCHARGE PLANNING	❏ PAC → PT/OT video ❏ Social services screening ❏ Patient path given ❏ Education booklet given ❏ Home visit referral		❏ ❏ Instruct re: PCA and pain scale 0–10 (2) ❏ ❏ Instruct post-op interventions per knee protocol ❏ ❏ Review patient path ___ ___ Verb understanding of pain control plan (2)	❏ ❏ Reinforce CPM instructions ❏ ❏ Review patient path

Symbol key: ❏ ❏ = Interventions   ___ ___ = Expected Outcomes

PATIENT IMPRINT   **SIGNATURE KEY:**

       _____   _____   _____   _____

       _____   _____   _____   _____

       _____   _____   _____   _____

       _____   _____   _____   _____

SYMBOL KEY: "Initials" on a line means done and findings as expected
                "✔" in a ❏ box means an intervention or item was completed
                "O" in a ❏ box or on a line indicates the item was not pertinent to that shift
                "*" in a ❏ box or on a line indicates an item was not done

continues

Exhibit 24–A–1 continued

ASSESSMENT	DATE: _____ DAY: _____ POST-OP DAY 2		DATE: _____ DAY: _____ POST-OP DAY 3		DATE: _____ DAY: _____ POST-OP DAY 4	
	W1 W2		W1 W2		W1 W2	
ASSESSMENT	❏ ❏ ❏ ❏ ❏ ❏ ___ ___ ___ ___ ___ ___	Call temp > 101 Lung sounds q12° ✔ CMS q4° Skin pressure points clear bid (1) Verb pain at tolerable level q4° (2) Verbalizes concerns	❏ ❏ ❏ ❏ ❏ ❏ ___ ___ ___ ___ ___ ___	Call temp > 101 Lung sounds q12° ✔ CMS q8° Skin pressure points clear bid (1) Verb pain at tolerable level q4° (2) Verbalizes concerns	❏ ❏ ❏ ❏ ❏ ❏ ___ ___ ___ ___ ___ ___	Call temp > 101 Lung sounds q12° ✔ CMS q8° Skin pressure points clear bid (1) Verb pain at tolerable level q4° (2) Verbalizes concerns
CONSULT					❏	OT to evaluate ADL needs PRN
TESTS/LABS	❏ ❏ ❏ ❏	Protime (5) _____ Hgb _____ CBC, lytes a.m. Call HO hgb < 8.5	❏	Protime (5) _____	❏	Protime (5) _____
MEDICATIONS	❏ ❏ ❏ ❏ ❏ ❏ ❏	? Transfuse blood (4) DC PCA (2) Coumadin (5) Bowel protocol (3) Pain controlled by PO pain meds (2)	❏ ❏ ❏ ❏ ❏ ❏ ___ ___	? Transfuse blood (4) Coumadin (5) Bowel protocol (3) Pain controlled by PO pain meds (2)	❏ ❏ ❏ ___ ___	Coumadin (5) Bowel protocol (3) Pain controlled by PO pain meds (2)
TREATMENTS	❏ ❏ ❏ ❏ ❏ ❏ ❏ ❏ ❏ ❏ ❏ ❏ ❏ ❏ ___ ___ ___ ___	Trapeze DC drain I.S. prn Teds/cuffs (off am/pm) (1) I&O DC IV CPM ___° flexion (2) Ice pack to knee (2) Dressing change Dressing dry & intact (6) Wound intact (6)	❏ ❏ ❏ ❏ ❏ ❏ ❏ ❏ ❏ ___ ___ ___ ___	Trapeze I.S. prn Teds/cuffs (off am/pm) (1) DC I&O CPM ___° flexion (2) Ice pack to knee (2) Dressing change Dressing dry & intact (6) Wound intact (6)	❏ ❏ ❏ ❏ ❏ ❏ ___ ___ ___ ___	Trapeze I.S. prn Teds/cuffs (off am/pm) (1) CPM ___° flexion (2) Ice pack to knee (2) Dressing change Dressing dry & intact (6) Wound intact (6)
MOBILITY/ADLs	❏ ___ ___	PT BID (2) BRP by chair (2)	❏ ___ ___	PT BID (2) BRP by chair (2)	❏ ___ ___	PT BID (2) BRP per walker (2)
NUTRITION	❏ ❏	Diet _____	❏ ❏	Diet _____	❏ ❏	Diet _____
ELIMINATION	❏ ❏ ❏	DC foley Straight cath prn q8°	___ ___ ___ ___	Voids adequately Maintain normal bowel function (3)	___ ___ ___ ___	Voids adequately Maintain normal bowel function (3)
TEACHING AND DISCHARGE PLANNING	❏ ❏ ❏ ❏	Instruct re: PO pain med (2) Review patient path	❏ ❏ ❏ ___ ___	Discharge to SNU if appropriate Review patient path Verbalizes understanding of pain control plan (2)	❏ ❏ ❏ ❏	Confirm home service Arrange discharge transportation Review patient path

Symbol key: ❏ ❏ = Interventions   ___ ___ = Expected Outcomes

PATIENT IMPRINT   **SIGNATURE KEY:**

_____ _____   _____ _____
_____ _____   _____ _____
_____ _____   _____ _____
_____ _____   _____ _____
_____ _____   _____ _____
_____ _____   _____ _____
_____ _____   _____ _____

SYMBOL KEY: "Initials" on a line means done and findings as expected
"✔" in a ❏ box means an intervention or item was completed
"O" in a ❏ box or on a line indicates the item was not pertinent to that shift
"*" in a ❏ box or on a line indicates an item was not done

continues

Exhibit 24–A–1 continued

	DATE: _____ DAY: _____ **POST-OP DAY 5** W1 W2	DATE: _____ DAY: _____ **POST-OP DAY 6** W1 W2	DATE: _____ DAY: _____ **POST-OP DAY 7** W1 W2	**Discharge Expected Outcomes**
ASSESSMENT	❑ ❑ Call temp > 101 ❑ ❑ Lung sounds q12° ❑ ❑ ✔ CMS q8° __ __ Skin pressure points clear bid (1) __ __ Verb pain at tolerable level q4° (2) __ __ Verb concerns	❑ ❑ Call temp > 101 ❑ ❑ Lung sounds q12° ❑ ❑ ✔ CMS q8° __ __ Skin pressure points clear bid (1) __ __ Verb pain at tolerable level q4° (2) __ __ Verb concerns	❑ ❑ Call temp > 101 ❑ ❑ Lung sounds q12° ❑ ❑ ✔ CMS q8° __ __ Skin pressure points clear bid (1) __ __ Verb pain at tolerable level q4° (2) __ __ Verb concerns	____ Temp < 100° for 24 hours ____ Skin integrity intact (1)
CONSULT				
TESTS/LABS	❑ Protime (5) _____	❑ Protime (5) _____	❑ Protime (5) _____	
MEDICATIONS	❑ Coumadin (5) ❑ ❑ Bowel protocol (3) __ __ Pain controlled by PO pain meds (2)	❑ Coumadin (5) ❑ ❑ Bowel protocol (3) __ __ Pain controlled by PO pain meds (2)	❑ Coumadin (5) ❑ ❑ Bowel protocol (3) __ __ Pain controlled by PO pain meds (2)	____ Pain controlled by PO meds (2) ____ Verb understanding of medications/schedules
TREATMENTS	❑ ❑ Trapeze ❑ ❑ I.S. PRN ❑ ❑ CPM ___° flexion (2) ❑ ❑ Ice pack to knee (2) ❑ ❑ TED/cuffs (off am/pm) (1) ❑ Dressing change __ __ Dressing dry & intact (6) __ __ Wound intact (6)	❑ ❑ Trapeze ❑ ❑ I.S. PRN ❑ ❑ CPM ___° flexion (2) ❑ ❑ Ice pack to knee (2) ❑ ❑ TED hose (off am/pm) (1) ❑ Dressing change __ __ Dressing dry & intact (6) __ __ Wound intact (6)	❑ ❑ Trapeze ❑ ❑ I.S. PRN ❑ ❑ CPM ___° flexion (2) ❑ ❑ Ice pack to knee (2) ❑ ❑ TED hose (off am/pm) (1) ❑ Dressing change __ __ Dressing dry & intact (6) __ __ Wound intact (6)	____ Wound edges well-approximated without redness or drainage (6)
MOBILITY/ADLs	❑ PT BID (2) __ __ Chair TID (2) __ __ BRP per walker (2)	❑ PT BID (2) __ __ Chair TID (2) __ __ BRP per walker (2)	❑ PT BID (2) __ __ Chair TID (2) __ __ BRP per walker (2)	____ In & out of bed/chair independently (2) ____ Ambulates with minimal assistance (2) ____ Verb/demonstrates ability to perform care at home and/or with support or community resources
NUTRITION	❑ ❑ Diet _____	❑ ❑ Diet _____	❑ ❑ Diet _____	____ Verb understanding of discharge/diet nutrition goal
ELIMINATION	__ __ Voids adequately __ __ Maintain normal bowel function (3)	__ __ Voids adequately __ __ Maintain normal bowel function (3)	__ __ Voids adequately __ __ Maintain normal bowel function (3)	____ Elimination patterns re-established (3)
TEACHING AND DISCHARGE PLANNING	❑ Instruct on wound care ❑ ❑ Review patient path	❑ ❑ Review patient path __ __ Received & verbalizes understanding of D/C meds, diet, & symptoms to report to Dr. & follow-up appointment	❑ ❑ Review patient path __ __ Received & verbalizes understanding of D/C meds, diet, & symptoms to report to Dr. & follow-up appointment	____ Verb/demonstrates understanding of safety measures & physical limitations ____ Identifies support systems ____ Verb signs & symptoms of infection & demonstrates preventive measures

Symbol key: ❑ ❑ = Interventions   __ __ = Expected Outcomes

PATIENT IMPRINT   **SIGNATURE KEY:**

_____   _____

_____   _____

_____   _____

_____   _____

Time: _____ Initials: _____
To: _____
Mode: ❑ w/c ❑ other
❑ Valuables with pt
❑ All nsg dx resolved
Accompanied by: _____

SYMBOL KEY: "Initials" on a line means done and findings as expected
"✔" in a ❑ box means an intervention or item was completed
"O" in a ❑ box or on a line indicates the item was not pertinent to that shift
"✱" in a ❑ box or on a line indicates an item was not done

Exhibit 24–A–2 Patient Path for Total Knee Replacement (Average Hospital Stay Range: 3–7 days)

SAINT LUKE'S HOSPITAL OF KANSAS CITY

Patient: _____

Saint Luke's Hospital's staff and doctors are dedicated to giving you the best possible care. This path is to help you and your family become more involved in your care. Since each person is an individual, your care may differ from this general guideline.

HEALTH CARE TEAM

Doctor(s):

Directs your medical and surgical care.

Registered Dietitian/Diet Technician:

Works with your doctor to meet your nutritional needs. They check your nutritional status and eating habits and teach you about your nutritional goals. They will also assist with any alternative feeding methods other than by mouth, if your doctor deems necessary.

Registered Nurse (RN):

Your nurses are responsible for your nursing care. They will plan your care with all health team members as well as with you and your family.

Occupational Therapist:

Will help you learn ways to manage at home and make arrangements for any special equipment you need.

Patient Care Technician (PCT):

Your PCT will help you with personalized care and daily needs. They may perform procedures such as blood pressure, temperature, drawing of blood, and EKGs.

Physical Therapist:

Will help you strengthen your muscles and get as much knee motion as possible. Our goal is to help you return to your normal daily activities.

Patient Service Associate (PSA):

Your PSA will assist you with meal selection and serving, maintain the cleanliness of your room, and respond to your "hospitality" needs.

Pre-Op Assessment Center:

The doctor may send you here before surgery for evaluation and teaching. If you have questions, call the Center at (816) 932-2390.

Patient Representative:

Your patient representative is your personal connection to the hospital system. If you have a question about your hospital stay, please call (816) 932-2328.

Social Worker:

Social workers provide counseling, information, education, support, and referrals to you and your family as you adjust to the impact of your illness or treatment. The social worker will assist you in making plans for your care after you leave the hospital.

continues

Exhibit 24–A–2 continued

CARE CATEGORY	DAYS BEFORE SURGERY	DAY 1 BEFORE SURGERY	DAY 1 AFTER SURGERY	DAY 2 POST-OP DAY 1	DAY 3 POST-OP DAY 2
TESTS AND TREATMENT	• Lab Work	• Start IV Fluids	• Blood test • IV Fluids • Cough and deep breathe • May have urinary catheter • Support stockings	• Blood test • IV Fluids • Cough and deep breathe • May have urinary catheter • Support stockings	• Blood test • No IV Fluids • Cough and deep breathe • No urinary catheter • Knee drain out • Support stockings
FOOD AND DRINK	• Regular food until night before surgery	• No food or drink after midnight before surgery	• Clear liquids	• Solid food • May have high protein snacks between meals	• Solid food • May have high protein snacks between meals
ACTIVITY	• ✔ with doctor	• Shower at home • Dress in hospital gown	• Bedrest • CPM (knee bending machine) • Bedpan • Ice pack to knee	• Chair twice—15 minutes • CPM machine • Bedpan • Ice pack to knee	• Chair twice—15 minutes • CPM machine • Bathroom by rolling chair • Start Physical Therapy • Ice pack to knee
MEDICATION AND PAIN	• ✔ with doctor		• IV Pain medicines • Coumadin • Antibiotics (IV)	• IV Pain medicines • Coumadin • Antibiotics (IV)	• Pain pills • Coumadin • Laxative if needed
DISCHARGE PLANNING AND LEARNING	• Pre-Op Assessment Center: Physical Therapy Visit • Read teaching packet • Discuss hospital care and home care plans • Learn and practice cough and deep breathing exercises	• Practice cough and deep breathing exercises • Learn pain scale	• Use pain scale	• Use pain scale • Discuss medicines (i.e., pain medicine—Coumadin)	• Use pain scale • Discuss arrangements for care after discharge

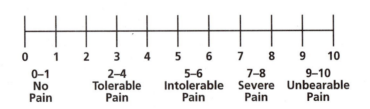

<table>
<tr><td>0–1
No
Pain</td><td>2–4
Tolerable
Pain</td><td>5–6
Intolerable
Pain</td><td>7–8
Severe
Pain</td><td>9–10
Unbearable
Pain</td></tr>
</table>

continues

Exhibit 24–A–2 continued

CARE CATEGORY	DAY 4 POST-OP DAY 3	DAY 5 POST-OP DAY 4	DAY 6 POST-OP DAY 5	DAY 7 POST-OP DAY 6	DAY 8 POST-OP DAY 7
TESTS AND TREATMENT	• Blood tests • Support stockings • Cough and deep breathe	• Blood tests • Support stockings • Cough and deep breathe	• Blood tests • Support stockings • Cough and deep breathe	• Blood tests • Support stockings • Cough and deep breathe	• Blood tests • Support stockings • Cough and deep breathe
FOOD AND DRINK	• Solid food • May have high protein snacks between meals	• Solid food • May have high protein snacks between meals	• Solid food • May have high protein snacks between meals	• Solid food • May have high protein snacks between meals	• Solid food • May have high protein snacks between meals
ACTIVITY	• Up in chair as much as possible • CPM machine • To physical therapy 2x day • Bathroom by rolling chair • Ice pack to knee	• Up in chair as much as possible • CPM machine • To physical therapy 2x day • Bathroom with walker • Ice pack to knee	• Up in chair as much as possible • CPM machine • To physical therapy 2x day • Bathroom with walker • Ice pack to knee	• Up in chair as much as possible • CPM machine • To physical therapy 2x day • Bathroom with walker • Ice pack to knee	• Up in chair as much as possible • CPM machine • To physical therapy 2x day • Bathroom with walker • Ice pack to knee
MEDICATION AND PAIN	• Pain pills • Coumadin • Laxative if needed	• Pain pills • Coumadin • Laxative if needed	• Pain pills • Coumadin • Laxative if needed	• Pain pills • Coumadin • Laxative if needed	• Pain pills • Coumadin • Laxative if needed
DISCHARGE PLANNING AND LEARNING	• Use pain scale • May move to Skilled Nursing at Saint Luke's	• Use pain scale • May move to Skilled Nursing at Saint Luke's • Home today?	• Use pain scale • Finalize home plans • Home today?	• Use pain scale • Home today?	• Use pain scale • Home today?
YOUR QUESTIONS AND COMMENTS					

The suggested plan represents the initial desired course of treatment and goals of recovery. These are representative or average guidelines only. They should be reviewed periodically by the attending physician and other involved care providers. Deviations are generally expected and revisions to the plan should be made as warranted.

continues

Exhibit 24–A–2 continued

LEARNING FOR DAILY LIVING
Knee Replacement Home Instructions

ACTIVITY

- Walk with crutches/walker
- Follow physical therapy exercises
- It is OK to resume sexual activity with special attention to your knee restrictions
- Do not drive until your surgeon says you can

- Do not put a pillow right under the knee joint. It is OK to raise your leg by placing the pillow from your knee to ankle.
- Wear support stockings all day and evening. It is OK to take them off at bedtime and put them back on in the morning. Wash stockings in mild detergent and hang to drip dry.

DIET

- No change in diet.
- Eat well-balanced meals.
- To avoid constipation, increase fiber by eating fruits, vegetables, and grains.
- If advised, increase protein calories as recommended. (See printed materials.)

INCISION CARE

- OK to shower after staples are removed. Do not take a tub bath (incision line should not be submerged). Do not scrub over incision line. Keep dressing clean and dry. Change it as needed.

WHEN TO CALL FOR MEDICAL ADVICE

- Pain in your calf, tenderness or increased swelling in calf
- An increase in drainage or pain from your knee
- Change in feeling in your affected leg
- Chilling or your temperature over 100°F
- Any episode of shortness of breath or chest pain or tightness
- Problems urinating, burning or itching on urination

MEDICATION INSTRUCTION

- Do not drink alcohol while taking medicine.
- Preventive (prophylactic) antibiotic coverage is needed before invasive procedures. Tell all your doctors and dentist.

MEDICATION AND DOSE

Take one coated/buffered aspirin two times a day as instructed by your doctor.

_____ One or two tablets every _____ hours as needed for pain

_____

_____

APPOINTMENT

- APPOINTMENTS: Your return appointment with Dr. _____ has been made for ___/___/___ at _____ AM/PM. If this is not convenient, please feel free to reschedule with the office. The phone number is _____ - _____.

◼ 25 ◼

Total Knee Arthroplasty: A Collaborative Project in Quality Improvement and Cost Reduction

Claire M. Paras and Patricia Brita-Rossi

In 1994, Beth Israel Deaconess Medical Center (BIDMC) of Boston's orthopaedic work team began a process to evaluate the orthopaedic program. The membership of the orthopaedic work team consisted of administrators, physicians, nurses, social workers, rehabilitation specialists, admitting department personnel, medical record department personnel, and clinical specialists. Baseline data, including program costs and clinical outcomes, were reviewed to identify opportunities for improvement. The goal of the process was to enhance the quality of the care delivered by the orthopaedic program while reducing overall costs. After the initial review of the program, it became evident that improvements could be made and that efficiencies could be gained in certain areas of orthopaedic care. Problems included lack of communication and coordination among caregivers, signficant variability in practice, lack of an evaluative process regarding practice patterns, and delivery of inconsistent messages to patients and families. The orthopaedic work team met on a regular basis to develop a comprehensive plan for redesign of the orthopaedic program and to monitor progress of other related improvement projects.

Some of the initiatives of the orthopaedic work redesign project included the development of a patient preoperative teaching session, perioperative redesign of the operating room staff and supplies, weekly orthopaedic case study rounds, development of a care management team, daily patient rounds of the care management team, and the development of clinical pathways, referred to as *interdisciplinary practice guidelines* (IPGs). As the work redesign project identified specific areas for change and improvement, subgroups of the larger work team evolved. Each subgroup was then delegated one of the initiatives. This chapter discusses the development

and implementation of a clinical pathway for management of the total knee arthroplasty patient.

DEVELOPMENT OF THE INTERDISCIPLINARY PRACTICE GUIDELINE

Formation of the Work Team

In June 1994, the organization initiated a hospitalwide clinical pathway program, formally known as the Interdisciplinary Practice Guideline Program. (On October 1, 1996, Beth Israel Hospital merged with New England Deaconess Hospital. The Interdisciplinary Practice Guideline Program refers to the former Beth Israel Hospital, now called the East Campus.) The orthopaedic work team was the first clinical team to integrate the IPGs into practice as a major component of an overall practice improvement strategy. A subgroup, the orthopaedic IPG work team, was formed to develop IPGs that would be used by the department as tools in the process of evaluating and improving practice on a continuous basis. The clinical pathway modality was identified as one that would provide a method to consistently and continuously communicate all improvement targets and expectations from which practice would be measured to all members of the health care team. The orthopaedic IPG work team focused its initial efforts on joint replacement patients because of the large volume of total hip replacement and total knee replacement procedures. Extended hospital lengths of stay (LOS) and variability in physician practices were also factors in the selection process.

The core membership of the orthopaedic IPG work team is interdisciplinary and includes

- orthopaedic surgeons
- nurse manager, inpatient unit

- nurse practitioners, inpatient unit and outpatient clinic
- clinical nurses, preadmitting clinic, postanesthesia care unit, inpatient unit
- physical therapist
- occupational therapist
- social worker
- facilitator, department of health care quality

Representation from and review by the department of pharmacy, laboratory medicine, and anesthesia are included on a consultative basis.

Goals and Objectives

By implementing IPGs as tools to support the overall improvement plan, the orthopaedic IPG work team expected to achieve the following goals:

- Improve the management of LOS and decrease process delays.
- Improve management of utilization.
- Decrease unnecessary and wasteful variations in practice.
- Provide a forum for developing patient/family educational materials and processes.
- Provide a forum for maximizing development and implementation of caregiver educational materials and processes.
- Increase patient satisfaction.
- Increase interdisciplinary communication.
- Provide a consistent and comprehensive system for continuous evaluation of the quality and cost of clinical performance and the care that is being delivered.

The IPG work team identified the following tasks for development:
- Develop an IPG for management of the acute care of total knee replacement patient.
- Create preprinted physician order sheets to be used in conjunction with the IPG.
- Prepare a patient version of the IPG as a tool for patient/family education.
- Expand the original inpatient version to include postcare management with rehabilitation facilities and home care.

To accomplish the goals and tasks that had been identified, the IPG work team initially met on a weekly basis to develop the IPG and to plan the implementation and evaluative process. The available data were analyzed and used to specify improvement targets:

- Decrease LOS from 6.7 days. Discharge targets were set at postoperative day 3 for patients who were

transferred to rehabilitation settings and at postoperative day 4 for patients who were sent home with services.
- Institute a preoperative nursing assessment by the expert inpatient unit nurses to provide an initial identification of actual or potential psychosocial and/or discharge planning needs. In necessary cases, the nurse would then facilitate an early referral to the social worker in a collaborative effort to plan more effective and efficient discharges.
- Provide a preoperative functional assessment by physical therapy.
- Provide a formalized and structured preoperative educational session by nursing and physical therapy to all patients undergoing total knee arthroplasty surgery. In view of the abbreviated LOS and rapid postoperative course, a preoperative patient educational program was identified as a way to support patients and families by building their comprehension and readiness.
- Decrease and standardize laboratory utilization.
- Decrease and standardize radiology utilization.

Development and Early Implementation

The total knee arthroplasty IPG for management in the acute care setting reflects the patient's progress through all aspects of the surgical experience (Exhibit 25–A–1 in Appendix 25–A). It begins with the preoperative phase, which includes both the physician's office and the preoperative assessment/teaching visit in the preadmitting clinic of the hospital. The IPG continues to track the patient's progress through hospitalization to discharge. Interdisciplinary interventions specifically related to the surgeons, nurses, social workers, and physical therapists, as well as patient outcomes, are articulated on a daily time frame. The standards of care for patients undergoing total joint surgery were used as a reference for the IPG. Targets for improvement that were identified by the orthopaedic work team, as well as projects such as the preoperative patient educational session, were incorporated into the guideline. The IPG work team chose to use the IPG as a reference tool for care and not as a documentation tool.

Once an initial draft was prepared, the IPG was put through an extensive endorsement process to build consensus and to expand the number of participants in the development and planning processes. Copies of the draft were distributed to all physicians in the department of orthopaedics. Each physician was asked to make comments, recommendations, or modifications on the actual draft and to return the document to the IPG work team. A follow-up review session was held in the orthopaedic physician staff meeting, a monthly meeting held by the attending physicians. Another review occurred in the orthopaedic walk rounds, a weekly case review session that includes attending physicians, residents, and representatives from all the disciplines involved with the care of the total knee arthroplasty patients. In addition, each discipline represented in the orthopaedic IPG work team was responsible for reviewing the IPG in department staff meetings or review sessions. For example, on the inpatient unit, the nurses posted a copy of the draft in a visible location and distributed a copy to each individual nurse as well. The draft was then discussed in nursing staff meetings. The IPG draft was reviewed by the pharmacy and therapeutics committee, the associate anesthesiologist in chief, the director of the division of laboratory medicine, and the medical director of the hematology laboratory. All recommendations and corrections were reviewed by the orthopaedic IPG work team and incorporated into the IPG.

Once the draft was approved, the total knee arthroplasty IPG underwent a 60-day pilot implementation period, beginning in November 1994. In some respects, the pilot served as an extension to the development process, since it enabled the work team to evaluate the content of the IPG as well as the effectiveness of the implementation and evaluation processes. Frequent educational sessions were provided to prepare all members of the health care team for use of the IPG in practice. Both the orthopaedic IPG work team and the orthopaedic work team obtained important information from the pilot implementation regarding how to proceed with long-term implementation and evaluation. Since the pilot implementation, the IPG has undergone several content and process revisions based on review of data and changes in the clinical environment.

Development of Preprinted Physician Order Sheets

Preprinted physician order sheets were developed for use in conjunction with the total knee arthroplasty IPG. Oxygen use, laboratory tests, physical therapy and social worker referrals, advancement of mobility, diet advancement, and medication use are stated on the orders in correlation with the IPG. Although the order sheets reflect the expectations of the IPG, all orders are activated pending the condition and needs of each patient and are subject to changes and modifications.

Development of "Patient Pathways"

A patient version of the total knee arthroplasty IPG has been developed and integrated as an important patient teaching tool in the formalized patient/family preoperative education session. The patient version has been created in the same format as the caregiver version and outlines the outcomes and interventions that the patient and family can expect throughout the hospitalization. This tool is serving as a prototype for other clinical work teams.

Development of a Transitional Pathway

The most recent phase in the continued development of the total knee arthroplasty IPG has focused on extending the IPG beyond the acute care setting and into the rehabilitative setting. In 1996, BIDMC and Youville Lifecare signed a memorandum of understanding. A major component of this joint venture is collaboration to increase the quality of care for all patients who are transferred from one facility to another. As one of the collaborative projects, an orthopaedic work team was established with interdisciplinary members from each facility. The focus of this work team was the total knee arthroplasty patient population, and the goal of the team was to create a guideline or pathway that would support a seamless transition of the patient from one location to the other. The work team analyzed the current transfer process and identified areas for improvement and consensus regarding changes in practice. A transitional guideline was developed that links the expectations regarding both patient outcomes and interdisciplinary interventions that are important to both facilities at the point of transfer (see Exhibit 25–A–2 in Appendix 25–A). Communication and information requirements are also clearly identified on the tool. The tool is partially completed at the time of the patient's discharge from BIDMC and is faxed to the Youville facility, where it is completed. When the patient is discharged from Youville Lifecare, the indicator section is completed and then returned to the clinical nurse specialist at BIDMC for data collection. The pilot for this guideline began in May 1997, and the data will be reviewed after six months.

IMPLEMENTATION OF THE INTERDISCIPLINARY PRACTICE GUIDELINE

Care Management

The IPG is used within a professional practice model of primary nursing. Within this model, the primary nurse is recognized as the patient's care manager. The primary nurse has 24-hour accountability for the care of the patient and is responsible for maintaining the patient's course on the practice guideline and achieving clinical outcomes. The primary nurse coordinates the treatment plan as described on the practice guideline. When the primary nurse is unavailable to provide direct care to the patient, the associate nurse continues to implement the plan of care to provide seamless care. The primary nurse also incorporates intimate knowledge of the patient's care needs into the practice guideline, making modifications appropriate to the patient's situation.

On the day of preadmission testing, the IPG is initiated during the patient's preoperative teaching session. During this time, an orthopaedic clinical nurse specialist and a physical therapist meet with the patient and family members to discuss the postoperative course, using the patient version of the practice guidelines. At this time, the clinical version of the practice guideline and corresponding preprinted physician order sheets are placed in the medical record.

The clinical nurse specialist assesses the patient's physical and psychosocial status, and the physical therapist assesses functional status. Together, they collect and share patient information while exchanging expectations with the patient as the guideline is implemented. Both care providers document in the medical record pertinent information that is identified on the practice guideline.

On the procedure day, the practice guideline is already a part of the medical record, with preprinted physician order sheets activated for the patient's postoperative course. The practice guideline tracks the patient through surgery to the postsurgical recovery room and the orthopaedic nursing unit. When the patient arrives on the orthopaedic nursing unit after surgery, a primary nurse is assigned to the patient's care through a voluntary process. The primary nurse then documents in the medical record the patient's treatment plan using the practice guideline as a reference. Patient information obtained at the time of the patient's preoperative teaching session by both the clinical nurse specialist and physical therapist is now passed along to the primary nurse. This information assists the primary nurse in developing the patient's personalized plan of care, using the clinical outcomes and interventions identified on the practice guideline.

Care management team rounds take place on the nursing unit on a daily basis to facilitate communication and coordination between the care team members. The care management team consists of primary nurses as the care managers, a physical therapist, a clinical nurse specialist, a social worker, and a nurse case manager. The nurse case manager is a new role within the organization and was formerly the utilization review coordinator. The nurse manager of the orthopaedic nursing unit functions not only as the leader of the care team but also as facilitator and integrator of team members. Orthopaedic surgeons make rounds with the primary nursing staff. They examine patients early in the morning so that information exchanged at this time can be added to the patient's plan of care before the care management team rounds.

The primary nurse presents the patient's plan of care to all team members. If the primary nurse has identified that the patient is in need of social work support, a referral is made to the social worker. The nurse case manager is available to the primary nurse to assist in ensuring that the patient meets acute care criteria and to communicate any delays or transitions of care to the payers. The primary nurse, in collaboration with the surgeons and physical therapist, facilitates identification of the patient's care needs within the context of the hospitalization and progression along the practice guideline. The primary nurse's in-depth knowledge of the patient gives direction to the nurse case manager, who then communicates the patient's care needs to the payers.

The IPGs are used by every member of the care management team. At care team rounds and morning physician rounds, the practice guidelines are used to identify and measure daily outcomes, to guide and coordinate the interventions of all team members, and to assist in the development of the patient's plan of care.

Measurement of Performance

A crucial element of a clinical pathway/practice guideline program is the institution of an evaluative process that continuously and consistently measures practice and systems performance. The orthopaedic health care providers recognize their accountability for managing outcomes, as well as their responsibilities for participating in the evaluation of their practice and for contributing to effective quality improvement in the delivery of care to their patients. Therefore, it becomes increasingly important to provide accurate and timely information to assist the caregivers in identifying problems and making corrective or improvement actions.

The data management component of the Interdisciplinary Practice Guideline Program has been undergoing a constant evolutionary process. However, the primary goals of the evaluative process have remained the same:

- to verify that all expected patient outcomes have been achieved as stated on the IPG
- to verify that all interdisciplinary interventions and actions have been completed as articulated on the IPG
- to identify variances from the expectations outlined on the IPG
- to facilitate identification of both negative and positive practice and systems issues that affect outcomes

The evaluative component of the Interdisciplinary Practice Guideline Program has proven to be the most challenging to develop and implement effectively. As a result, the whole process has gone through several phases of development and is still in a state of transition. The total knee arthroplasty IPG has experienced each phase in the evolution of the outcomes and variance measurement. Several key factors determine how successfully a measurement system functions:

- selection and identification of appropriate, realistic, and meaningful indicators
- ability to obtain data either via data collection or from other existing databases
- data entry and analysis processes and resources
- organizationwide feedback/response loop

As the Interdisciplinary Practice Guideline Program and the individual clinical work teams were striving to develop a valid and practical system for measuring performance, the organization's plan and structure for providing an infrastructure to support quality and cost improvement were also maturing. The organizational evolution contributed to the processes and tools that are now incorporated into the Interdisciplinary Practice Guideline Program.

Data Management Tools and Processes

The pilot implementation phase offered the orthopaedic IPG work team an opportunity to identify key indicators for measurement. A variance tracking tool was considered, but there was concern that a nontailored approach would result in many useless details with a minimum of useful information for the caregivers. Ultimately, each discipline identified a few focused indicators for measurement, while also leaving some room for limited free-form variance tracking. Some of the indicators that were originally tracked included

- LOS
- discharge disposition
- comorbidities
- completion of a preoperative patient education session by nursing
- documentation by nursing of a discharge plan by postoperative day 1
- patient out of bed and ambulating by postoperative day 1
- devices required by patients (e.g., walker or crutches)
- patient able to tolerate oral pain medication and have epidural discontinued by postoperative day 2

The specific variances that were tracked included

- pain management problems
- urinary retention
- delays in discharge (as related to patient condition, caregiver decisions, systems issues)

- indicators for utilization of laboratory tests, radiology, medications, and transfusion

Over approximately 24 months, it became clear that many of the indicators were related to practices that were well established and had reached 100 percent compliance (e.g., number of patients attending a preoperative educational session, documentation by nursing of a preliminary discharge plan before postoperative day 1, patients out of bed by postoperative day 1). The IPG work team identified the need for more focused information. As this team and others within the Interdisciplinary Practice Guideline Program were striving to develop a practical and valid system for measuring performance, the organization's plan for providing an infrastructure to support quality improvement was also maturing.

The organizational quality improvement plan contributed to the processes and tools that are now incorporated into the Interdisciplinary Practice Guideline Program. In conjunction with the organizational plan for quality improvement, the orthopaedic IPG work team has streamlined the type of indicators that will be tracked and is integrating the IPG evaluative process more comprehensively into the department's quality improvement structure.

The basis of the organizational quality improvement plan is the development of "instrument panels" from which each clinical department can immediately identify the necessary improvements and initiate actions. Instrument panels can be tailored to address total system performance as well as specific clinical processes.[1] Using the concept of the instrument panel, the organization has identified several broad areas for measurement: patient characteristics and clinical care, utilization and financial, customer satisfaction, and "other dimensions" of performance. "Patient characteristics and clinical care" encompasses patient demographics, clinical care processes, and clinical outcomes. "Utilization and financial" includes volume, utilization, and costs. "Customer satisfaction" includes patient satisfaction, access to care, referring physician satisfaction, and employee satisfaction. "Other dimensions" includes research and teaching. As used for the IPGs, the basic premise is to focus more on obtaining data that can be used immediately and effectively for quality improvement and less on obtaining data for research or for academic projects. The Interdisciplinary Practice Guideline Program has standardized a set of reports that will be available on a quarterly basis for each work team. The orthopaedic IPG work team was instrumental in determining the indicators that would provide meaningful reports, and the total knee arthroplasty IPG will be the first to undergo a data analysis via this set of reports. Although the reports may seem limited in scope, these data elements have been identified as containing the most useful information about outcomes and variances in relation to the IPGs. The outcomes reports that will now be available via the Interdisciplinary Practice Guideline Program are

1. Patient Characteristics and Clinical Care
   - age
   - sex
   - principal diagnosis
   - secondary diagnosis
   - primary procedure
   - secondary procedure(s)
   - comorbidities
   - preadmission setting
   - admission type
   - discharge disposition
   - complications
   - delays in discharge
   - readmission rate (especially within 31 days of previous discharge)
   - ZIP code
2. Financial and Utilization
   - LOS (preoperative and postoperative)
   - variable costs (e.g., laboratory testing, pharmacy, radiological testing)
3. Customer Satisfaction
   - patient satisfaction
   - caregiver satisfaction

For the orthopaedic IPG work team, as well as other IPG teams, the ability to obtain data, either via a data collection process or by downloading from existing databases, has significantly influenced the type of indicators that are selected and the validity and usefulness of the data analysis. This component of the program is also undergoing continual development.

During planning for pilot implementation of the total knee arthroplasty IPG, a preprinted data collection sheet was designed for each discipline, listing the critical indicators identified by each discipline as well as a brief variance-tracking section. This outcomes and variance documentation method proved to be cumbersome and problematic, requiring frequent assessments for compliance and completion. The data collected were often incomplete or unclear. For the postpilot implementation of the IPG, the data were collected by the utilization review nurse assigned to the inpatient unit. The data collection sheet was redesigned to include critical outcomes, critical processes, demographics, clinical complications, nonclinical variances/systems problems, and delays in discharge. The sheet was completed for every total knee arthroplasty patient and submitted for collating and analysis to the office of the director of the Interdisciplinary Practice Guideline Program.

Currently, the organization is undergoing a massive budget reduction as well as a postmerger restructuring. Many roles and departments are experiencing significant changes. To control cost, efforts were maximized to institute a system for quality and cost measurement and improvement. One example is the transition of the role of utilization review nurse to case manager. Their responsibilies include more aggressive interaction and monitoring between caregivers and payers and less data collection activity.

Another problem regarding data availability was the inability of the clinical and financial databases to "talk" to each other. There was also a clear recognition and consensus that to implement improvement activities effectively, the clinical teams, such as the orthopaedic IPG work team, required a reliable and accessible system for retrieving data and information.

An important collaborative effort involving the department of health care quality, the fiscal department, and information services has resulted in a comprehensive database, accessible via the Interdisciplinary Practice Guideline Program and other quality/cost improvement forums. For the first time, clinical and financial data will be available in a common database. Training for utilization of this database and production of reports is currently underway, and the first reports according to the revised list of indicators and regarding the total knee arthroplasty will be distributed to the orthopaedic IPG work team in September 1997. This system will eliminate the need for separate resources to manage data collection and data entry functions.

A separate outcomes measurement is underway for the collaborative transitional pathway/IPG that was developed with Youville Lifecare. The data elements include demographic information, LOS, discharge disposition, unexpected clinical events such as unplanned readmission to BIDMC, and nonclinical delays in discharge.

Review of Outcomes and Variances

Historically, the data analysis for the total knee arthroplasty IPG has occurred twice a year. The data are reviewed by the chief of orthopaedics and the IPG work team. Problems with achievement of outcomes or interventions are analyzed. In some circumstances, the data, especially variance data, may require deeper analysis by the team to determine exact causes and ultimately identify the most effective response. Positive and negative variances are also analyzed to determine new trends or issues. If it is indicated by the data, the IPG work team plans corrective actions, revises the IPG, and plans ongoing implementation and evaluation of the guideline. The information is then disseminated by each discipline in respective staff meetings. The data and subsequent action plan are also reviewed and discussed in orthopaedic walk rounds and physician staff meetings.

Organizationwide Feedback/Response Loop

The department of health care quality, under the direction of the vice president and medical director for the department, provides support and education for all quality improvement projects. Each department has assigned a physician or nurse as its quality improvement representative. These representatives have attended a quality improvement retreat and have been included in an ongoing group with the objective to identify performance indicators and data requirements. The representative for the department of orthopaedics has become an active member of the orthopaedic IPG work team as well. The outcomes and variances information that is tracked for the total knee arthroplasty IPG will now become a part of a much more cohesive and structured departmental quality improvement program.

An important goal for the department of health care quality, administration, fiscal services, and all clinical departments is to structure an ongoing systematic and formalized process for reviewing action plans that are developed in response to data.

Until this point, there has been no centralized, organizational forum or responsibility for presenting IPG data and response plans. However, in 1997, a new entity, the committee for cost and care, was convened and is performing several functions. The committee is collating and presenting financial and utilization data to each department chief regarding specific high-cost procedures or diagnoses within that service. One goal is to identify areas in which cost and utilization are out of line with expectations. The second goal is to identify which cases require development of an IPG.

In addition, the budget process for this year has been very aggressive, and each department has been required to make a presentation regarding the current status and future plans for the budget. Discussions regarding the implementation of the IPGs and the achievement of targeted practice improvements have been incorporated into these sessions. Maintaining accountability for cost and quality improvement has become a priority within the organization.

OUTCOMES OF THE INTERDISCIPLINARY PRACTICE GUIDELINE

The overall outcomes to utilization of the total knee arthroplasty IPG have been consistently positive over time.

Length of Stay

LOS for fiscal year (FY) 1993 for the total knee arthroplasty was 6.7 days. In FY 1994, it had decreased to 5.7 days. In FY 1995, it had decreased to 4.9 days, and in FY 1996, it had decreased to 4.4 days. By July 1997, LOS had decreased to 4.0 days. The goal of discharging patients to rehabilitative settings by postoperative day 3 and to home with services by postoperative day 4 remains a successful plan. However, in view of rapidly changing payer requirements, it will be necessary to consider abbreviating this LOS even further in the very near future (see Figure 25–1).

VARIABLE COSTS

One of the indicators that has been monitored since the implementation of the IPG has been variable costs (see Figure 25–2). The variable costs have been directly affected by the management of LOS, radiological tests, laboratory tests, and overall utilization as recommended on the IPG and have been consistently decreasing since 1994. A recent slight increase in variable costs has led to increased vigilance by the orthopaedic work team in the areas of laboratory utilization, operating room management, and blood product utilization.

Patient Education and Assessment

Since late 1994, every patient who was scheduled to undergo surgery for a total knee replacement has had a preoperative assessment and teaching session with nursing and physical therapy (see Figure 25–3).

Patient Satisfaction

Patient satisfaction surveys are conducted by Picker Institute, a close affiliate of the East Campus of BIDMC. Survey data are collected and benchmarked to other Picker Institute clients. Picker Institute bases the survey questions on specific dimensions of care: coordination of care, continuity and transition, involvement of family and friends, respect for patient preferences, information and education, emotional support, and physical comfort. In 1994, the patient survey for the entire organization revealed that patients had concerns regarding continuity and transition as well as information and education. In response to the survey, the orthopaedic work team instituted the formalized patient preoperative education program. The 1995 survey demonstrated positive results for the unit in the dimensions of coordination of care and information and education.

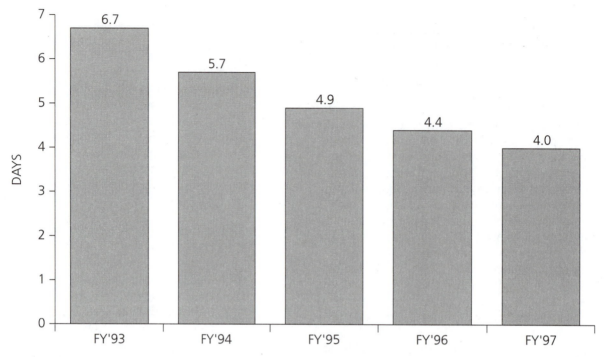

Figure 25–1 Total Knee Arthroplasty Length of Stay. *Source:* Copyright © Beth Israel Medical Center.

Staff Satisfaction

A survey has been developed and distributed to the orthopaedic caregivers (see Exhibit 25–A–3 in Appendix 25–A). The survey asks specific questions about the caregiver's response to the utilization of IPGs in practice. The initial results of the survey indicate a positive response to the IPG as a support to quality improvement and cost containment (see Table 25–1). The majority of responders indicate that the IPG

- reflects their practice
- does not interfere with their professional autonomy
- is an educational tool for caregivers as well as patients and families

INTEGRATION OF THE IPG INTO PRACTICE

Acceptance

The total knee arthroplasty IPG has been well accepted into the practice structure of the orthopaedic program. Several key elements made this a reality:

- The chief of orthopaedics is a major champion of the IPGs and supports all efforts to manage quality and cost in a rapidly changing environment. Several of the other surgeons in the department also serve as champions and supporters.

- The orthopaedic work team is a group of dedicated interdisciplinary critical thinkers who are committed to improvement of outcomes and practice.
- The orthopaedic department is composed of motivated interdisciplinary caregivers who recognize that the environment is turbulent and that responding to the environment appropriately is very important to the care provided to the patients.
- The orthopaedic IPG work team works steadily to review the IPG and to keep it updated in accordance with patient care requirements and in response to changes in the environment, practice, and systems.
- Demonstrable cost reductions and quality improvement have been shared with all members of the patient care team.
- The tool itself has been implemented as a guideline and not as a mandatory standard. Also, the tool was not implemented as a documentation tool. At the outset, early input from a variety of staff indicated a negative reaction to using the IPG as a documentation tool, and this was supported.

Throughout the development, implementation, and evaluation of the IPG, every effort was made by the IPG work team to avoid isolating the IPG from the caregivers who would be responsible for using it in practice. All revisions are disseminated throughout all the surgeons

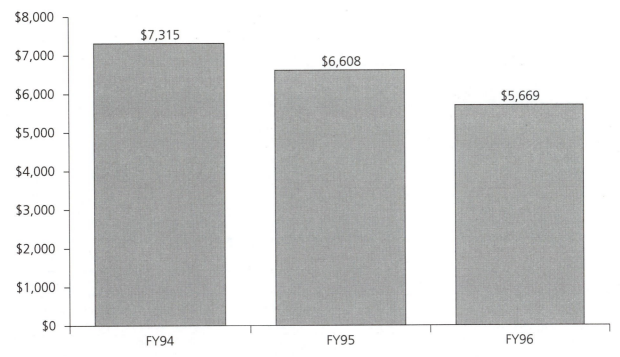

Figure 25–2 Total Knee Arthroplasty Variable Costs (Based on Principal Procedure 81.54/DRG 209). *Source:* Copyright © Beth Israel Medical Center.

and other disciplines for approval. The orthopaedic IPG work team presents itself as a core team of representatives for each discipline and not as a governing team for decision making. Interdisciplinary endorsement of the guideline is actively sought to avoid the perception of

mandating clinical and professional behaviors. All guideline revisions and data are reviewed and shared with each physician, as well as with each discipline, to emphasize that the IPG and its successful implementation belong to every participant in the patient's care. Each

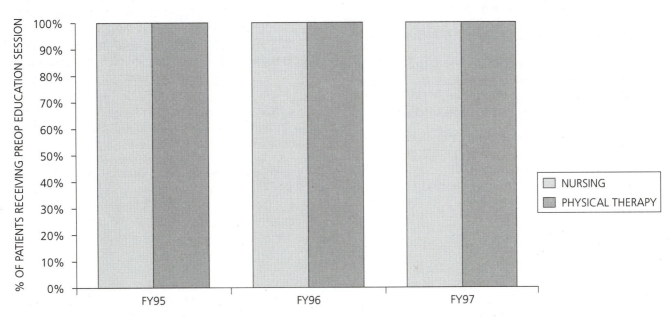

Figure 25–3 Total Knee Arthroplasty Preoperative Patient Education Compliance. *Source:* Copyright © Beth Israel Medical Center.

Table 25–1 Staff Survey Outcomes

	Strongly Disagree	Disagree	Neutral	Agree	Strongly Agree
The clinical content of the IPG is sound.				43%	57%
The implementation process of the IPG is acceptable.			7%	57%	36%
The IPG reflects my practice.			7%	50%	43%
The IPG increases my paperwork.	29%	57%	14%		
The IPG interferes with my autonomy in determining treatment.	36%	50%	14%		
I have been involved in the development of the IPG.		14%	29%	36%	21%
In practice, I use the IPG as:					
a. an operational guide			21%	50%	29%
b. an educational tool for myself		7%	14%	57%	21%
c. an educational tool for other caregivers			14%	65%	21%
d. an educational tool for patient/families		7%	14%	43%	36%

Note: Responders were 14 percent faculty MDs, 19 percent registered nurses, and 7 percent physical therapists.
Source: Copyright © Beth Israel Medical Center.

caregiver is viewed as a stakeholder in the successful implementation of the IPG.

Future Directions

Measurement and management of outcomes by caregivers have become a critical element in the delivery of health care. Today's health care environment has become increasingly competitive, with an emphasis on reducing costs while improving quality. IPGs or clinical pathways have facilitated the ability of clinical work teams to achieve positive outcomes in both cost reduction and quality improvement.

The total knee arthroplasty IPG enables caregivers to work together from one set plan of care, and this has produced several important advantages:

- Interdisciplinary communication and coordination have increased.
- LOS and costs of care have been reduced markedly.
- A system for providing continuous monitoring of performance has been established.
- Patient and family receive consistent information.

The orthopaedic IPG work team continues to meet weekly and has expanded the IPGs to include total hip arthroplasty, decompressive laminectomy, and fractured hip. Patient versions are also being created to be used as patient/family educational tools. The next version of the total knee arthroplasty IPG will include home care after discharge from acute care or rehabilitation.

Quality improvement, cost reduction, and management of care within the constraints of managed and/or capitated care will continue to present challenges to health care providers. The orthopaedic work team will continue to accept and meet each challenge with innovation and ingenuity.

REFERENCE

1. Nelson EC, Batalden PB, Plume SK, Mihevc NT, Swartz WB. Report cards or instrument panels: who needs what? *J Qual Improvement.* 1995;21:155–166.

SUGGESTED READING

Berwick D. Quality comes home. *Ann Intern Med.* 1996;125:839–843.

Hofmann P. Critical path method: an important tool for coordinating care. *J Qual Improvement.* 1993;19:235–246.

James M. Critical path implementation requires review and revision. *Inside Case Management.* 1996;3:1–4.

Pearson S, Goulart-Fisher D, Lee T. Critical pathways as a strategy for improving care: problems and potential. *Ann Intern Med.* 1995;123:941–948.

■ Appendix 25–A ■

Practice Guideline, Transitional Guideline and Data Collection Tool, and Staff Survey

The following is a key to the abbreviations used in the exhibits of this appendix:

ABX—antibiotics
BI—Beth Israel
BM—bowel movement
BR—bathroom
CBC—complete blood count
coags—coagulation levels
CPM—continuous passive motion
CSM—color, sensation, movement
CXR—chest X-ray
DAT—diet as tolerated
DB&C—deep-breathe and cough
D/C—discontinue, discharge
DTV—due to void
D5½ NS—5½% dextrose in normal saline
EKG—electrocardiogram
F/E—fluid and electrolyte
FHPA—functional health pattern assessment
F/U—follow-up
HCT—hematocrit
HL—heparin lock
H&P—history and physical
IPG—interdisciplinary practice guideline
IV—intravenous
KCL—potassium chloride
LOS—length of stay
MOM—milk of magnesia
NPO—nothing by mouth
N/V—nausea/vomiting
OOB—out of bed
OR—operating room

O_2 Sat—oxygen saturation
OT—occupational therapy
PO—oral
PCA—patient-controlled analgesia
POD—postoperative day
PT—physical therapy, prothrombin time
PTT—partial thromboplastin time
prn—as needed
qs—quantity sufficient
ROM—range of motion
SCD—sequential compression device
SMA 7—blood chemistry-7 tests (sodium, potassium, chloride, carbon dioxide, glucose, blood urea nitrogen, creatinine)
SNF—skilled nursing facility
S&S—signs and symptoms
SW—social work(er)
SWAPP—Social Work Advanced Planning Program
T&C—type and cross
TCDB—turn, cough, deep-breathe
TID—three times daily
TKA—total knee arthroplasty
tx—therapy
U/O—urinary output
UPC—unit packed cells
URI—upper respiratory infection
US—ultrasound
UTI—urinary tract infection
VCC—vital signs and symptoms
WNL—within normal limits

Exhibit 25–A–1. Total Knee Arthroplasty Interdisciplinary Practice Guideline

BETH ISRAEL DEACONESS MEDICAL CENTER: INTERDISCIPLINARY PRACTICE GUIDELINE
PROCEDURE: TOTAL KNEE ARTHROPLASTY

The Interdisciplinary Practice Guideline is not a mandatory standard, but is a guideline for care. All patient care requires adjustment via clinical judgment and decision making to meet the needs of each patient.

Sequence	Pre-Op	OR Day	POD 1	POD 2	POD 3—Discharge to Rehab	POD 4—Discharge to Home
Patient Need Categories			**Intermediate and Discharge Patient Outcomes**			
Physiological Parameters	• VSS • No pre-op S&S of URI or UTI	• VSS • No pre-op S&S of URI or UTI • O₂ Sat ≥ 90%	• Independent DB&C • VSS • CSM intact • F/E WNL • U/O 30 cc or greater every hour • O₂ Sat ≥ 90%	• Breath Sounds Clear • VSS • CSM intact • F/E WNL • No N/V • DTV 8 hours after foley removed	• Breath Sounds Clear • VSS • Afebrile • CSM intact • F/E WNL • No N/V • Voiding qs	• Breath Sounds Clear • VSS • Afebrile • CSM intact • Voiding qs • Tolerating diet • BM x 1
Pain		• States pain relieved in response to epidural/PCA	• States pain relieved in response to epidural/PCA	• States/demonstrates pain relieved in response to po pain med	• States/demonstrates pain relieved in response to PO pain med	• States/demonstrates pain relieved in response to PO pain med • Increase activity without intolerable pain
Mobility		• Bedrest	• Patient gets OOB to chair with minimal assist (one person) • Ambulate with walker at least 5 feet with minimal assistance	• Patient tolerating being OOB TID (Meals) • Ambulating to door with walker/crutches at least 25 ft. with contact guard assistance	• Patient ambulating with walker/crutches with supervision at least 50 feet	• Independent ambulation with assistive device on levels and stairs (at least 75 feet)
Patient Education	• Verbalizes concerns and asks questions • Patient/family demonstrates full knowledge re: postoperative course, expected LOS, discharge disposition	• Patient demonstrates awareness of surroundings, unit routine and immediate post-op course	• Patient demonstrates independent exercise program • Patient demonstrates understanding re: — Medications (dose, route, purpose, schedule, side effects) — Mobility — Emergency indicators (i.e., fever, incision changes) — F/U Visit — Coumadin/Blood Tests • Patient demonstrates knowledge of Rehab or Home Plan	• Patient demonstrates independent exercise program • Patient demonstrates understanding re: — Medications (dose, route, purpose, schedule, side effects) — Mobility — Emergency indicators (i.e., fever, incision changes) — F/U Visit — Coumadin/Blood Tests • Patient demonstrates knowledge of Rehab or Home Plan	• Patient demonstrates independent exercise program • Patient demonstrates understanding re: — Medications (dose, route, purpose, schedule, side effects) — Mobility — Emergency indicators (i.e., fever, incision changes) — F/U Visit — Coumadin/Blood Tests • Patient demonstrates knowledge of Rehab or Home Plan	• Patient demonstrates independent exercise program • Patient demonstrates understanding re: — Medications (dose, route, purpose, schedule, side effects) — Mobility — Emergency indicators (i.e., fever, incision changes) — F/U Visit — Coumadin/Blood Tests • Patient demonstrates knowledge of Rehab or Home Plan
Skin Integrity			• Dressing dry and intact	• Wound edges approximated • No significant surrounding erythema • Decreasing or no serious drainage • No purulent discharge	• Wound edges approximated • No significant surrounding erythema • Decreasing or no serious drainage • No purulent discharge	• Wound edges approximated • No significant surrounding erythema • Decreasing or no serious drainage • No purulent discharge

continues

The Interdisciplinary Practice Guideline is not a mandatory standard, but is a guideline for care. All patient care requires adjustment via clinical judgment and decision making to meet the needs of each individual patient.

Sequence	Pre-Op	OR Day	POD 1	POD 2	POD 3—Discharge to Rehab	POD 4—Discharge to Home
Intervention Categories			Interdisciplinary Interventions			
Referrals/Consults	• Anesthesia Evaluation • Preoperative Education Group: PT/Nursing/SW—SWAPP • MD Consult • Information faxed to BI Home Care as indicated	• Verify PT order				
Assessments	• Pre-admitting Testing: — FHPA — Verify autologous blood • Pre-op assessment	• Initial and ongoing assessment per TKA standards	• Ongoing assessment per TKA standards • Initial post-op assessment • Daily follow up by Social Worker as needed	• Ongoing assessment per TKA standards • Ongoing assessment • Daily follow up by SW as needed	• Ongoing assessment per TKA standards • Ongoing assessment • Daily follow up by SW as needed	• Ongoing assessment per TKA standards • D/C Assessment
Specimens/Tests	• If indicated: — CBC, SMA7; PT/PTT; CXR; EKG per anesthesia guidelines — T&C 2 UPC or per blood bank agenda	• X-rays to OR	• HCT • Lytes if indicated • PT	• PT	• PT • Discharge knee X-rays for patients being d/c'd per MD • US per MD	• PT • Discharge knee X-rays for patients being d/c'd per MD • US per MD
Treatments		• Sequential compression device on unaffected limb • Teds per MD • CPM • Drain per MD • Knee immobilizer per MD • Towel roll under heel as indicated • Foley • TCDB q 2 hours • Electricool per MD • Dressing • O₂ therapy to maintain O₂ Sat ≥ 90%	• Discontinue sequential compression device • Teds per MD • Discontinue CPM • Discontinue drain per MD • Incision care • Knee immobilizer per MD • Towel roll under heel as indicated • Foley • TCDB q 2 hours • Psychosocial intervention as needed by MSW • Electricool per MD • Dressing chg'd by MD • D/C O₂ if O₂ Sat ≥ 90%	• Discontinue SCD when patient ambulates • Teds per MD • Discontinue Foley • Straight cath as ordered prn • Knee immobilizer per MD • Towel roll under heel as indicated • Psychosocial intervention as needed by MSW • Electricool per MD	• Teds per MD • Knee immobilizer per MD • Towel roll under heel as indicated • Incision care • Psychosocial intervention as needed by MSW • Electricool per MD	• Teds per MD • Knee immobilizer per MD • Towel roll under heel as indicated • Electricool per MD • Incision care
Medications		• IV: D5½NS w/20 KCL at 80cc/hr • Epidural • IV ABX x 3 doses (cefazolin 1 gm q 8 hrs) or ((Cefoxitin 1 gm q 8 hrs) • Antiemetics prn • Anticoagulants	• IV: D5½NS w/20 KCL at 80 cc/hr • Change to Hep lock when patient tolerates 400cc PO fluids/shift • Cap epidural post PT eval & tx • Discontinue ABX • Antiemetics prn • Anticoagulants • Colace TID • MOM prn • Dulcolax prn	• H.L. or D/C IV • Change to PO pain med (Percocets 1–2 tabs PO q 3 hrs prn) • Check further requirements for antiemetics • Antiemetics prn • Anticoagulants • Colace TID • MOM prn • Dulcolax prn • D/C Epidural	• PO pain med (Percocets 1–2 tabs PO q 3 hrs prn) • Anticoagulants • Colace TID • MOM prn • Dulcolax prn	• PO pain med (Percocets 1–2 tabs PO q 3 hrs prn) • Anticoagulants • Colace TID • MOM prn • Dulcolax prn

continues

Exhibit 25-A-1 continued

The Interdisciplinary Practice Guideline is not a mandatory standard, but is a guideline for care. All patient care requires adjustment via clinical judgment and decision making to meet the needs of each individual patient.

Sequence / Intervention Categories	Pre-Op	OR Day	POD 1	POD 2	POD 3—Discharge to Rehab	POD 4—Discharge to Home
			Interdisciplinary Interventions			
Nutrition/Diet	• NPO after midnight	• NPO • Advance DAT	• DAT	• DAT	• DAT	• DAT
Safety/Activity		• Bedrest	• OOB to chair • Ambulate patient with walker • Transfer/gait training • ROM/strengthening exercises • Reinforce activity	• Transfer/gait training • ROM/strengthening exercises • Stationary bike/pedaler • Up for meals and BR • Reinforce activity	• Transfer/gait training • ROM/strengthening exercises • Stationary bike/pedaler • Up for meals and BR • Reinforce activity	• Up for meals and BR • Reinforce activity
Teaching	• Pre-op Teaching: — Expected post-op course — Potential discharge disposition including range of options for rehab/home — LOS — Autologous blood transfusion — H&P	• Orientation to unit and post-op routine	• Provide instruction re: — Medications (dose, route, purpose, side effects) — Post-op progression — Tubes — CPM — F/U visits — Mobility Progression — Exercise program — Positioning	• Provide instruction re: — Medications (dose, route, purpose, side effects) — Post-op progression — Tubes — CPM — F/U visits — DTV — Mobility Progression — Exercise program — Positioning	• Provide instruction re: — Medications (dose, route, purpose, side effects) — Post-op progression — F/U visits — CPM — Emergency indicators — Mobility Progression — Exercise program — Positioning	• Provide instruction re: — Medications (dose, route, purpose, side effects) — Post-op progression — F/U visits — CPM — Emergency indicators — Contact people — DVT prophylaxis x 6 weeks (including medication, follow up blood drawing, possible medialert bracelet)
D/C Planning	• Plans/assessment for D/C planning initiated • Assessment of home supports, insurance and psychosocial issues		• Patient screened for rehab as necessary		• Interdisciplinary: 3 page referral completed • MD Discharge Summary dictated • Post-discharge equipment ordered as indicated for home use • For all patients *d/c'd* to a facility, compile: — OR Reports — X-ray findings — Most recent lab results (3 days of PTs & Coumadin doses) — When to schedule F/U appt with MD	• Verify that patient has all prescriptions, *in particular* pain medication prescription • Complete D/C summary and emphasize who to contact with problems or questions • Discharge with necessary services • Complete 3 page referral • Discharge Summary dictated

Authors: Stephen Lipson, MD; Patty Brita-Rossi, RN, MSN; Cheryl Totte, RN, BSN; Karen Wasserman, LICSW; Deborah Adduci, PT; Claire Paras, RN, MBA; Donald Reilly, MD; Stephen Murphy, MD.
References: Coultron, CJ, Evaluating screening and early intervention: a puzzle with many pieces. *Social Work in Health Care.* 1988;13:65–72. Beth Israel Standard of Patient Care: Total Hip Arthroplasty.
Source: Copyright © Beth Israel Medical Center.

Exhibit 25–A–2 Beth Israel Deaconess Medical Center—East Campus/Youville LifeCare Transitional Interdisciplinary Practice Guideline/Clinical Pathway for Total Knee Replacement

	Beth Israel Deaconess Medical Center—East Campus/Youville LifeCare Transitional Interdisciplinary Practice Guideline/Clinical Pathway for TOTAL KNEE REPLACEMENT	
	Discharge from BIDMC	**Admission to Youville**
Assessments	• *Preoperatively:* —Assess patient/family awareness of Youville as a potential rehabilitative option	• *During hospitalization:* —Screening by Youville screeners by POD 1 —Verification of bed approval with insurance company
Specimens/Tests	• *At discharge:* —Provide most recent knee X-Ray at discharge —Provide most recent CBC results —Pending coags will be called/faxed by 2 PM —Provide available CXR results or indicate if patient did not require preoperative CXR	• *At admission:* —Determine necessity of CXR based on pre-surgical requirements (if necessary prior to surgery, a repeat CXR should be considered at this point)
Treatments	• *At discharge:* —CPM; provide patient range & recommendations for continued utilization —Knee immobilizer; provide specific recommendations per surgeon for utilization x 4 weeks	• *Upon admission:* —Discontinue CPM at 90° —Provide towel roll under heel to extend knee
Medications	• *At discharge:* —Coumadin management per surgeon recommendations —Provide copy of Coumadin Protocol & Anticoagulation Documentation Sheet —Provide information re: oral pain medication management regimen	• *Upon admission* —Verify recommendations for anticoagulation —Continue & advance pain medication management regimen
Safety/Activity	• *At discharge:* —Provide recommendations for bike therapy —Provide summary of patient's functional status with recommendations for advancement (page 3) —*If patient transferred after 12 PM, PT treats patient at BIDMC* —*If patient is transferred on a weekend, PT treats patient at BIDMC & informs patient that the next treatment will occur at Youville on the following Monday*	• *Upon admission* —Continue bike therapy • *If patient is transferred weekdays between 10 AM–12 PM, PT treats patient at Youville*
Teaching	• *Preoperatively:* —Provide information to build patient/family awareness of Youville as a rehabilitation option (brochure, contact numbers to arrange a tour, etc.) • *During hospitalization:* —Provide realistic expectation of rehab course & LOS —Prepare patient/family for transfer	• *Pre-admission:* —Provide tour • *Upon admission:* —Provide orientation & introduction to patient/family
Discharge Check Off *Must be completed at discharge*	• **BIDMC provide:** ❑ Most recent CXR report ❑ Most recent CBC result ❑ Most recent level reached on CPM ❑ Coumadin Protocol ❑ Anticoagulation Documentation Sheet ❑ Completed 3-page referral ❑ Operative report ❑ Discharge Summary ❑ CXR report, if applicable ❑ Recommendations for exercise bike ❑ Recommendations/orders for knee immobilizer/towel roll ❑ Recommendations/orders for CPM ❑ Indications for contacting surgeon	• **Youville Lifecare provide:** ❑ Discharge Summary to surgeon and BIDMC inpatient unit ❑ Preferred home care provider for management post-discharge from Youville Lifecare: _____ ❑ Coumadin information transferred to BIDMC at discharge from Youville Lifecare

Signature: _____ Date ____/____/____

continues

Exhibit 25–A–2 continued

Beth Israel Deaconess Medical Center/Youville LifeCare
Data Collection

1. **Youville Admission Date:** _____/_____/_____

2. **Youville Discharge Date:** _____/_____/_____

3. **Transfer from Youville Hospital to Youville Healthcare ❑ Date:** _____/_____/_____

4. **Discharge Disposition from Youville:**

Home with Services 1		Home without Services 2	
Nursing Home 3		SNF 4	
Other (specify): _____			

4. Unexpected Clinical Events	Yes	No
A. Unplanned readmission to BIDMC	1	2
B. Surgical Site Infection	1	2
C. DVT	1	2
D. At any time during the admission, did ROM drop below 90°?	1	2
E. If the answer to "D" is "yes," was the surgeon notified?	1	2
F. Other (specify):	1	2

5. Non-Clinical Delays in Discharge:	Yes	No
A. Patient/family decision	1	2
B. Family issue	1	2
C. MD order	1	2
D. Information/results delay	1	2
E. Home care arrangement problems	1	2
F. Supplies/equipment delay	1	2
G. Insurance/payer	1	2
H. Other (specify):	1	2

Signature: _____ Date _____/_____/_____

Source: Copyright © Beth Israel Medical Center.

Exhibit 25–A–3 Staff Survey, Department of Healthcare Quality

Interdisciplinary Practice Guideline Staff Survey
Department of Healthcare Quality
Beth Israel Deaconess Medical Center—East Campus

This survey intends to assess the staff's perceptions on the acceptability and efficacy of the Interdisciplinary Practice Guidelines (IPGs). Please answer each question by circling the appropriate number. Thank you for your participation and cooperation.

1. Date: _____ 2. Dept/Unit: _____ 3. Time at BIDMC _____ Yrs _____ Mons

4. Your job at BIDMC (please circle):
   (1) Faculty MD   (7) Registered Nurse
   (2) Attending MD   (8) Rehab Services (PT/OT)
   (3) Fellow   (9) Respiratory Therapist
   (4) Resident   (10) Social Worker
   (5) Intern   (11) Nutritionist/Dietitian
   (6) Medical Student   (12) Other (specify): _____

	Strongly Disagree	Disagree	Neutral	Agree	Strongly Agree
5. The clinical content of the IPG is sound.	1	2	3	4	5
6. The implementation process of the IPG is acceptable.	1	2	3	4	5
7. The IPG reflects my practice.	1	2	3	4	5
8. The IPG increases my paperwork.	1	2	3	4	5
9. The IPG interferes with my autonomy in determining treatment.	1	2	3	4	5
10. I have been involved in the development of the IPG.	1	2	3	4	5
11. I believe that the IPG is an effective way to reduce cost and improve clinical quality.	1	2	3	4	5
12. In practice, I use the IPG as:					
a. a daily operational guide	1	2	3	4	5
b. an educational tool for myself	1	2	3	4	5
c. an educational tool for other caregivers	1	2	3	4	5
d. an educational tool for patients/families	1	2	3	4	5
e. other (specify): _____	1	2	3	4	5

13. Specific IPG(s) referred to when answering the above questions: _____

_____

14. Additional comments & suggestions: _____

_____

Source: Copyright © Beth Israel Medical Center.

■ Part VI ■
Geriatric

■ 26 ■

Pathways for Managing the Frail Older Adult across the Continuum

Becky Trella and H. Scott Sarran

From 1991 through 1996, as part of an ongoing, grant-funded project to establish and test the effectiveness of a chronic care management system for older adults, Lutheran General HealthSystem (LGHS) developed six disease-specific care management tools. These tools, dealing with the management of stroke, dementia, depression, hip fracture, hip and knee replacements, and congestive heart failure, were designed to help patients and providers provide optimum care to the frail older adult population. A major goal of the original design was to enhance the coordination of care across sites for frail seniors, with particular emphasis on managing transitions between acute care, skilled nursing facilities, home care, community-based services, and physician offices. By the end of 1996, 600 patients had been enrolled and followed for a three-year period. This chapter will discuss how the implementation of these pathways led to the development of a generic geriatric pathway to manage frail older adults across multiple sites of care.

In 1990, LGHS joined the National Chronic Care Consortium (NCCC), an organization of 28 health care systems committed to develop and implement integrated systems of care (Chronic Care Networks) for patients with chronic conditions. The work from the pathway project was a result of efforts to meet the goals of the NCCC and develop a customer-centered, systems-oriented approach to the integration of chronic care across time, place, and profession. Although the NCCC had three areas of focus for integration (finance mechanisms, information systems, and care management), the work from the pathway project developed out of the NCCC care management committee. The first task of the LGHS care management effort was to develop common protocols that would enable network providers to collectively prevent, delay, or reduce the ongoing effects of disability throughout a patient's life.

PATHWAYS AND OTHER TOOLS

LGHS was one of the first NCCC sites to develop common care protocols across sites. The system formed a committee composed of physicians (both primary care and specialists), nurses, therapists, social workers, and other health care team members. These members represented multiple settings of care, including acute care, home care, nursing facilities, community-based services, and physician offices. Although some sites were owned by LGHS at the time, other sites (e.g., nursing homes and home care agencies) were invited to the committee, since they provided care in the community to many of the same patients.

LGHS had adopted the philosophy of continuous quality improvement in 1990 and had provided staff with the training and tools to use this approach to care. Many process improvement teams had successfully developed and implemented improvement projects, including the development of critical pathways in the hospital. Since the hospital had developed critical pathways, a similar process and format were used to develop the common protocols, which later became known as extended care pathways.

During the developmental phase of the pathways, clinicians found that regular meetings with other sites helped them better understand each other's roles and dramatically improved communication and coordination across these sites. Clinicians, instead of blaming each other for what had not been done at the last site of care, were calling each other to discuss their patients' needs.

After the standard format and process were agreed upon, the larger committee divided into subcommittees for the three originally chosen conditions: depression, dementia, and stroke. Subgroups used literature review and consensus from system experts to develop the pathways. At that time, little work had been done to develop pathways or guidelines outside acute care, making it difficult to use tools developed by other health care systems as examples.

The first three sets of pathways took approximately a year to develop. They were detailed documents with a separate tool for each site that followed the same standard format. The time frames listed on the pathways varied according to the sites: days for acute care, weeks and even months for nursing facilities and outpatient rehabilitation, and number of visits for home care and physicians' offices. The interventions listed on the pathways included all disciplines to improve interdisciplinary care and communication. When using the pathways, clinicians checked off the care provided and signed their name. A variance record was included to document the reasons for not providing expected care and to add interventions specific to a patient's individual needs.

While the pathways allowed for standardized best-practice planning and coordination, they did not provide a record of the patient's history that could travel with the patient from site to site. A common assessment tool containing clinical information required by all sites was therefore developed. The tool could be updated and used to track a patient's progress over time. It included demographic information; a systems review; history of service utilization; assessments of living situation, support systems, functional status, nutritional status, and mental status; advance directives; goals of treatment; emergency contacts; and a medication list.

A condensed version of the pathways and assessment tool was developed for the physician's office. This one-page document contained a brief synopsis from home care, community-based programs, or nursing facilities describing any changes in the patient's status or plans since the last visit to the physician. This was designed to be faxed to the physician's office before an office visit. All the tools (pathways, variance records, common assessment tool, and physician communication tool) were kept in a patient record designed to travel with the patient from site to site.

In an attempt to gain input and sponsorship, the tools were widely distributed. The committee realized that it would take a significant effort to educate clinicians and to implement processes to review pathways regularly at each site. In October 1992, a grant was obtained from the Retirement Research Foundation to develop a process to implement and evaluate the use of the pathways across sites. Multiple sites agreed to participate in the original project: the hospital, three nursing facilities, a home care agency, older adult day care, a comprehensive outpatient facility, and four physician offices (including a geriatric clinic, an internal medicine clinic, and two ortho-

paedic clinics). By the time the grant ended in 1996, 17 sites had participated.

In January 1995, LGHS merged with Evangelical Health Systems to form Advocate HealthCare. This new organization is one of the largest integrated delivery systems in the Midwest, with 20,000 employees providing care at 180 diverse sites (e.g., eight hospitals, four nursing homes, physician offices, older adult housing units, home care, adult day care, hospice, and congregations) throughout greater metropolitan Chicago. The merger gave the Lutheran project team the opportunity to share their experiences and project tools with other Advocate sites of care and to help determine an infrastructure to expand the older adult care management system across all of Advocate HealthCare.

As part of the process to evaluate, refine, and disseminate tools to the larger Advocate system, experience and feedback from the initial five pathways were reviewed. Although the feedback from patients and providers was good, the most commonly heard concern was that many patients did not "fit well" on one pathway or tool. Some of the most difficult management challenges came from patients whose functional disabilities resulted from more than just a summation of individual diseases. Many of these frail complex patients represented significant management challenges, with combinations of medical, cognitive, and psychosocial issues that stressed their caregivers and medical providers.

Several key interventions, such as evaluating changes in functional status and discussing goals of treatment and life-prolonging measures, were present on every pathway. In fact, 75 percent of the interventions were listed on all the pathways, regardless of disease or the site. Given these issues, providers and the committee came to agree that a basic geriatric pathway could be used as the primary tool for all patients, with one-page disease-specific interventions added to address the unique needs for that population. A patient with a stroke, congestive heart failure, and dementia would therefore be managed using a generic geriatric pathway with a few additional interventions targeted at preventing another stroke, managing the cardiac medications, and providing caregiver support.

To help providers grapple with this challenge, work was begun in 1995 on a Pathway for Managing the Frail Older Adult. The target population for this pathway was older adults with multiple chronic conditions and functional impairments, and the overall goal was to improve the coordination of care across sites. Like all other clinical pathways designed in Advocate, the development was the result of a multidisciplinary effort involving primary care physicians, geriatricians, nurses, therapists, and other health care providers.

With this framework, the generic pathway was written fairly quickly and completed in April 1995. The challenge was implementing the new tool with the targeted population. To clarify appropriate patients, the committee members developed the following criteria: multiple chronic conditions, needing assistance with three or more activities of daily living, more than two hospital admissions and/or emergency room visits in a six-month period, taking more than five medications, and lacking social support. If a patient met two or more of these criteria, he or she was considered appropriate for the generic geriatric pathway.

A total of 72 patients were managed in two years with the generic pathways. Most of these patients were enrolled while on the geriatric acute care unit. Many of these were subsequently transitioned to a nursing facility for long-term or home care. The physician office was the second most common site for patient recruitment. There was no typical pattern of site utilization for these patients, since most did not have a single primary diagnosis driving service use or cost of care.

It soon became clear that clinicians in home care (Exhibit 26–A–1 in Appendix 26–A), day care (Exhibit 26–A–2 in Appendix 26–A), and nursing facilities felt the generic pathway was the tool of choice. In contrast, many of the providers in the acute care setting, especially on cardiac and surgical units, felt that patients were primarily admitted and treated on the basis of a disease and did not want to change to a more generic tool.

THE CARE MANAGEMENT SYSTEM

The majority of systems have implemented case managers to follow patients across sites and to plan and coordinate care. Typically, much of a case manager's time is spent communicating the patient's history, status, and plan of care to the next site of care.

Instead of using case managers to follow patients across sites, the pathway project provided a means for sites to share clinical information and plan for a patient's care across multiple sites. Since it was not practical to bring clinicians together to plan for each individual patient's care, the goal was to establish standardized plans of care and common assessment data that would be used to track and manage patients with chronic conditions.

It was clear that educating and encouraging the sites to use the pathways was inadequate to implement the tools. A system was needed to provide for accountable clinicians at each site who would ensure that the tools were being used appropriately. To provide this system of accountability, a care coordinator role was established at each site. Care coordinators were responsible for ensuring that the care outlined by the pathway was implemented (or a variance was documented), the assessment tool was completed or updated, the physician communication tools were completed, and the assessment, pathways, and variance forms were sent to the next site of care when the patient was transferred. The care coordinators also provided staff education regarding the tools and the process. Monthly cross-site care coordinator meetings were established to review problems, develop solutions, and provide continuing education.

Most care coordinators were nurses who already had a major role at their sites coordinating care, although they had never been made accountable for coordination and communication with other sites. This project helped establish this accountability and provided these clinicians with an avenue to address issues and improve communication across sites.

The care coordinator role was implemented as a result of this project. It demonstrated to site management that having an accountable, designated clinician responsible for coordinated hand-offs to other sites of care for each patient is important to effective care. This integrated systems approach to case management (Figure 26–1) was one of the most important learning experiences from the project.

Project managers provided consultative services to sites, training, data collection, and problem solving regarding the use of the tools. The project managers developed the client and provider satisfaction instruments and collected the satisfaction data and the client utilization data. They conducted monthly cross-site meetings to maintain open lines of communication between sites and to resolve transition problems. These meetings were also a forum to address variance trending and to update and improve pathways.

Besides coordinating the implementation of the project, the project manager became the case manager for the highest risk patients who did not fit easily on any pathways. It was clear that a minority of patients would always "fall out" of the standard framework and need a much more individualized approach.

Tracking variances as patients transitioned across sites provided insight as to the barriers to care coordination across the continuum. By tracking these barriers, care coordinators were able to improve the process of coordinated transitions. One such barrier was the difficulty that home care and nursing home staff had in obtaining timely physician orders to institute care for a newly transitioned patient. To eliminate this barrier, pretyped physician orders were developed that were completed and signed by the physician and sent with the patient to home care and the nursing facilities. Another frequent variance was the inconsistency of teaching material across sites. For example, patients often received multiple

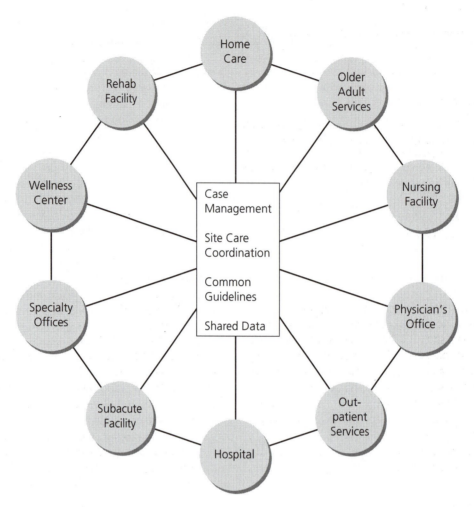

Figure 26–1 Integrated Care Management System

brochures on hip precautions before and after their hip replacement surgery. To eliminate this barrier, clinicians from across sites agreed on one set of teaching materials for each pathway condition. The same information was then reinforced at each site of care.

OUTCOMES

The key outcomes measured as part of this project were utilization patterns (including sites of care and lengths of stay), costs, and satisfaction (both patient and provider). Unfortunately, no baseline data are available for this population.

With the exception of cost data (which were derived from sites' various cost accounting systems), these outcomes were manually tracked by the project team, especially the project managers. The project managers, in conjunction with the site-based care coordinators, were also responsible for variance tracking.

Tables 26–1 and 26–2 show the three most common "paths" that patients experienced during their management under this project. Combined with disease-specific data (for cerebrovascular accidents, depression, dementia, and major joint replacements), these outcome data have led to numerous internal discussions regarding the appropriate role of lengthy stays in inpatient settings (especially acute rehabilitation and psychiatric services) or skilled nursing facilities. There appeared to be a relatively poor ability to predict those patients likely to benefit from these stays, since many of these patients went on to poor outcomes (such as death or permanent disability requiring custodial care) without obvious improvements from earlier interventions. These issues are receiving significant attention in the context of a dramatic increase in our population of globally capitated Medicare patients.

Patient satisfaction was assessed via monthly telephone interviews of a sample of five patients. For patients who were deemed incapable of participating in the

Table 26–1 Site Utilization for Comprehensive Geriatric Clients

	1995	1996
No. of Clients	21	51
No. with multiple-pathway diagnosis	4 (19%)	11 (21%)
No. followed by pathway MD	18 (85%)	30 (58%)
No. transitioned from medical unit to skilled nursing facility, became long-term care residents	5 (23%)	11 (21%)
No. transitioned from medical unit to skilled nursing facility for short-term care	4 (19%)	7 (13%)
Average length of stay in acute care	5 days	4 days
No. transitioned from medical unit to skilled nursing facility to home care	1 (4%)	4 (13%)
No. transitioned from medical unit to home care	3 (14%)	6 (11%)
Average length of stay in acute care	9 days	9 days
No. transitioned from MD office, to home care	3 (14%)	6 (11%)
No. followed at MD office only	2 (9%)	6 (11%)
No. followed at adult day care only	2 (9%)	2 (3%)
Miscellaneous transitions	1 (4%)	13 (25%)

survey process, the identified caregiver was interviewed. Since there were no tools that adequately dealt with care coordination across sites for frail elderly, we developed an instrument (Exhibit 26–A–3 in Appendix 26–A). Key feedback received was patients' continued frustration at being asked for the same information by multiple caretakers at multiple sites on multiple occasions. This has led to changes in how patients' clinical and demographic information is processed and communicated between sites and caregivers. During the first year of the project, patients frequently reported poor communication between their physicians and nursing facility staff; after feedback to providers, this concern was resolved.

Providers were surveyed twice yearly using specific project-developed tools targeted at physicians (via telephone interview; see Exhibit 26–A–4 in Appendix 26–A), care coordinators, and other staff (the latter two via a written questionnaire). In general, physicians felt neutral (i.e., the pathway was frequently "transparent" to the physician) or positive (because of perceived improvements in the coordination of their patients' care). Nonphysician providers commonly reported concerns with the length and com-

Table 26–2 Utilization Patterns and Costs for 19 Comprehensive Geriatric Clients

Sites	Path A	Path B	Path C
Outpatient MD office	n = 6		
Inpatient medical unit		n = 6 Length of stay = 9 days Charges = $13,905	n = 7 Length of stay = 4 days Charges = $6,180
Extended care facility			Length of stay = 35 days No. of physical therapy visits = 37 No. of occupational therapy visits = 30 No. of speech therapy visits = 19 Charges = $17,680
Home care	No. of RN visits = 8 No. of physical therapy visits = 5 No. of occupational therapy visits = 1 Charges = $1,680	No. of RN visits = 15 No. of physical therapy visits = 8 No. of occupational therapy visits = 6 Charges = $3,480	
Adult day care			
Estimated average charges/client	$1,680 plus MD office visits	$17,385	$23,860

Note: Path A, MD office to home care; Path B, inpatient medical unit to home care; Path C, inpatient medical unit to extended care facility.

plexity of the tools—criticisms that, in the long term, can probably be best dealt with by the incorporation of the pathway into an electronically based plan of care. Unfortunately, at this time, the only component of the pathway that is automated is the assessment tool.

Outcomes were reviewed in an ongoing fashion by the project managers, and with care coordinators and site administrators at regular meetings. The most common variance noted was difficulty in getting physician orders to "travel" with the patient upon transfer of care between sites (especially upon transfer from the inpatient setting to a nursing facility). There was particular difficulty in ensuring the continuity of care for specific medical conditions (especially anticoagulation management) for patients transferring across sites.

CONCLUSIONS AND LESSONS LEARNED

Tools' Simplicity and Documentation

The original pathways and common assessment tool were very extensive and cumbersome. While the goal was to use them to replace the current care plans and assessment tools, many sites were unable to make that commitment. This meant that clinicians were required to complete two sets of paperwork. With this in mind, we have since shortened and simplified our pathways and guidelines. Instead of trying to list 30 or more interventions, the goal is to target the two or three most important interventions across the entire continuum or episode of illness. For example, current hip and knee replacement pathways are one page for all sites of care (Exhibit 26–A–5 in Appendix 26–A).

Since documentation on the pathways did not replace the need to document in accordance with a site's specific process, this also created a dual documentation system. The new pathways do not require that the sites use them as a documentation system. It is up to the individual site to determine how clinicians document the completion of an intervention on the pathway. When possible, sites are encouraged to use the pathways for their care plan and their documentation tool, since this does encourage compliance if it is in place of and not in addition to old tools.

Access to Information Systems

The pathway project attempted to implement an information system that had a common assessment tool on line. Though the tool was a fairly simple database, many sites (especially the nursing facilities and home care) had limited availability of personal computers for clinician use. Therefore, clinicians had to write the information on the paper tool to be later entered into the database. This created more work and therefore affected compliance. Until clinicians have access to terminals where they are assessing and providing care, systems will not be well accepted. Although some of our acute care sites have started to implement in-room clinical databases, nursing facilities, physician offices, and home care are much farther behind.

The Care Coordination Role in Physician Offices

Two primary care and three specialty offices were involved in the pathway project. Care coordinators (nurses or social workers) were available in all but one office. They found the tools cumbersome at times (Exhibit 26–A–6 in Appendix 26–A) but appreciated the enhanced communication with the other sites of care. The office lacking a care coordinator was a busy internal medicine clinic with many frail geriatric patients. The staff was largely composed of physician assistants and aides who did not feel comfortable with the care coordinator role. To compensate, one of the project managers provided care coordination three half-days per week. In a short time, the physicians became very dependent on the project manager to deal with their most complex geriatric patients. When the project ended, a job description was written for a full-time primary care case manager with sponsorship from the physicians in the group. The physicians felt that the funding of the new position was easily justified by freeing up physician time from dealing with care coordination and allowing them to take more patients. The nurse had also prevented hospital admissions and other unnecessary care, a task that was crucial if the group was to successfully manage its Medicare health maintenance organization (HMO) population.

Managing Frail Complex Geriatric Patient with Pathways

Frail geriatric patients do not follow narrowly defined treatment pathways that are disease specific. Most frail seniors have issues related to multiple chronic conditions, functional impairment, and social support. The disease-specific pathways developed by this project were broad and focused on comprehensive assessment, functional issues, and other geriatric concepts. A generic pathway with additional brief disease-specific attachments as needed was a more practical solution to incorporating key geriatric concepts into care. In addition, the use of a common assessment tool that follows a patient across sites can prevent fragmented inappropriate care by assisting clinicians to follow the patient's history and progress over time.

Role of the Case Manager for Complex Patients

Even with appropriate tools, a much more individualized approach is required for very complex patients. If case managers are located in the primary care physician's offices, they can target patients before an acute episode. If their patients move to other sites, the case managers can refer clinicians to the generic pathway and also consult when the patient does not fit the pathway. The common assessment tool can be completed in the physician's office and sent to the other sites of care.

Forming a System of Coordinated Care

Clinicians from various sites need to work together to resolve problems and improve coordination. Care planning should not stop at one door and start at the next. Clinicians need to be held accountable to coordinate care with the next site and demonstrate outcomes beyond one site of care. Developing the tools for the entire patient experience and reporting outcomes accordingly forced our sites to examine the whole network of care and improve care coordination. This focus proved to be more important than the actual tools. Although some sites have not used the pathways on a regular basis, having a care coordinator who is accountable at the sites for coordination and cost-effectiveness has proved invaluable when managing our Medicare HMO patients. The Advocate System has 10,000 Medicare HMO members that it has taken on global capitation and managed within budget.

These important lessons are being implemented throughout Advocate, including the establishment of accountable, clear, formalized roles for care coordination within and across sites. Advocate is slowly moving to standardized roles and responsibilities for care coordinators and has now formed a systemwide entity to address ongoing issues related to care management. Each site is responsible for achieving certain common goals, and there is significant accountability at the system executive level for the adoption of coordinated best practices across the system. Although much work continues in the area of pathways, guidelines, and other tools, they are considered a means to support an integrated, common, patient-centered care management process.

■ Appendix 26–A ■
Clinical Pathways and Satisfaction Surveys

The following key is for abbreviations used in the exhibits of this appendix:

ADC—adult day care
ADL—activities of daily living
AHA—American Heart Association
C—client
CBC—complete blood count
CCP—Community Care Provider
CG—caregiver
D/C—discontinue
DM—diabetes mellitus
DME—durable medical equipment
DVT—deep-vein thrombosis
ECF—extended care facility
EKG—electrocardiogram
FBS—fasting blood sugar
Hgb—hemoglobin
HHA—home health aide
IDOA—Illinois Department on Aging
INR—anticoagulation ratio

I&R—information and referral
LOS—length of stay
MM—Mini Mental Status Exam
OT—occupational therapy/therapist
PM&R—physical medicine and rehabilitation
PO—oral
PRN—as needed
PT—physical therapy/therapist
ROM—range of motion
SE—side effect(s)
SMAC—blood chemistry
S/S—signs and symptoms
ST—speech therapist
TB—tuberculosis
TD—tetanus diptheria
TTWB—toe touch weight bearing
WB—weight bearing
W/C—wheelchair

Exhibit 26-A-1 Comprehensive Geriatric Pathway for Home Health Care

Patient's Name: _____

Date of Admission: _____

Site: _____

Primary Physician: _____

Health Status Goals:

☒ 1.0 Patient will achieve optimal level of health, as indicated by:
☒ Vital Signs
☒ Lab Values
☒ Wound Healing
☒ Pain Management
☒ Control of Physical Symptoms
☒ Other _____
☒ 2.0 Pt will achieve optimal level of
2.1 muscle strength
2.2 coordination
2.3 balance
3.0 _____

1. *Medical instability*
2. *Health status declining*
3. *Health status the same or stable*
4. *Health status improving*
5. *Optimal status for pt*

Functional Status Goals:

☒ 4.0 Pt/CG performs ADL at max level of independence (includes use of assistive devices)
☒ 5.0 Pt is mobile within their environment (includes use of assisted devices)
☒ transfer
☒ ambulance/maneuver W/C
☒ stairclimb
☒ 6.0 Pt/CG perform procedures related to care
☒ 7.0 Pt/CG manage equipment related to care

☒ 8.0 Pt demonstrates ability to:
☒ 8.1 communicate
☒ 8.2 swallow
☒ 8.3 recall information to perform activities
☒ Other
☒ 9.0 *

1. *Dependent—0% participation*
2. *Max assist—25% participation*
3. *Mod assist—50% participation*
4. *Min assist—75% participation*
5. *Independent—100% participation*

Knowledge Goals:

☒ 10.0 Pt/CG verbalizes understanding of
☒ 10.1 disease process
☒ 10.2 medication/therapy regimen
☒ 10.3 diet
☒ 10.4 activity management/exercise program
☒ 10.5 S/S of complications
☒ 10.6 Other*
☒ 11.0 Pt/CG identify necessary lifestyle changes which may improve health status. Mandatory for chronic illness.
☒ 12.0 Pt/CG Identify home and environment safety measures.

1. *No knowledge*
2. *Minimal knowledge*
3. *Basic knowledge*
4. *Adequate knowledge*
5. *Knowledgeable and takes appropriate action*

Behavioral Adaptation Goals:

☒ 14.0 Pt/CG demonstrates compliance with
☒ 14.1 medication regimen/therapy
☒ 14.2 diet
☒ 14.3 activity management/exercise program
☒ 14.4 appropriate response to complications
☒ 14.5 Other _____
☒ 15.0 Pt/CG demonstrates coping mechanism that enhances participation in care

1. *Non-compliant*
2. *Rarely appropriate*
3. *Inconsistently appropriate*
4. *Usually appropriate*
5. *Follows treatment plan consistently*

Components	Admission—48 Hrs	Day 2–5	Week 2–3	Week 4–5	Week 6–8	Transition
Assessment/ Monitoring	☒ RN assessment within 24 hrs of discharge from institution ☒ Order verified with physician(s) to include MSW assessment, therapies, HHA ☒ History & discharge instruction received from prior institution ☒ Initial visit done by RN Case Manager during which time the RN will do or complete: ■ assessment form update ■ complete history ■ physical exam ■ assess client's living environment ☒ Assess client/caregiver support systems	☒ Ongoing assessment and confirmation of baseline data	☒ Ongoing assessment and confirmation of baseline data ☒ Assess for cause in change from baseline ☒ Assess readiness for discharge	☒ Ongoing assessment and confirmation of baseline data ☒ Assess readiness for discharge, complete forms and turn in to supervisor.	☒ Ongoing assessment and confirmation of baseline data ☒ Assess readiness for discharge, complete forms and turn in to supervisor.	☒ Assess appropriateness of discharge plan ☒ Update Pathway Project Assessment forms and communicate status to new health care providers

continues

Exhibit 26–A–1 continued

Components	Admission—48 Hrs	Day 2–5	Week 2–3	Week 4–5	Week 6–8	Transition
Consults	☒ MSW, PT, OT, ST, HHA per physician order	☒ Confer that treatment is in place ☒ Ongoing evaluation for consults PRN	☒ Send communication tool for office visits ☒ Assist physician with determining need for transition with alternative arrangements when client goals have been met or client no longer homebound	☒ Send communication tool for office visits ☒ Assist physician with determining need for recertification or transition with alternative arrangement when client goals have been met or client no longer homebound	☒ Send communication tool for office visits. ☒ Assist physician with determining need for recertification or transition with alternative arrangement when client goals have been met or client no longer homebound	☒ Consult with physician for approval of transition plan
Tests	☒ Labs as defined by physician ☒ Notify physician of all lab results ☒ Therapeutic medication levels as indicated (i.e., digoxin, dilantin, theophylline, anticoagulants, etc.)	☒ Labs as defined by physician ☒ Notify physician of all labs	☒ Labs as defined by physician ☒ Notify physician of all labs	☒ Labs as defined by physician ☒ Notify physician of all labs	☒ Labs as defined by physician ☒ Notify physician of all labs	☒ Assist client in arranging for follow-up lab work
Functional-Rehab	☒ Information regarding limitations, activities permitted, and rehab needs as given by referral source at time initial referral is being given ☒ RN to assess patient functional status, with ADLs and complete initial assessment sheet on functional abilities. ☒ Consult therapies if appropriate and consider HHA to assist with ADLs ☒ Initial client care plan & goals determined ☒ All therapy evaluations completed & frequency established ☒ Evaluate need for DME & safety equipment. Provide referrals to caregiver	☒ Encourage and increase activity as tolerated ☒ Instruct on safety and fall precaution prn ☒ Ongoing assessment of functional status & rehab needs ☒ Assess risk for skin breakdown—initiate protocol prn ☒ All therapy evaluations completed & frequency established	☒ Reassess need for ongoing therapy or D/C if no progress or change to outpatient therapy when available ☒ Reassess client/caregiver's ability in performing ADLs, consider HHA if needed ☒ Reinforce safety & fall precautions ☒ Assess risk for skin breakdown, initiate protocol prn ☒ RN to complete HHA supervisory visit @ 2 weeks with appropriate forms	☒ Reassess need for ongoing therapy or D/C if no progress or change to outpatient therapy when able ☒ Reassess client/caregiver ability in performing ADLs, consider HHA if needed ☒ Ongoing assessment of functional status & rehab needs ☒ Reinforce safety & fall precautions ☒ Assess risk for skin breakdown, initiate protocol prn	☒ Reassess need for ongoing therapy or D/C if no progress or change to outpatient therapy when able ☒ Reassess client/caregiver ability in performing ADLs, consider HHA if needed ☒ Ongoing assessment of functional status & rehab needs ☒ Reinforce safety & fall precautions ☒ Assess risk for skin breakdown, initiate protocol prn	☒ Arrange for follow-up therapies as indicated or other supportive services to maintain function (i.e., adult day care)
Nutrition	☒ Assess client/caregiver knowledge of diet ☒ Assess ability to chew, feed & swallow ☒ Verify diet with physician ☒ Baseline weight ☒ Assess general appearance	☒ Diet instruction to client/caregiver ☒ Confer with physician for poor nutritional status ☒ Assess general appearance ☒ Weekly weight prn	☒ Weekly weight PRN ☒ Confer with physician for poor nutritional status ☒ Assess general appearance	☒ Weekly weight PRN ☒ Confer with physician for poor nutritional status ☒ Assess general appearance	☒ Weekly weight PRN ☒ Confer with physician for poor nutritional status ☒ Assess general appearance	☒ Final instruction to client/caregiver ☒ Client/caregiver able to provide & plan proper diet without assistance

Components	Admission—48 Hrs	Day 2–5	Week 2–3	Week 4–5	Week 6–8	Transition
Medication	☒ RN to check client's medication at home with the medication on discharge list ☒ Medication regimen validated with doctor ☒ Assess educational needs regarding medication, indication, dose, time, SE	☒ Monitor response to medication ☒ Confer with physician if need to adjust medication ☒ Provide education of instruction sheets regarding medication, indication, dose, time, SE ☒ Update medical sheet to include new medications	☒ Monitor response to medication ☒ Ongoing assessment of C/CG ability to state use, dosage, and administration of medication for transition ☒ Continue education needs regarding medication, indication, dose, time, SE ☒ Confer with physician if need to adjust medication	☒ Client/caregiver can state use, dosage & administration of medication for discharge or continue to re-educate ☒ Review drug regimen prior to discharge	☒ Client/caregiver can state use, dosage & administration of medication for discharge ☒ Review drug regimen prior to transition	☒ Instruct client on medication regimen ☒ Written discharge instructions provided to client and caregiver
Treatments/ Teaching	☒ Provide appropriate teaching materials to C/CG ☒ Assess client/caregiver teaching needs regarding condition, care & treatment plan ☒ Agency information given ☒ Mutual goals discussed ☒ Review advance directives, code status & client's goals for treatment ☒ Obtain any specific treatment procedure or client teaching need at time referral is made from physician/prior site	☒ Provide assistance to manage current problems of: –falls & safety –sleep –agitation –wandering ☒ Review mutual goals	☒ Reinforce appropriate teaching materials ☒ Provide assistance to manage current problems of: –falls –sleep –agitation –wandering ☒ Review mutual goals ☒ Client/caregiver verbalizes understanding of instruction for transition	☒ Client/caregiver verbalizes understanding of instruction for transition or need to recertify ☒ Client/caregiver demonstrates ability to provide for care needs for transition or need to recertify	☒ Client/caregiver verbalizes understanding of instruction for transition or need to recertify ☒ Client/caregiver demonstrates ability to provide for care needs for discharge or need to recertify	☒ Client/caregiver can verbalize/demonstrate knowledge of condition, care & treatment plan
Spiritual/ Psychosocial	☒ Assess spiritual needs, values & preferences ☒ Assess impact of condition & coping in client/caregiver ☒ Supportive counsel PRN ☒ Assess any specific concerns that may adversely affect client's ability to live at home ☒ Begin building trust in relationship	☒ Offer options of spiritual support ☒ Assess impact of condition & coping in client/caregiver ☒ Assess for signs & symptoms of depression ☒ Supportive counseling PRN ☒ Determine client's ability to make decisions and/or need for surrogate decision maker	☒ Offer options of spiritual support ☒ Assess impact of condition & coping in client/caregiver ☒ Supportive counseling PRN	☒ Offer options of spiritual support ☒ Assess impact of condition & coping in client/caregiver ☒ Supportive counseling PRN	☒ Offer options of spiritual support ☒ Assess impact of condition & coping in client/caregiver ☒ Re-evaluate for signs & symptoms of depression ☒ Supportive counseling PRN ☒ Re-evaluate client's ability to make decisions and/or need for surrogate decision maker	☒ Encourage client to follow up with religious organization ☒ Referrals for ongoing counseling & support counseling PRN

continues

Exhibit 26–A–1 continued

Components	Admission—48 Hrs	Day 2–5	Week 2–3	Week 4–5	Week 6–8	Transition
Continuing Care Needs	☒ Assess current living situation, financial needs, & ability to care for client at home ☒ Multidisciplinary team recommend to physician for estimated LOS ☒ Transition plans discussed with client/caregiver	☒ Appropriate community resource referrals made ☒ Social worker evaluation prn ☒ Review Pathway at Interdisciplinary care conference ☒ Consult with benefits case manager for DME/supply approval	☒ Follow-up with doctor as scheduled ☒ Client/caregiver conference to discuss medical treatment, care plan & goals, ongoing needs & transition plans ☒ Review Pathway at Interdisciplinary care conference ☒ Consult with benefits case manager for DME/supply approval	☒ Referrals made for continued care needs if patient to be transitioned; day care, outpatient follow-up labs, nursing home, hospice or community parish nurse ☒ Recertification per physician order ☒ Review Pathway at Interdisciplinary care conference ☒ Consult with benefits case manager for DME/supply approval	☒ Referrals made for continued care needs if patient to be transitioned; day care, outpatient follow-up labs, nursing home, hospice or community parish nurse ☒ Recertification per physician order ☒ Review Pathway at Interdisciplinary care conference ☒ Consult with benefits case manager for DME/supply approval	☒ Client transitioned with appropriate care ☒ Transition plan & paperwork provided to next site ☒ Notify Pathway Project Manager of discharge
Outcomes						☒
Signature						
Date						

Source: Copyright © Lutheran General Hospital–Advocate, Park Ridge, Illinois, used with permission.

Exhibit 26–A–2 Comprehensive Geriatric Pathway for Adult Day Care

Patient's Name: _____

Care Coordinator: _____

Date of Admission: _____

Site: _____

Primary Physician: _____

Components	Admission—72 Hrs	Week 1	Week 2	90 days	Quarterly Review
Monitoring	☒ Initiate & complete medical and psychosocial assessment forms ☒ IDOA (CCP) papers completed as needed	☒ Ongoing observation and confirmation of baseline data ☒ Assess for change in behavior and/or cognition since admission ☒ Assess for sensory (vision, hearing) loss	☒ Ongoing monitoring of client for deviation from initial assessment ☒ Assess client's overall adjustment and integration into day care	☒ Ongoing monitoring of client for deviation from initial assessment ☒ Assess client appropriateness for continued ADC.	☒ Ongoing monitoring of client for deviation from initial assessment ☒ Assess client appropriateness for continued ADC.
Consults	☒ Contact referral source ☒ Consults as indicated	☒ PT/OT/PRN ☒ Send letter to medical physician requesting updated status and medication if not already received ☒ Evaluate sensory loss and encourage family to consult appropriate services (i.e., audiology, ophthalmology) ☒ Consults as needed	☒ Evaluate sensory loss and encourage family to consult appropriate services (i.e., audiology, ophthalmology) ☒ Consults as needed	☒ Consult physician if new problems develop and send communication tool for office visits	☒ Consult physician if new problems develop and send communication tool for office visits
Tests	☒ Request family to arrange for periodic lab draws if indicated	☒ Request family to arrange for periodic lab draws if indicated	☒ Request family to arrange for periodic lab draws if indicated	☒ Request family to arrange for periodic lab draws if indicated	☒ Request family to arrange for periodic lab draws if indicated
Functional Rehab	☒ Assess for physical condition & functional status ☒ Therapies PRN ☒ Safety & fall precautions PRN ☒ Bowel & bladder program PRN ☒ Skin care PRN ☒ Assess need for elopement risk	☒ Ongoing evaluation & update of functional status ☒ Therapies PRN ☒ Safety & fall precautions PRN ☒ Bowel & bladder program PRN ☒ Skin care PRN ☒ Continue to assess need for elopement risk	☒ Ongoing evaluation & update of functional status ☒ Assess client/caregiver's ability to meet rehab/self care needs ☒ Therapies PRN ☒ Safety & fall precautions PRN ☒ Bowel & bladder program PRN ☒ Skin care PRN ☒ Continue to assess need for elopement risk	☒ Ongoing evaluation & update of functional status ☒ Assess need for arranging ongoing therapies outside adult day care setting ☒ Therapies PRN ☒ Safety & fall precautions PRN ☒ Bowel & bladder program PRN ☒ Skin care PRN ☒ Continue to assess need for elopement risk	☒ Ongoing evaluation & update of functional status ☒ Assess need for arranging ongoing therapies outside adult day care setting ☒ Therapies PRN ☒ Safety & fall precautions PRN ☒ Bowel & bladder program PRN ☒ Skin care PRN ☒ Continue to assess need for elopement risk
Nutrition	☒ Assess nutritional status ☒ Assess ability to chew, feed & swallow ☒ Nutrition consult PRN (outpatient nutrition 696-7770)	☒ Monitor PO intake ☒ Assess client/caregiver knowledge of diet ☒ Confer with physician if nutritional needs not being met PRN ☒ Nutrition consult PRN (outpatient nutrition 696-7770)	☒ Monitor PO intake ☒ Supplemental nutrition PRN ☒ Dietary instruction to client/caregiver PRN ☒ Confer with physician if nutritional needs not being met PRN	☒ Monitor PO intake ☒ Supplemental nutrition PRN ☒ Dietary instruction to client/caregiver PRN ☒ Confer with physician if nutritional needs not being met PRN	☒ Monitor PO intake ☒ Supplemental nutrition PRN ☒ Dietary instruction to client/caregiver PRN ☒ Confer with physician if nutritional needs not being met PRN

continues

Exhibit 26-A-2 continued

Components	Admission—72 Hrs	Week 1	Week 2	90 days	Quarterly Review
Medications	☒ Review medications ☒ Assess education needs regarding medications, indication, dose, time, SE ☒ Review medication policies with client	☒ Monitor response to medication ☒ Instruct client/caregiver regarding medication ☒ Client to provide a 2 week supply of meds in appropriate container	☒ Monitor response to medication ☒ Confer with physician if poor response to medication ☒ Instruct client/caregiver regarding medication ☒ Monitor for common side effects ☒ Reinforce medication policy	☒ Monitor response to medication ☒ Instruct client/caregiver regarding medication PRN ☒ Review appropriateness of drug regime every three months ☒ Reinforce medication policy PRN	☒ Monitor response to medication ☒ Instruct client/caregiver regarding medication PRN ☒ Review appropriateness of drug regime every three months ☒ Reinforce medication policy PRN
Teaching	☒ Assess client/caregiver knowledge of condition, care, day care program & treatment plan ☒ Orient to day care & introduce to staff, client, volunteers	☒ Formulate care plan ☒ Orient to day care & introduce to staff, client, volunteers	☒ Instruct client/caregiver regarding condition, care and treatment plan as needed	☒ Instruct client/caregiver regarding condition, care and treatment plan as needed	☒ Instruct client/caregiver regarding condition, care and treatment plan as needed
Treatment	☒ Evaluate for treatments including dressing changes, foley catheter changes, tube feeding, wound care, skin care needs PRN	☒ Provide treatments as indicated ☒ Baseline vital signs ☒ Baseline weight	☒ Provide treatments as indicated ☒ Vital signs as indicated, a minimum of monthly	☒ Provide treatments as indicated ☒ Vital signs as indicated, a minimum of monthly ☒ Weigh monthly	☒ Provide treatments as indicated ☒ Vital signs as indicated, a minimum of monthly ☒ Weigh monthly
Spiritual/ Psychosocial	☒ Assess spiritual needs, values & preferences ☒ Assess impact of condition & coping in client/caregiver ☒ Determine client's ability to make decisions & need for surrogate decision making ☒ Supportive counseling ☒ Provide information on advanced directives, code status	☒ Offer options of spiritual support ☒ Monitor adjustment to day care ☒ Initiate contact with social worker for supportive counseling PRN ☒ Monitor support caregiver adjustment to day care ☒ Offer support groups to caregiver ☒ Support counseling ☒ Advanced directives, code status in place	☒ Offer options of spiritual support ☒ Monitor adjustment to day care ☒ Weekly session with social worker PRN ☒ Encourage caregiver to participate in counseling & support groups PRN	☒ Offer options of spiritual support ☒ Encourage continued participation in day care ☒ Weekly session with social worker PRN ☒ Encourage caregiver to participate in counseling & support groups PRN	☒ Offer options of spiritual support ☒ Encourage continued participation in day care ☒ Weekly session with social worker PRN ☒ Encourage caregiver to participate in counseling & support groups PRN
Continuing Care Needs	☒ Assess current living situation & need for other services ☒ Assess need for alternative funding	☒ Assess appropriateness of day care, plan for alternative services PRN ☒ Adjust day care schedule if needed ☒ Assess caregiver's ability to care for client at home	☒ Assess appropriateness of day care, plan for alternative services PRN ☒ Adjust day care schedule if needed	☒ Caregiver conference PRN to discuss care plan, client goals, ongoing needs	☒ Caregiver conference PRN to discuss care plan, client goals, ongoing needs
Activity	☒ Identify for client/caregiver the available programs	☒ Leisure survey completed ☒ Initiate group involvement & integrate into setting	☒ Ongoing evaluation of appropriate group placement ☒ Reassign to group PRN ☒ Provide for a variety of group experiences	☒ Ongoing evaluation of appropriate group placement ☒ Reassign to group PRN ☒ Provide for a variety of group experiences	☒ Ongoing evaluation of appropriate group placement ☒ Reassign to group PRN ☒ Provide for a variety of group experiences

Components	Admission—72 Hrs	Week 1	Week 2	90 days	Quarterly Review
Outcomes	☒ Client has been assessed appropriate for adult day care ☒ Client/caregiver verbalizes understanding of adult day care procedures & practices ☒ If using transportation client/caregiver verbalizes understanding of transportation policies & practices ☒ Client/caregiver understands payment procedure	☒ Client is oriented to primary areas of adult day care, ie, bathroom, dining room ☒ Client beginning to socialize with staff & other clients ☒ Client remains in the activity room for at least half of the activity ☒ Client/caregiver adhere to medication policy ☒ Assessment forms completed	☒ Client is oriented to adult day care ☒ Client follows appropriate diet ☒ Client tolerating medication ☒ Client remains in the activity room for the full activity ☒ Client participates in outings ☒ Client appears to be reaching a comfort level with staff & other clients	☒ Client is oriented to adult day care ☒ Client follows appropriate diet ☒ Client tolerating medication ☒ Client participating in all recommended activities ☒ Client attending adult day care on all scheduled days ☒ Client interacting with staff & other clients	☒ Client is oriented to adult day care ☒ Client follows appropriate diet ☒ Client tolerating medication ☒ Client participating in all recommended activities ☒ Client attending adult day care on all scheduled days ☒ Client interacting with staff & other clients
Signature					
Date					

Source: Copyright © Lutheran General Hospital–Advocate, Park Ridge, Illinois, used with permission.

Exhibit 26–A–3 Client Satisfaction Survey

Hello _____, my name is _____ of the Lutheran General Pathway Project. The Pathway Project's goal is to improve care management, communication and coordinate your care as you move from one health site to another. I would like to ask you a few questions about your past illness/surgery.

1. Do you feel your care (or your family member's) was well planned during this past episode of illness/surgery?

    _____ Yes   _____To some extent   ____No

2. Do you feel there was adequate communication between the physicians and other staff members at:
    ____Hospital   _____Yes   ____To some extent   ____No
    ____Nursing Home   _____Yes   ____To some extent   ____No
    ____Home Health Care   _____Yes   ____To some extent   ____No
    ____Adult Day Care   _____Yes   ____To some extent   ____No
    ____Physician Office   _____Yes   ____To some extent   ____No
    ____Pre-Admission Testing   _____Yes   ____To some extent   ____No
    ____Comprehensive Outpatient Rehabilitation   _____Yes   ____To some extent   ____No

3. Do you feel that the physician and other staff members at each site were well informed about you (or your family member?)

    _____Yes   ____To some extent   ____No

4. Did you have to repeat information about your (family member's) health history at each site?

    _____Yes   ____To some extent   ____No

5. Were you included in decisions regarding your care (or your family member's care)?

    _____Yes   ____To some extent   ____No

6. What did you like most about the care you received regarding your past illness/surgery?

7. What one thing could we improve on?

Exhibit 26–A–4 Physician Satisfaction Survey (Structured Interview—Interviewer Makes All Entries)

1. Are you aware there are pathways for managing frail seniors with:

Dementia	____Yes	____No	____Does not apply
Depression	____Yes	____No	____Does not apply
Stroke	____Yes	____No	____Does not apply
Hip Fracture	____Yes	____No	____Does not apply
Hip Replacement	____Yes	____No	____Does not apply
Knee Replacement	____Yes	____No	____Does not apply

2. Do you feel the pathways have been beneficial to your patients?

   1. Yes   2. No   3. To some extent   4. I don't know

   (If 1 or 2) In what way? _____

   _____

   _____

3. Have the pathways assisted you in caring for your patients?

   1. Yes   2. No   3. To some extent   4. I don't know

   (If 1 or 2) In what way? _____

   _____

   _____

4. Have the pathways facilitated clinicians at other sites in caring for your patients?

   1. Yes   2. No   3. To some extent   4. I don't know

   (If 1 or 2) In what way? _____

   _____

   _____

5. Who do you feel is the most appropriate person in your office to use the pathways to coordinate care for your patients?

   1. Physician   2. RN   3. Practice Manager   4. Other staff

Exhibit 26-A-5 Orthopaedic Total Joint Replacement Extended Path

	Discharge to 13 Days Post-Op	2-4 Weeks Post-Op	4-6 Weeks Post-Op	6-8 Weeks Post-Op	3 Months Post-Op
Outcomes	Patient functioning safely in least restrictive environment with no complications	Patient progressing in independent ADL and mobility. Patient independent/compliant with exercise program.	Patient is independent and functioning safely in home environment. Patient reports pain does not limit functional performance.	Patient no longer homebound.	
Consults/Office Visits	Follow-up office visit with orthopaedic surgeon. Home Care or post-acute facility to fax/send Orthopaedic Communication Tool to physician office. If HMO patient, verify with Case Manager approval process for continual services.			Follow-up office visits with orthopaedic surgeon. Home Care or post-acute facility to fax/send status report to physician.	Follow-up office visits with orthopaedic surgeon.
Assessments	Re-evaluate the need for further services (i.e., HHA) and equipment. Ongoing assessment for DVT, ROM, pain, strength, function incision.	↑	↑	↑	
Tests	Hip/knee X-ray. Protime or INR twice weekly, if patient on coumadin.	Hip/knee X-ray. Protime or INR weekly, if patient on coumadin.	Protime or INR weekly until anticoagulation d/c'ed.	Hip/knee X-ray PRN.	Hip/knee X-ray PRN.
Interventions	Remove staples at office visit. Refer to Level Of Care grid for recommended intensity/frequency of OT/PT. For uncemented joints—TTWB. For cemented joints—WB as tolerated.	D/C OT/PT when patient meeting goals. ↑ ↑	↑ Progressive WB as ordered by physician. ↑	↑	↑
Medications	Discuss with physician need for pain control. Iron supplements PRN. Continue anticoagulant, if ordered by physician.	↑ ↑ ↑	Continue iron if Hgb not acceptable. Anticoagulant D/C as ordered by physician.		
Education	Review teaching material with patient (i.e., sexual function, driving, dental work). Reinforce hip/knee precautions. Reinforce ted hose until d/c'ed by physician. *Continue to reinforce use of abductor pillow.*	↑ ↑ *Instruct patient to use regular pillow to maintain abduction.*	↑ ↑	↑ ↑	Physician to discuss with patient need for ongoing hip/knee precautions. *D/C use of pillow with approval of physician.*
Transitional Plan	Continue to evaluate appropriate transition to next level of care (see flowchart).	ECF to notify Home Care 2 days prior to transition.	D/C services when patient meeting goals.		

Exhibit 26–A–6 Physician's Office Continuing Care Pathway for the High-Risk Older Adult

Patient's Name: _____

Care Coordinator: _____

Date of Admission: _____

Site: _____

Attending Physician: _____

Date Components	Initial Visit	Follow up Visits	Follow up Visits	Follow up Visits
Assessment/ Monitoring	☒ History and physical, focus on functional and cognitive status ☒ Assess bowel and bladder, skin and hygiene ☒ Blood pressure ☒ Assess for hearing and vision loss ☒ Care coordinator to complete assessment forms	☒ Complete physical every year, concentrate on changes from baseline ☒ Follow up visits as needed to focus on specific problems ☒ Care coordinators to update assessment form every 6 months to 1 year	☒ Complete physical every year, concentrate on changes from baseline ☒ Follow up visits as needed to focus on specific problems ☒ Care coordinators to update assessment form every 6 months to 1 year	☒ Complete physical every year, concentrate on changes from baseline ☒ Follow up visits as needed to focus on specific problems ☒ Care coordinators to update assessment form every 6 months to 1 year
Consults	☒ PT, OT, PM&R for functional changes ☒ Speech prn for swallowing and/or communication concerns ☒ Psych prn for signs and symptoms of depression ☒ Incontinence clinic prn (696-8014) ☒ Nutrition prn for dietary issues ☒ Other _____ ☒ indicated for _____	☒ If client's status changes, reconsider need for consult ☒ Periodic communication with other doctors involved in care re: status and treatment plans	☒ If client's status changes, reconsider need for consult ☒ Periodic communication with other doctors involved in care re: status and treatment plans	☒ If client's status changes, reconsider need for consult ☒ Periodic communication with other doctors involved in care re: status and treatment plans

Tests	☒ Preventive screening if consistent with goals
	  — mammogram every 1–2 years
	  — glaucoma testing by eye specialist
	  — dipstick urinalysis
	  — CBC, SMAC, non-fasting total blood cholesterol
	  — thyroid function test
	  — for high risk group—FBS, TB test, EKG, pap smear, fecal occult
	  — blood
	☒ Audiology evaluation prn
	☒ Vision evaluation prn
	☒ Dementia work up for cognitive changes (see protocol)
	☒ Other _____
	☒ indicated for _____
		☒ Mammogram every 1–2 years
		☒ Follow up any abnormalities
			☒ Mammogram every 1–2 years
			☒ Follow up any abnormalities
				☒ Mammogram every 1–2 years
				☒ Follow up any abnormalities

continues

Exhibit 26–A–6 continued

Date Components	Initial Visit	Follow up Visits	Follow up Visits	Follow up Visits
Functional Rehab	☒ Assess for changes in function, cognition, behavior, gait, falls, sleep (helpful tools: Folstein MM, Katz, ADL) ☒ Institute PT, OT prn for functional decline ☒ Home evaluation for history of falls (LifeEase 696-7051) ☒ Encourage activity and exercise based on client's tolerance ☒ Injury prevention: prevention of falls, safety belts, smoke detector, smoking near bed or furniture, hot water heater	☒ Assess each visit for changes in function, cognition and behavior ☒ Re-educate to benefit of activity/exercise	☒ Assess each visit for changes in function, cognition and behavior ☒ Re-educate to benefit of activity/exercise	☒ Assess each visit for changes in function, cognition and behavior ☒ Re-educate to benefit of activity/exercise
Nutrition	☒ Baseline weight ☒ Compare changes in last 4 months to 1 year ☒ Assess nutritional status and ability to chew, swallow and feed ☒ Determine need to change consistency of diet (pureed, finger food, etc.) ☒ Explore goals of long term nutritional support (comfort vs. life prolonging methods) ☒ If consistent with goals, educate for low fat, low cholesterol, low sodium, high fiber, variety of foods, increase calcium for women (AHA diet pamphlet) ☒ Nutrition consult for weight loss, weight gain, DM diet or other prescribed diets (696-7777)	☒ Assess each visit for changes in weight, nutritional status, appetite and ability to feed, chew and swallow ☒ Reconsider need for nutritional consult prn	☒ Assess each visit for changes in weight, nutritional status, appetite and ability to feed, chew and swallow ☒ Reconsider need for nutritional consult prn	☒ Assess each visit for changes in weight, nutritional status, appetite and ability to feed, chew and swallow ☒ Reconsider need for nutritional consult prn
Medication	☒ Vaccine: influenza every year, pneumococ-cal, TD booster every 10 years ☒ Review current and past medications including over the counter ☒ Assess for compliance and ability to safely take medication ☒ Evaluate effectiveness, side effects and/or drug toxicity ☒ Assess understanding and provide teaching re: medication, indication, dose, time, side effects ☒ If client/caregiver unable to follow medication regime: arrange visiting nurse for medication monitoring	☒ Influenza every year ☒ Update medication list including medication prescribed by consults and over the counter ☒ Evaluate each visit: compliance, ability to take medication, drug effectiveness, side effects or drug toxicity ☒ Re-adjust medication as indicated ☒ Review knowledge and teaching needs re: medication	☒ Influenza every year ☒ Update medication list including medication prescribed by consults and over the counter ☒ Evaluate each visit: compliance, ability to take medication, drug effectiveness, side effects or drug toxicity ☒ Re-adjust medication as indicated ☒ Review knowledge and teaching needs re: medication	☒ Influenza every year ☒ Update medication list including medication prescribed by consults and over the counter ☒ Evaluate each visit: compliance, ability to take medication, drug effectiveness, side effects or drug toxicity ☒ Re-adjust medication as indicated ☒ Review knowledge and teaching needs re: medication

Date / Components	Initial Visit	Follow up Visits	Follow up Visits	Follow up Visits
Treatment/ Teaching	☒ Assess and review client/caregivers teaching needs re: diagnosis, care and treatment plan ☒ Provide instruction for fall and safety precautions, skin care, bowel and bladder program prn ☒ Tobacco cessation, limiting alcohol consumption ☒ Provide education on advance directives and goals of treatment ☒ Encourage completion of advance directives and copy to records	☒ Re-evaluate each visit teaching needs re: diagnosis, care and treatment plan ☒ Encourage completion of advance directives if not done, copy to records ☒ Re-establish goals, especially if client's status changes	☒ Re-evaluate each visit teaching needs re: diagnosis, care and treatment plan ☒ Encourage completion of advance directives if not done, copy to records ☒ Re-establish goals, especially if client's status changes	☒ Re-evaluate each visit teaching needs re: diagnosis, care and treatment plan ☒ Encourage completion of advance directives if not done, copy to records ☒ Re-establish goals, especially if client's status changes
Psychosocial	☒ Determines client's ability to make decisions and need for surrogate decision maker ☒ Assess for depression signs and symptoms, suicide risk factors, abnormal bereavement, physical abuse or neglect (see depression scale) ☒ Assess caregiver burden—refer for counseling prn (696-7770)	☒ Assess each visit impact of condition on client/caregiver ☒ With changes in health or social situation (i.e., losses) be alert for depression or caregiver burden ☒ Psych evaluation and/or counseling prn	☒ Assess each visit impact of condition on client/caregiver ☒ With changes in health or social situation (i.e., losses) be alert for depression or caregiver burden ☒ Psych evaluation and/or counseling prn	☒ Assess each visit impact of condition on client/caregiver ☒ With changes in health or social situation (i.e., losses) be alert for depression or caregiver burden ☒ Psych evaluation and/or counseling prn
Continuing Care Needs	☒ Assess current living situation—ability for client to live alone or ability of caregiver to provide care ☒ Family meeting prn to address issues and goals ☒ Consider referral for adult day care, visiting nurses for skilled needs, other in home support services, nursing home placement or legal counseling ☒ Social work evaluation to assist with assessment, decision and referrals (Senior I&R 696-7770)	☒ Reassess need for support services especially if client status changes or support network changes ☒ Family meeting prn to address issues and goals especially if status changes ☒ Periodic communication with other health professionals providing care (visiting nurses, day care) especially when client status, medication or treatment plan changes	☒ Reassess need for support services especially if client status changes or support network changes ☒ Family meeting prn to address issues and goals especially if status changes ☒ Periodic communication with other health professionals providing care (visiting nurses, day care) especially when client status, medication or treatment plan changes	☒ Reassess need for support services especially if client status changes or support network changes ☒ Family meeting prn to address issues and goals especially if status changes ☒ Periodic communication with other health professionals providing care (visiting nurses, day care) especially when client status, medication or treatment plan changes
Outcomes	☒ Etiologies responsible for decline identified ☒ Client/caregiver have established goals of treatment (i.e., treat all, comfort only, etc.) and will complete advance directives ☒ Appropriate diagnosis and intervention(s) consistent with client's goals, established ☒ Appropriate support and education provided to client and caregiver ☒ Current living situation and care maximized client's independence while maintaining a safe environment ☒ Appropriate referrals made	☒ Client status unchanged/improved ☒ Etiologies responsible for changes in baseline identified ☒ Change in treatment plan and care provided consistent with client's goals	☒ Client status unchanged/improved ☒ Etiologies responsible for changes in baseline identified ☒ Change in treatment plan and care provided consistent with client's goals	☒ Client status unchanged/improved ☒ Etiologies responsible for changes in baseline identified ☒ Change in treatment plan and care provided consistent with client's goals

continues

Exhibit 26–A–6 continued

Date								
Components	Initial Visit		Follow up Visits		Follow up Visits		Follow up Visits	
Referrals	Date	Agency	Contact Person			Reason		
Signature								
Date								

Source: Copyright © Lutheran General Hospital–Advocate, Park Ridge, Illinois, used with permission.

■ 27 ■

Achieving Quality and Fiscal Outcomes for Older Adults with Depression

Janice K. Bultema, Marcia A. Colone, Debra J. Livingston, and Mary Jane Strong

The Norman and Ida Stone Institute of Psychiatry is part of Northwestern Memorial Hospital, a not-for-profit 750-bed academic medical center affiliated with Northwestern University Medical School. It is located in downtown Chicago. Stone Institute of Psychiatry is composed of 56 inpatient beds and ambulatory programs, including intake and crisis intervention, outpatient treatment center, rehabilitation, residential care, and partial hospitalization. These programs have treatment tracks for adolescent, adult, chemical dependence, and older adult care. Northwestern Memorial's patients primarily come from the immediate geographic locale, which includes a diversity of income levels.

In an effort to effectively manage the care of these patients across the continuum, Northwestern Memorial initiated a clinical pathway endeavor. The use of pathways to improve quality care was consistent with the mission of being "an academic medical center where the patient comes first." Moreover, the pathways paved the way for cost reduction efforts that changing payer mixes mandated. In 1992, the hospital established a department of case management and hired a director to lead the clinical pathway initiative throughout the hospital. Since psychiatry was not the first department within the hospital to develop pathways, they were able to learn from the experiences of their peers. Stone Institute learned the importance of selecting appropriate case types for pathways and involving interdisciplinary team members in all phases of pathway development and implementation.

In psychiatry, staff members deliver patient care as a comprehensive multidisciplinary team, so including this team in the pathway initiative was a natural extension of current practice. Since each clinical area had its own multidisciplinary team and culture, the Stone Institute leadership decided to start the pathway work by focusing on the inpatient component of care for one patient population. Stone administrators recognized that eventually the pathway work would need to extend to other inpatient populations, the ambulatory programs, and home health. But they believed that there would be a higher likelihood of success if they initially focused the work in one area.

This chapter describes the first clinical pathway initiated in psychiatry and its longitudinal outcomes over three years. The pathway was developed for older adults with depression because this patient population was high volume, high risk, and problem prone.[1] Additionally, although patients are treated individually, the course of treatment is fairly predictable.[2] The medical and nursing leadership and the multidisciplinary team on the unit treating this patient population were strong, cohesive, and committed to quality improvements. This laid a strong foundation for the required work.

DEVELOPING THE PATHWAY

The multidisciplinary work group on the 20-bed geriatric acute care unit developed the first psychiatric pathway in the late fall of 1993. This group, composed of the medical director, a clinical nurse manager, a psychiatrist, a psychiatry resident, a clinical nurse specialist, three nurses, two social workers, two occupational therapists, a unit secretary, a nursing graduate student, and a geriatric nurse practitioner, developed the pathway and or-

We acknowledge the support and assistance of Anne Bolger, Senior Vice President, Hospital Operations; Julie Creamer, Vice President, Patient Care; Mary Kay Getzfrid, Manager, Transition Planning and former Clinical Nurse Manager of the older adult unit and initiator of this pathway; Jeremy Cohen; and Jason Moore in the preparation of this manuscript.

chestrated its implementation. They began their work by reviewing the literature, practice guidelines,[3] and standards for treating older adults with depression. In an effort to understand and incorporate the treatment culture at the Stone Institute, the clinical nurse specialist interviewed key psychiatrists and clinical staff to determine their standard treatment interventions and reviewed the medical records of previously treated patients. The clinical nurse specialist used these data to draft the pathway, which was then revised by the multidisciplinary team during a series of meetings.

While this work was in progress, the unit leaders organized a full-day retreat for multidisciplinary staff members to learn about systems change theory, the impact of change, and how to commit to change. These activities, combined over a three-month period, resulted in the development of a pathway and established multidisciplinary ownership of the treatment program.[1]

The clinical pathway lists daily interventions required for the projected 12-day hospitalization and defines who is accountable for these actions (Exhibit 27–A–1 in Appendix 27–A). It includes a section that delineates consultations and diagnostic procedures that may be indicated at strategic points during the hospitalization. The pathway also defines expected outcomes for the patient and significant others throughout the hospitalization. Process improvements built into the pathway-defined treatment program included early discharge planning and emphasis on collaboration with patients and their significant others. The pathway also incorporated the use of quantitative and qualitative measures of patient outcomes—namely, the Mini Mental Status Exam,[4] the Geriatric Depression Scale,[5,6] and the specified target symptoms and discharge criteria for inpatient care.

In tandem with the pathway, the team developed a quality improvement monitoring tool referred to as the *variance report*. Variances are discrepancies between expected and actual interventions and outcomes.[7] Positive variances indicate that the patient achieved an outcome or that an intervention occurred sooner than expected. Negative variances indicate that the patient did not achieve expected outcomes either at specific intervals along the pathway or at discharge or that staff members did not carry out the interventions as recommended by the pathway. Monitoring these variances is integral to successful implementation of pathways, since it ensures that the team makes appropriate modifications in the patient's treatment course.

The systematic monitoring of variances is critical for two reasons. First, patients are individuals and do not follow the pathway identically. If a variance occurs, the team can promptly adapt the pathway interventions to better meet this patient's needs. Second, variance monitoring allows for retrospective improvements through system changes. For example, the team may immediately resolve a variance that is impeding the patient's progress, yet they may need to continually monitor other variances to ascertain the root cause and appropriate actions. Variances that occur for several patients over time may point to issues that require planning and resolution at a system or clinical practice level. To maximize the quality improvements indicated by the variance analysis, the organization must create an infrastructure to ensure continuous monitoring.

MANAGING THE PATHWAY

The primary purpose of clinical pathways is to manage groups of patients effectively by identifying common multidisciplinary interventions and patterns of outcomes occurring throughout the treatment episode, and ideally throughout the care continuum. Organizations can meet this goal only when they define, measure, and use outcomes and variances to improve patient care. Without the identification and monitoring of variances and outcomes, a clinical pathway becomes a static template of care as opposed to a dynamic patient care tool that provides ongoing information regarding effective treatment interventions. Furthermore, successful pathway implementation is only possible when the caregivers who manage the day-to-day activities of patients assigned to clinical pathways are integrally involved in all aspects of managing the pathway.

The older adult unit assigned two geriatric patient care coordinators (PCCs) to manage the pathways, including the monitoring of pathway compliance and variances. These registered nurses provide direct patient care to assigned patients and are responsible for utilization management. This has proven to be an effective model for case management, since the PCC knows the clinical and management information regarding the case. By having one identified information source, the unit improved the efficiency of its contacts with managed-care companies and the satisfaction of these companies. The PCC's utilization management role also has decreased payment denials and increased reimbursement.

Although the PCC is responsible for the management of the pathways, all staff nurses administer the measurement tools, monitor the criteria, and initiate the pathways. The admitting nurse, after completing the nursing assessment, initiates the pathway as appropriate, places the pathway in the patient's chart, and documents the initiation date on the cardex. The pathway becomes part of the patient's individual treatment plan and is used by the nurses in shift report. Specifically, in the shift-to-shift report, the nurse identifies the pathway day, the interventions completed and not completed, and variances that occurred. The oncoming nursing staff members use the pathway to plan the care of the patient for that shift. The pathway also guides documentation of patient care and is incorporated into the medical record.

Similarly, all disciplines use the pathway to structure their daily treatment and documentation. The team schedules a multidisciplinary treatment planning conference within the first 24 hours of a patient's hospitalization. In this meeting, the treatment team determines if the pathway initiated is appropriate and uses it to guide prescribed individualized interventions. Each discipline-specific team member has a section of the pathway to complete daily and is expected to document on the variance report. Additionally, the treatment team meets weekly to evaluate the effectiveness of the prescribed interventions. The team uses the pathway to guide these meetings, evaluate the variances, and refine the treatment plan.

The multidisciplinary team uses the pathway on a daily basis to direct patient education. Patients do not receive the actual pathway document. Instead, the nursing staff developed a pathway-based patient education tool. This tool is easier for the older adults to read than the pathway document because it has larger print and is written in book format. It is also easier to understand because the nurses wrote it at a sixth-grade reading level. The booklet divides the pathway into days and provides the patients and their significant others with information about what to expect and do throughout the course of hospitalization. This helps patients gain mastery over the experience of hospitalization.

Patients need to know what the team expects of them and how they can measure their improvement. They also need to understand how the multidisciplinary team and their significant others will be measuring their improvements and readiness for discharge. When the team member shares the pathway education document with the patients' significant others, it communicates the importance of their role in the patients' progress and eventual discharge. This lays the foundation for the significant others' participation throughout the course of treatment.

The final piece of managing the pathway is the effective use of the variance data. At discharge, the discharging nurse completes the variance report, notes the discharge date, and computes the total length of stay. The PCC reviews the pathway and variances for that patient, compiles the data, and sends the data to the management engineering department for length-of-stay and cost-per-case compilation. At an aggregate level, the PCC summarizes and tracks all patient variances, which include system and practice issues. This monitoring of variances and outcomes over time provides important trend information that the PCC routinely shares with the multidisciplinary team during monthly quality improvement meetings.

The multidisciplinary team reviews specific cases and overall trends indicated by the pathway variances and identifies possible treatment implications. After reviewing the pathway and barriers encountered in its use, the team initiates a variety of strategies to address negative variance trends. These strategies may include refining a component of clinical practice, initiating staff education, or revising the pathway. Making the pathway and variance analyses a major focus in these quality meetings underscores that the pathway process is a method of continually improving patient care.

MEASURING OUTCOMES

Both quality and fiscal outcomes are measured. The team measures quality outcomes by evaluating whether patients and their significant others demonstrate the expected outcomes identified in the pathway. Expected outcomes can be defined as anticipated responses to treatment. By measuring these daily, the team is able to determine if the patient is responding as anticipated at key intervals of the treatment regimen. These benchmarks also help the patients and their significant others to understand the goals of the treatment episode.

The team measures its own performance against the predetermined standard of care that the pathway defines. This measuring process provides direct feedback to the caregivers and ongoing positive reinforcement of their practice. It may suggest ways to improve clinical practice or streamline care. The pathway challenges some caregivers to intervene ahead of schedule as appropriate for that patient. For example, the social worker may set a personal goal to complete assessments on day 1 rather than waiting for day 2 as designated on the pathway.

Additionally, the team measures the satisfaction quality outcomes. This includes the satisfaction of multidisciplinary team members and of patients and their significant others. The satisfaction of referral sources, discharge placement facilities, and other departments within the organization can also offer a helpful perspective to the treatment team.

Fiscal outcomes that are measured include cost-per-case, reimbursement, and length-of-stay data. Measuring and interpreting these outcomes over time often lead to opportunities for improvement. For example, if over time patients are discharged sooner than expected, this may indicate the need to shorten the prescribed length of stay. Conversely, if the trend analysis demonstrates that most patients are discharged after the prescribed length of stay, the pathway may need to be longer, or work processes may need to be revised to be more efficient.

ACHIEVING LONGITUDINAL QUALITY OUTCOMES

Initial Quality-of-Care Outcomes

Quality outcomes achieved with the older adult depression pathway included improvements in patient care as well as patient, staff, and physician satisfaction. The multidisciplinary team experienced the initial quality outcomes upon completion of the pathway development phase. After working together for three months to develop the pathway, team members better understood each other's professional contribution and became more willing to hold each other accountable for the completion of the defined work. Team members also found that there was less controversy regarding how to care for patients because they could reference this predetermined care standard. This focused on the needs of the patient rather than on personality- and discipline-related differences. Collaboration and negotiation significantly increased among team members.[1]

Following a two-month pilot of the pathway with the patients of two psychiatrists, multidisciplinary staff members found that the pathway was consistent with the unit's values. They also found that it incorporated the standards of care and standard operating procedures of the unit. Patients on the pathway and their significant others expressed improved satisfaction. They liked knowing what was going to happen each day of the hospitalization. They also found it helpful to know the expected length of stay, target symptoms, and discharge criteria. As a result of these quality improvements, additional psychiatrists chose to include their patients on the pathway and expressed interest in developing more pathways throughout the Stone Institute.

Six months after pathway implementation, additional quality outcomes were achieved. As shown in Figure 27–1, the pathway resulted in more patients receiving medical consultation and physical examination by a geriatric specialist. This improved the care of these patients, who usually presented with multiple medical conditions. Further improvements demonstrated were an increased level of involvement of the patient's significant other and enhanced and earlier contact with aftercare agencies.

Quality Outcomes

* Prepath:
    * 83% of patients had medical consultation
    * 8% of patients examined by nurse practitioner or internist in first 24 hours
    * 25% of significant others involved in first 24 hours
    * 33% of aftercare agencies contacted by day 14

* Postpath:
    * 100% of patients had medical consultation
    * 92% of patients examined by nurse practitioner or internist in first 24 hours
    * 83% of significant others involved in first 24 hours
    * 83% of aftercare agencies contacted by day 14

Figure 27–1 Six-Month Quality Outcomes

During this first six-month period, the team suspended some patients from the pathway. The most common cause was a medical condition that interfered with the patient's response to psychiatric intervention. In response, the nursing and medical staff developed guidelines to help the nurse respond appropriately to abnormal laboratory values and untoward patient symptoms. For each abnormality, the guidelines defined what is urgent compared to nonurgent and listed necessary actions. After implementation of these guidelines, physicians and nurses substantiated that the guidelines improved the quality of care provided for medically ill older adults.

Longitudinal Patient Satisfaction Outcomes

The organization compared its patient satisfaction survey results to those of 48 peer hospitals from a national benchmark database of 457 institutions. On the basis of these results, the management engineering department completed statistical analysis from September 1995 through February 1997, comparing patient satisfaction of pathway and nonpathway patients. Since the sample size for patients on diagnosis-specific pathways throughout the organization was too small for statistically sound analysis, they performed *t* tests and logistic regressions, using the congregate grouping of all patients on pathways. They found that clinical pathway patients were significantly more satisfied overall than nonpathway patients ($p < .05$). The researchers also found that privacy, staff sensitivity, pain control, communication among physicians and nurses, and visitor and family service dimensions influenced overall satisfaction almost twofold for pathway patients compared to nonpathway patients.

Additional longitudinal patient satisfaction outcomes have been demonstrated by the callback surveys that nursing conducts with patients who return home. For example, for the first quarter of this fiscal year, the satisfaction results were positive, with 85 percent very satisfied, 12 percent fairly satisfied, and 3 percent unsatisfied. The questions related to communication with the patient and the various disciplines were also positive. For one quarter, the nurses added pilot questions regarding the patient's perception of how well the staff met their needs. The response to this question indicated that the patients perceived that the staff cared about them. The team believes that this response is in part due to the patient and significant other's increased understanding of the course of treatment and the staff's intent to meet their specific needs. The pathway educational tool communicates both of these concepts.

Longitudinal Staff Satisfaction Outcomes

Staff satisfaction has improved over the last three years with pathway use. Initially, staff voiced concern about in-creased work and additional documentation requirements. The unit leadership listened and responded to the concerns. They demonstrated that long term, the pathways would decrease staff members' work by standardizing interventions and replacing other documentation.[1(p.34)] The continued multidisciplinary involvement in the management of the pathway has sustained a patient-focused, team commitment. Additionally, the pathway provided staff with common language related to providing care and helped support clinical decision making. This increased staff confidence, competence, and satisfaction.

In 1996, more than two years after implementation of the older adult pathway, the nursing development department conducted a research study on caregiver morale and satisfaction.[8] This unit scored significantly higher than other units on morale, goals, and satisfaction. This included the perception of team spirit. The unit also scored significantly higher than other units on work environment, which included questions on involvement, autonomy, and innovation. The unit scored significantly lower than other units in caregiver stress, including conflict with physicians, perceptions of inadequacy, and lack of social support. Although the pathway implementation is not the sole reason for this, the process of developing and managing the pathway as a multidisciplinary team improved the collaboration that contributed to these high satisfaction scores.

Longitudinal Physician Satisfaction Outcomes

The pathway has also improved physician satisfaction, although initially it was difficult to integrate the pathway with physician practice. The complex delivery system in an academic health care setting, the unfamiliarity with clinical pathways, and the large number of admitting physicians were barriers to the implementation process. A compounding complication was presented by the division of private physicians, faculty practice physicians, hospital-employed physician medical directors, and house staff who admitted patients to the unit. It was a challenge to communicate and gain consensus from the various groups to ensure consistent implementation of the pathway.

The support and endorsement of the department chairman and hospital-employed medical directors were integral in overcoming these barriers. Moreover, the unit leaders, who strongly believed in the benefits of pathways, encouraged as many physicians as possible to participate in the pathway development and monitoring process. Initially, only two physicians piloted the pathway, one a private attending and the other a faculty practice physician who was also a hospital-employed medical director. As the other physicians watched their peers pilot the pathway and heard their endorsement, they volunteered to have their patients assigned to the pathway. The unit leaders provided individual support for the physicians during their first pathway implementation case. Additionally, the physician-chaired psychiatry utilization management committee, under the direction of the department chairman, shared with the physicians their length-of-stay and resource utilization data compared to the data of their peers. This was new information for most. It challenged the physicians to practice according to the predetermined standard of care defined by the pathway. Combined, these efforts resulted in a predominantly positive perception of the pathway initiative among the entire physician group.

Physicians have been pleased that the same standard of care is being provided and monitored by all physicians. Multidisciplinary team members use the pathway as a teaching tool for residents and medical students, since it represents cumulative best practice. The physicians have found that this standardization ensures a predictable treatment plan and consistent interventions from multidisciplinary caregivers.

ACHIEVING LONGITUDINAL FISCAL OUTCOMES

Although many health care providers consider quality and cost outcomes to be separate variables, improved fiscal outcomes can also be interpreted as quality improvements. Patients today are paying more of their health care costs out of pocket than ever before. When they are discharged, patients often experience stress related to these bills.[9] This becomes an even more difficult problem for mentally ill patients who have limited coping skills and resources and are trying to reenter the community following an acute episode. Furthermore, the rate of nosocomial infection declines with decreased length of stay. Thus, decreasing length of stay improves quality care and decreases cost per case.

The Stone Institute was pleased with the significant fiscal improvements that resulted from pathway implementation. Length of stay and financial improvements occurred after six months of implementation (Figure 27–2). Specific improvements included a 39 percent decrease in length-of-stay, a 40 percent decrease in cost-per-case, a 3.26 percent decrease in cost-per-day, and a 104.1 percent increase in net revenue. These improvements have continued over time, as demonstrated in Figure 27–3 and Figure 27–4. Both average length-of-stay and cost-per-case gradually decreased except for a brief rise in the third quarter of fiscal year 1995. During this time period, the unit's medical director and clinical nurse manager resigned. This emphasizes the importance of unit leadership in pathway maintenance. When the unit recovered from this loss and the new medical director and clinical

Financial & LOS Outcomes

Prepath comparison group:
- *n* = 153
- geriatric patients
- DRG 430
- ICD-9-CM 296.20–296.39

Postpath comparison group:
- *n* = 58
- geriatric patients
- DRG 430
- ICD-9-CM 296.20–296.39

❖ 58% of patients discharged within 14 days
❖ 39% decrease in length-of-stay
❖ 40% decrease in cost-per-case
❖ 3.26% decrease in cost-per-day
❖ 104.1% increase in net revenue

Figure 27–2 Six-Month Financial and Length-of-Stay (LOS) Outcomes

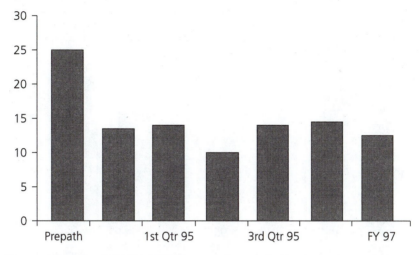

Longitudinal ALOS Outcomes

Figure 27–3 Longitudinal Average-Length-of-Stay (ALOS) Outcomes

nurse manager established themselves, the average length-of-stay and cost-per-case decreased, surpassing prior achievements.

These improvements cascaded to nonpathway patients as well, as shown in Figure 27–5. Cost-per-case decreased dramatically over a two-year period for the 13 diagnostic related groupings most commonly admitted to the Stone Institute, even though only five pathways were in use. This demonstrates the halo effect of improvements that occur as staff members focus on maximizing the benefit of each day of the hospitalization. It may be that with managed care influences, cost-per-case would have decreased without the implementation of pathways. However, only with the use of pathways can caregivers ensure quality improvement while reducing cost and length-of-stay.

LOOKING BACK

Certain findings stand out as the Stone Institute looks back over the last three years. The pathways resulted in significant care improvements and clinical advantages. However, issues and concerns related to pathway implementation also emerged.

Clinical Advantages

As described earlier, the pathway raised the standard of care for patients, and there were fewer variations in care. Stone Institute achieved this because the development process included a combination of the caregivers' best practices and published research findings. The variance tracking provided a consistent quality feedback mechanism. Aggregate information about key variances made the need for changes in care processes more obvious and targeted where these changes were needed. Over time, variance data uncovered groups of patients with specific needs that were not being met by the interventions delineated on the pathway. For example, the older adults who received electroconvulsive therapy were not able to stay on the pathway. This pointed to the need to decide about electroconvulsive therapy earlier in the course of treatment and made clear the need to develop an overlay pathway detailing the guidelines and outcomes for care of these patients.

The older adult depression pathway served as an important communication tool between caregivers, patients, and their significant others. The common language used in the pathway facilitated an understanding among all disciplines of the goals, activities, and outcomes targeted for the patient. Using both written and verbal communication, team members discussed the overall goals and key activities associated with the pathway with patients and their significant others. This provided information about each discipline's activities and anticipated the responses of patients and significant others throughout the hospitalization. This process in-

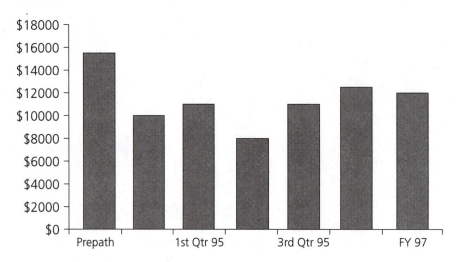

Figure 27–4 Longitudinal Cost-per-Case Trend

creased the locus of control for patients and their significant others and facilitated the treatment progress.

The role of the unit leadership was paramount in setting the tone of acceptance for the pathway process. The ability to view the process as one of evolution enabled everyone to understand the need for ongoing revisions to ensure continual improvements. The use and value of the pathway were directly related to the degree that the staff integrated the pathway into daily unit activities. Referring to the pathway during treatment planning conferences, in review rounds, and when giving and receiving report encouraged familiarity with the planned interventions and established the clinical pathway as a vital patient care tool.

While psychiatry has a long history of interdisciplinary practice, the pathways fostered consensus building related to cost-effective treatment processes and common goals. Feedback to the multidisciplinary team members regarding cost per case helped all disciplines understand their responsibility in ensuring cost containment while sustaining or improving quality.

Another important advantage to using clinical pathways was the positive impression it communicated to managed-care companies. These companies prefer to contract with providers who are cost-efficient and quality focused and who organize care according to outcomes. Managed-care companies have demonstrated a genuine interest in building relationships with providers who establish best-practice standards and track outcomes to improve care.

Clinical Path Issues and Concerns

Disadvantages to patient care have not been evident in this clinical pathway process. However, there have been traps to avoid. The pathway is intended to establish best-practice guidelines; it is not a template of mandatory activities that team members must follow from start to finish with every patient who meets the clinical criteria. Moreover, individuals placed on the pathway may not stay on it for the entire hospitalization episode. The decision to remove a patient from the pathway may be made for appropriate reasons that stem from his or her medical or psychiatric condition. For example, an elderly person who continues to display severe short-term memory loss and confusion needs a clinical assessment for dementia and should not continue on the depression pathway.

Variance tracking provides important information about systems' and clinicians' performance. Staff members must carefully communicate this information, ensuring confidentiality where appropriate. The team can use the variance analysis findings to educate clinical staff about strategies to improve treatments and care for patients. For example, repeated delays in illness teaching should prompt a review of why the intervention is not being done on day 3. Is the patient too ill and not ready for learning? Do the clinical staff need help finding time to teach? Team members should view the variance data as information that provides direction for planned improvements in care. Using the data to point a finger at poor performers does not motivate caregivers to track and report variances or continue to use the clinical pathways.

Psychiatrists, nurses, social workers, and occupational therapists must work with the patient to make sure they identify and meet individual needs. The pathway activities are guidelines and are written in language that allows for individualization. For example, the nursing intervention "Determine schedule for weighing patient two-three times per week" suggests the need to be flexible within the three-times-per-week standard. The wording of this intervention encourages the nurse, when planning the weekly weighing schedule, to take into account the individual patient's activities of daily living and sleep schedule.

If not well understood, clinical pathways can easily become a reactionary tool for clinicians. In some cases, physicians who needed to exert their independence and underscore their control over patient care insisted that their patients stay an extra day. If the patient has no medical necessity to stay, these patients become outliers and fall off the path. Unit leaders should establish the expectation that physicians will document an extra day of stay as a variance and record the reason. They can then use this

information to discuss trends related to different clinical practices. Outcome data can be used to identify whether patients who frequently stay one day longer do better, worse, or the same as patients who have successfully completed the pathway within the prescribed length-of-stay. In most situations, data provide clinicians with information leading to more careful consideration of the need to change practice.

Pathway maintenance takes time and demands careful attention from multidisciplinary staff members. Someone must oversee the pathway implementation and variance reports. Successful organizations constantly restate their commitment to pathways. Support for data collection, graphic display of quarterly data, assistance with the formatting and printing of the pathways, and information systems support for on-line documentation of the interventions are necessary components of successful implementation.

LOOKING FORWARD

To date, the focus of outcome measurement at the Stone Institute has been on quality of care, satisfaction, length-of-stay, cost-per-case, and reimbursement-related data. Over time, this information provided important feedback regarding the success of the older adult depression pathway in reducing costs and making care more efficient. Having established this, phase 2 of track-

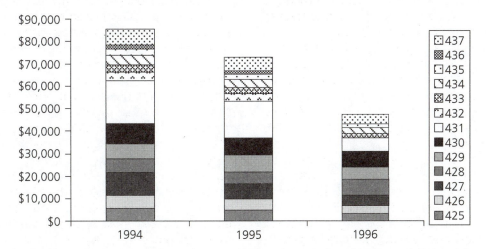

Figure 27–5 Longitudinal Cost-per-Case Trend for Pathway and Nonpathway Patients by Diagnostic-Related Groups

ing outcome data will focus on patient-specific outcomes. This will include the measurement of changes in the admission and discharge scores on the Geriatric Depression Scale and Mini-Mental Status Exam, measurement of medication tolerance, tracking decreases in target symptoms, and maintenance of low to no risk for suicide.

As the depression pathway has become more accepted and used in the inpatient setting, it is time to develop and implement pre- and posthospitalization segments of the pathway. Day 0, which will occur in psychiatry's intake setting, will define interventions that can and should occur before admission to the inpatient unit. For example, preliminary teaching of the significant other, measurement scales, and blood levels can be initiated. After hospitalization, the older adult unit refers many of the inpatients to Stone's outpatient treatment center or partial hospital program or to Northwestern's home health agency. Pathway extensions guiding care in these settings will improve the efficiency and effectiveness of care across the entire continuum.

Furthermore, as relationships continue to develop with long-term care facilities, the opportunity may arise to collaboratively develop and implement pathways external to Northwestern. This would maximize the utilization of best-practice guidelines, increase early identification of relapse potential in this patient population, and capture the collective delineation of what interventions should happen when and where. Everyone involved would be more clear about what the patient and family know about the illness, how to manage the depression in the acute phases, how best to prevent relapses, and how

to identify a relapse early. These pathway extensions would also lay the foundation for cumulative cost-per-case analysis that could define the best course of treatment for the illness episode.

REFERENCES

1. Bultema JK, Mailliard L, Getzfrid MK, Lerner RD, Colone M. Geriatric patients with depression: improving outcomes using a multidisciplinary clinical path model. *J Nurs Adm.* 1996;26:31–38.

2. Rush AJ. Clinical practice guidelines. *Arch Gen Psychiatr.* 1993;50:483–490.

3. Depression Guideline Panel. *Depression in Primary Care: Volume 2. Treatment of Major Depression. Clinical Practice Guideline, Number 5.* AHCPR Publication No. 93-0551. Rockville, MD: US Department of Health and Human Services, Public Health Service, Agency for Health Care Policy and Research; 1993.

4. Folstein MF, Folstein SE, McHugh PR. Mini-Mental State: a practical method for grading the cognitive state of patients for the clinician. *J Psychiatr Res.* 1975;12:189–198.

5. Brink TL, Yesavage JA, Lum O, Heersema PH, Adey M, Rose TL. Screening tests for geriatric depression. *Clin Gerontol.* 1982;1:37–44.

6. Yesavage JA, Brink TL, Rose TL, et al. Development and validation of a geriatric depression screening scale: a preliminary report. *J Psychiatr Res.* 1983;17:37–49.

7. Rudisill PT, Phillips M, Payne CM. Clinical paths for cardiac surgery patients: a multidisciplinary approach to quality improvement outcomes. *J Nurs Care Qual.* 1994;8:27–33.

8. Gaynor SE, Verdin JA, Bucko JP. Peer social support: a key to care giver morale and satisfaction. *J Nurs Adm.* 1995;25:23–28.

9. Crummer MB, Carter V. Critical pathways: the pivotal tool. *J Cardiovasc Nurs.* 1993;7:30–37.

■ Appendix 27–A ■
Clinical Pathway

Exhibit 27–A–1 Clinical Pathway for Older Adults with Depression

Northwestern Memorial Hospital: Stone Institute of Psychiatry			
Admission Date _____ Date _____ Patient Care Coordinator _____			
CLINICAL PATHWAY TEMPLATE FOR OLDER ADULTS WITH DEPRESSION			
This clinical pathway presents guidelines for patient care. Individual patients and specific clinical events, as determined by the patient's attending physician, may require modification to these guidelines.			

	1	**2**	**3**
DATE:			
Psychiatry	❑ Complete initial assessment ❑ Assess for advance directives ❑ Obtain consents ❑ Sign up for treatment planning conference and family meeting ❑ Request medical consult ❑ Address discharge needs ❑ Complete certificate/petition if needed ❑ Preliminary specifications of endpoints for inpatient treatment	❑ Decision regarding starting antidepressants and/or antipsychotics ❑ Review lab results ❑ Contact internist for abnormal lab values ❑ Monitor for medication side effects	❑ Determine primary diagnosis and co-morbid conditions ❑ Treatment planning conference #1 ❑ Determine plan for management of non-primary active co-morbid conditions ❑ Document medical internist plan ❑ Final specification of endpoints for inpatient treatment
Nursing	❑ Complete initial assessment ❑ Initiate precautions prn ❑ Initiate schedules; e.g., toileting, out of room, exercise ❑ Determine group and activity schedule ❑ Complete initial care plan ❑ Review pathway and expected outcomes with patient and/or significant key other	❑ Medication teaching to patient and/or significant key other ❑ Complete Mini Mental Status Exam ❑ Complete Geriatric Depression Scale ❑ Determine schedule for weighing patient 2–3 times per week ❑ Assist with ADLs prn while fostering independence ❑ Sleep assessment	❑ Begin to correct nutritional status prn ❑ Begin teaching regarding illness, expected outcomes and likely discharge needs to patient and/or significant key other following treatment planning conference #1
Social Work	❑ Begin assessment ❑ Contact family and/or nursing home to collaborate regarding key issues	❑ Complete assessment with input from family and/or nursing home ❑ Identify family member's own stressors and understanding of patient's illness ❑ Assess existing supports ❑ Set collaborative goal with patient/significant key other	❑ Adequacy of safety at home determined ❑ Initiate referral; e.g., Department on Aging, home health aide, etc. as identified in treatment planning conference ❑ Provisional significant key other identified

continues

Exhibit 27–A–1 continued

	1	2	3
DATE:			
Occupational Therapy	❏ Begin assessment ❏ Involve patient in 1 activity (group or individual)	❏ Determine appropriate set of occupational therapy groups (i.e., skills vs. Leisure Exploration)	❏ Initial functional assessment complete ❏ Identify collaborative goal with patient ❏ Collaborate with nursing on ADL status
Tests, Procedures, and Consults	❏ H & P completed ❏ Medical internist contacted ❏ Orthostatic BPs ordered ❏ Consider Chest X-ray ❏ SMA-20 ❏ B-12 and Folate ❏ CBC and UA ❏ Thyroid Profile ❏ EKG	❏ Consider the following consults and order prn: ❏ Physical Therapy ❏ Psychology/Neuropsychology testing ❏ Audiology ❏ Ophthalmology ❏ Dental ❏ Chaplain ❏ Internal Medicine	
Patient Outcome	❏ Does not harm self ❏ Attends scheduled milieu groups and activities ❏ Demonstrates readiness to learn ❏ Demonstrates no barriers to learning ❏ Verbalizes understanding of pathway	❏ Does not harm self ❏ Attends scheduled milieu groups and activities ❏ Demonstrates readiness to learn ❏ Demonstrates no barriers to learning ❏ Verbalizes understanding of illness and treatment	❏ Attends scheduled milieu groups and activities ❏ Demonstrates readiness to learn ❏ Demonstrates no barriers to learning ❏ Can state symptoms which require treatment
Significant Key Other Outcome	❏ Voices goals/expectation for treatment	❏ Speaks with social worker to provide psychosocial information	
Signatures	Day RN _____ PM RN _____ NOC RN _____	Day RN _____ PM RN _____ NOC RN _____	Day RN _____ PM RN _____ NOC RN _____

	4–5	6–7	8–10	11–12
DATE:				
Psychiatry	❏ Continue evaluation of progress toward inpatient endpoint ❏ Monitor for medication side effects ❏ Day 5—Family meeting ❏ Day 5—Set discharge date	❏ Arrange psychiatric follow-up ❏ Ongoing consultation with internist as needed	❏ Continue evaluation of progress toward inpatient endpoint ❏ Monitor for medication side effects ❏ Ongoing consultation with internist as needed	❏ Continue evaluation of progress toward inpatient endpoint ❏ Monitor for side effects ❏ Ongoing consultation with internist as needed
Nursing	❏ Reassess plan and treatment outcomes/progress toward inpatient endpoint ❏ Reinforce teaching ❏ Day 5—Family meeting ❏ Day 5—Collaborate in setting discharge date	❏ Continued education and support to patient and significant key other regarding illness, medications, discharge needs ❏ Begin Continuity of Care form if indicated ❏ Initiate aftercare contacts	❏ Monitor for medication side effects ❏ Reassess treatment outcomes ❏ Reinforce teaching with patient/significant key other	❏ Repeat Geriatric Depression Scale ❏ Summarize assessment of target symptom resolution and progress toward inpatient endpoint ❏ Reinforce teaching ❏ Complete discharge instructions/continuity of care instructions ❏ Provide report to referral agency
Social Work	❏ Day 5—Family meeting ❏ Address aftercare needs and confirm information from family ❏ Collaborate in setting discharge date	❏ Ongoing family interventions: counseling, support, education ❏ Provide significant key other with resources as additional needs are identified	❏ Finalize referrals such as Older Adult Program, Partial Hospitalization Program, Geriatric Evaluation Service, nursing home	❏ Family support and education

continues

Exhibit 27–A–1 continued

	4–5	6–7	8–10	11–12
DATE:				
Occupational Therapy	❑ Bathing and dressing assessment prn ❑ Day 5—Family meeting	❑ Continue to treat identified occupational therapy problems in appropriate groups (i.e., Skills, Leisure, Volunteer, Grooming, Reminiscence II, Horticulture) ❑ Review discharge needs with patient	❑ Skill-specific ADL assessment ❑ Make recommendations for services or equipment required for discharge	❑ Review progress toward collaborative goal with patient
Tests, Procedures, and Consults			❑ Internal medicine follow-up appointment set ❑ Home Health contacted ❑ Other outpatient post-discharge appointments set	
Patient Outcome	❑ Attends scheduled milieu groups and activities ❑ Demonstrates readiness to learn ❑ Demonstrates no barriers to learning ❑ Tolerates medications	❑ Attends scheduled milieu groups and activities ❑ Demonstrates readiness to learn ❑ Demonstrates no barriers to learning ❑ Decrease in target symptoms	❑ Attends scheduled milieu groups and activities ❑ Demonstrates readiness to learn ❑ Demonstrates no barriers to learning ❑ Suicide risk low to absent ❑ Agrees to discharge plan	❑ Attends scheduled milieu groups and activities ❑ Demonstrates readiness to learn ❑ Demonstrates no barriers to learning ❑ Reflects the previously identified specifications for ending inpatient treatment
Significant Key Other Outcome	❑ Attends family meeting ❑ Demonstrates readiness to learn ❑ Demonstrates no barriers to learning	❑ Accepts discharge recommen-dations	❑ Makes initial contacts with aftercare providers prn	❑ Demonstrates readiness to learn ❑ Demonstrates no barriers to learning
Signatures	**Day 4** Day RN _____ PM RN _____ NOC RN _____ **Day 5** Day RN _____ PM RN _____ NOC RN _____	**Day 6** Day RN _____ PM RN _____ NOC RN _____ **Day 7** Day RN _____ PM RN _____ NOC RN _____	**Day 8** Day RN _____ PM RN _____ NOC RN _____ **Day 9** Day RN _____ PM RN _____ NOC RN _____ **Day 10** Day RN _____ PM RN _____ NOC RN _____	**Day 11** Day RN _____ PM RN _____ NOC RN _____ **Day 12** Day RN _____ PM RN _____ NOC RN _____

PRIVILEGED AND CONFIDENTIAL UNDER THE ILLINOIS MEDICAL STUDIES ACT

Note: ADLs, activities of daily living; BP, blood pressure; CBC, complete blood count; EKG, electrocardiogram; H&P, history and physical; prn, as needed; SMA, sequential multiple analysis; UA, urinalysis.

Source: Copyright © Northwestern Memorial Hospital.

Index

About the Authors

Suzanne Smith Blancett, EdD, RN, FAAN, has been the Editor-in-Chief of the *Journal of Nursing Administration* and *Nurse Educator* since 1981. She is the author of numerous articles and two books that received Book of the Year Awards from the *American Journal of Nursing,* and she is the founder of *Nursing SCAN in Research: Application for Clinical Practice,* which received two Sigma Theta Tau Media Awards. She is coauthor/editor, along with Dr. Flarey, of *Reengineering Nursing and Health Care: The Handbook for Organizational Transformation,* the *Handbook of Nursing Case Management: Health Care Delivery in a World of Managed Care,* and *Case Studies in Nursing Case Management: Health Care Delivery in a World of Managed Care,* all published by Aspen Publishers, Inc.

A former baccalaureate faculty member and director of a federal manpower planning project, Dr. Blancett earned her degrees from Simmons College and Boston University. She is a member of the American Organization of Nurse Executives, the National Advisory Board of the Center for Medical Ethics and Mediation, the American Academy of Nursing, the National League for Nursing, and the American Nurses Association. A Sigma Theta Tau International Distinguished Lecturer, Virginia Henderson Fellow, and 1991 to 1995 member of the Board of Directors, Dr. Blancett chairs the Sigma Theta Tau Foundation.

Dr. Blancett's public service has included being chair of a city board of health and an elected member of a board of aldermen and school board. She was also a member and president of a human services council.

Dominick L. Flarey, PhD, MBA, RN, CS, CNAA, FACHE, is a Health Care Management Consultant and Owner and President of Dominick L. Flarey & Associates and The Center for Medical-Legal Consulting. He holds a BSN, an MBA in health care administration, and doctorates in nursing administration and management. He has held positions as a certified nurse practitioner, an associate administrator of patient services, a chief operating officer, and an administrator in acute care organizations. He is certified as an adult nurse practitioner and in nursing administration advanced by the American Nurses Credentialing Center. He is also board certified as a health care executive and is a Fellow in the American College of Health Care Executives.

Dr. Flarey is the author/editor of *Redesigning Nursing Care Delivery: Transforming Our Future,* published by Lippincott/Raven. He is coauthor/editor, along with Dr. Blancett, of *Reengineering Nursing and Health Care: The Handbook for Organizational Transformation,* winner of a 1995 *American Journal of Nursing* Book of the Year Award; the *Handbook of Nursing Case Management: Health Care Delivery in a World of Managed Care,* and *Case Studies in Nursing Case Management: Health Care Delivery in a World of Managed Care,* all published by Aspen Publishers, Inc.

Dr. Flarey is a member of the editorial boards of the *Journal of Nursing Administration, Seminars for Nurse Managers, Nursing Case Management,* and *Outcomes Management in Nursing Practice.* He speaks nationally on the topics of case management, reengineering, delivery systems redesign, and organizational transformation.